T0185361

ENVIRONMENTAL AND ECOLOGICAL STATISTICS WITH R

Second Edition

CHAPMAN & HALL/CRC APPLIED ENVIRONMENTAL STATISTICS

Published Titles

Michael E. Ginevan and Douglas E. Splitstone, **Statistical Tools for Environmental Quality**

Timothy G. Gregoire and Harry T. Valentine, **Sampling Strategies for Natural Resources and the Environment**

Daniel Mandallaz, **Sampling Techniques for Forest Inventory**

Bryan F. J. Manly, **Statistics for Environmental Science and Management, Second Edition**

Bryan F. J. Manly and Jorge A. Navarro Alberto, **Introduction to Ecological Sampling**

Steven P. Millard and Nagaraj K. Neerchal, **Environmental Statistics with S Plus**

Wayne L. Myers and Ganapati P. Patil, **Statistical Geoinformatics for Human Environment Interface**

Nathaniel K. Newlands, **Future Sustainable Ecosystems: Complexity, Risk and Uncertainty**

Éric Parent and Étienne Rivot, **Introduction to Hierarchical Bayesian Modeling for Ecological Data**

Song S. Qian, **Environmental and Ecological Statistics with R, Second Edition**

Thorsten Wiegand and Kirk A. Moloney, **Handbook of Spatial Point-Pattern Analysis in Ecology**

ENVIRONMENTAL AND ECOLOGICAL STATISTICS WITH R

Second Edition

Song S. Qian

The University of Toledo
Ohio, USA

CRC Press is an imprint of the
Taylor & Francis Group, an informa business
A CHAPMAN & HALL BOOK

CRC Press
Taylor & Francis Group
6000 Broken Sound Parkway NW, Suite 300
Boca Raton, FL 33487-2742

First issued in paperback 2020

ISBN-13: 978-1-4987-2872-0 (hbk)
ISBN-13: 978-0-367-73675-0 (pbk)

Visit the Taylor & Francis Web site at
http://www.taylorandfrancis.com

and the CRC Press Web site at
http://www.crcpress.com

In memory of my grandmother 张一贯，mother 仲泽庆, and father 钱拙.

Contents

Preface xiii

List of Figures xvii

List of Tables xxiii

I Basic Concepts 1

1 Introduction 3

1.1 Tool for Inductive Reasoning 3
1.2 The Everglades Example 7
1.2.1 Statistical Issues 10
1.3 Effects of Urbanization on Stream Ecosystems 14
1.3.1 Statistical Issues 15
1.4 PCB in Fish from Lake Michigan 16
1.4.1 Statistical Issues 16
1.5 Measuring Harmful Algal Bloom Toxin 17
1.6 Bibliography Notes 18
1.7 Exercise 18

2 A Crash Course on R 19

2.1 What is R? 19
2.2 Getting Started with R 20
2.2.1 R Commands and Scripts 21
2.2.2 R Packages 22
2.2.3 R Working Directory 22
2.2.4 Data Types 23
2.2.5 R Functions 25
2.3 Getting Data into R 27
2.3.1 Functions for Creating Data 29
2.3.2 A Simulation Example 31
2.4 Data Preparation 34
2.4.1 Data Cleaning 35
2.4.1.1 Missing Values 36

2.4.2 Subsetting and Combining Data 36
2.4.3 Data Transformation 38
2.4.4 Data Aggregation and Reshaping 38
2.4.5 Dates 42
2.5 Exercises 44

3 Statistical Assumptions 47

3.1 The Normality Assumption 48
3.2 The Independence Assumption 54
3.3 The Constant Variance Assumption 55
3.4 Exploratory Data Analysis 56
3.4.1 Graphs for Displaying Distributions 57
3.4.2 Graphs for Comparing Distributions 59
3.4.3 Graphs for Exploring Dependency among Variables . . 61
3.5 From Graphs to Statistical Thinking 69
3.6 Bibliography Notes 72
3.7 Exercises 73

4 Statistical Inference 77

4.1 Introduction 77
4.2 Estimation of Population Mean and Confidence Interval . . . 78
4.2.1 Bootstrap Method for Estimating Standard Error . . . 86
4.3 Hypothesis Testing 90
4.3.1 t-Test 91
4.3.2 Two-Sided Alternatives 98
4.3.3 Hypothesis Testing Using the Confidence Interval . . . 99
4.4 A General Procedure 101
4.5 Nonparametric Methods for Hypothesis Testing 102
4.5.1 Rank Transformation 102
4.5.2 Wilcoxon Signed Rank Test 103
4.5.3 Wilcoxon Rank Sum Test 104
4.5.4 A Comment on Distribution-Free Methods 106
4.6 Significance Level α, Power $1 - \beta$, and p-Value 109
4.7 One-Way Analysis of Variance 116
4.7.1 Analysis of Variance 117
4.7.2 Statistical Inference 119
4.7.3 Multiple Comparisons 121
4.8 Examples 127
4.8.1 The Everglades Example 127
4.8.2 Kemp's Ridley Turtles 128
4.8.3 Assessing Water Quality Standard Compliance 134
4.8.4 Interaction between Red Mangrove and Sponges . . . 137
4.9 Bibliography Notes 142

4.10 Exercises 142

II Statistical Modeling **147**

5 Linear Models **149**

5.1 Introduction 149
5.2 From t-test to Linear Models 152
5.3 Simple and Multiple Linear Regression Models 154
5.3.1 The Least Squares 154
5.3.2 Regression with One Predictor 156
5.3.3 Multiple Regression 158
5.3.4 Interaction 160
5.3.5 Residuals and Model Assessment 162
5.3.6 Categorical Predictors 170
5.3.7 Collinearity and the Finnish Lakes Example 174
5.4 General Considerations in Building a Predictive Model . . . 185
5.5 Uncertainty in Model Predictions 189
5.5.1 Example: Uncertainty in Water Quality Measurements 191
5.6 Two-Way ANOVA 193
5.6.1 ANOVA as a Linear Model 193
5.6.2 More Than One Categorical Predictor 195
5.6.3 Interaction 198
5.7 Bibliography Notes 200
5.8 Exercises 200

6 Nonlinear Models **209**

6.1 Nonlinear Regression 209
6.1.1 Piecewise Linear Models 220
6.1.2 Example: U.S. Lilac First Bloom Dates 226
6.1.3 Selecting Starting Values 229
6.2 Smoothing 240
6.2.1 Scatter Plot Smoothing 240
6.2.2 Fitting a Local Regression Model 243
6.3 Smoothing and Additive Models 245
6.3.1 Additive Models 245
6.3.2 Fitting an Additive Model 248
6.3.3 Example: The North American Wetlands Database . . 250
6.3.4 Discussion: The Role of Nonparametric Regression Models in Science 254
6.3.5 Seasonal Decomposition of Time Series 259
6.3.5.1 The Neuse River Example 261
6.4 Bibliographic Notes 267
6.5 Exercises 269

7 Classification and Regression Tree **271**

7.1 The Willamette River Example 272
7.2 Statistical Methods 275
7.2.1 Growing and Pruning a Regression Tree 277
7.2.2 Growing and Pruning a Classification Tree 285
7.2.3 Plotting Options 289
7.3 Comments 293
7.3.1 CART as a Model Building Tool 293
7.3.2 Deviance and Probabilistic Assumptions 297
7.3.3 CART and Ecological Threshold 298
7.4 Bibliography Notes 300
7.5 Exercises 300

8 Generalized Linear Model **303**

8.1 Logistic Regression 305
8.1.1 Example: Evaluating the Effectiveness of UV as a Drinking Water Disinfectant 306
8.1.2 Statistical Issues 307
8.1.3 Fitting the Model in R 308
8.2 Model Interpretation 309
8.2.1 Logit Transformation 310
8.2.2 Intercept 310
8.2.3 Slope 311
8.2.4 Additional Predictors 312
8.2.5 Interaction 314
8.2.6 Comments on the Crypto Example 315
8.3 Diagnostics 316
8.3.1 Binned Residuals Plot 316
8.3.2 Overdispersion 316
8.3.3 Seed Predation by Rodents: A Second Example of Logistic Regression 319
8.4 Poisson Regression Model 332
8.4.1 Arsenic Data from Southwestern Taiwan 332
8.4.2 Poisson Regression 333
8.4.3 Exposure and Offset 340
8.4.4 Overdispersion 341
8.4.5 Interactions 344
8.4.6 Negative Binomial 351
8.5 Multinomial Regression 353
8.5.1 Fitting a Multinomial Regression Model in R 354
8.5.2 Model Evaluation 358
8.6 The Poisson-Multinomial Connection 361
8.7 Generalized Additive Models 367

8.7.1 Example: Whales in the Western Antarctic Peninsula 369
8.7.1.1 The Data 371
8.7.1.2 Variable Selection Using CART 371
8.7.1.3 Fitting GAM 374
8.7.1.4 Summary 378
8.8 Bibliography Notes 380
8.9 Exercises 381

III Advanced Statistical Modeling 385

9 Simulation for Model Checking and Statistical Inference 387

9.1 Simulation 388
9.2 Summarizing Regression Models Using Simulation 390
9.2.1 An Introductory Example 390
9.2.2 Summarizing a Linear Regression Model 392
9.2.2.1 Re-transformation Bias 396
9.2.3 Simulation for Model Evaluation 397
9.2.4 Predictive Uncertainty 405
9.3 Simulation Based on Re-sampling 408
9.3.1 Bootstrap Aggregation 410
9.3.2 Example: Confidence Interval of the CART-Based Threshold 411
9.4 Bibliography Notes 414
9.5 Exercises 414

10 Multilevel Regression 417

10.1 From Stein's Paradox to Multilevel Models 417
10.2 Multilevel Structure and Exchangeability 421
10.3 Multilevel ANOVA 425
10.3.1 Intertidal Seaweed Grazers 426
10.3.2 Background N_2O Emission from Agriculture Fields . . 431
10.3.3 When to Use the Multilevel Model? 434
10.4 Multilevel Linear Regression 436
10.4.1 Nonnested Groups 447
10.4.2 Multiple Regression Problems 453
10.4.3 The ELISA Example—An Unintended Multilevel Modeling Problem 464
10.5 Nonlinear Multilevel Models 465
10.6 Generalized Multilevel Models 469
10.6.1 Exploited Plant Monitoring—Galax 470
10.6.1.1 A Multilevel Poisson Model 471
10.6.1.2 A Multilevel Logistic Regression Model . . . 474

10.6.2 Cryptosporidium in U.S. Drinking Water—A Poisson Regression Example 478
10.6.3 Model Checking Using Simulation 482
10.7 Concluding Remarks 486
10.8 Bibliography Notes 489
10.9 Exercises 489

11 Evaluating Models Based on Statistical Significance Testing 493

11.1 Introduction 493
11.2 Evaluating TITAN 495
11.2.1 A Brief Description of TITAN 496
11.2.2 Hypothesis Testing in TITAN 498
11.2.3 Type I Error Probability 499
11.2.4 Statistical Power 503
11.2.5 Bootstrapping 511
11.2.6 Community Threshold 512
11.2.7 Conclusions 513
11.3 Exercises 514

Bibliography 515

Index 529

Preface

I learned statistics from Bayesian statisticians. As a result, I do not pay attention to hypothesis testing and p-values in my work. Likewise, I do not emphasize the use of them in my teaching. However, most students from my classes remember the term "statistically significant" (or $p < 0.05$) better than anything and check the R^2 value when evaluating a regression model. I have talked to many of them on their experiences in learning and using statistics to understand why they seem to be naturally drawn to these numbers that few can explain clearly in plain language. I came to a satisfactory explanation around 2007 when I read slides of a presentation given by Dick De Veaux of Williams College entitled "Math is Music; Statistics is Literature." (This presentation is now available on YouTube.) According to Dr. De Veaux, statistics is challenging to both students and instructors alike, because we want to teach not only the mechanical part of statistics, but also the process of making a judgment. As a statistics course is always counted as a quantitative methods class, students naturally view statistics as a mathematics class. But statistics is not mathematics. In a typical statistical class for environmental/ecological graduate students, we typically use very simple (but often tedious) mathematics. Students expect to learn statistics as they learn mathematics. However, the mode of inference in mathematics is deduction while the mode of inference of statistics is induction. As a result, statistics cannot be learned by remembering rules and formulae. The process of making a judgment requires putting the analysis in the context, combining information from multiple sources, using logic and common sense. Learning statistics is not about learning rules (as in mathematics) but more about interpretation and synthesis, which requires experience (as in literature). When deciding to write this book, I wanted to put together some examples to illustrate the process of making a judgment and integrate these examples to illustrate the iterative process of statistical inference. This process will inevitably include more than one statistical topic. As a result, many examples included in this book are used in multiple chapters. For example, I used the PCB in fish example as an example of a two-sample t-test in Chapter 4, simple and multiple regressions in Chapter 5, and an example of nonlinear regression in Chapter 6. With these examples, I try to illuminate the difference between how we learn statistics and how we use statistics. In learning statistics, we learn by topics (e.g., from t-test to ANOVA to linear regression, and so on). By the end of the class, students often see statistics as a collection of unrelated methods. When using statistics, we first must decide what is the nature of the problem before deciding what statistical tools to use. This first step is not always taught in a statistics class.

Using the PCB in fish example, I want to illustrate the iterative nature of a statistical inference problem. We may not be able to identify the most appropriate model at first. Through repeated effort on proposing the model, identifying flaws of the proposed model, and revising the model, we hope to reach a sensible conclusion. As a result, a statistical analysis must have subject matter context. It is a process of sifting through data to find useful information to achieve a specific objective. The basic problem of the PCB in fish example is the risk of PCB exposure from consuming fish from Lake Michigan. The initial use of the data showed a large difference between large and small fish PCB concentrations. However, Figure 5.1 suggests that the difference between small and large fish PCB concentrations cannot be adequately described by the simple two sample t-test model. Throughout Chapter 5, I used this example to discuss how a linear regression model should be evaluated and updated. In Chapter 6, some alternative models are presented to summarize the attempts made in the literature to correct the inadequacies of the linear models. But I left Chapter 6 without a satisfactory model. In Chapter 9, I used this example again to illustrate the use of simulation for model evaluation. While writing Chapter 9, I discovered the length imbalance. In a way, this example shows the typical outcome of a statistical analysis — no matter how hard we try, the outcome is always not completely satisfactory. There are always more "what if"s. However, the ability to ask "what if" is not easy to teach and learn, because of the "seven unnatural acts of statistical thinking" required by a statistical analysis: think critically, be skeptical, think about variation (rather than about center), focus on what we don't know, perfect the process, and think about conditional probabilities and rare events [De Veaux and Velleman, 2008]. By examining the same problem from different angles, I hope to bring home the essential message: statistical analysis is more than reporting a p-value.

Since the publication of the first edition, I have learned more about the problem of using statistical hypothesis testing. One part of these problems lies in the terminology we use in statistical hypothesis testing. The term "statistically significant" is particularly corruptive. The term has a specific meaning with respect to the null hypothesis. But by declaring our result to be "significant" without further explanation, we often mislead not only the consumer of the result but also ourselves. In this edition, I removed the term "statistically significant" whenever possible. Instead, I try to use plain language to describe the meaning of a "significant" result. As I explained in a guest editorial for the journal *Landscape Ecology*, a statistical result should be measured by the MAGIC criteria of Abelson [1995]: a statistical inference should be a principled argument and the strength of the inference should be measured by Magnitude, Articulation, Generality, Interestingness, and Credibility, not just a p-value or R^2 or any other single statistic. Throughout the book, I emphasize the interpretation of a fitted model and making conclusions based on the context of the problem. I have followed the following rules in all examples:

- Verbal description of a model – a clear description of the model using nonstatistical terms should be a first step. When describing the model in clear scientific terms, we can better judge whether the model is sensible and whether the real world can be reasonably represented by the model. Even for a simple model such as a t-test or ANOVA, a verbal description can be helpful.

- Verifying model assumptions – plots, plots, and more plots.

- Verbal description of estimated model coefficients – before finalizing the model, we should describe the estimated model coefficients in words. This should be done even in a simple two-sample t-test.

The American Statistical Association issued a statement on p-values [Wasserstein and Lazar, 2016]. The statement emphasizes that the use of statistics should include the context of the problem, the process of data collection and model formulation, and the purpose of the analysis. I will use the statement as a required reading in my class during the first and last weeks of the semester.

Major changes made in this edition include:

- New and revised Chapters and Sections
  - Sections 1.2–1.5 describe main examples used in more than one chapter.
  - Chapter 2 is rewritten with a brief introduction to R and the use of R for data manipulation.
  - Section 5.1 is rewritten to use the PCB in fish example as the lead for linear regression model.
  - New section 5.3.1 introduces the ELISA data collected during the Toledo water crisis in 2014.
  - New section 6.1.3 presents the use of a self-starter function for nonlinear regression.
  - Sections 8.5–8.6 present the multinomial regression and the connection between multinomial and Poisson models.
  - Section 9.2 is revised to include nonlinear regression simulation.
  - Two-way ANOVA is removed from section 10.3.
  - Section 10.4.3 is added to introduce the ELISA example as a multilevel modeling problem.
  - Section 10.5 is added to introduce nonlinear multilevel models.

- Section 10.6.1 uses new examples for generalized multilevel models.
- Chapter 11 is added to discuss the use of simulation in evaluating hypothesis testing based methods. This chapter demonstrates the importance of putting a statistical test in the context of a real-world problem. We should ask: what is the scientific problem at hand, what is the null hypothesis in the context of the problem, what alternatives are supported when the null is rejected? Once these questions are answered, we often have a better understanding of the problem and can be better prepared for making a sound judgment.

- Exercises are added to the end of each chapter.
- Online materials (data and R code) are at GitHub (`https://github.com/songsqian/eesR`).

Song S. Qian
Sylvania, Ohio, USA
July 2016

List of Figures

1.1	Map of WCA2A	10
2.1	RStudio screenshot when opened for the first time.	20
2.2	RStudio screenshot with the R script file of this book open.	22
2.3	An example stream networks	40
3.1	The standard normal distribution	49
3.2	Everglades background TP concentration distribution	50
3.3	Normal Q-Q plot of the Everglades background TP concentration	53
3.4	TP in Lake Erie as a function of distance to Maumee	55
3.5	Comparing standard deviations using S-L plot	56
3.6	Histograms of Everglades TP concentrations	57
3.7	An example quantile plot	58
3.8	Explaining the boxplot	59
3.9	Additive versus multiplicative shift in Q-Q plot	60
3.10	Bivariate scatter plot	62
3.11	Scatter plot matrix	63
3.12	Scatter plot of North American Wetland Database	64
3.13	Power transformation for normality	65
3.14	Daily PM2.5 concentrations in Baltimore	67
3.15	Seasonal patterns of daily PM2.5 in Baltimore	68
3.16	Conditional plot of the air quality data	68
3.17	The iris data	71
4.1	Simulating the Central Limit Theorem	83
4.2	Distribution of sample standard deviation	84
4.3	Distribution of the 75th percentile of Everglades background TP concentration	85
4.4	The t-distribution	93
4.5	Relationships between α, β, and p-value	94
4.6	A two-sided test	99
4.7	Factors affecting statistical power	110
4.8	Residuals from an ANOVA model	120
4.9	S-L plot of residuals from an ANOVA model	121
4.10	ANOVA residuals	122

4.11 Normal quantile plot of ANOVA residuals 123
4.12 Annual precipitation in the Everglades National Park 128
4.13 Yearly variation in Everglades TP concentrations 129
4.14 Statistical power is a function of sample size. 136
4.15 Boxplots of the mangrove-sponge interaction data 138
4.16 Normal Q-Q plots of the mangrove-sponge interaction data 139
4.17 Pairwise comparison of the mangrove-sponge data 140

5.1 Q-Q plot comparing PCB in large and small fish 153
5.2 PCB in fish versus fish length 154
5.3 Temporal trend of fish tissue PCB concentrations 157
5.4 Simple linear regression of the PCB example 159
5.5 Multiple linear regression of the PCB example 160
5.6 Normal Q-Q plot of PCB model residuals 166
5.7 PCB model residuals vs. fitted 167
5.8 S-L plot of PCB model residuals 168
5.9 Cook's distance of the PCB model 169
5.10 The rfs plot of the PCB model 170
5.11 Modified PCB model residuals vs. fitted 173
5.12 Finnish lakes example: bivariate scatter plots 175
5.13 Conditional plot: chlorophyll *a* against TP conditional on TN (no interaction) 178
5.14 Conditional plot: chlorophyll *a* against TN conditional on TP (no interaction) 179
5.15 Finnish lakes example: interaction plots (no interaction) . . 180
5.16 Conditional plot: chlorophyll *a* against TP conditional on TN (positive interaction) 182
5.17 Conditional plot: chlorophyll *a* against TN conditional on TP (positive interaction) 183
5.18 Finnish lakes example: interaction plots (positive interaction) 184
5.19 Finnish lakes example: interaction plots (negative interaction) 184
5.20 Box–Cox likelihood plot for response variable transformation 188
5.21 ELISA standard curve and prediction uncertainty 193

6.1 Nonlinear PCB model 211
6.2 Nonlinear PCB model residuals normal Q-Q plot 212
6.3 Nonlinear PCB model residuals vs. fitted PCB 213
6.4 Nonlinear PCB model residuals S-L plot 214
6.5 Nonlinear PCB model residuals distribution 214
6.6 Four nonlinear PCB models 219
6.7 Simulated % PCB reduction from 2000 to 2007 219
6.8 The hockey stick model 222
6.9 The piecewise linear regression model 223
6.10 The estimated piecewise linear regression model for selected years 225

6.11 First bloom dates of lilacs in North America 227
6.12 All first bloom dates of lilacs in North America 230
6.13 Toledo ELISA standard curve data 231
6.14 Toledo ELISA model diagnostics 1 238
6.15 Toledo ELISA model diagnostics 2 239
6.16 A moving average smoother 242
6.17 A loess smoother 244
6.18 Graphical presentation of a multiple linear regression model 246
6.19 Graphical presentation of a multiple linear regression model with log-transformation 247
6.20 Graphical presentation of a multiple linear regression model with log-transformation 247
6.21 Additive model of PCB in the fish 248
6.22 Effects of smoothing parameter 250
6.23 The North American Wetlands Database 252
6.24 The effluent concentration–loading rate relationship 253
6.25 Fitted additive model using `mgcv` default 253
6.26 Contour plot of a two-variable smoother fitted using `gam` . . 256
6.27 Three-dimensional perspective plot of a two variable smoother fitted using `gam` 257
6.28 The one-gram rule model 258
6.29 Fitted additive model using user-selected smoothness parameter value 258
6.30 CO_2 time series from Mauna Loa, Hawaii 259
6.31 Fecal coliform time series from the Neuse River 264
6.32 STL model of fecal coliform time series from the Neuse River 265
6.33 STL model of total phosphorus time series from the Neuse River 266
6.34 Long-term trend of TKN in the Neuse River 268

7.1 A classification tree of the iris data 274
7.2 Classification rules for the iris data 275
7.3 Diuron concentrations in the Willamette River Basin 278
7.4 First diuron CART model 280
7.5 Cp-plot of the diuron CART model 282
7.6 Pruned diuron CART model 1 283
7.7 Pruned diuron CART model 2 284
7.8 Quantile plot of diuron data 286
7.9 First diuron CART classification model 288
7.10 Cp-plot of the diuron classification model 289
7.11 Pruned diuron classification model 290
7.12 CART plot option 1 291
7.13 CART plot option 2 292
7.14 CART plot option 3 294
7.15 Alternative diuron classification models 296

8.1 A dose-response curve 310
8.2 Logit transformation 311
8.3 Mice infectivity data 313
8.4 Logistic regression residuals 317
8.5 The binned residual plot 317
8.6 Seed predation versus seed weight 320
8.7 Seed predation over time 323
8.8 Time varying seed predation rate 324
8.9 Probability of predation by time and seed weight 325
8.10 Probability of seed predation as a function of seed weight . 328
8.11 Seed weight and topographic class interaction 330
8.12 Binned residual plot of the seed predation model 331
8.13 Arsenic in drinking water data 1 336
8.14 Arsenic in drinking water data 2 337
8.15 Arsenic in drinking water data 3 338
8.16 Arsenic in drinking water data 4 339
8.17 Raw versus standardized residuals of an additive Poisson model 343
8.18 Fitted overdispersed Poisson model 347
8.19 Fitted overdispersed Poisson model with age as a covariate . 350
8.20 Residuals of a Poisson model 351
8.21 Tolerance group multinomial model 1 357
8.22 Tolerance group multinomial model 2 359
8.23 Multinomial residual plot 361
8.24 The Poisson-multinomial connection 364
8.25 Independent Poisson models for tolerance groups 365
8.26 Independent Poisson models of mayfly taxa 366
8.27 Comparing mayfly taxa models 367
8.28 Antarctic whale survey locations 370
8.29 Antarctic whale survey data scatter plots 372
8.30 Antarctic whale survey CART model Cp plot 373
8.31 Antarctic whale survey CART (regression) model 373
8.32 Antarctic whale survey CART (classification) model 374
8.33 Antarctic whale survey Poisson GAM 376
8.34 Residuals from GAM show overdispersion 378
8.35 Antarctic whale survey logistic GAM 379

9.1 Fish tissue PCB reduction from 2002 to 2007 398
9.2 Fish size versus year 398
9.3 Residuals as a measure of goodness of fit 400
9.4 Simulation for model evaluation 401
9.5 Tail areas of selected PCB statistics 402
9.6 Cape Sable seaside sparrow population temporal trend . . . 403
9.7 Cape Sable seaside sparrow model simulation 404
9.8 ELISA test uncertainty 409

9.9 Bootstrapping for threshold confidence interval 412

10.1 Seaweed grazer example comparing `lm` and `lmer` 430
10.2 Comparisons of three data pooling methods in the N_2O emission example 432
10.3 Logit transformation of soil carbon 434
10.4 N_2O emission as a function of soil carbon 435
10.5 The EUSE example data 437
10.6 EUSE example linear model coefficients 440
10.7 Comparison of linear and multilevel regression 443
10.8 Multilevel model with a group level predictor 446
10.9 Antecedent agriculture land-use as a group level predictor . 448
10.10 Antecedent agriculture land-use and temperature as group-level predictors 450
10.11 Antecedent agriculture land-use and temperature interaction 452
10.12 Lake type-level multilevel model coefficients 455
10.13 Conditional plots of oligotrophic lakes (TP) 456
10.14 Conditional plots of oligotrophic lakes (TN) 457
10.15 Conditional plots of eutrophic lakes (TP) 458
10.16 Conditional plots of eutrophic lakes (TN) 459
10.17 Conditional plots of oligotrophic (P limited) lakes (TP) . . . 460
10.18 Conditional plots of oligotrophic (P limited) lakes (TN) . . 461
10.19 Conditional plots of oligotrophic/mesotrophic lakes (TP) . . 462
10.20 Conditional plots of oligotrophic/mesotrophic lakes (TN) . . 463
10.21 Random effects of ELISA model coefficients using `SSfpl2` . 467
10.22 Random effects of ELISA model coefficients using `SSfpl` . . 469
10.23 Random effects (sites) of the Galax model 473
10.24 Large leaf density of the Galax model 474
10.25 Large leaf proportion random effects 476
10.26 Large leaf proportions 477
10.27 System means of cryptosporidium in U.S. drinking water systems 482
10.28 System mean distribution of cryptosporidium in the United States 483
10.29 Simulating cryptosporidium in U.S. drinking water systems 485

11.1 *IV* and z-score under the null model 502
11.2 Permutation μ and σ under the null model 502
11.3 TITAN's underlying models 505
11.4 *IV* and z-score under a linear model 507
11.5 *IV* and z-score under a hockey stick model 508
11.6 *IV* and z-score under a step function model 509
11.7 *IV* and z-score under a sigmoidal model 509
11.8 *IV* and z-score under a sigmoidal model 510
11.9 *IV* and z-score under a sigmoidal model 510

List of Tables

2.1 An example data file 39
2.2 An example data frame 39
2.3 Date formats in R date-time classes 43

3.1 Model-based percentiles versus data percentiles 51

4.1 ANOVA table 119
4.2 Everglades data sample sizes 128

5.1 ANOVA table of a linear model 164
5.2 Linear model coefficients with two categorical predictors . . 197
5.3 Galton's peas data 203

6.1 Estimated piecewise linear model coefficients (and their standard error) for the data used in Figure 6.11 228

8.1 Seed predation model intercepts 326
8.2 The arsenic in drinking water example data 334
8.3 The arsenic standard effect in cancer death rates 341
8.4 Interactions between gender and cancer type 345

10.1 Finnish lake type definition 455

Part I

Basic Concepts

Chapter 1

Introduction

1.1	Tool for Inductive Reasoning	3
1.2	The Everglades Example	7
	1.2.1 Statistical Issues	10
1.3	Effects of Urbanization on Stream Ecosystems	14
	1.3.1 Statistical Issues	15
1.4	PCB in Fish from Lake Michigan	16
	1.4.1 Statistical Issues	16
1.5	Measuring Harmful Algal Bloom Toxin	17
1.6	Bibliography Notes	18
1.7	Exercise	18

1.1 Tool for Inductive Reasoning

We learn from data, both experimental and observational data. Scientists propose hypotheses about the underlying mechanism of the subject under study. These hypotheses are then tested by comparing the logic consequences of these hypotheses to the observed data. A hypothesis is a model about the realworld. The logical consequence is what the model predicts. Comparing model predictions and observations is to decide whether the proposed model is likely to produce the observed data. A positive result provides evidence supporting the proposed model, while a negative result is evidence against the model. This process is a typical scientific inference process. The proper handling of the uncertainty in data and in the model is often the difficulty in this process. The role of statistics in scientific research is to provide quantitative tools for bridging the gap between observed data and proposed models.

The foundation of modern statistics was laid down partly by R.A. Fisher in his 1922 paper "On the Mathematical Foundations of Theoretical Statistics" [Fisher, 1922]. In this paper, Fisher launched "the first large-scale attack on the problem of estimation" [Bennett, 1971], and introduced a number of influential new concepts, including the level of significance and the parametric model. These concepts and terms became part of the scientific lexicon routinely used in environmental and ecological literature. The philosophical contribution of the 1922 essay is Fisher's conception of inference logic, the "logic of inductive inference." At the center of this inference logic is the role of "models" – what is to be understood by a model, and how models are to be embedded in the logic of inference. Fisher's definition of the purpose of statistics is perhaps the best description of the role of a model in statistical inference:

> In order to arrive at a distinct formulation of statistical problems, it is necessary to define the task which the statistician sets himself: briefly, and in its most concrete form, the object of statistical methods is the reduction of data. A quantity of data, which usually by its mere bulk is incapable of entering the mind, is to be replaced by relatively few quantities which shall adequately represent the whole, or which, in other words, shall contain as much as possible, ideally the whole, of the relevant information contained in the original data.
>
> This object is accomplished by constructing a hypothetical infinite population, of which the actual data are regarded as constituting a random sample. The law of distribution of this hypothetical population is specified by relatively few parameters, which are sufficient to describe it exhaustively in respect of all qualities under discussion.

In other words, the objective of statistical methods is to find a parametric model with a limited number of parameters that can be used to represent the information contained in the observed data. A model serves both as a summary of the information in the data and a representation of a mathematical generalization of the real problem. Once a model is established, it replaces the data. Also in his 1922 essay, Fisher divided statistical problems into three types:

1. Problems of specification – how to specify a model
2. Problems of estimation – how to estimate model parameters
3. Problems of distribution – how to describe probability distributions of statistics derived from data.

Applications of statistics can be summarized as a three-step process of addressing these three problems.

The first step of problem solving is to propose a working model (or hypothesis). The model inevitably has unknown parameters, which will be estimated based on collected data. Once these parameters are estimated, we have a quantified model to describe the variation of the variable of interest. As there can be many alternative models, the quantified model must be verified.

Model specification requires scientific knowledge. Applications of statistical methods cannot be isolated from the real-world problems. Consequently, applications of statistical methods must consider the characteristics of the real-world problem and data on the one hand, and the mathematical properties of the model on the other hand. Problems of specification are difficult because a model must serve as an intermediate between the real-world problem and the mathematical formulation. On the one hand, a scientist's conception about the real world can only be tested when predictions based on the conception can be made. Therefore, building a quantitative model is a necessary step. On the other hand, we will always be confined by those model forms which we know how to handle. But a mathematically tractable model is not necessarily the best model. Because any specific model formulation is likely to be wrong, an important statistical problem is how to test a model's goodness of fit to the data. Models which passed the test are more likely to be (or closer to) the true model than are models which failed the test. Therefore, a "good" model is a model that can be tested.

In statistics, model specification is to propose a probability distribution model for the variable of interest. Although the number of probability distributions can be large, probability distributions are grouped into families each with a unique mathematical form and the number of probability distribution families is limited. In a model specification problem, we select a family of distribution to approximate the variable of interest. Parameters of the selected distribution model are estimated. It is important to know that a proposed model is a hypothesis, not a known fact. The objective of statistical inference is to assess the proposed hypothesis based on data.

Problems of estimation are mainly mathematical problems: given the model formulation how to best calculate model parameters from the data. Various optimization methods are used for estimating model parameters, such that the resulting model is "optimal" – the resulting model is the most likely to have produced the observed data.

Problems of distribution are theoretical ones: what is the theoretical distribution of the statistic we estimated? Finding the theoretical distribution of an estimated parameter is the first step of model assessment. The estimated statistics is a function of the data, that is, the estimated parameter value is determined by a specific data set. Because data are random samples of the variable of interest, the estimated statistics is also random – a different data set will lead to a different estimate. The theoretical distribution of an estimated statistics (known as the sampling distribution of the statistics) summarizes how the estimate will vary. This distribution is contingent on the validity of the model specified in the first step. With the sampling distribution, we can make a quick assessment on the proposed model. If the estimated statistics value fits the sampling distribution well, we have a first confirmation of the proposed model. Otherwise we will likely question and reexamine the proposed model.

In this book I present statistics from a scientist's perspective, that is, statistics as a tool for dealing with uncertainty. We are forced to deal with uncertainty in our daily life, especially in our professional life. Environmental scientists face uncertainty in every subject and every experiment. However, we are trained to ignore uncertainty under an academic setting where the pursuit of knowledge is often equivalent to the discovery of underlying mechanisms of natural phenomena. Once the mechanism is known, the outcome can be predicted with certainty. Uncertainty is dealt with as untidiness to be cleaned up with more data and more measurements or more advanced technology. Unfortunately, this untidiness is inevitable. As a result, understanding how to deal with uncertainty and how to learn and draw conclusions from imprecise data are important. Furthermore, policy and management decisions must be made based on imperfect knowledge. Decision making under uncertainty forces us to think carefully about the consequences of all possible strategies/decisions under all possible circumstances. Ignoring randomness will inevitably bring consequences.

Statistics is the science about randomness. Statistics has been part of the core curriculum of biological and life sciences since the time of R. A. Fisher. The traditional biostatistics curriculum is, however, focused mainly on the analysis of experimental data. Environmental and ecological studies often must rely on observational data. The distinction between experimental and observational data is not entirely obvious if viewed purely as a data analysis problem. The issue is whether statistical analysis can be used for causal inference. Data from well-designed experiments are considered appropriate for causal inference because a well-designed experiment can estimate the chance that the estimated effect is due to chance. This capability is because of the randomization in treatment assignment. Although nothing prevents the use of the same statistical techniques for observational data, such analysis cannot be directly used for causal inference, because the estimated treatment effect can be the result of one or more confounding factors that were not observed. But in practice, observational data are often the main source of information for studies of the environment. As a result, researchers are often either not entirely confident about the use of statistics or unable to recognize the spurious correlation due to confounding factors or lurking variables.

This book is aimed at providing a link between environmental and ecological scientists and the commonly used statistical modeling techniques. The emphasis is on the proper application of statistics for analyzing observational data. When possible, mathematical details are omitted and examples are used for illustrating methods and concepts.

Examples used in this book come from published journal papers and books. These examples are typical of many environmental and ecological studies. Most of the examples use observational data, and they were selected to demonstrate the statistical methods as well as to provide a critical review of many practices in the current environmental and ecological literature. The critiques presented in this book represent a hindsight review after many new techniques in statistics are available. The Everglades example is of particular interest because of its large sum of data and the complexity of the problems. In the remaining pages of this chapter, environmental and ecological backgrounds of selected examples are introduced, highlighted by the Everglades example.

1.2 The Everglades Example

The Florida Everglades is one of the largest freshwater wetlands in the world. At the beginning of the twentieth century, the Everglades spanned nearly one million hectares covering almost the entire area south of Lake Okeechobee [Davis, 1943]. The region was mostly undisturbed by humans until the 1940s when a small portion of the land was drained for agriculture and settlement. Then, in 1948, the federal "Central and Southern Florida Project for Flood Control and Other Purposes" was enacted, leading to today's large scale system of canals, pumps, water storage areas, levees, and large agricultural tracts within the Everglades [Light and Dineen, 1994]. The Florida Everglades is a phosphorus limited ecosystem. Therefore, the increased agricultural production, achieved with phosphorus enhanced fertilizers, led to increasing amounts of phosphorus in the water and soil, extensive shifts in algal species, and altered community structure.

In 1988, the federal government sued the South Florida Water Management District (SFWMD, a state agency) and the then Florida Department of Environmental Regulation (now the Department of Environmental Protection) (*U.S. vs. South Florida Water Management District*, Case No. 88-1886-CIV-HOEVELER, U.S.D.C.), for violations of state water quality standards, particularly phosphorus, in the Loxahatchee National Wildlife Refuge (LNWR) and the Everglades National Park (ENP). The United States alleged that the Park and Refuge were losing native plant and animal habitat due to increased phosphorus loading from agricultural runoff. Moreover, according to pleadings filed by the United States, for more than a decade Florida regulators had ignored evidence of worsening conditions in the Park and Refuge, thereby avoiding confrontation with powerful agricultural interests.

In 1991, after two-and-one-half years of litigation, the United States and the State of Florida reached a settlement agreement that recognized the severe harm ENP and LNWR had suffered and would continue to suffer if remedial steps were not taken. The 1991 Settlement Agreement, entered as a consent decree by Judge Hoeveler in 1992, sets out in detail the steps the State of Florida would take over the next ten years to restore and preserve water quality in the Everglades. Among the steps are a fundamental commitment by all parties to achieve the water quality and quantity needed to preserve and restore the unique flora and fauna of ENP and LNWR, a commitment to construct a series of storm-water treatment areas, and to implement a regulatory program requiring agricultural growers to use best management practices to control and cleanse discharges from the Everglades Agricultural Area (EAA).

In 1994, Florida passed the Everglades Forever Act (EFA) which differs from the settlement agreement and consent decree. The EFA included the entire Everglades and changed the time lines for implementing project components, requiring compliance with all water quality standards in the Everglades by 2006. The EFA authorized the Everglades Construction Project including schedules for construction and operation of six storm-water treatment areas to remove phosphorus from the EAA runoff. The EFA created a research program to understand phosphorus impacts on the Everglades and to develop additional treatment technologies. Finally, the EFA required a numeric criterion for phosphorus to be established by the Florida Department of Environmental Protection (FDEP), and a default criterion be created in the event final numerical criterion is not established by 2003.

In studying an ecosystem, ecologists measure various parameters or biological attributes that represent different aspects of the system. For example they might measure the relative abundance of certain species among a particular group of organisms (e.g., diatoms, macroinvertebrates) or the composition of all species in a particular group. Different attributes may represent ecological functions at different trophic levels. (A trophic level is one stratum of a food web, comprised of organisms which are the same number of steps removed from the primary producers.) Algae, macroinvertebrates, and macrophytes form the basis of a wetland ecosystem. Therefore, attributes representing the demographics of these organisms are often used to study the state of wetlands. Changes in these attributes may indicate the beginning of changes of habitat for other organisms. Because of the large redundancy at low trophic levels (the same ecological function is carried out by many species), collective attributes may remain stable even though individual species flourish or disappear when the environment starts to change. When collective attributes do change, the changes are apt to be abrupt and well approximated by step functions. In other words, an ecosystem is capable of absorbing a certain amount of a pollutant up to a threshold without significant change in its function. This capacity is often referred to as the assimilative capacity of an ecosystem [Richardson and Qian, 1999]. The phosphorus threshold is the highest phosphorus concentration that will not result in significant changes in ecosystem functions. The EFA defined this threshold as the phosphorus concentration that will not lead to an "imbalance in natural populations of aquatic flora or fauna."

FDEP is charged with setting a legal limit or *standard* for the amount of phosphorus that may be discharged into the Everglades. The standard should be set so the threshold is not exceeded. Two studies were carried out in parallel – one by the FDEP and one by the Duke University Wetland Center (DUWC) – to determine what the total phosphorus standard should be. The two studies reached different conclusions. The Florida Environmental Regulation Commission (ERC) must consider the scientific and technical validity of the two approaches, the economic impacts of choosing one over the other, and the relative risks and benefits to the public and the environment. The role of the ERC is to advise the FDEP which does the actual adoption of the standards.

Generally, there are two different approaches to study an ecosystem: experimental and observational. Ecological experiments are usually conducted in enclosures within the ecosystem of interest. These enclosures are referred to as mesocosms, within which ecologists can alter the specific aspects of the environment and measure the response of the ecosystem. As in the familiar agricultural experiments with different treatment levels assigned to multiple plots to quantify the treatment effects, a mesocosm must be designed in order to discern the changes in ecosystem due to the "treatment" (or the main factor of interest) from changes due to other uncontrolled factors. Mesocosm experiments are typically conducted in the field by altering certain conditions in isolated small plots of the ecosystem understudy. Because a mesocosm is conceptually appealing and because many statistical methods are available for analyzing its results, a mesocosm has always been very popular in ecological studies. Compared to a mono-culture agricultural plot in an agricultural experiment, a mesocosm of a wetland ecosystem is more complex. Interactions among species in an ecosystem often depend on the spatial and temporal scale. In other words, what happened in the ecosystem is not guaranteed to happen in a mesocosm simply because of the reduced spatial scales and the limited duration of an experiment we can afford. As a result, there are questions about the contribution of mesocosm studies to our understanding of complex ecological systems (see, for example, Daehler and Strong [1996]).

Ecologists are often not satisfied until their mesocosm results can be validated by observational evidence. Observational studies often consist of collecting long-term observational data from a sequence of otherwise similar sites with varying levels of the factor of interest. The natural variation of the factor of interest provides different "treatment" levels. Observational studies are often limited by the difficulty of finding sites that are similar in all aspects but the factor of interest. In fact, ecologists always expect to see differences between any two ecosystems.

In the Everglades, FDEP established 28 permanent sites in a wetland just south of LNWR known as the Water Conservation Area – 2A (WCA2A), a 44,000-ha diked wetland in the northern Everglades (Figure 1.1). WCA2A is isolated from the rest of the Everglades, with its water (both inflow and outflow) controlled by several water-control structures. Water flow inside WCA2A generally follows a north to south direction. After nearly a half-century of receiving outflow from Lake Okeechobee and phosphorus-enriched runoff from EAA, a steep eutrophication gradient has been established along the general north to south direction, resulting in three relatively distinct zones of impact. If we travel in an air boat from the water-control structures (the three arrows in the northern border of WCA2A) southward, the impacted area is within the first 3 km, where we would see dense stands of cattail and other invasive macrophytes and high phosphorus concentrations in both surface water and soil. Further south ($\sim$ 3–7 km from water-control structures), we would see mixed cattail and sawgrass vegetation and infrequent open water sloughs (freshwater marshes dominated by floating aquatic plants, such as white water lily and bladderworts, with some emergent plant, such as spike rush). We

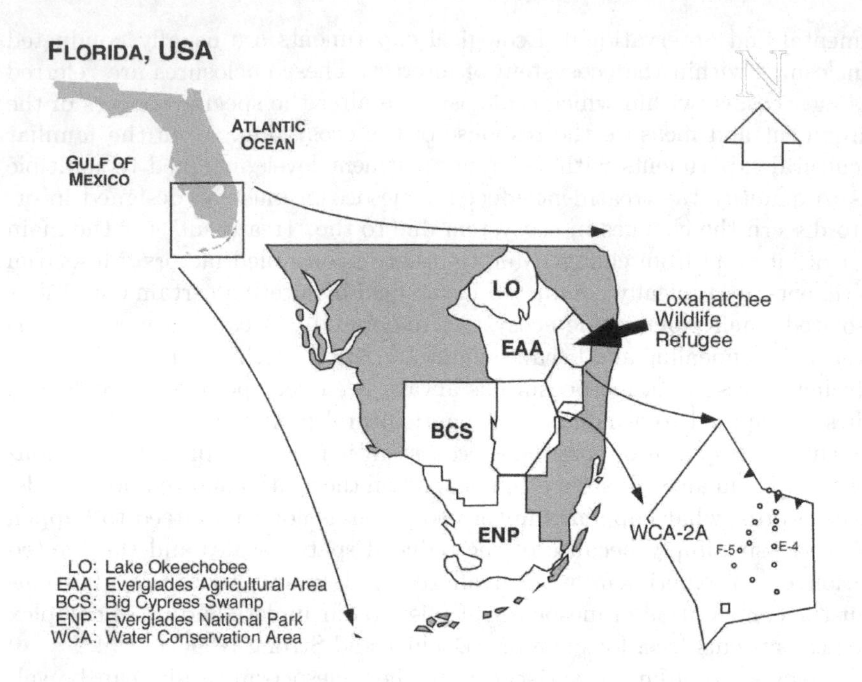

FIGURE 1.1: Location of WCA2A and sampling sites maintained by FDEP.

expect the phosphorus concentrations to be lower. The un-impacted area is about 7 km south of the water-control structure. Here the water chemistry would represent the pristine Everglades, with vegetation structure as a mosaic of sawgrass stands laced with open-water sloughs.

FDEP's monitoring sites are aligned from north to south to capture the phosphorus gradient. Water samples were taken from these sites by various research teams dating back to 1978. FDEP collected biological samples 16 times from 1994 to 1998, each time from a subset of the 28 sites. Among the 28 sites, 13 (shown in the map) were the regularly sampled "main" sites and the other 15 were sampled irregularly for verification purposes. Biological data were used to determine which of the 28 locations are reference sites. The water quality data (TP concentrations) from those reference sites were then used to set the TP standard.

1.2.1 Statistical Issues

In setting an environmental standard, statistics plays an important role. Water quality changes naturally; so do ecological conditions. FDEP used the *reference condition* approach for setting an environmental standard for phosphorus. This method requires the estimation of the probability distribution of total phosphorus (TP) concentrations measured in areas known to be free of anthropogenic impact, known as the *reference areas.* The distribution is often called the *reference* distribution. The U.S. Environmental Protection Agency (EPA) recommended that the 75th percentile of the reference distribution be used as the numerical environmental standard [U.S. EPA, 2000]. This process involves important statistical concepts that cover the basis of statistics.

1. *Probability Distribution* is the first key concept in the setting of an environmental standard. A probability distribution is often defined in introductory texts as an urn with a potentially infinite number of balls inside. A random variable is defined as the process of drawing a ball from the urn; each time the value of the random variable is what is painted on the ball. If the balls inside are written with numbers between 1 and 100, we know a randomly drawn ball would have a number between 1 and 100. Furthermore, if we know that 10% of the balls have numbers less than 3 or greater than 97, we would expect a 1 in 10 chance of picking up a ball with a number less than 3 or greater than 97. Drawing a ball from the urn and recording the number written on it is conceptually the same as taking a water sample from the Everglades wetlands and sending the sample to a lab for measuring the TP concentrations. If we know the contents of the urn, we can calculate the probability of any randomly picked ball with a number in a certain range. By the same token, if we know the probability distribution, we know the probability of a TP measurement exceeding a certain value. The TP concentration distribution in the reference sites is now the direct connection between the classical definition as the urn with a potentially infinite number of balls inside and the physical feature important in environmental management. A probability distribution can be used to describe the scatter of data, parameter values (e.g., the TP threshold of an ecosystem), and error. The most frequently used probability distribution in statistics is the normal or Gaussian distribution. This is because (1) when a random variable can be described using a normal distribution, we need only two parameters, mean and variance, to describe the distribution, and (2) the central limit theorem (see Section 4.1) ensures that many quantities (sum or mean of many independent random variables) are approximately normal. The probability distribution commonly used to describe an environmental concentration variable is the log-normal distribution. If a variable follows a log-normal distribution, the logarithm of this variable follows a normal distribution. Consequently, the first rule of thumb in analyzing environmental and ecological data is to take the logarithm of the data before any analysis. The two parameters of a log-normal distribution are the log mean (μ) and log standard deviation (σ). (In statistics literature, the term "log" refers to natural log.) The exponential of μ (e^{μ}) is called the *geometric mean.* The TP concentration standard for the Everglades is defined in terms of the annual geometric mean. When we know μ and σ of a log-normal distribution, the mean and standard deviation on the original scale are $e^{\mu+\frac{1}{2}\sigma^2}$ and $e^{\mu+\frac{1}{2}\sigma^2}\sqrt{e^{\sigma^2}-1}$, respectively. The stan-

dard deviation of a log-normal distribution is proportional to its mean, and the proportional constant $\sqrt{e^{\sigma^2}-1}$ is known as the coefficient of variation (cv).

2. *Representative* samples of TP concentrations must be obtained to estimate the reference TP concentration distribution. This is a sampling design problem. When using a fraction of the population (here a small volume of water from a small number of locations in the Everglades) for estimating population characteristics (TP concentration distribution), we encounter *sampling error*. Statistical inference is the process of learning the characteristics of a distribution from samples. If the underlying probability distribution is log-normal or normal, statistical inference about the distribution is the same as estimating the distribution model parameters (mean and standard deviation). Because a sample is only a fraction of the population, the estimated model parameters are inevitably dependent on the data included in the sample. Each time a new sample is taken, a new set of estimates will be generated. In other words, the estimated model parameters are random variables. Representative samples are samples taken at random from the population. When a sample is not taken at random, the sample will likely lead to *biased* estimate. Examples of nonrandom samples in the context of this example are samples from only summer, samples from only one site, samples taken only from a particularly wet year. Once the sample is obtained, it is usually difficult to assess the randomness directly from the sample itself. Other information is necessary to properly identify potential bias.

3. *Statistical inference* not only provides estimates of the parameters of interest, but also provides information on the uncertainty associated with the estimated parameters. In practice, both sampling error and measurement error are present in any given data. Sampling error describes the difference between the estimated population characteristics and the true one. For example, the difference between the average of 12 monthly measurements of TP concentration and the true mean concentration is such an error. A sampling error occurs because we use a fraction of the population to infer the entire population. Sampling error is the subject of the *sampling model*, and a sampling model makes no direct reference to measurement error. Measurement error occurs even when the entire population (or complete data) are observed. *Measurement error model* is the tool for this uncertainty. Usually, we combine these two approaches in making a statistical model. Statistical inference focuses on the quantification of the errors.

4. *Statistical assumptions* are the basis for statistical inference. The most frequently used statistical assumption is the normality assumption on measurement error. Measurement error is assumed to have a normal distribution with mean 0 and standard deviation σ. When these basic

assumptions are not met, the resulting statistical inference about uncertainty can be misleading. All statistical methods rely on the assumption that data are random samples of the population in one way or the other.

The reference condition approach for setting an environmental standard relies on the capability of identifying reference sites. In South Florida, identification of a reference site is through statistical modeling of ecological variables selected by ecologists to represent ecological "balance." This process, although complicated, is a process of comparing two populations – the reference population and the impacted population.

Once an environmental standard is set, assessing whether or not a water body is in compliance of the standard is frequently a statistical hypothesis testing problem. Translating this statement into a hypothesis testing problem, we are testing the null hypothesis that the water is in compliance against the alternative hypothesis that the water is out of compliance. In the United States, the definition of "in compliance" used to be "less than 10% of the observed data exceeding the standard." When this definition was translated into a statistical hypothesis testing problem by Smith et al. [2001], "10% of the observed concentration values" was equated to "10% of the time." As a result, many states require that a water body is to be declared in compliance with a water quality standard only if the water quality standard is exceeded by no more than 10% of the time. Therefore, a specific quantity of interest is the 90th percentile of the concentration distribution. When the 90th percentile is below the water quality standard the water is considered in compliance, and when the 90th percentile is above the standard the water is considered in violation.

In addition, numerous ecological indicators (or metrics) are measured for studying the response of the Everglades ecosystem to elevated phosphorus from agriculture runoff. These studies collect large volumes of data and often require sophisticated statistical analysis. For example, the concept of ecological threshold is commonly defined as a condition beyond which there is an abrupt change in a quality, property, or phenomenon of the ecosystem. Because ecosystems often do not respond smoothly to gradual change in forcing variables, instead, they respond with abrupt, discontinuous shifts to an alternative state as the ecosystem exceeds a threshold in one or more of its key variables or processes, materials covered in this book are unable to tackle the problem easily. However, this book will provide the reader with a basic understanding of statistics and statistical modeling in the context of ecological and environmental studies. Data from the Everglades case study will be repeatedly used to illustrate various aspects of statistical concepts and techniques.

1.3 Effects of Urbanization on Stream Ecosystems

The U.S. Geological Survey (USGS) is responsible for monitoring the country's natural resources. In 1991, USGS started a program to develop long-term consistent and comparable information on water quality and factors affecting aquatic ecosystems. The program is designed to understand the conditions of streams, rivers, and groundwater in the U.S. and their temporal trends. The program, known as the National Water Quality Assessment program (NAWQA), has both long-term monitoring networks and short-term topical studies. The project on the effects of urbanization on stream ecosystem (EUSE) is a topic study with an emphasis on the effects of various urbanization-induced changes in the landscape on water quality and aquatic ecosystem. The project, started in 1999, consists of a series of studies with a common design to examine the regional effects of urbanization on aquatic biota (fish, invertebrates, and algae), water chemistry, and physical habitat in nine metropolitan areas in different environmental settings. These studies are known as "urban gradient studies" because they were conducted on watersheds selected along urban gradients within their respective study areas. These study areas are in the metropolitan areas of Atlanta, Georgia (ATL); Boston, Massachusetts (BOS); Birmingham, Alabama (BIR); Denver, Colorado (DEN); Dallas-Fort Worth, Texas (DFW); Milwaukee-Green Bay, Wisconsin (MGB); Portland, Oregon (POR); Raleigh, North Carolina (RAL); and Salt Lake City, Utah (SLC).

During the initial stage of EUSE, researchers developed a multimetric urban intensity index (UII) to identify representative gradients of urbanization within relatively homogeneous environmental settings associated with each urban area. Within each study area, 30 watersheds were selected to represent the urbanization gradient. These watersheds are of similar size and other natural characteristics, so that researchers can address several main questions:

- Do physical, chemical, and biological characteristics of streams respond to urban intensity?
- What are the rates of such response?
- What are typical indicators of changes caused by urbanization?
- What are typical characteristics of biological response to increased urban intensity?
- Do biological responses to urbanization vary by region?

Although these questions are ecological and environmental in nature, statistics played an important role in analyzing the data collected in the subsequent years.

1.3.1 Statistical Issues

In both the Everglades and the EUSE examples, we are interested in how ecosystems respond to changes due to human activities. In the Everglades, the change is represented in the elevated phosphorus concentration, whereas the change in the EUSE example is the urbanization level of a watershed. Ecologists used a similar approach by measuring ecological metrics along the gradient of the factor of interest. The main difference between the two examples lies in how study units were selected.

The Everglades mesocosm study is a typical experimental study, where study units are built so that they have the same conditions except the phosphorus conditions. For example, the mesocosm experiment reported in Richardson [2008] consists of 12 experimental flumes constructed in an open water slough in the Everglades. These flumes are identical except for the levels of phosphorus dosed into them. As a result, ecological differences observed in these flumes can be attributed to the different phosphorus concentrations. The EUSE study is, however, an *observational study*. The main difference between an observational study and an experimental study is how the main factor of interest (urban intensity in the watershed, the treatment) is "assigned" to each study unit. In the Everglades example, the treatment (various levels of phosphorus concentrations) is randomly assigned to each flume. In the EUSE example, the treatment (urban intensity) is associated with the study unit without the interference from the researchers. Because ecosystem health can be affected by many factors, we cannot definitely attribute changes observed along an urban gradient to urbanization. Other factors may also change along the urban gradient. As a result, the main statistical issue in analyzing observational data is to understand whether the observed correlation can be interpreted as causal. Consequently, researchers often take extra measures to ensure that study units are as similar as possible except for the factor of interest. In the EUSE study, study units are individual watersheds. They were carefully selected based on many natural and cultural factors.

The EUSE study seeks a general understanding of the urbanization effect throughout the United States, not just in one watershed or one region. As a result, conditions that are locally constant (e.g., annual mean precipitation and temperature) become important. Specifically, each observation in the EUSE data has a number of attributes to represent a spatial or temporal hierarchy, a common feature of many environmental data. How to properly address the hierarchical nature of the data is the topic of Chapter 10.

1.4 PCB in Fish from Lake Michigan

Human exposure to polychlorinated biphenyls (PCBs) from Great Lakes fish consumption has been a health concern and an area of contention for many decades. Early studies reported birth weight anomalies and other enduring problems among babies born to women who consumed large amounts of Lake Michigan fish. The production of PCB was banned in 1970s, resulting in substantial declining in PCB concentrations in Lake Michigan fish over time. Though fish PCB concentrations have dropped since the 1970s, declines after the early to mid-1980s have been minimal in Lake Michigan, and concentrations remain relatively high, particularly in Lake Michigan and Lake Ontario fishes. States bordering Lake Michigan issued fish consumption advisories cautioning the public of possible risks associated with eating contaminated fish. However, varying standards arising from myriad jurisdictions have caused a confusing array of warnings. In 1993, an advisory task force drafted a protocol to standardize consumption advisories. In response, the state of Wisconsin issued an advisory for Lake Michigan fishes based on standards established in the protocol. Because anglers cannot readily know the PCB concentration of their catch, the advisory translates these concentration-based consumption categories into fish size ranges for the important recreational species. However, PCB concentrations among individual fish are highly variable, even among similar sized fish of the same species. In a series of studies, various statistical models were used to provide probabilistic assessment of PCB exposure from consumption of five Lake Michigan salmonids, based on fish size. In this book, we use the data for lake trout (*Salvelinus namaycush*) collected by the Wisconsin and Michigan Departments of Natural Resources from 1974 to 2003.

1.4.1 Statistical Issues

PCB concentrations in lake trout from Lake Michigan will be used in a number of statistics topics. Initially, the data will be used as exercises of two-sample t-test, comparing mean concentrations between small and large fish. The data will then be used to illustrate simple and multiple linear regressions. In Chapter 6, the data will be used again as an example of a nonlinear regression model. Applications of various statistical techniques on the same data illustrate the hypothetical deductive nature of statistics. In addition, these applications illustrate the iterative nature of statistical inference.

1.5 Measuring Harmful Algal Bloom Toxin

Harmful algal blooms are increasingly becoming a common environmental problem. Algal blooms are overgrowths of algae in water. Some algal species produce toxins in fresh or marine water that can sicken or kill people and animals. Even nontoxic blooms can result in environmental degradation (e.g., depletion of dissolved oxygen) and economic losses (e.g., increased cost in treating drinking water). One particular harmful algal bloom is the overgrowth of cyanobacteria (commonly known as blue-green algae), which can produce microcystin, a group of toxins that can cause liver damage in humans. On August 1, 2014, one microcystin concentration value from a treated water sample in Collins Park Water Treatment Plant of City of Toledo, Ohio, exceeded 1 μg/L, the World Health Organization (WHO) drinking water quality criterion [World Health Organization, 1998] for microcystin adopted by the State of Ohio. Three more tests were carried out on the same day using the same water sample and each time at least one replicate showed a concentration above the criterion. These results prompted the City of Toledo to issue a "Do Not Drink" advisory on the morning of August 2, 2014, affecting over a half million residents. Additional tests were conducted on drinking water from the water treatment plant and throughout the distribution system until all samples consistently showed microcystin concentration below detectable levels (< 0.30 μg/L) during the water advisory, which lasted nearly 3 days.

While the harmful effects of microcystin have been reported widely [Carmichael, 1992], the risk of exposure to harmful levels of the toxin has not been adequately communicated. This is largely because of the lack of information on the uncertainty associated with concentrations measured using common laboratory equipment. This example describes the measurement process commonly used by drinking water facilities for measuring microcystin concentrations. The measurement method requires a regression model to quantify the relationship between microcystin concentration and the variable measuring the color change. The regression model can be linear or nonlinear. When a linear regression model is used, the response variable is the microcystin concentration. As a result, the measurement uncertainty of microcystin concentration is the regression model's predictive uncertainty (Chapter 5). When the nonlinear regression model is used, the response variable is the color variable. The estimation uncertainty is not directly available. In Chapter 6, the nonlinear regression model from this example is used for illustrating the self-starter function. Measurement uncertainty is quantified using simulation methods described in Chapter 9. In Chapter 10, an alternative approach is discussed for reducing the estimation uncertainty. This example uses data collected between August 1 and 3, 2014, during the Toledo water crisis.

1.6 Bibliography Notes

The Everglades example is discussed in detail in two books [Davis and Ogden, 1994, Richardson, 2008], and a brief summary of the statistical issues by Qian and Lavine [2003]. Litigations on the Everglades issue is summarized by Rizzardi [2001]. The EUSE example is derived largely from McMahon and Cuffney [2000], Cuffney and Falcone [2008]. The PCB in fish example is developed based on numerous papers by C.A. Stow and others [Stow, 1995, Stow et al., 1994, Stow and Qian, 1998]. Details of the ELISA test data and the Toledo water crisis can be found in Qian et al. [2015a].

1.7 Exercise

1. Data story. The success of statistical analysis is dependent on our understanding of the underlying science. Find a data set and tell the "story" behind the data – why the data were collected (the hypothesis) and whether the data support the hypothesis.

Chapter 2

A Crash Course on R

2.1 What is R? 19
2.2 Getting Started with R 20
2.2.1 R Commands and Scripts 21
2.2.2 R Packages 22
2.2.3 R Working Directory 22
2.2.4 Data Types 23
2.2.5 R Functions 25
2.3 Getting Data into R 27
2.3.1 Functions for Creating Data 29
2.3.2 A Simulation Example 31
2.4 Data Preparation 34
2.4.1 Data Cleaning 35
2.4.1.1 Missing Values 36
2.4.2 Subsetting and Combining Data 36
2.4.3 Data Transformation 38
2.4.4 Data Aggregation and Reshaping 38
2.4.5 Dates 42
2.5 Exercises 44

2.1 What is R?

R is a computer language and environment for statistical computing and graphics, similar to the S language developed at the Bell Laboratories by John Chambers and others. Initially, R was developed by Ross Ihaka and Robert Gentleman in the 1990s as a substitute teaching tool to the commercial version of S, the S-Plus. The "R Core Team" was formed in 1997, and the team maintains and modifies the R source code archive at R's home page (`http://www.r-project.org/`). The core of R is an interpreted computer language. It is a free software distributed under a GNU-style copyleft,[1] and an official part of the GNU project ("GNU S"). Because it is a free software developed for multiple computer platforms by people who prefer the flexibility and power of typing-centric methods, R lacks a common graphical user interface (GUI). As

[1] Copyleft is a general method for making a program or other work free, and requiring all modified and extended versions of the program to be free as well.

a result, R is difficult to learn for those who are not accustomed to computer programming.

2.2 Getting Started with R

There are many books, documents, and online tutorials on R. The best teaching notes on R are probably the lecture notes by Kuhnert and Venables [2005] (*An Introduction to R: Software for Statistical Modelling & Computing*). The data sets and R scripts used in the notes are available at R's home page. Instead of repeating the materials already discussed elsewhere, this section describes the basic concepts of R objects and syntax necessary for the example in the next section. The best way to learn R depends on the background of each user. For most users, a good user-interface, such as RStudio, is helpful.

For those with a good computer programming background, Kuhnert and Venables [2005] may be the best place to start. For readers with little programming experience, I recommend Zuur et al. [2009]. In this chapter, I will introduce the basic setup of R using RStudio followed by a quick summary of data management using R.

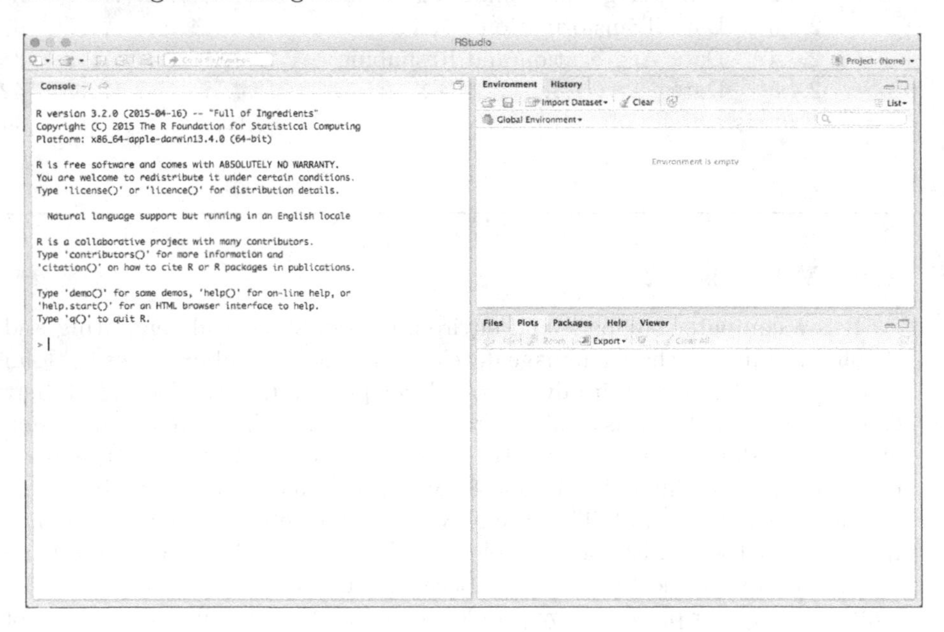

FIGURE 2.1: RStudio screenshot when opened for the first time.

Once R and RStudio are installed, we can start an R session by opening

RStudio. When opened for the first time, RStudio will open a window with three panels, one on the left half of the window and two on the right (Figure 2.1).

The left panel, the R command window (known as the `R Console`), opens with a message about the installed R (version, copyright, how to cite R, and a simple demo).

At the prompt (`>`), we can enter R command to carry out specific operations. R commands or code are better typed into a *script* file. We can open a script panel by using the pull-down menu (clicking `File > New File > R Script` to open a new script file or `File > Open File...` to open an existing script file). The script file will typically be in the top left panel. Figure 2.2 shows the RStudio screenshot with the script file for this book.

A command can take one or more lines of code. To run a command, we can either place the cursor at the line of the command or highlight the lines of one or more commands and click the `Run` button on top of the command window. Once a command (e.g., reading a file) is executed, an R object (imported data) is created in R environment (current memory). The contents of the R environment is shown in the top right panel under the `Environment` tab. During an R session, all commands we run (both those from the script file and typed directly into the R console) are recorded in a log file shown under the `History` tab. The bottom right panel has the following tabs: `File` (showing local files), `Plots` (showing generated graphics), `packages` (listing available packages), `Help` displaying help messages, and `Viewer` (showing local web content).

2.2.1 R Commands and Scripts

The larger than sign (`>`) in the R console is the R prompt, indicating that R is ready to receive a command. For example:

```
> 4 + 8 * 9
[1] 76
```

The line `4 + 8 * 9` is a command telling R to perform an arithmetic operation. R returns a result after we hit the `Enter` key.

By default, the result is displayed on the screen. We can also store the result into an *object*, a named variable with specific values:

```
> a <- 4 + 8 * 9
```

In this line `a` is the object receiving value of the arithmetic operation on the back of the arrow. The arrow (`<-`, the less than sign followed by the minus sign) is the assignment operator, assigning the value from the expression (or object) behind the arrow to the object in front of the arrow. The content of the object can be displayed by typing the object name at the prompt:

```
> a
[1] 76
```

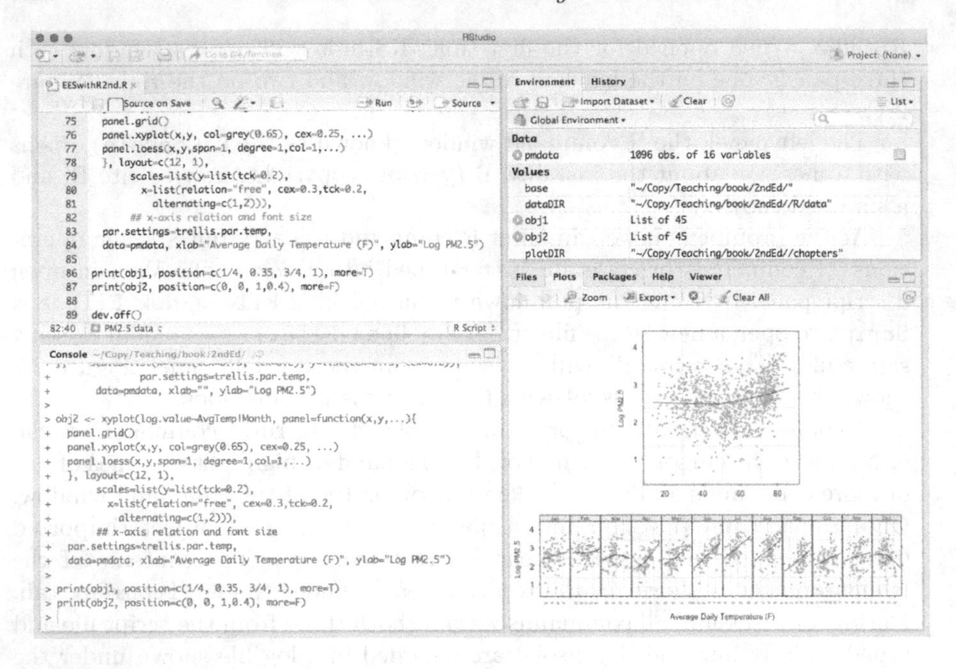

FIGURE 2.2: RStudio screenshot with the R script file of this book open.

We can type these commands in the script panel and save them into a script file.

2.2.2 R Packages

A package is a collection of functions for specific tasks. Some packages come with the R distribution. These packages are listed under the `Package` tab in the lower right panel when RStudio is opened for the first time. Other packages must be installed manually by using the pull-down menu (`Tools > Install Packages...`) or use the function `install.packages` in the commend console. Once a package is installed, it will appear in the list of packages when the package list is refreshed or RStudio starts the next time. An installed package must be loaded into the R memory before its functions are available. Loading an installed package can be as easy as clicking on the unchecked box to the left of the package name. In the scripts for this book, I wrote a small function for loading a package. The function (named `packages`) will first check if the package is installed, then install (if necessary) and load the package.

2.2.3 R Working Directory

The R working directory is the default location where R will search for and save files. It is advisable not to save your data to the R default working directory. In my work, I always set the working directory first by using the function `setwd`. For example, on a computer with OSX or Unix operating system, I use the following scripts

```
base <- "~/MyWorkDir/"
setwd(base)
```

On a Windows computer:

```
base <- "C:\\Users\\Song\\MyWorkDir"
setwd(base)
```

where `MyWorkDir` is an existing file directory.

2.2.4 Data Types

The four atomic data types in R are: numeric, character, logical, and complex number. A numeric data object, such as `a`, contains numeric values. A character object is to store a character string:

```
> hi <- "hello, world"
> hi
[1] "hello, world"
>
```

A logical object contains results of a logical comparison. For example,

```
> 3 > 4
```

is a logical comparison ("is 3 larger than 4?") and the answer to a logical comparison is either "yes" (`TRUE`) or "no" (`FALSE`):

```
> 3 > 4
[1] FALSE
> 3 < 5
[1] TRUE
```

and the result of a logical comparison can be assigned to a logical object:

```
> Logic <- 3 < 5
> Logic
[1] TRUE
```

Data type is known as "mode" in R. The R function `mode` can be used to get the data type of an object:

```
> mode(hi)
[1] "character"
```

A data object can be a vector (a set of atomic elements of the same mode), a matrix (a set of elements of the same mode appearing in rows and columns), a data frame (similar to matrix but the columns can be of different modes), and a list (a collection of data objects). The most commonly used data object is the data frame, where columns represent variables and rows represent observations (or cases).

A logic object is coerced into a numeric one when it is used in a numeric operation. The value TRUE is coerced to 1 and FALSE to 0:

```
> (3<4) + (3>4)
[1] 1
```

This feature can be very useful when calculating the frequency of certain events. For example, the U.S. Environmental Protection Agency (EPA) guidelines once required that a water body be listed as impaired when greater than 10% of the measurements of water quality conditions exceed numeric criteria [Smith et al., 2001]. Suppose we have a sample of 20 observed total phosphorus concentration values stored in the object TP. We can calculate the percentage of observations exceeding a known numeric criterion (e.g., 10) as follows:

```
> TP
 [1]  8.91  4.76 10.30  2.32 12.47  4.49  3.11  9.61  6.35
[10]  5.84  3.30 12.38  8.99  7.79  7.58  6.70  8.13  5.47
[19]  5.27  3.52
> violation <- TP > 10
> violation
 [1] FALSE FALSE  TRUE FALSE  TRUE FALSE FALSE FALSE FALSE FALSE
[11] FALSE  TRUE FALSE FALSE FALSE FALSE FALSE FALSE FALSE FALSE
> mean(violations)
[1] 0.15
```

Three of the 20 TP values exceed the hypothetical numerical criterion of 10. These three are converted to TRUE and the rest to FALSE. When these logical values are put into the R function mean, they are converted to 1s and 0s. The mean of the 1s and 0s is the fraction of 1s (or TRUEs) in the vector.

To access individual values of a vector, we add a square bracket behind the variable name. For example, TP[1] points to the first value in the vector TP and TP[c(1,3,5)] selects three (first, third, and fifth) values. The order of the numbers inside the bracket is the order of the result:

```
> TP[c(1,3,5)]
[1]  8.91 10.30 12.47
> TP[c(5,3,1)]
[1] 12.47 10.30  8.91
```

We can sort the data using this feature. The function order returns the ordering permutation:

```
> order(TP)
 [1]  4 7 11 20 6 2 19 18 10 9 16 15 14 17 1 13 8 3 12 5
```

which means that the fourth element in vector TP is the smallest, the seventh is the second smallest, and so on. Naturally, sorting the the vector is as easy as putting the above result into the square bracket:

```
> TP[order(TP)]
 [1]   2.32  3.11  3.30  3.52  4.49  4.76  5.27  5.47  5.84
 [10]  6.35  6.70  7.58  7.79  8.13  8.91  8.99  9.61 10.30
 [19] 12.38 12.47
```

We can also select values satisfying certain conditions. If we want to list the values exceeding the standard of 10, we use TP[violation]. Here, the logic object violation has the same length as TP. The expression TP[violation] keeps only TP concentrations at locations where the vector violation takes value TRUE, which are the third, the fifth, and the twelfth:

```
> TP[violation]
[1] 10.30 12.47 12.38
```

2.2.5 R Functions

To calculate the mean of the 20 values, the computer needs to add the 20 numbers together and then divide the sum by the number of observations. This simple calculation requires two separate steps. Each step requires the use of an operation. To make this and other frequently used operations easy, we can gather all the necessary steps (R commands) into a group. In R, a group of commands bounded together to perform certain calculations is called a *function*. The standard installation of R comes with a set of commonly used functions for statistical computation. For example, we use the function sum to add all elements of a vector:

```
> sum(TP)
[1] 137.00
```

and the function length to count the number of elements in an object:

```
> length(TP)
[1] 20
```

To calculate the mean, we can either explicitly calculate the sum and divide the sum by the sample size:

```
> sum(TP)/length(TP)
```

or create a function such that when needed in the future we just need to call this function with new data:

```
> my.mean <- function(x){
+   total <- sum(x)
+   n <- length(x)
+   return(total/n)
+ }
```

These lines of commands create an object named `my.mean`, which consists of three lines. This object has a mode of `function`. To run this function, we need to supply a numeric vector `x`. Suppose we want to calculate the mean of `TP`. We type `my.mean(x=TP)`. The function takes the 20 values in the vector `TP` and passes them to object `x` and runs the three lines of commands. The value calculated in the last line is returned to the R console:

```
> my.mean(x=TP)
[1] 6.9
```

R comes with many functions for standard statistical procedures and mathematical functions. For example, the function `mean` calculates the mean of a numeric vector:

```
> mean(TP)
[1] 6.9
```

Users can create new functions to simplify their work. When using an existing function, we need to know the required arguments. That is, we need to tell the function, e.g., `mean`, to perform the operation on which vector, and under what condition. To learn about a function, we can consult the built-in helps:

```
> help(mean)
```

The help file will be displayed on the lower right panel under the `Help` tab in RStudio. For the function `mean`, there are three arguments to specify: `x, trim=0, na.rm=FALSE`. The first argument `x` is a numeric vector. The argument `trim` is a number between 0 and 0.5 indicating the fraction of data to be trimmed at both the lower and upper ends of `x` before the mean is calculated. This argument has a default value of 0 (no observation will be trimmed). The other argument `na.rm` takes a logical value `TRUE` or `FALSE`) indicating whether missing values should be stripped before proceeding with the calculation. The default value of `na.rm` is `FALSE`. For each function, examples of using the function are listed at the end of the help file. Often these examples are very helpful. These examples can be viewed in the R console directly by using the function `example`:

```
> example(mean)

mean> x <- c(0:10, 50)
mean> xm <- mean(x)
mean> c(xm, mean(x, trim = 0.10))
```

```
[1] 8.75 5.50

mean> mean(USArrests, trim = 0.2)
  Murder  Assault UrbanPop     Rape
    7.42   167.60    66.20    20.16
```

The argument `na.rm` has a default of `FALSE`. When there is one or more missing values in the data and the default for `na.rm` is left unchanged, the calculated mean is also missing (`NA`). To remove missing values before calculating the mean, we must change the value of `na.rm` to `TRUE`:

```
> mean(x, na.rm=T)
```

If we must use the function repeatedly, we may create a new function by simply changing the default setting:

```
> my.mean <- function(x)
+    return(mean(x, na.rm=T))
```

2.3 Getting Data into R

Small data sets can be typed into R in many different ways. The TP data used in the previous section was typed using the function `c` (concatenation):

```
> TP <- c(8.91, 4.76, 10.30, 2.32, 12.47, 4.49, 3.11, 9.61,
+         6.35, 5.84, 3.30, 12.38, 8.99,  7.79, 7.58, 6.70,
+         8.13, 5.47, 5.27, 3.52)
```

This line creats a vector `TP` of length 20. Suppose that these TP concentration values were measured at five locations; we can type in this information by creating a vector of site names:

```
> Site <- c("s1","s2","s3","s4","s5","s1","s2","s3","s4","s5",
+           "s1","s2","s3","s4","s5","s1","s2","s3","s4","s5")
```

We now have a numeric vector `TP` and a character vector `Site`. To associate each TP concentration value to its respective site, we can use the function `names` to set the site names as names of `TP`:

```
> names(TP) <- Site
> TP

   s1    s2    s3    s4    s5    s1    s2    s3    s4    s5
 8.91  4.76 10.30  2.32 12.47  4.49  3.11  9.61  6.35  5.84
   s1    s2    s3    s4    s5    s1    s2    s3    s4    s5
 3.30 12.38  8.99  7.79  7.58  6.70  8.13  5.47  5.27  3.52
```

Alternatively, we can combine the two vectors to create a data frame using the function `data.frame`:

```
> TPdata <- data.frame(Conc=TP, Loc=Site)
> TPdata
    Conc Loc
1   8.91  s1
2   4.76  s2
3  10.30  s3
4   2.32  s4
5  12.47  s5
6   4.49  s1
7   3.11  s2
8   9.61  s3
9   6.35  s4
10  5.84  s5
11  3.30  s1
12 12.38  s2
13  8.99  s3
14  7.79  s4
15  7.58  s5
16  6.70  s1
17  8.13  s2
18  5.47  s3
19  5.27  s4
20  3.52  s5
```

The resulting object `TPdata` has two columns with names `Conc` and `Loc` and 20 rows. Elements of a two-dimensional data frame can be accessed using the same squared bracket method using two numbers separated by a comma. For example, `TPdata[4,1]` extracts the value on the fourth row and first column and `TPdata[,1]` extracts the first column. A negative number removes the corresponding row or column: `TPdata[,-2]` removes the second column.

Most data sets we work with are data frames because we commonly store data in a spreadsheet format with each row representing an observation and each column representing a variable. Variables can be numeric and categorical such as the data frame we just created. Importing a spreadsheet like data file into R is commonly done in two steps:

1. Preparing and exporting a spreadsheet file into a text file.

   In this step, a spreadsheet file is processed and cleaned. For example, if missing values are labeled in different ways we can change them to a single label (e.g., `NA`); we should rename columns if their names are too long or have symbols such as $, %, &. The cleaned data file is exported into a comma or tab delimited text file. If a comma delimited file is used, we should check if a comma exists in one or more variables (e.g., location name). Commas in a variable may be confused with the commas used for separating columns.

2. Importing a text file into R using function `read.table` or `read.csv`.

   The function `read.table` reads a text file in table format and creates a data frame from it. Many options are available when using this function. But for most applications, we need only tell this function (1) where is the file and what is its name, (2) whether the file contains the names of the variables as its first line (the default is no), and (3) what is the field separator character (the default is white space). When we use comma delimited format, the field separator character is a comma. The function `read.csv` is essentially the same as `read.table` except that the default of field separator character is a comma (and the file has a header line).

2.3.1 Functions for Creating Data

A number of R functions are handy in creating data with specific patterns. To illustrate the use of these functions, let us import the monthly precipitation data for the Evergaldes National Park, available from the U.S. Southeast Regional Climate Center (`http://www.sercc.com/cgi-bin/sercc/cliMONtpre.pl?fl2850`). It reports monthly total precipitation from 1927 to 2012. A cleaned up version is included in the data folder of this book's GitHub page (`EvergladesP.txt`), a tab-delimited text file. When the data file is imported using:

```
> everg.precip <- read.table(file="EvergladesP.txt", header=T)
```

we have a data frame with 86 rows (years) and 14 columns (12 months plus the leading column of year and the ending column of annual total). Before moving further, I will discuss the concept of data structure first.

When making an observation, we record the values of the variables of interest and attributes used to identify each measurement. In this data set, the measurement is the monthly precipitation – numbers occupy the middle 12 columns. Each measurement is associated with a column (month) and row (year). That is, monthly total precipitation is the variable of interest and month and year are attributes. This data set includes a total of 1032 (86×12) observations. In statistical analysis, we often organize data in a data frame where each row represents an observation and each column represents a variable. For this data set, the reorganized data frame should have 1032 rows and 3 columns. We will not consider the annual total column in the current data frame for now.

The columns representing the monthly precitation measurements can be extracted from the current data frame by discarding the first and the 14th columns and transforming them into a vector. A data frame in R is a list of several vectors of the same length. We can use the function `unlist` to simplify the list into a vector, in this case, putting one column after another.

```
unlist(everg.precip[,-c(1,14)])
```

We can use the resulting vector as the first variable of the new data frame. The two-attribute variable (year and month) can be created using the function `rep`, which replicates a vector in certain ways. When using `rep(x, n)`, the vector `x` will be repeated `n` times. For example,

```
rep(everg.precip[,1], 12)
```

replicates the column `YEAR` 12 times. As there are 86 rows, the resulting vector is of length 1032. The attribute month is the middle 12 values of the names of the data frame `everg.precip`. To match each month's name to a data value, we need to replicate 86 times:

```
rep(names(everg.precip)[-c(1,14)], each = 86)
```

Alternatively, we can use numeral month names (1 to 12) by using the function `seq` or simply `1:12`:

```
seq(1, 12, 1)
## or
1:12
```

Putting these steps together, we use the function `data.frame`:

```
> EvergData<-data.frame(Precip = unlist(everg.precip[,-c(1,14)]),
+                        Year = rep(everg.precip[,1], 12),
+                        Month = rep(1:12, each=86))
> head(EvergData)
     Precip Year Month
JAN1   0.28 1927     1
JAN2   0.02 1928     1
JAN3   0.28 1929     1
JAN4   1.96 1930     1
JAN5   6.32 1931     1
JAN6   1.47 1932     1
```

In Chapter 9, we will use random number generators to generate random variates from known probability distributions. Drawing random numbers from a known distribution is the first step of a simulation study, where we mimic the process of repeated sampling so that we can understand the statistical features of a model. With the advent of fast personal computers, simulation is increasingly becoming the most versatile tool for characterizing a distribution and quantities derived from the distribution. As a preview, I will introduce an example of evaluating a method used by the EPA for assessing a water's compliance to an environmental standard.

2.3.2 A Simulation Example

The U.S. Clean Water Act requires that states report water quality status of all water bodies regularly and submit a list of waters that do not meet water quality standards. EPA is responsible for developing rules for water quality assessment. Smith et al. [2001] described that EPA's guidelines once required that a water body be declared as "impaired" if 10% or more of the water quality measurements exceed the limit of a numeric criterion (or water quality standard). Smith et al. [2001] interpreted that this rule was intended to ensure that the water quality standard is violated at most 10% of the time and discussed potential problems with this rule. They concluded that the 10% rule performs poorly – using the rule we frequently wrongly declare a water as impaired when the water is in compliance and we often fail to identify a water as impaired when it is truly impaired. To learn why the rule may be flawed, we can conduct simulations to see how often this rule makes mistakes – declaring a water in violation of the standard when the water is in compliance and failed to identify an impaired water when we know it is in violation of a standard.

Because water quality measurements are random, results from any sampling study of the water quality of a lake or a river segment are subject to sampling error. Using simulation we can see how often we will make mistakes using the 10% rule. To do this, the easiest way is to repeatedly sample a water that is known to be in compliance and measure the concentration and determine how often we will declare the water to be impaired using this rule. Obviously the easiest method is impossible in practice. But if we know the distribution of the water quality concentration variable, we can let the computer simulate the actual sampling process. Taking a water sample and measuring the concentration is simulated by drawing a random number from the known distribution. Repeatedly drawing random numbers using a computer is easy to do.

Because most water quality concentration variables can be approximated by the log-normal distribution (hence the logarithm of the concentration variable will be approximately normal), we will use the logarithm of a concentration variable and assume a normal distribution. For this example, suppose the (natural) logarithm of a water quality standard is 3, and we know that the distribution of the log concentration of the pollutant is $N(2, 0.75)$ (mean 2, standard deviation 0.75). This distribution's 90th precentile or (0.9 quantile) is 2.96, below the hypothetical standard of 3. In other words, the chance of a random variable from this distribution exceeding 3 is less than 0.1. The 0.9 quantile of a normal distribution is calculated by the function `qnorm`:

```
> qnorm(p=0.9, mean=2, sd=0.75, lower.tail=TRUE)
[1] 2.961164
```

With the 90th percentile being less than 3, if we repeatedly draw random numbers from this distribution, on average about 90% of samples will be less than 2.96. In other words, if a pollutant log concentration distribution is

$N(2, 0.75)$, the water body is in compliance with the water quality standard of 3. This conclusion is based on the law of large numbers. When we collect a small number of samples, we may see more or less than 10% of the values above 3. Suppose that we take a sample of 10 measurements, or draw 10 random numbers from this distribution using function `rnorm`:

```
> set.seed(123)
> samp1 <- rnorm(n=10, mean=2, sd=0.75)
> samp1
 [1] 1.579643 1.827367 3.169031 2.052881 2.096966
 [6] 3.286299 2.345687 1.051204 1.484860 1.665754
```

Because we are drawing random numbers from a distribution, no two runs should have the same outcome. But with a computer, random numbers are drawn with a fixed algorithm. These algorithms usually start a random number sequence from a random starting point (a seed). We will set the random number seed to be 123 using the function `set.seed`, so that the outcome printed in this book should be the same as the outcome from your computer.

We can count how many of these 10 numbers exceed 3 (which is 2, more than 10% of the total). Based on the 10% rule, if two or more measurements exceed 3, the water will be listed as impaired. This process is simulated in R in three steps:

```
## 1. compare each value to the standard
> viol <- samp1 > 3
## 2. calculate the number of samples exceeding the standard
> num.v <- sum(viol)
## 3. compare to allowed number of violations (1)
> Viol <- num.v > 1
```

The object `Viol` takes value `TRUE` (the water is declared to be impaired) or `FALSE` (not impaired). To assess the probability of wrongly listing this water as impaired, we can repeat this process of sampling and counting many times and record the total number of times we wrongly declared the water as impaired. To repeat the same computation process many times, we can use the `for` loop:

```
> Viol <- numeric() ## creating an empty numeric vector
> for (i in 1:1000){
    samp <- rnorm(10, 2, 0.75)
    viol <- samp > 3
    num.v <- sum(viol)
    Viol[i] <- num.v > 1
}
```

This script can be further simplified:

```
> Viol <- numeric() ## creating an empty numeric vector
> for (i in 1:1000){
```

```
    Viol[i] <- sum(rnorm(10, 2, 0.75) > 3) > 1
}
```

After simulating the process 1000 time, the vector `Viol` has 1000 logic values. When imposing a numeric operation, `TRUE` becomes 1 and `FALSE` becomes 0. Therefore, the estimated probability of wrongly declaring the water to be impaired can be calculated by `mean(Voil)`, which is 0.21 if the random seed is set at 123.

The example illustrates several frequently used procedures in statistics. Random number generation is an important aspect of statistics. It is the basis of a simulation study. In applied statistics, simulation is often the best way to understand the behavior of a model or of an assumption. We will use simulation repeatedly in this book. The basic idea of simulation is the use of the long-run frequency definition of probability and the use of a computer to replicate a process repeatedly. The resulting random numbers can be directly used to calculate the quantity of interest and to evaluate probabilities.

As an exercise to complete this section, let us further study the problems of the 10% rule by making two more simulations.

First, let us suppose the distribution of the variable is $N(2, 1)$. The 90th percentile of this distribution is `qnorm(0.9, 2, 1)` (=3.28). The standard will be violated more than 10% of the time (in fact `1-pnorm(3, 2, 1)` or 15.9% of the time).

The water body should be declared as "impaired." We can estimate the probability that the water is declared to be "not impaired" under the EPA's rule (assuming that we are still using a sample size of 10, not impaired means that there is 1 or 0 observations exceeding 3).

```
> Viol2 <- numeric()
> for (i in 1:1000){
    Viol2[i] <- sum(rnorm(10, 2, 1) > 3) > 1
}
```

Because the water is impaired, a mistake is made when we conclude that the water is not impaired. The average of the vector `Viol2` is the probability of making the correct decision. The probability of wrongly declaring the water is in compliance is `1 - mean(Viol2)` or 0.52.

Often, the high error rate can be a result of the small number of observations (10). We can use the same simulation method to see whether the error rate reduces when the sample size is increased to 100. A water is in compliance if no more than 10 observations exceed the standard of 3. We can make the program more flexible so that we don't have to make changes to the code every time when we change the sample size:

```
> Viol <- numeric() ## creating an empty numeric vector
> n <- 100 ## sample size
> nsims <- 1000 ## number of simulations
> mu <- 2
```

```
> sigma <- 1
> cr <- 3
> for (i in 1:nsims){
+    Viol2[i] <- sum(rnorm(n, mu, sigma) > cr) > 0.1*n
}
```

More efficiently, we can group these lines into a function:

```
> viol.sim <- function(n=10,nsims=1000,mu=2,sigma=0.75,cr=3){
+   temp <- numeric()
+   for (i in 1:nsims)
+      temp[i] <- sum(rnorm(n, mu, sigma) > cr) > 0.1*n
+   return(mean(temp))
+ }
```

Once executed, we have a function named `viol.sim`, which can be used to perform simulations using different sample size, different number of simulations, and different underlying distribution. The function returns the probability of declaring that a water is in violation of the water quality standard (`cr`). Using this function, our first simulation can be done using just one line of code:

```
> pr.sim1 <- viol.sim()
```

When we want to increase the sample size from 10 to 100, we change the input of `n` from the default of 10 to 100:

```
> pr.sim2 <- viol.sim(n=100)
```

For the second simulation:

```
> pr.sim3 <- 1-viol.sim(sigma=1)
```

2.4 Data Preparation

A large part of a statistical analysis project is preparing data for statistical analysis. In this section, I use two data sets as examples of typical tasks in data preparation. Both data sets are available from the book's webpage. The first task of data preparation is cleaning – detecting/correcting obvious data entry errors, missing data coding method, and basic summary statistics as a means for data checking. In my work, I convert spreadsheet data files into comma separated text files before reading them into R. The conversion also includes renaming variable names to conform with R convention.

2.4.1 Data Cleaning

Before importing data to R, a quick examination of the data is necessary to avoid some common errors. For example, many variables have a method reporting limit (MRL), the lowest value the measurement method can reliably measure. When a value is below the MRL, the value should be reported as censored – it is below the MRL and otherwise unknown. There is no standard method for recording a censored value. If a censored value is recorded as the MRL with a leading less than sign (e.g., < 0.01), the column will be read into R as a character string without any indication of a problem. But when the variable is used later in computation, we will encounter error. Another example of a potential problem that should be handled before importing data is how missing values are represented in a data file. Many old data sets use a large negative number (e.g., -999, -9999). It is advisable to use the same value or character string (e.g., NA) to represent missing values, such that when using the function `read.table`, we can set the option `na.string` properly.

Once data are imported to R, obvious entry errors often can be detected by using simple plots and summary statistics. The function `summary` displays basic summary statistics (minimum, maximum, mean, first and third quartiles, and the number of missing values) for each numeric variable and the number of unique values of character variables. From the summary statistics, we can tell if all numeric variables are correctly imported as numeric and whether numeric variable values are within ranges of their definition or reasonable bounds. For example, after importing the long-term Lake Erie harmful algal bloom monitoring data from the Great Lakes Environmental Research Laboratory (GLERL), the summary statistics show that the maximum value for variable `Latitude` was 875.73. It is an obvious data entry mistake.

Some mistakes are less obvious. Cleveland [1985] discussed the data set on animal intelligence published in a book by Carl Sagan. In the book, Carl Sagan defined a measure of the intelligence of an animal species as the ratio of average brain weight over average body weight to the 2/3 power:

$$Int = \frac{brain}{body^{2/3}}$$

In a log-log scale, the definition can be expressed as:

$$\log(brain) = \log(Int) + 2/3\log(body)$$

Sagan showed a figure of log brain weight plotted agains log body weight. He suggested that it is "obvious" that the human is the most intelligent species. But the figure in his book (a scatter plot of brain weight against body weight, both in logarithmic scales) is difficult to visualize the conclusion. As an example of data cleanup, we will return to this data set in the Exercise to uncover a data entry error.

2.4.1.1 Missing Values

Many studies use environmental monitoring data collected over a long time. An unavoidable problem in such data is the presence of missing data. Data can be missing because of, for example, a missed sampling date, a bad sample, or a botched analytic process. These missing values can be attributed to random factors (missing at random). When using R for data analysis, we can encode such missing values with NA.

Another kind of missing values is often problematic in environmental data. Most analytic methods have "method reporting limits" (MRLs), values below which the reported values are unreliable. These values are known in statistics as "censored data." In environmental data analysis, censoring occurs most commonly because the measured value is below the respective MRL, or left censored. In the data cleaning step, we need to know how a censored value is recorded. When an observed datum is censored, we know only that the value is below a certain number. It is, in a way, missing. But a censored value, unlike a data point missing at random, still has some information. Consequently, we need to record a censored value using a different method. In the past, left-censored values were either recorded as 0 or the MRL (or half of MRL). As lab analytic techniques improve, MRLs change over time. If left-censoring is present, I always use the MRL as the reported value and add a column of 0s (uncensored values) and 1s (censored) to indicate which observations are censored.

2.4.2 Subsetting and Combining Data

Subsetting and merging data are frequently used in data preparation. In R, the basic concept in subsetting is the use of the square brackets. As discussed earlier in this chapter, numbers inside brackets identify the elements of an object. In this section, we discuss subsetting a vector and a data frame. Subsetting a data object requires defining a condition and then asking whether this condition is met for each observation. For example, if we want to select data with values less than 3 in a vector x, we ask

```
> x < 3
```

The result is a vector of TRUE and FALSE. Using this vector of logic values, we can extract elements in x meeting the condition:

```
> x[x<3]
```

When working with data frames, subsetting can be by rows or by columns. If we want to use data collected in 1993 in a large data frame xframe, we need to use the column for year to specify the condition – year==1993:

```
> xframe_sub <- xframe[xframe$year==1993,]
```

Note that it is necesary to use `xframe$year` because the name `year` is inside the data frame. The logic expression `xframe$year==1993` returns a vector of `TRUE` and `FALSE` with a length of the number of rows of the data frame. Putting the expression in the first position, only those rows corresponding to positions of `TRUE`s in the column `year` will be kept. To keep a subset of columns, we can enter a vector of either column numbers (e.g., `xframe[,c(1,3,5)]`) or column names (e.g., `xframe[,c("y","year","z")]`). When a subset of rows and columns are needed, we specify both conditions in the bracket separated by a comma. For example, `xframe[xframe$year == 1993, c(1,3,5)]` will keep only observations in columns 1, 3, and 5 from 1993.

In many R commands for graphing and fitting statistical models, we have an option to select a subset of data for the operation. This is usually to select a subset of rows of a data frame. As a result, when specifying the `subset = year==1993`, we want to plot (or fit a model) using data from 1993 only.

In some cases, we want to expand a data frame by adding additional attributes. For example, in the EUSE example, we have about 30 observations (watersheds) for each of the nine regions. When the data are compiled, we have a data file of watershed-level observations (`euse_all`) and a data file (`esue_env`) of regional environmental conditions such as mean temperature, precipitation, soil characteristics, etc. The 30 observations of a specific region share the same regional environmental condition attributes. Because the two data sets share the same "region" attribute (a character variable with same values), we can use the square bracket operation to add, e.g., annual mean precipitation to the watershed-level data set:

- Add a numeric column representing the region to each data set:

  ```
  > euse_all$reg <- as.numeric(ordered(euse_all$Region))
  > euse_env$reg <- as.numeric(ordered(euse_env$Region))
  ```

  The column `reg` in both data sets is an integer vector with values 1 through 9 representing the nine regions. In `euse_all`, each value is repeated about 30 times for respective watersheds in each region. The data frame `euse_env` has 9 rows, one for each region.

- Sort `euse_env` by region:

  ```
  > oo <- order(euse_env$reg)
  > euse_env <- euse_env[oo,]
  ```

- We can now add the mean precipitation column to `euse_all`:

  ```
  > euse_all$precip <- euse_env$precip[euse_all$reg]
  ```

2.4.3 Data Transformation

In statistical analysis, data transformtion is a frequently used approach to simplify analysis and to make the transformed data closer to the normal distribution. The most commonly used transformation is the log-transformation.

Most of the time, we don't have to create a separate variable for the transformed data. We can make the operation directly in a plot or a model. To create a new variable (or add a new column to an existing data frame) use the "$" operation:

```
> my.data$logY <- log(my.dats$Y)
```

Another frequently used transformation is to center a variable around its mean:

```
> my.data$x.cen <- my.data$x - mean(my.data$x)
```

2.4.4 Data Aggregation and Reshaping

In a typical statistical analysis, we frequently spend a lot of time to get the data into the format required by specific analyses. When collecting data, we use the most efficient means for data entry or use the default structure of an instrument. These formats are not necessarily the ones required for data analysis in R. In this section, I will use two examples to illustrate data aggregation and data reshape.

In general, a data file, in the form of a data frame, consists of variables (in columns) and observations (in rows). Variables can be grouped into measured variables and identification variables. A measured variable is a variable of interest (e.g., total phosphorus in a water sample), and an identification variable identifies each measured variable value (e.g., sampling time, location). When recording measured TP concentration data, we may use a large table with rows representing sampling location and columns representing sampling time, a convenient format for a field/lab setup. In this format, the measured variable is TP (the main body of the table) and the two identification variables are sampling location and sampling time (the two margins of the table). Before analyzing the data, we need to convert the data into a data frame with three columns: one measured variable (TP) and two identification variables (Time and Site). For example, Table 2.1 shows a table of measured TP values from 4 sites in three days.

This format can be convenient for data summary. For example, we can use the function `apply` to calculate row or column means:

```
> site.mean <- apply(TPdata, 1, mean)
> day.mean <- apply(TPdata, 2, mean)
```

The function `apply(X, MARGIN, FUN, ...)` performs a specific calculation (specified by `FUN`) on a matrix (`X`) by row (`MARGIN=1`) or column (`MARGIN=2`).

TABLE 2.1: An example data file

	Day 1	Day 2	Day 3
Site 1	20.1	21.5	30
Site 2	15.2	31.0	12
Site 3	20	25	19
Site 4	11	14	21

However, for most statistical analysis, the table should be converted into a data frame as in Table 2.2.

TABLE 2.2: An example data frame

TP	Site	Day
20.1	Site 1	Day 1
21.5	Site 1	Day 2
30	Site 1	Day 3
15.2	Site 2	Day 1
31	Site 2	Day 2
12	Site 2	Day 3
20	Site 3	Day 1
25	Site 3	Day 2
19	Site 3	Day 3
11	Site 4	Day 1
14	Site 4	Day 2
21	Site 4	Day 3

Instead of using `apply`, we now use `tapply(X, INDEX, FUN, ..., simplify=T)` for aggregation:

```
> attach(TPdataframe}
> site.mean <- tapply(TP, Site, mean)
> day.mean <- tapply(TP, Day, mean)
```

The function `tapply` can be used for more complicated data reshaping tasks. In Qian et al. [2005b], I used a nonlinear regression model for estimating watershed nutrient loading to downstream receiving water. The model uses variables representing nutrient source as the main input. Data are typically collected using a Geographical Information System (GIS) software, where streams in a large watershed are divided into reaches and pollution sources are allocated to each reach. For example, Figure 2.3 is a simplified watershed with five reaches (represented by arrows). In this example watershed, we have two monitoring sites where we collect nutrient flux data (represented by shaded ovals).

From GIS, we have source variables for each reach and the input data file is typically in the following format:

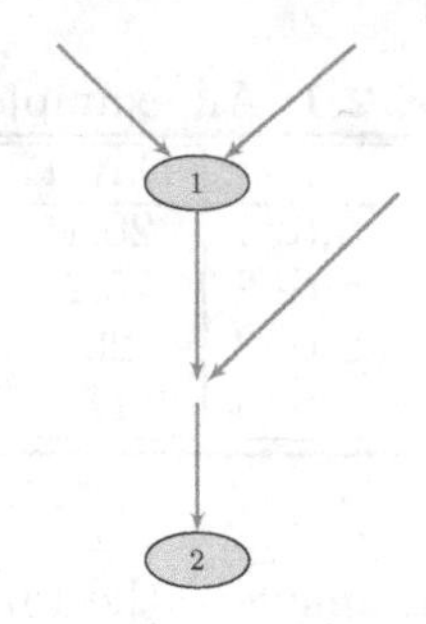

FIGURE 2.3: An example stream networks

DWNSTID	Load	X1	X2	Z1
1	3	10	3	0.2
1	NA	14	5	0.7
2	10	20	1	0.4
2	NA	40	2	0.3
2	NA	10	3	0.2

The column `DWNSTID` identifies the immediate downstream monitoring site for a reach, `X1` and `X2` are two source variables, `Z1` is a variable describing the potential loss of nutrients before reaching to the next monitoring site, and the column `Load` is the measured nutrient flux (recorded as missing, `NA`, for reaches without a monitoring site). Because we only have observations of nutrient fluxes at a small number of reaches, the observed nutrient flux at, for example, node 1 represents the sum of flux from the two upstream reaches. The nonlinear regression model is based on a mass-balance equation:

$$Y_i = \sum_{j=1}^{J_i} \left((\beta_1 X_{1.j} + \beta_2 X_{2.j}) e^{-\alpha Z_{1.j}} \right) \tag{2.1}$$

where i is the ith monitoring site, J_i is the number of reaches upstream of monitoring site i, $X_{1.j}$ and $X_{2.j}$ are the two sources from reach j, and βs and α are parameters to be estimated. In other words, the data should be transposed to:

DWNSTID	Load	X1.1	X1.2	X1.3	X2.1	X2.2	X2.3	Z.1	Z.2	Z.3
1	3	10	14	0	3	5	0	0.2	0.7	0
2	10	20	40	10	1	2	3	0.4	0.3	0.2

Instead of using `X1` as a source variable for a reach, we want to create three source variables, one for each of the maximum number of three reaches above each monitoring site. For monitoring sites with less than the maximum number of reaches, reaches of 0 source value are added to simplify the subsequent coding. With this data frame, we can implement the nonlinear regression.

If the number of reaches above each monitoring sites is the same (let the number be 3), we can use the function `tapply`, one source variable at a time:

```
> X1.temp <- tapply(GISdata$X1, GISdata$DWNSTID, as.vector)
```

which goes through each unique value of `DWNSTID` and lists the corresponding values of `X1` as vectors. When the number of reaches is the same, we will have a vector combining the source term for all six reaches. We can convert the resulting vector into a matrix.

In our case, the number of reaches is not the same, the above line will return a list of two vectors, one of length 2 and the other of length 3. When using `tapply`, we can write our own function, instead of using existing functions such as `mean` and `as.vector`. In this case, we want the function to return a fixed length (maximum number of reaches) vector. When the number of reaches is smaller than the maximum, we fill the remaining with 0s:

```
id <- as.numeric(ordered(GISdata$DWNSTID))
idtbl <- table(id) ## tabulate
ns <- max(idtbl) ## maximum number of reaches
nr <- max(id) ## number of monitoring sites

temp <- tapply(GISdata$X1,GIDdata$DWNSTID,FUN=function(x,ns=nc){
               tt <- as.vector(x)
               if (length(tt) < ns)
                  tt <- c(tt, rep(0, ns-length(tt)))
               return(tt)})
temp <- as.data.frame(mxtrix(unlist(temp),nrow=nr,
                       ncol=nc,byrow=T))
```

Because there are multiple source terms, we can put these lines into a function so that we do not have to repeatedly type and change them:

```
oo <- order(GISdata$DWNSTID) ## sort by monitoring site
GISdata <- GISdata[oo,]
Y <- GISdata$Load[!is.na(GISdata$Load)] ## Load data
id <- as.numeric(ordered(GISdata$DWNSTID))
idtbl <- table(id)
ns <- max(idtbl)
nr <- max(id)

my.unstack <- function(X, ID, nc, x.names){
  temp <- tapply(X, ID, FUN = function(x, ns=nc){
                   tt <- as.vector(x)
                   if (length(tt < ns))
                      tt <- c(tt, rep(0, ns-length(tt)))
```

```
                    return(tt)})
  temp <- as.data.frame(matrix(unlist(temp), nrow=nr, ncol=ns,
         byrow=T))
  names(temp) <- x.names
  return(temp)
}
X1.names <- paste("X1", 1:ns, sep="_")
X2.names <- paste("X2", 1:ns, sep="_")
Z1.names <- paste("Z1", 1:ns, sep="_")
X1 <- my.unstack(GISdata$X1, GISdata$DWNSTID, nc=ns, X1.names)
X2 <- my.unstack(GISdata$X2, GISdata$DWNSTID, nc=ns, X2.names)
Z1 <- my.unstack(GISdata$Z1, GISdata$DWNSTID, nc=ns, Z1.names)
GISdata_reshaped <- cbind (Y, X1,X2, Z1)
```

2.4.5 Dates

A commonly used method for processing dates and time in computer programming is the POSIX standard. It measures dates and times in seconds since the beginning of 1970 in UTC time zone. In R, `POSIXct` is the R date class for this standard. The `POSIXlt` class breaks down the date object into year, month, day of the month, hour, minute, and second. The `POSIXlt` class also calculates day of the week and day of the year (Julian day). The `Date` class are similar but with dates only (without time).

Typically, dates are entered as characters. For example, dates are typically entered in the U.S. using numeric values in a format of `mm/dd/yyyy` (e.g., 5/27/2000) or with month name plus numeric day and year (e.g., December 31, 2013). When reading into R, the date column becomes a factor variable. We can use the function `as.Date` to convert the factor variable into dates:

```
> first.date <- as.Date("5/27/2000", format="%m/%d/%Y")
> second.date<-as.Date("December 31, 2003",format="%B %d, %Y")
> second.date - first.date
```

The first two lines convert two character strings to date class objects. As date objects are numeric (days since January 1, 1970), we can use them to calculate days eclipsed between two dates. A more general function for converting date-time objects is `strptime`, which converts a date-time character string to a POSIXlt class object, measuring time in seconds since the beginning of 1970.

```
first.d <- strptime("5/27/2000 22:15:00",
                    format="%m/%d/%Y %H:%M:%S")
second.d <- strptime("December 31, 2003, 4:25:00",
                     format="%B %d, %Y, %H:%M:%S")
second.d - first.d
```

The format of a date object is defined by the POSIX standard, consisting of a "%" followed by a single letter. Table 2.3 lists some of them.

TABLE 2.3: Date formats in R date-time classes

Format	Description
%a	Abbreviated weekday name in the current locale on this platform
%A	Full weekday name in the current locale
%b	Abbreviated month name in the current locale on this platform
%B	Full month name in the current locale
%c	Date and time (%a %b %e %H:%M:%S %Y)
%C	Century (00-99)
%d	Day of the month as decimal number (01-31)
%D	Date format %m/%d/%y
%e	Day of the month as decimal number (1-31)
%F	Equivalent to %Y-%m-%d (the ISO 8601 date format)
%G	The week-based year as a decimal number
%h	Equivalent to %b
%H	Hours as decimal number (00-23)
%I	Hours as decimal number (01-12)
%j	Day of year as decimal number (001-366)
%m	Month as decimal number (01-12)
%M	Minute as decimal number (00-59)
%n	New line on output, arbitrary whitespace on input
%p	AM/PM indicator in the locale
%r	The 12-hour clock time (using the locale's AM or PM)
%R	Equivalent to %H:%M
%S	Second as decimal number (00-61)
%t	Tab on output, arbitrary whitespace on input
%T	Equivalent to %H:%M:%S
%u	Weekday as a decimal number (1-7, Monday is 1)
%U	Week of the year as decimal number (00-53) using Sunday as the first day 1 of the week
%V	Week of the year as decimal number (01-53) as defined in ISO 8601
%w	Weekday as decimal number (0-6, Sunday is 0)
%W	Week of the year as decimal number (00-53) using Monday as the first day of week
%y	Year without century (00-99)
%Y	Year with century
%z	Signed offset in hours and minutes from UTC, so -0800 is 8 hours behind UTC.

Once a date object is created, we can extract relevant information associated with dates using function `format`. We now create a data frame with a date column:

```
> mytime <- data.frame(x = rnorm(100),
+                      date=as.Date(round(runif(100)*5000),
+                        origin="1970-01-01"))
```

We can add a column of month and a column of week days to the data frame:

```
> mytime$Month <- format(mytime$date, "%b")
> mytime$weekday <- format(mytime$date, "%a")
```

We can also store date object as a POSIXlt class object, which is a list of nine elements: (1) seconds, (2) minutes, (3) hours, (4) day of month (1-31), (5) month of the year (0-11), (6) years since 1900, (7) day of the week (0, Sunday, through 6), (8) day of the year (0-365), and (9) daylight savings indicator. If we want to extract day of the year, month, and year as numeric vectors, we can simply assign the eighth (Julian day), fifth (month), and sixth (year) elements:

```
mytime$Julian <- as.POSIXlt(mytime$date)[[8]]+1
mytime$Month <- as.POSIXlt(mytime$date)[[5]]+1
mytime$Year <- as.POSIXlt(mytime$date)[[6]]+1900
```

2.5 Exercises

1. Using R as a calculator to perform the following operations:

    (a) Calculate the area of a circle $A = 2\pi r^2$ with $r = 2$;

    (b) Calculate the density of the normal distribution $x \sim N(2, 1.25)$ (mean and standard deviation) at `x <- seq(0,4,0.5)` by using the normal density formula ($\frac{1}{\sqrt{2\pi}\sigma}e^{-\frac{(x-\mu)^2}{2\sigma^2}}$), and verify your result by using the function `dnorm`.

2. The 10% rule:

    (a) In the example discussing the 10% rule, we used two hypothetical waters with pollutant concentration distributions $N(2, 0.75)$ (indicating that the water is in compliance) and $N(2, 1)$ (indicating that the water is impaired), respectively. Use simulation to evaluate the performance of the 10% rule by estimating the probability of making mistakes when sample size is low (`n=10`) and large (`n=100`) and discuss the implications of the poor performance.

(b) Use the function `viol.sim` to calculate the error rates for the impaired water (with pollutant concentration distribution $N(2,1)$) with sample sizes `n = 6, 12, 24, 48, 60, 72, 84, 96` and present the result in a plot. Repeat the same for a water that is not impaired.

(c) Write a short essay on the 10% rule, discussing the consequences of the rule in terms of the probability of making two type of mistakes – declaring a water to be impaired while the water is in compliance and vice versa.

3. Carl Sagan's intelligence data. In his book, *The Dragons of Eden*, Carl Sagan presented a graph showing the brain and body masses, both on log scale, of a collection of animal species. The purpose of the graph was to describe an intelligence scale: the ratio of the average brain weight over the average body weight to the power of 2/3.

   (a) Read the data (in file `Intelligence.csv`) into R and graph brain weight against body weight, both in logarithmic scales. The brain weight is in grams and body weight is in kilograms. Can you tell which species has the highest intelligence from this figure?

   (b) Calculate the intelligence measure (call it `Int`) and add the result as a new column to the data frame.

   (c) Use the function `dotplot` from package `lattice` to plot the intelligence measure directly:
   `dotplot(Species~Int, data=Intelligence)`

   (d) The dot plot orders the species alphabetically. Reorder the column based on the intelligence scale using function `ordered` and redraw the dot plot so that the species are sorted based on their intelligence scales. Is there a problem in the data? If so, what might be the cause of the problem?

4. The Heidelberg University in Tiffin, Ohio, U.S.A., maintains a long-term monitoring program of several Lake Erie tributaries. The water quality and flow data from the Maumee River station near Waterville, Ohio, can be found at `http://www.heidelberg.edu/sites/default/files/dsmith/files/MaumeeData.xlsx`.

   (a) Convert the date column into R dates.

   (b) Summarize flow and total phosphorus by year and by month.

   (c) Plot flow and TP over time.

Chapter 3

Statistical Assumptions

3.1	The Normality Assumption	48
3.2	The Independence Assumption	54
3.3	The Constant Variance Assumption	55
3.4	Exploratory Data Analysis	56
	3.4.1 Graphs for Displaying Distributions	57
	3.4.2 Graphs for Comparing Distributions	59
	3.4.3 Graphs for Exploring Dependency among Variables	61
3.5	From Graphs to Statistical Thinking	69
3.6	Bibliography Notes	72
3.7	Exercises	73

Statistical inference is about probabilistic distributions of data, model error, and model parameters. Because statistical thinking is inductive in nature, distributional assumptions are the basis of statistical analysis and inference. When using a statistical procedure, exploratory analysis of the data should be performed to check whether these assumptions are met. Although some statistical procedures are robust against the departure of probabilistic assumptions, understanding these underlying assumptions is an important part of learning statistics. Especially when applying statistics in ecological and environmental studies, an understanding of these assumptions will help us avoid some commonly seen mistakes in applications of statistics. A thorough exploratory analysis in terms of statistical assumptions often relies on graphical presentation of data. Graphical presentation of the data often serves as the bridge linking an ecological or environmental problem and an abstract statistical representation of the problem. I emphasize graphical procedures for checking important assumptions. In this chapter, three commonly used assumptions are briefly discussed. In the rest of the book, assumptions of each statistical procedure will be discussed in detail with graphical methods for checking the compliance of these assumptions.

3.1 The Normality Assumption

The most commonly used distributional assumption is the normality assumption which assumes a normal distribution for the quantity of interest. In 1809, Carl Friedrich Gauss published a monograph commonly known as *Theoria Motus* (see, for example, *Theory of Motion of the Heavenly Bodies Moving About the Sun in Conic Sections: A Translation of Theoria Motus*, Dover Phoenix Editions, ISBN 0486439062). In it, Gauss derived the probability law of measurement error as a justification for using the least squares method for estimating a mean. This probability law was later known as the normal or Gaussian distribution. Pierre-Simon Laplace published the central limit theorem in 1812 [Stigler, 1975], which states that the distribution of sample averages of independent random variables can be approximated by the normal distribution, regardless of the original distribution from which these random variables were drawn. The closer the original distribution is to normal, the better the approximation will be, particularly with small sample sizes (see Figure 4.1 on page 83).

In environmental studies, the normal distribution is particularly important because many environmental variables (concentration variables in particular) can be approximated by the log-normal distribution [Ott, 1995]. Thus, a rule of thumb in environmental statistics is that we should log-transform concentration variables before statistical analysis [van Belle, 2002], so that properties of normal distributions can be used advantageously.

The normal distribution is defined by two parameters, the mean (μ) and the standard deviation (σ). The probability density function of a normal random variable Y is

$$\frac{1}{\sqrt{2\pi}\sigma}e^{-\frac{(y-\mu)^2}{2\sigma^2}} \tag{3.1}$$

In the rest of this book, this distribution will be denoted as $N(Y|\mu, \sigma)$ or simply $N(\mu, \sigma)$ when the random variable in question is unambiguous, indicating the random variable Y has a normal distribution with mean μ and standard deviation σ. Once the distribution parameters μ and σ are known, we can make a statistical inference about Y using the density function. The standard (or unit) normal distribution ($\mu = 0, \sigma = 1$) density function is shown in Figure 3.1. Three quantities are of interest in statistical inference: the *quantile* or *percentile* (y), the *cumulative probability* or the lower tail area, and the density of y also known as the *likelihood* of observing y.

The density of y is calculated using the density function in equation 3.1. For example, the density for $y = 0.5$ is $\frac{1}{\sqrt{2\pi}1}e^{-\frac{(0.5-0)^2}{2\times1^2}} = 0.352$. The density value is the height of the curve at y, and it is meaningless by itself. However, this quantity is of crucial importance in statistical estimation. It represents the relative likelihood of observing the value in question. For example, the density of $y = 0$ is 0.399, indicating that it is more likely to observe a value of

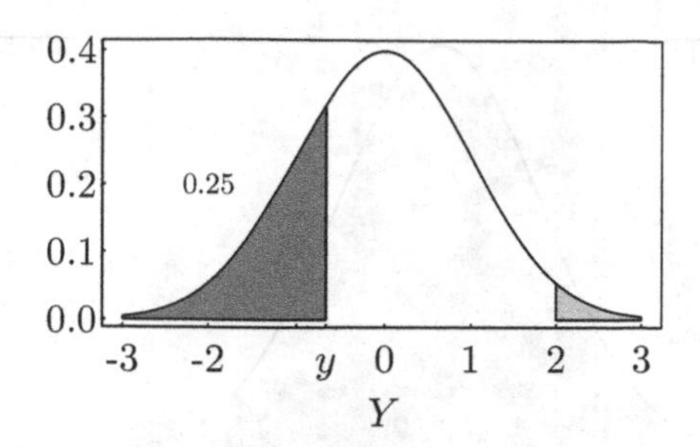

FIGURE 3.1: The standard normal distribution – the dark shaded area to the left has an area of 0.25 (y is the 0.25 quantile or 25th percentile) and the size of the light shaded area is the probability of seeing a value larger than 2 (~ 0.023).

0 than to observe a value of 0.5 when the underlying probability distribution of the random variable is the standard normal distribution. Likewise, if we don't know the mean μ (and $\sigma = 1$), the likelihood of observing $y = 0$ is 0.399 if $\mu = 0$, and the likelihood is 0.242 if $\mu = 1$, indicating that μ is more likely to be 0 than to be 1. The density value can be calculated using the R function `dnorm`:

```
> dnorm(0.5, mean=0, sd=1)
  [1] 0.3520653
```

The cumulative probability of y is the area under the density curve to the left of y (the dark shaded area in Figure 3.1) or $\Phi(y) = \int_{-\infty}^{y} \frac{1}{\sqrt{2\pi}\sigma} e^{-\frac{(y-\mu)^2}{2\sigma^2}} dy$, the probability of observing a value less than or equal to y. The R function `pnorm` is used to calculate the cumulative probability:

```
> pnorm(0.5, mean=0, sd=1)
  [1] 0.691462
```

To calculate the probability of observing a value larger than or equal to $y = 0.5$:

```
> 1 - pnorm(0.5, mean=0, sd=1)
  [1] 0.308538
```

The percentile is the opposite of the cumulative probability – calculating the y value from a known cumulative probability value. It is calculated using R function `qnorm`:

```
> qnorm(0.25, mean=0, sd=1)
  [1] -0.67449
```

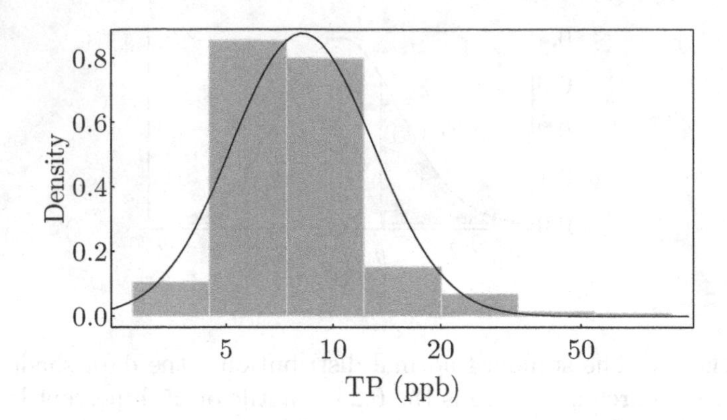

FIGURE 3.2: Everglades background TP concentration distribution – TP concentrations from reference sites in the Everglades are shown in the histogram. The dark line is the log-normal distribution based on the mean and standard deviation of the data.

```
> qnorm(0.05, mean=0, sd=1)
  [1] -1.64485

> qnorm(0.95, mean=0, sd=1)
  [1] 1.64485
```

That is, 25% of the values from $N(y|0, 1)$ are less than -0.674 (y in Figure 3.1), 5% less than -1.645, and 95% less than 1.645, or the middle 90% of the values are bounded between ± 1.645.

When checking whether a sample is from a normal distribution, we often use two graphical methods. First, a histogram of the data can be used to show whether the distribution of the data is approximately symmetric. Figure 3.2 shows the histogram of annual geometric means of TP observed from several reference sites in the Everglades. The data are shown in the logarithm scale. This data set was used by the Florida Department of Environmental Protection for estimating the reference distribution when setting the TP standard for the Everglades. The distribution shown in Figure 3.2 is apparently asymmetric, atypical of an environmental concentration variable. Because the measured variable is a mean variable, the central limit theorem suggests that the distribution of the mean should be close to normal. Why are the observed data showing a skewed distribution? To answer this question, we must dig into the details of the data set. The data set consists of data points from several sampling sites in the Everglades collected over many years. When combining these annual log averages into a single data set, we imply that these sites have TP concentration distributions similar to each other and the TP concentration distributions for these sites are the same over the entire sampling period. We will get to the answer to the question later in Chapter 4. For now,

TABLE 3.1: Model-based percentiles versus data percentiles – Selected percentiles estimated using the log normal model are compared to the same calculated from the data

	5%	10%	25%	50%	75%	90%	92%	95%
Data	4.00	5.00	6.00	8.00	10.00	14.00	15.00	20.00
Lognormal	3.88	4.58	6.04	8.21	11.17	14.73	15.58	17.38

we comment on the consequences of the normality assumption applied to this data set.

The log mean and standard deviation of these annual averages are 2.11 and 0.46, respectively, representing the estimated population mean and standard deviation of the background annual mean TP concentration in the Everglades. From this distribution, the 75th percentile is estimated to be `qnorm(0.75, mean=2.11, sd=0.46) = 2.42` (or 11.25 ppb). Table 3.1 lists selected percentiles estimated from the estimated normal distribution and percentiles directly calculated from the data. The estimated normal distribution (Figure 3.2) cannot adequately capture the distribution of the observed data. As a result, estimated parameter of interest (the 75th percentile, Table 3.1) is potentially biased. In this case, the estimated normal distribution underestimates the chance of observing large TP concentrations.

The second method for checking the normality assumption is the normal quantile-quantile (or Q-Q) plot. A normal Q-Q plot is constructed using the relationship between the q-quantile (Section 3.4.1) of a normal distribution $N(\mu, \sigma)$ (y_q) and the same quantile of the standard normal distribution (z_q), that is, $y_q = \mu + \sigma z_q$. If the sample is from a normal distribution, a straight line will form when plotting the quantiles from the data against the same quantiles from the standard normal distribution. The line should have an intercept of μ and a slope of σ. Although exact percentiles are unknown, quantiles estimated directly from data should be close to the true values if the data are drawn from a normal distribution. The y-axis of the Q-Q plot is the estimated quantiles of data, and the x-axis is the quantiles of the standard normal distribution. A reference line with an intercept of $\bar{y}$ (sample average of the data) and slope of $\hat{\sigma}$ (sample standard deviation of the data) is superimposed. Quantiles can be estimated by first sorting the data in ascending order: $y^{(1)}, y^{(2)}, \cdots, y^{(n)}$, and assigning an approximate quantile

$$\frac{i - 0.5}{n}$$

to the ith value $y^{(i)}, i = 1, \cdots, n$, or in R:

```
yq <- ((1:n) - 0.5)/n
```

The respective quantiles of the standard normal distribution can be estimated by using `qnorm`:

```
#### R Code ####
y <- rnorm(100)
n <- length(y)
yq <- ((1:n) - 0.5)/n
zq <- qnorm(yq, mean=0, sd=1)
plot(zq, sort(y), xlab="Standard Normal Quantile", ylab="Data")
abline(mean(y), sd(y))
```

The calculated data quantiles are approximations. As a result, a normal Q-Q plot may not show a perfect straight line even when the data are from a normal distribution. To gain a sense of how much deviation of the normal Q-Q plot is acceptable, we can repeatedly run the above R scripts. Instead of typing these lines, we can put them into a function:

```
#### R Code ####
my.qqnorm <- function (y=rnorm(100)){
  n <- length(y)
  yq <- ((1:n) - 0.5)/n
  zq <- qnorm(yq, mean=0, sd=1)
  plot(zq,sort(y),xlab="Standard Normal Quantile",ylab="Data")
  abline(mean(y), sd(y))
  invisible()
}
```

and call the function repeatedly:

```
> my.qqnorm()
> my.qqnorm(rnorm(20))
```

This is equivalent to calling the R functions `qqnorm` and `qqline`:

```
> qqnorm(y)
> qqline(y)
```

Quite often we see data points on both ends of the plot deviate from the straight line when running the above scripts repeatedly, which is in part because the estimated data quantiles on both ends are likely to be inaccurate. As such, when comparing real data to a normal distribution, we should take this behavior into consideration. However, a systematic pattern of departure from the straight line should be considered as evidence of normality violation. The histogram in Figure 3.2 indicates that the log TP concentration distribution is not symmetric; there are more large values than expected if the underlying distribution is a normal distribution. This pattern is apparent in a normal Q-Q plot (Figure 3.3).

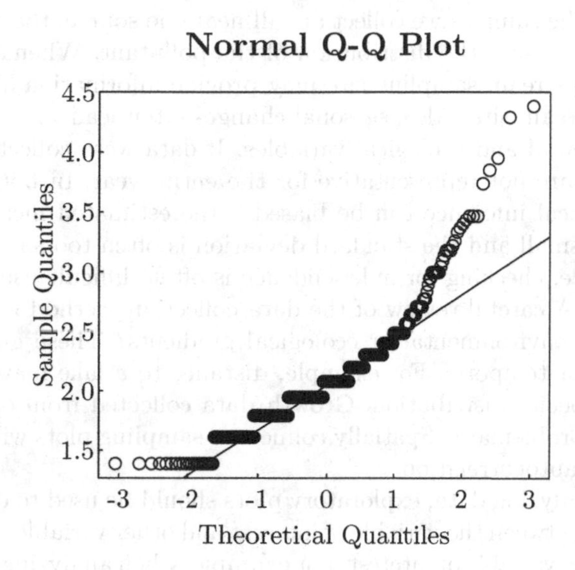

FIGURE 3.3: Q-Q normal plot of TP concentrations from reference sites in the Everglades shows a systematic departure from the normal distribution. The observed data points are more likely to deviate away from the reference line when the TP values are high. This is a typical right-skewed data distribution (see Figure 3.2), which has more high values than expected.

3.2 The Independence Assumption

Intuitively independence of two events means that the occurrence of one event makes it neither more nor less probable that the other occurs. Applied to environmental and ecological studies, the independence assumption is often used to describe observations that are random and uncorrelated. Knowing the value of one observation will give us no information on the next observations. Two situations may lead to dependency among observations: clustering and serial correlation. When high pollutant concentrations are clustered around sources and the samples we collect are all near one source, the samples cannot be used to represent the distribution of the pollutant. When sampling along a river, an upstream sampling site may provide information about an immediate downstream site. Also, seasonal changes often lead to seasonal patterns in environmental and ecological variables. If data were collected only in one season, they are not representative for the entire year. In both cases, subsequent statistical inference can be biased – the estimated means may be too large or too small and the standard deviation is often too small.

In practice, checking for independence is often difficult using only the observed data. A careful review of the data collection method is necessary. Are there known environmental or ecological gradients? These gradients can be spatial and/or temporal. For example, distance to a lake may determine the pattern of species distribution. Growth data collected from one animal over time can be problematic. Spatially connected sampling plots will result in data with spatial autocorrelation.

Before analyzing data, exploratory plots should be used to detect potential correlations between the variable of interest and other variables that may cause changes in the variable of interest. For example, when analyzing the Everglades TP concentration data, researchers often plot the observed TP concentrations against the distance between the sampling location and the pumping stations that distribute water (including agriculture runoff) in the area. These pumping stations (the three arrows on the northern border of WCA2A in Figure 1.1) are the main anthropogenic sources of phosphorus in the study area. Likewise, when studying harmful algal blooms in the western basin of Lake Erie, where the main nutrient source is the Maumee River entering the lake near Toledo, Ohio, to the west of the lake, distance to the Maumee River mouth is often a good indicator of nutrient concentration levels (Figure 3.4).

Plots like Figure 3.4 are often used to show spatial correlation. When spatial or temporal correlations are present, we need to introduce variables to model the correlation. This is because the independence assumption is not always imposed on the *raw* data. In most statistical modeling situations, the independence assumption is imposed on model residuals. Consequently, introducing distance as a covariate can often reduce the spatial correlation in residuals.

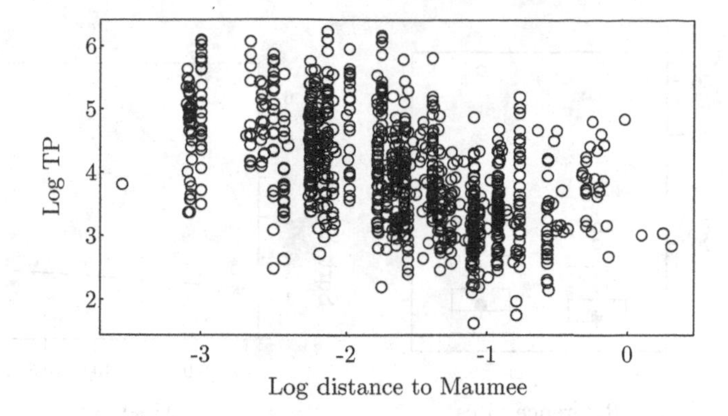

FIGURE 3.4: Scatter plot of log TP concentration against log distance to Maumee River mouth (calculated using latitude and longitude) shows a spatial pattern.

3.3 The Constant Variance Assumption

A constant variance (or standard deviation) among populations is a necessary condition when comparing means among populations. If the standard deviations are different, comparing means is less meaningful. When the difference among the populations is in their means only, comparing means will be sufficient to reveal the nature of the difference. When applying t-test or analysis of variance, we imply that the differences in different populations are in their means only. The constant variance assumption is also imposed on residuals of a linear or nonlinear model.

Conceptually, standard deviation is a measure of spread, representing the "typical" distance between each data point and the population mean. Because this "typical" distance can never be observed, graphical display is difficult. One effective graphical tool for checking this assumption is the S-L plot suggested by Cleveland [1993]. In this plot, standard deviation is represented by the median of the distances between data points and their mean. For example, when comparing the standard deviation of two data sets $X : (x_1, \cdots, x_n)$ and $Y : (y_1, \cdots, y_m)$, we calculate the distances between each data value minus their respective mean ($|\varepsilon_{x_i}| = |x_i - \bar{x}|$ and $|\varepsilon_{y_j}| = |y_j - \bar{y}|$). The median of these differences is called the *median absolute deviances* (*mads*). Because *mads* is the median of distances to the mean, we can consider it as a typical distance. As a result, *mads* can be considered as a measure of standard deviation. We can compare the *mads* of Y and *mads* of X as a substitute for directly comparing standard deviations because the *mads* are relatively easy to visualize: plotting $|\varepsilon_{x_i}|$ against $\bar{x}$ and $|\varepsilon_{y_j}|$ against $\bar{y}$, and connecting the

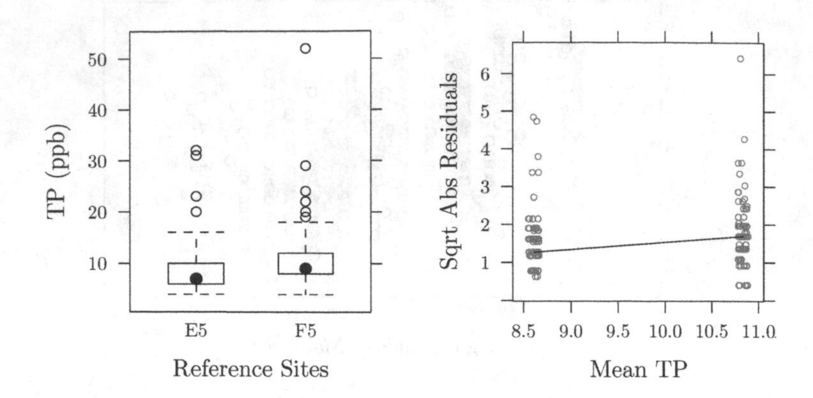

FIGURE 3.5: Comparing standard deviations using S-L plot – The left panel shows boxplots of TP concentrations from reference sites in the Everglades. The right panel shows the S-L plot from the same TP concentrations. The data points in the S-L plots are the *mads* and the dark line connects the two medians.

two *mads* with a straight line. Cleveland [1993] recommended that we take the square root of $|\varepsilon_{x_i}|$ and $|\varepsilon_{y_j}|$ before plotting because the distribution of these *mads* is highly skewed. The S-L plot converts a measure of spread (standard deviation) into a measure of location (*mads*), thereby facilitating visual comparisons of standard deviations.

Figure 3.5 (left panel) shows boxplots of TP concentrations measured in two of the five reference sites in the Everglades Wetland. Although the height of a boxplot is a measure of spread, we can hardly tell whether the spread of the two distributions is different because the boxes are not lined up for easy comparison. But when using the S-L plot (Figure 3.5, right panel), the difference, although small, is obvious from the plot. One feature stands out from the data is that the standard deviation of TP increases as the mean increases. This feature, known as *monotone spread*, is common in environmental and ecological data. S-L plot is possibly the best tool for detecting a monotone spread.

3.4 Exploratory Data Analysis

All statistical models are based on one or more assumptions about the data distribution and the nature of association among variables. When learning statistical methods, a key skill is the awareness of statistical assumptions and

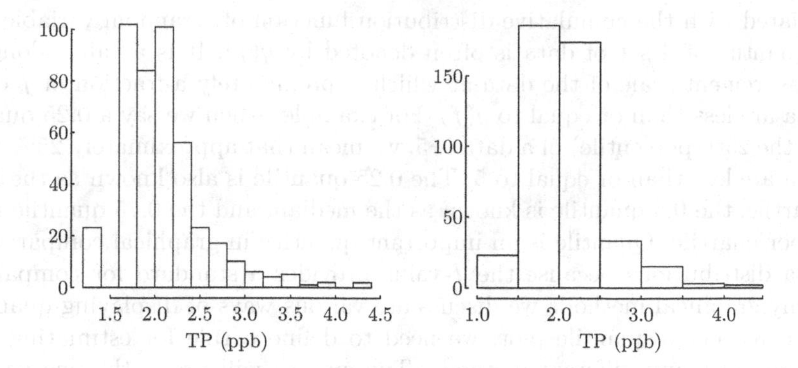

FIGURE 3.6: Histograms of Everglades TP concentrations – Two histograms constructed from the same data show that the shape of the graph is dependent on the number of bins used.

knowledge in how to assess the compliance of these assumptions. Many mistakes in applications of statistics stem from assumption violations. The first step in a typical data analysis work should be the exploratory data analysis for assessing data distribution, potential correlations, and possible problems with the data. We will focus on graphical analysis of the data. These graphs are designed to display data distribution and other specific features. Most of these graphical tools are discussed in Cleveland [1993]. In this section, some of the frequently used graphs are introduced.

3.4.1 Graphs for Displaying Distributions

Graphically displaying a distribution is the most commonly used approach for checking the normality assumption. Many alternatives can be used. The most frequently used method is the *histogram*. A histogram shows the data distribution by dividing the range of the data into bins and displaying the number (or fraction) of data points falling into each bin. A histogram is the most straightforward method for showing whether a distribution is approximately symmetric. Figure 3.6 shows two histograms of the Everglades TP concentration data used by the FDEP for estimating the background TP concentration distribution. The difference between the two histograms, the numbers of bins, shows the limitation of a histogram as a means for judging normality. The shape of a histogram depends on the number of bins. Often we rescale the height of the bars in a histogram such that the areas of the bars sum to 1. This rescaling will not change the shape of the histogram, but it allows us to superimpose an estimated probability distribution (or density) function onto the histogram (black lines in Figure 3.2).

While the shape of a histogram depends on the number of bins we use, a *quantile plot* can accurately reflect the data distribution. Quantiles are associated with the cumulative distribution function of a random variable. The f quantile of a set of data is often denoted by $q(f)$. It is a value along the measurement scale of the data at which approximately a fraction of f of the data are less than or equal to $q(f)$. For example, when we say a 0.25 quantile (or the 25th percentile) of a data is 5, we mean that approximately 25% of the data are less than or equal to 5. The 0.25 quantile is also known as the lower quartile, the 0.5 quantile is known as the median, and the 0.75 quantile is the upper quartile. Quantile is an important quantity in graphical comparison of data distributions because the f-value provides a standard for comparison. Many graphical methods we discuss are various ways of displaying quantiles. To construct a quantile plot, we need to define a rule for estimating $q(f)$. There are many different methods. The one we will use is the one used by R. The data points $x_{(i)}$, for $i = 1$ to n, are ordered from the smallest to the largest, ($x_{(1)}$ is the smallest and $x_{(n)}$ is the largest). For each data point, record what percentile of the data each value occupies:

$$f_i = \frac{i - 0.5}{n} \tag{3.2}$$

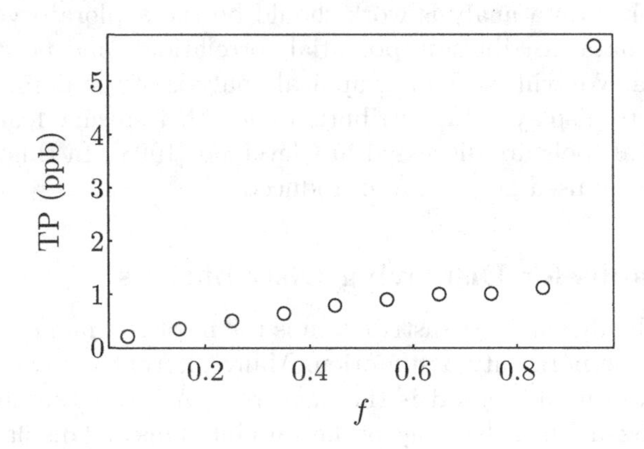

FIGURE 3.7: An example quantile plot – A quantile plot of the example TP concentration data shows all data points and their quantiles.

These numbers increase in equal steps of $1/n$ beginning with $\frac{1}{2n}$, which is slightly above zero, and ending with $1 - \frac{1}{2n}$, which is slightly below one. We will take $x_{(i)}$ to be $q(f_i)$. For example, a subset of 10 TP concentration values has the following f-values:

f	TP	f	TP	f	TP
0.05	0.21	0.45	0.79	0.75	1.01
0.15	0.35	0.55	0.90	0.85	1.12
0.25	0.50	0.65	1.00	0.95	5.66
0.35	0.64				

The 0.35 quantile is 0.64. The f-values not calculated from this definition (e.g., 0.10 and 0.99) are extended through linear interpolation or extrapolation. On a quantile plot, $x_{(i)}$ is graphed against f_i (Figure 3.7).

A histogram shows the general shape of a distribution and a quantile plot displays quantiles of all data points. But some times we are interested in certain statistics of a distribution. The Tukey's *box-and-whisker plot* (or *boxplot*) is such a device. In a boxplot, we display the mean (and/or median), 25th and 75th percentiles, and the outlying adjacent values on both directions. A boxplot shows the range of the middle 50% of the data, and from the location of the median line we can judge whether the distribution is approximately symmetric. A boxplot is not intended for checking the normality assumption. It is a general purpose graphical device for summarizing a data set. Figure 3.8 shows the relationship between a boxplot and a quantile plot, originally shown in Cleveland (1993).

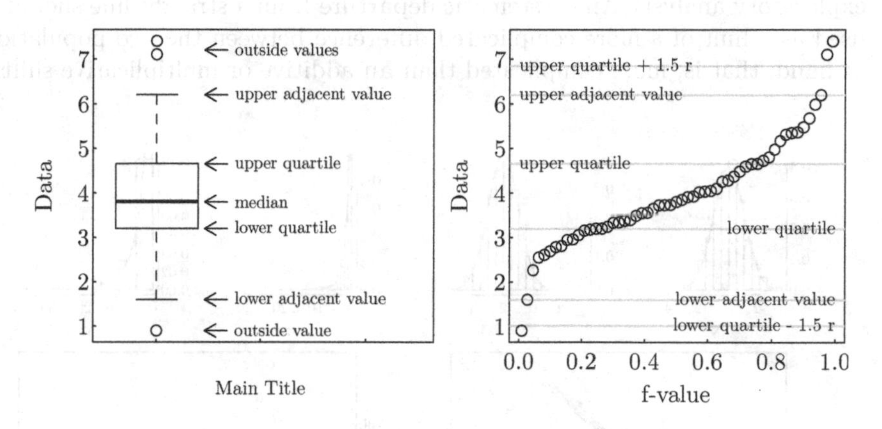

FIGURE 3.8: Explaining the boxplot – The boxplot (left panel) is explained by the quantile plot on the right panel. Both graphed using a generated data set. (Used with permission from Cleveland [1993])

3.4.2 Graphs for Comparing Distributions

When comparing distributions of two or more data sets, we use the *quantile-quantile plot* or *Q-Q plot*. The Q-Q plot is constructed by pairing data points from two data sets that have the same quantiles and graphing these pair on a bivariate scatter plot. The objective of the plot is to understand how the distributions shift in going from one data set to the next. If the two distributions are the same, a Q-Q plot will consist of points that fall around a straight line with slope 1 and intercept 0. If these points fall around a line with a nonzero intercept (but still has a slope of 1), the difference between the two distributions is *additive*. That is, the two distributions differ by a constant. The constant is the difference between the same quantiles from the two distributions. Figure 3.9 (left panel) shows a Q-Q plot comparing two data sets from normal distributions with different means but the same standard deviation. If these points fall around a line with a slope not equal to 1, the two distributions differ both in location and spread, but the distributions have similar shapes. If the intercept is 0, the difference between the two distributions is *multiplicative*. That is, the two distributions differ by a multiplicative factor. Figure 3.9 (right panel) shows a Q-Q plot of two data sets from log-normal distributions with different means and standard deviations. When these points do not fall around a straight line, the difference between the two distributions is more complicated. Data used for generating Figure 3.9 were random samples from normal (left panel) and log normal distributions (right panel). Even when the difference between the underlying distributions are strictly additive or multiplicative, departure from a straight line should be expected, especially near both ends of the data range. A Q-Q plot is a tool for exploratory analysis. Any systematic departure from a straight line should be used as a hint of a more complicated difference between the two populations at hand, that is, more complicated than an additive or multiplicative shift.

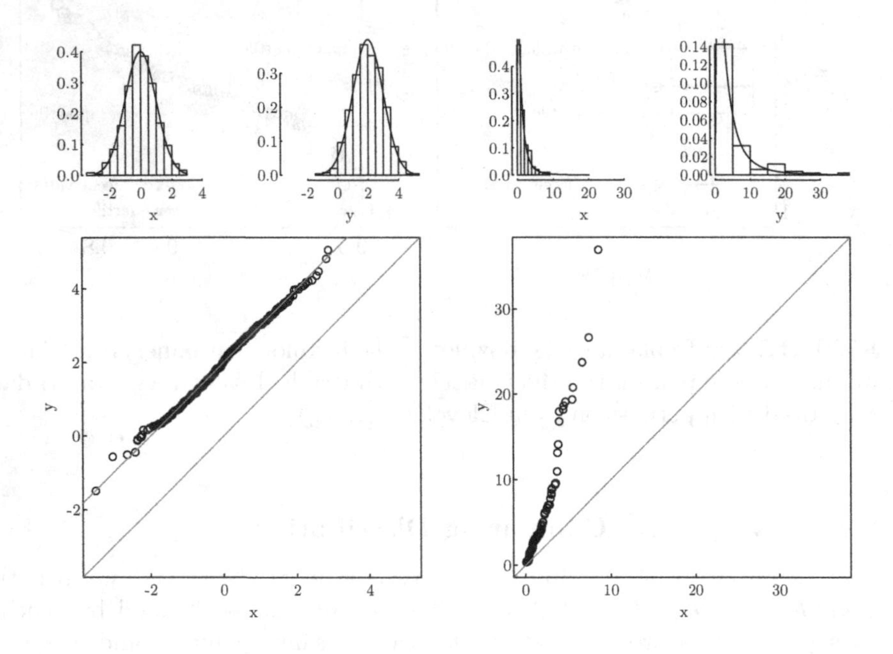

FIGURE 3.9: Additive versus multiplicative shift in Q-Q plot – The left panel shows an example of an additive shift. The Q-Q plot consists of data points falling around a straight line parallel to the reference 1-1 line. The right panel shows an example of multiplicative shift. The Q-Q plot consists of data points falling around a straight line intercepting the reference 1-1 line at 0.

A Q-Q plot graphs quantiles of one distribution against the corresponding quantiles of the other. Suppose that we have data set 1: $x_{(1)}, \cdots, x_{(n)}$ and data set 2: $y_{(1)}, \cdots, y_{(m)}$ and $m \leq n$. If $m = n$, then $y_{(i)}$ and $x_{(i)}$ are both $(i - 0.5)/n$ quantiles of their respective data sets, so on the Q-Q plot, $y_{(i)}$ is graphed against $x_{(i)}$; that is, ordered values of one set of data are graphed against the ordered values of the other set. If $m < n$, then $y_{(i)}$ is the $(i-0.5)/m$ quantile of the y data, and we graph $y_{(i)}$ against the $(i - 0.5)/m$ quantile of the x data, which typically must be computed by interpolation. With this method, there are always m points on the graph, the number of values in the smaller of the two data sets. Of course, if m is a big number, say 10^3, then we select fewer quantiles for comparison.

Normal Q-Q plot is a special Q-Q plot comparing a data distribution to the standard normal distribution. The purpose of a normal Q-Q plot is to visually assess whether the underlying distribution of a data set is likely a normal distribution. The plot is produced by plotting the data quantiles (calculated by $(i - 0.5)/n$) against their corresponding standard normal distribution quantiles. For example, the estimated quantile for $x_{(4)}$ from a data set with $n = 100$ is $(4 - 0.5)/100 = 0.035$, and the 0.035 quantile of a unit normal distribution is `qnorm(0.035)` or -1.812. In a normal Q-Q plot, the datum $x_{(4)}$ is located with the x-axis value of -1.812 and y-axis value is $x_{(4)}$ itself.

In R, the Q-Q plot is implemented in functions `qq` (from the `lattice` package) and `qqplot`. The `lattice` function `qqmath` can be used to compare the data to a number of distributions.

3.4.3 Graphs for Exploring Dependency among Variables

Bivariate scatter plots are the most commonly used graphical tool for displaying dependency between two variables. In showing a scatter plot, we try to convey the information that the two variables displayed in the figure are either correlated or independent of each other. In presenting a scatter plot, there are two considerations. One is the loess curve. The other is variable transformation.

The term loess, from the German *löss*, is short for local regression. It is one of the curve fitting methods that is often referred to as nonparametric regression. When plotting a bivariate scatter plot, fitting a loess curve can help us detect a nonlinear relationship. For example, the 1990 April issue of *Consumer Reports* provided information for new cars. Figure 3.10 shows the scatter plot of fuel consumption (miles per U.S. gallon, or mpg) against weight from the data set. The left panel shows an often used method in showing a scatter plot: fitting a straight line. The right panel shows a loess line, indicating that the relationship between `Mileage` and `Weight` is likely nonlinear, which is not obvious when only the straight line is shown. We will discuss the details of nonparametric curve fitting in Chapter 6. For now, we simply state that a loess line is a line that traces the center of the scatter plot data cloud. It is used to help us better judge the nature of the bivariate relationship, especially to detect departures from a linear relationship.

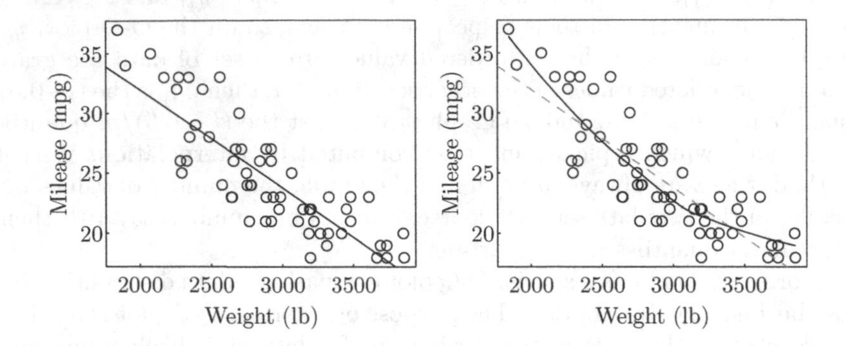

FIGURE 3.10: Bivariate scatter plot – A scatter plot displays bivariate data: new car fuel consumption and weight. The left panel shows the data and the best fit line. The right panel shows the best fit line (dashed line) and the loess line.

When including a straight line in a scatter plot, the line coerces the linear relationship to the data and can be misleading. Including a loess line in a scatter plot, instead of a straight line, is always a good idea.

When there are more than two variables, a bivariate scatter plot matrix is a good starting point of an exploratory analysis. Figure 3.11 shows a scatter plot matrix of the daily air quality measurements in New York from May to September 1973, a data set included in R. The data set includes the ground level ozone concentration in parts per billion from 1:00 to 3:00 PM at Roosevelt Island and three meteorological variables – solar radiation (`Solar.R`) in Langleys in the frequency band 4000–7700 Angstroms from 8:00 AM to 12:00 noon at Central Park, average wind speed (`Wind`) in miles per hour at 7:00 and 10:00 AM at LaGuardia Airport, and maximum daily temperature in degrees Fahrenheit at LaGuardia Airport (`Temp`). In each scatter plot, a loess line is superimposed. For each variable, a histogram is plotted in the diagonal panel.

Each panel of the matrix in Figure 3.11 is a bivariate scatter plot. The x-axis variable is the variable shown in the diagonal panel in the same column and the y-axis variable is the variable shown in the same row. For example, the scatter plot in the upper-right corner has `Ozone` as the y-axis variable and `Temp` as the x-axis variable. For this data set, we are interested in the effects of meteorological variables on ground level ozone concentration. The three plots in the first row show the ozone concentration as the response variable (plotted on the y-axis). From these three scatter plots we can draw some initial observations. First, the effect of solar radiation (`Solar.R`) is somewhat ambiguous. The loess line suggested that ozone concentration increases as solar radiation increases until the solar radiation reaches a value close to 200 Langleys, then the ozone concentration decreases as radiation increases beyond

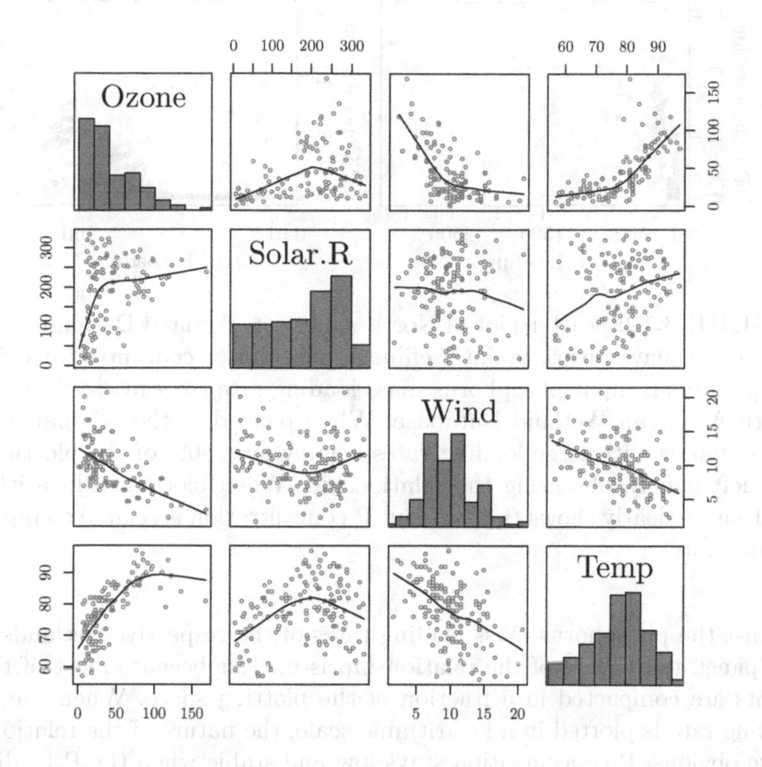

FIGURE 3.11: Scatter plot matrix – A scatter plot matrix displays bivariate scatter plots of four variables.

200 Langleys. This seems to contradict our understanding of the mechanism of smog formation. The figure also shows that the ozone concentration variation is much higher when the solar radiation is above 150 Langleys. The relationship between ozone and wind speed can be easily interpreted. Wind disperses groundlevel pollutants, therefore, the stronger the wind, the lower the ozone concentration. When the wind speed reaches about 10 mile per hour (16 km per hour), ozone concentration will stay at a low and relatively constant level. Lastly, ozone concentration increases as air temperature increases. But when temperature is below 75 °F ($\sim$ 24°C), the effect of temperature is not obvious. When reading this figure, we must be careful not to be overconfident on the observed relationship. The three meteorological variables are correlated (e.g., high temperature is usually associated with calmer winds). Their "interaction" effect on ozone is not displayed by bivariate scatter plots.

Variable transformation is often an important part of data visualization. For example, Figure 3.12 shows the scatter plot of phosphorus concentrations measured at outlets of a number of constructed wetlands in North America

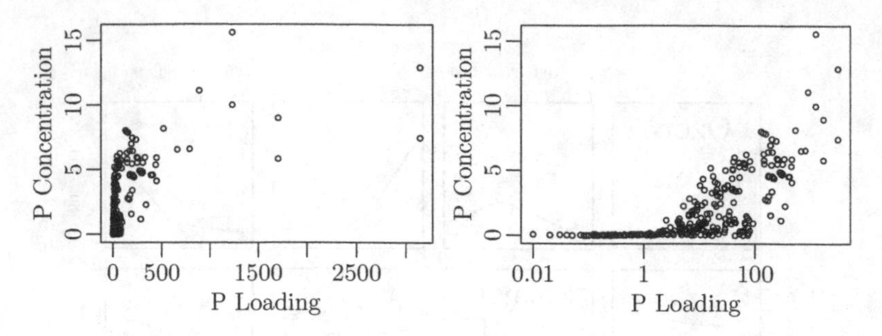

FIGURE 3.12: Scatter plot of North American Wetland Database – A scatter plot displays bivariate data: effluent phosphorus concentrations (in μg/L or ppb) versus input phosphorus mass loading rate (in g m^{-2}-yr^{-1}) from the North American Wetland Database. When plotted in the original scale, the few wetlands with large loading rates occupy over 80% of the plotting space (the left panel), obscuring the relationship. When plotted in logarithm, the right panel clearly shows that effluent P concentration is related to input mass loading rate.

against the phosphorus mass loading rates of the respective wetlands. In the left panel, the nature of the relationship is unclear because most of the data points are compacted in a fraction of the plotting space. When the P mass loading rate is plotted in a logarithmic scale, the nature of the relationship is quite obvious: P concentration stays low and stable when the P loading rate is below ˜1 g m^{-2}yr^{-1}, and P concentration and its variance increase as a function of the loading rate when the rate is above 1. This presentation was first shown in Qian [1995].

In general, the purpose of variable transformation is to make data points more or less evenly spread out in the plotting area to avoid overcrowding. In Figure 3.12 (left panel), most wetlands included in the data set had relatively low mass loading rate. The few large loading rate wetlands stretch the plotting region, but most data points in the plot are in a very small area. Variable transformation changes the scale of a variable. For example, in the original scale, the majority of the data have loading rates below 500 g/m^2-yr (only about 10 wetlands, or less than 10%, had loading rate above 500). With the largest loading rate to be above 3000, over 90% of the data points are crowded in less than 15% of the plotting area (to the left of 500 in the left panel). When transforming the loading data using logarithm ($\log(500) = 6.2$ and $\log(3000) = 8.0$), the distance between 500 and 3000 is less than 2 in natural logarithm scale, and the distance between 0.01 ($\log(0.01) = -4.6$) and 500 is larger than 10. In the logarithmic scale, the same 10% of data points with large loading rates are now taking only less than 20% of the space.

Logarithm transformation is a special case of the power transformation, which is a class of transformation of the form x^λ. By using different values of λ,

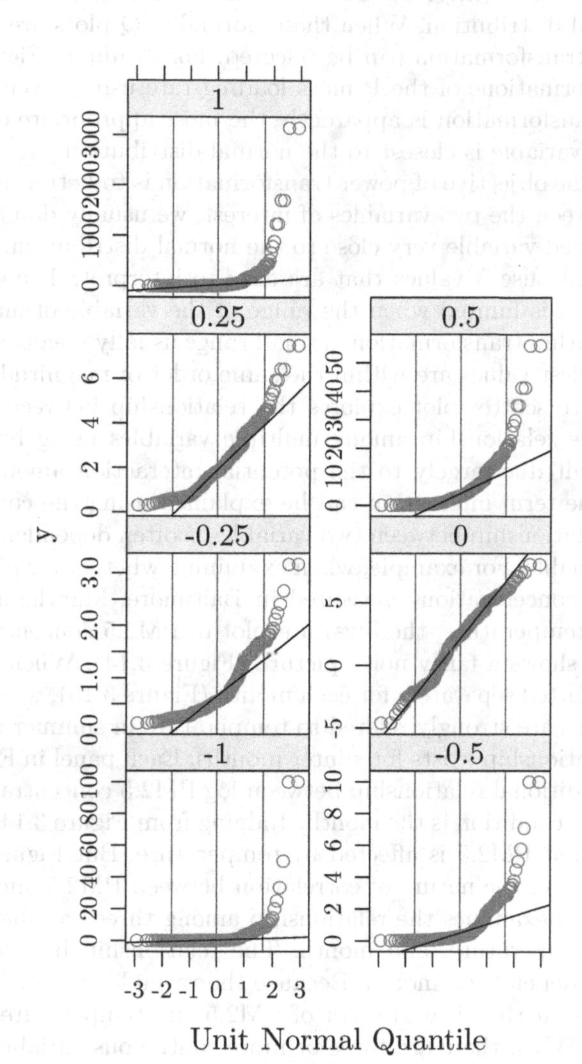

FIGURE 3.13: Power transformation for normality – Power transformations of the P mass loading rate are displayed in normal Q-Q plots. The number on top of each panel is the power.

the variable distribution after transformation can be made close to symmetric. To select the proper value of λ, we can use the normal Q-Q plot to determine whether the transformed variable distribution is close to a normal distribution. That is, we can select a few values of λ, e.g., $z = x^2$ (square), $z = x^{0.5}$ (square root), $z = \log(x)$ (defined by $\lambda = 0$) (logarithm), $z = x^{-0.5}$ (inverse square root), and $z = x^{-1}$ (inverse). The transformed variable z is then compared to the normal distribution. When these normal Q-Q plots are compared, the appropriate transformation can be selected. For example, Figure 3.13 shows power transformations of the P mass loading rate using seven λ values. The logarithm transformation is apparently the most appropriate one in that the transformed variable is closest to the normal distribution.

Because the objective of power transformation is to better display the relationship between the two variables of interest, we usually don't have to make the transformed variable very close to the normal distribution. Consequently, we usually only use λ values that are easy to interpret. The effect of power transformation is limited when the range of the variable of interest is small. For the logarithm transformation, a small range usually means that the largest and the smallest values are within the same order of magnitude.

A bivariate scatter plot explores the relationship between two variables. Exploring the relationship among multiple variables using bivariate scatter plot is difficult due largely to the potential interaction among the multiple variables. The term interaction can be explained using the concept of conditioning. A relationship between two variables is often dependent on the values of a third variable. For example, when examining whether air particulate matter (PM2.5) concentrations measured in Baltimore, Maryland, is related to ambient air temperature, the bivariate plot of PM2.5 concentrations against temperature shows a fairly noisy picture (Figure 3.14). When the same relationship is plotted separately for each month (Figure 3.15), we see that PM2.5 concentrations are strongly related to temperature in summer months and no apparent relationship exists for winter months. Each panel in Figure 3.15 represents a conditional relationship between log PM2.5 concentrations and temperature. The condition is the month. Judging from Figure 3.14 alone, we may not believe that PM2.5 is affected by temperature. But Figure 3.15 changes our perception of the nature of correlation between PM2.5 and temperature.

Figure 3.15 examines the relationship among three variables: PM2.5 concentration, temperature, and month. The relationship between PM2.5 and temperature depends on month. Because the variable "month" is categorical, it is natural that the bivariate plot of PM2.5 and temperature be plotted for each month. When there are three or more continuous variables, the concept of conditioning can still apply. For example, to fully understand the effect of solar radiation on ground level ozone concentration, we can plot a series of bivariate scatter plots of ozone and solar radiation conditional on different ranges of temperature and wind. That is, divide the data into subsets each representing a different wind and temperature condition. A conditional plot can be easily constructed using the R function `coplot` or the generic `lattice` function `xyplot`.

Figure 3.16 shows a multiple panel plot produced using the function `xyplot`. Each panel is a bivariate scatter plot of square root of ozone concentration against solar radiation. On top of each panel are two strips defining the conditions of wind speed and temperature for the plot. The ranges of

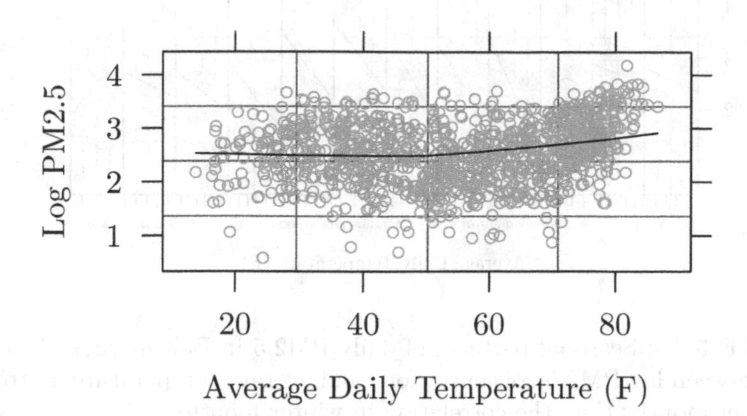

FIGURE 3.14: Daily PM2.5 concentrations in Baltimore – A bivariate scatter plot shows a weak correlation between log PM2.5 concentrations and average temperature.

the conditioning variables are marked by the shaded bar inside the respective strips. The wind speed increases when moving from left to right within each row of plots, while the temperature increases when moving from bottom to top within each column. Compared to the scatter plot of ozone against solar radiation for the entire data set (Figure 3.11), we can see that these conditional plots suggest a monotonic relationship, that is, the higher the solar radiation, the higher the ozone concentration, while the plot with the entire data set suggests that the effect of solar radiation peaks at a radiation value between 200 and 250 Langleys. Additionally, the conditional plots also suggest the following:

1. At low wind speed (left column), the effect of solar radiation intensifies as temperature increases (from bottom to top), reflected in the increasingly steeper slope.

2. The effect of radiation weakens as wind speed increases (from left to right).

Summarizing the findings in statistical terms, we say that the effect of radiation depends on the values of temperature and wind speed, or an interaction effect among the three meteorological variables.

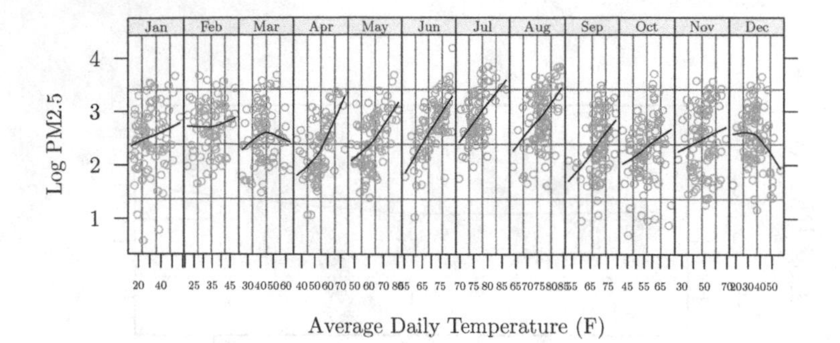

FIGURE 3.15: Seasonal patterns of daily PM2.5 in Baltimore – The correlation between log PM2.5 concentrations and average temperature is stronger in summer months than the correlation in winter months.

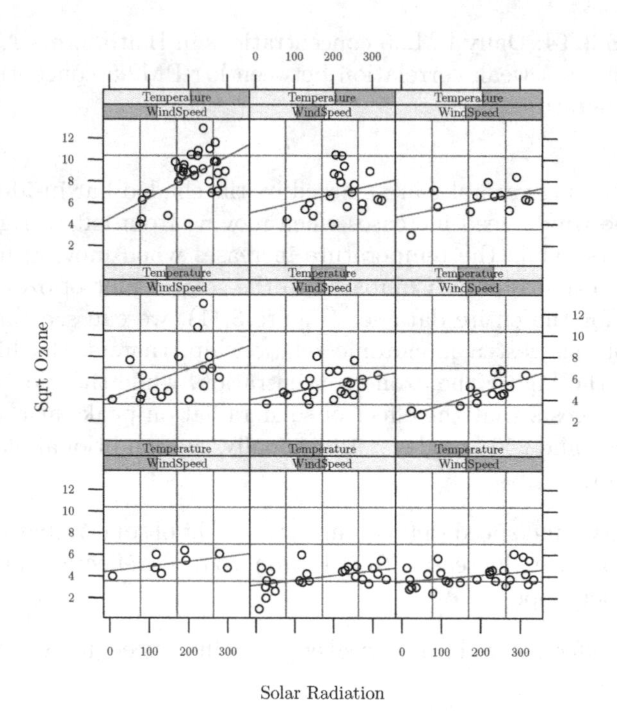

FIGURE 3.16: Conditional plot of the air quality data – The correlation between the square root transformed ozone concentration and solar radiation is generally positive. The strength of the relationship is conditional on wind speed and temperature. Calm wind (left column) and high temperature (top row) will enhance the correlation.

3.5 From Graphs to Statistical Thinking

Good data presentation is important for effective communication, and graphical presentation is more effective than tables of summary statistics. Because statistics is about the variation, good graphs are those that can help us think about the data in terms of variability. It is easy to calculate the mean, but difficult to think about the variance. Some of the graphs we studied in this chapter are specifically designed for displaying variability. Furthermore, graphs may show unexpected features in the data, thereby helping us clearly express our ideas and uncover relationships that may otherwise be missed. Exploratory data analysis is the first step of a sound statistical analysis. Statistical thinking is thinking scientifically, which requires us to think critically and to be skeptical. Graphs help us explore and communicate.

The main objective of data analysis and modeling is to find mathematical descriptions of the structure in data. Because we can never expect to know the correct mathematical formula and many competing models can potentially produce the data we observed, any model we develop is likely wrong [Box, 1976]. To make the possibly wrong model useful, the model we develop should be able to explain the discrepancy between the model prediction and the observed data. A critical aspect of statistical thinking is the capability of evaluating the discrepancy between the model and the reality reflected in data. For example, when comparing two data sets, the concept of additive or multiplicative shift describes a structural property of univariate data. For two distributions differing only in location and not in spread or shape (or $y = x + a$), an additive shift is fitted by estimating location – computing a location measure (mean or median) for each distribution. The difference between the two distributions is described by the difference in their mean (or median). If this description is correct (or useful), we expect that the Q-Q plot of x and y will form a line parallel to the 1-1 reference line. When two distributions differ by a multiplicative factor, or $y = ax$, comparing the means of the two distributions will no longer provide us with a meaningful description of the difference. The difference between the means is no longer an accurate description of the difference between the two distributions. However, in the logarithm scale (i.e., $\log(y) = \log(a) + \log(x)$), the two distributions differ only in location. So, if we suspect that the difference between two distributions is a multiplicative shift (Figure 3.9, right panel), we should take logarithm of both data sets and plot the Q-Q plot of the log-transformed data. If the Q-Q plot of the log-transformed data resembles an additive shift (Figure 3.9, left panel), we should describe the difference in terms of the log difference or the proportional factor.

Once we have determined that distributions from two data sets, x and y, differ only in location, we can divide each data set into two parts:

$$y_j = \bar{y} + \varepsilon_j$$

and

$$x_i = \bar{x} + \epsilon_i$$

The estimated means $\bar{x}$ and $\bar{y}$ are examples of "fit," the estimate of the parameter that describes the main feature of the distribution. The differences ε_j and ϵ_i are called *residuals* (or "misfits," frequently used in marine modeling literature). Residuals are important in statistical analysis because they provide information on variability. When we establish that the distributions of variables x and y differ only in location, we know that the distributions of ϵ and ε are the same. Consequently, combining the residuals from the two data sets will increase the sample size and improve the reliability of the estimated common variance of the two data sets. In fact, statistical assumptions described in this chapter are almost always made with respect to residuals. We will come back to the analysis of residuals repeatedly when discussing statistical models in Part II.

For example, in presenting the famous iris data, a data set originally collected by Anderson [1935] and used by Fisher [1936], Cleveland [1993] used the scatter plot matrix. This famous data set gives the measurements in centimeters of the variables, sepal length and width and petal length and width, respectively, for 50 flowers from each of 3 species of iris: *Iris setosa*, *versicolor*, and *virginica*.

An interesting question is that whether these measurements can be used to differentiate the three species of iris. This data set has been used repeatedly to illustrate different modeling approaches. Figure 3.17 uses three plotting characters and shading to represent the three species. From the plot of petal length against petal width, we see that all three species have petal width proportional to petal length. To separate the three species, we need only to define a new variable, for example, petal size as the sum of petal length and petal width. Iris setosa has a petal size range from 1.2 to 2.3 cm, versicolor has a size between 4.1 and 6.7 cm, and virginica has a size larger than 6.2 cm. We can translate this figure into a classification rule; a species is setosa if the petal size is less than 3 cm, versicolor if the size is between 3 and 6.5 cm, and virginica if the size if larger than 6.5. By doing so, we translate an exploratory plot into a classification model.

Exploratory data analysis as introduced by Tukey [1977] is an integral part of statistical inference. Properly manipulating and summarizing data through graphics can make the data more comprehensible to human minds, thus providing hints on the underlying structure in the data. A big intellectual leap required in probabilistic inference is that we must treat data as realizations of a random variable with a certain distribution function that cannot be directly observed. The objective of data analysis and statistical modeling is to find an approximation to this distribution. The result of a statistical analysis must be evaluated based on the likelihood that the observed data are generated from the resulting distribution. Because nothing in the real world follows the normal (or any theoretical) distribution strictly, all models we propose are wrong in one way or the other. As a result, statistics focuses on the discrepancy be-

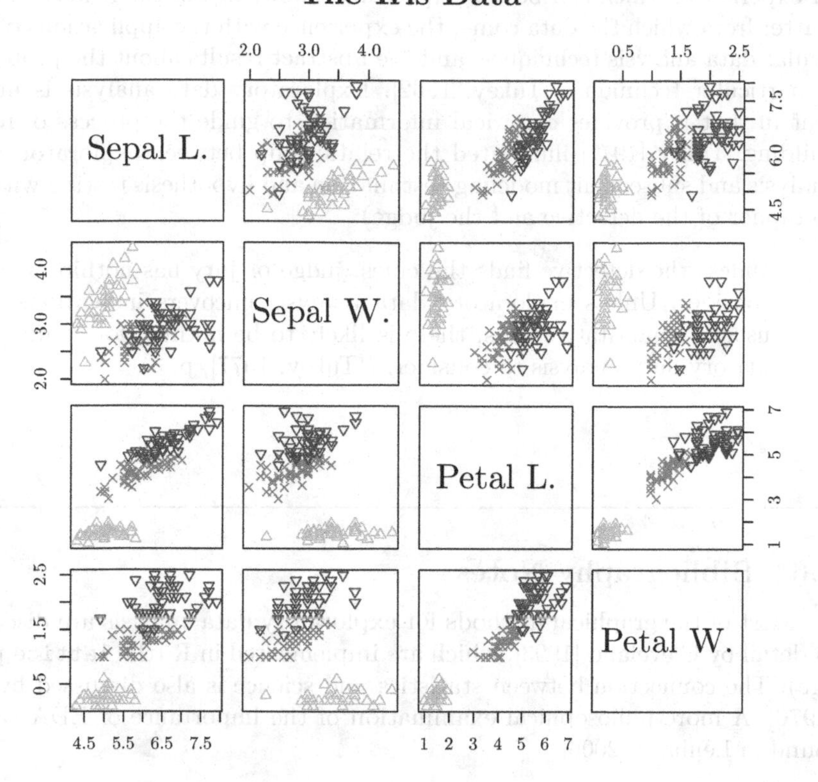

FIGURE 3.17: The iris data – A scatter plot matrix displays the famous iris data: measurements of sepal length (`Sepal L`), sepal width (`Sepal W.`), petal length (`Petal L.`), and petal width (`Petal W.`) are graphed against each other. The three species are represented by different plotting characters: Iris setosa (△), versicolor (×), and virginica (▽).

tween the proposed model and the data, represented by residuals or misfits. Although most students treat statistics as something similar to mathematics, statistical thinking is quite different from mathematics. In mathematics, we perform deductive reasoning. That is, we start from a set of premises and use a set of rules to derive what would result. The conclusion drawn from deductive reasoning contains no more information than the premises taken collectively. In statistics, we observe the results (data) and try to discover the cause. Although mathematics is an important part of statistics, statistical thinking is largely inductive, consistent with (empirical) scientific methods. Because of this difference, statistical inference relies heavily on the judgment about the underlying model or assumptions. This judgment is largely based on experience, which can be the experience/knowledge of the field of subject matter from which the data come, the experience with the applications of particular data analysis techniques, and the abstract results about the properties of particular techniques [Tukey, 1962]. Exploratory data analysis is important in that it provides empirical information to guide the process of model building. Tukey [1977] illustrated the relationship between exploratory data analysis and subsequent modeling (estimation and hypothesis testing with the metaphor of the detective and the judge):

> Unless the detective finds the clues, judge or jury has nothing to consider. Unless exploratory data analysis uncovers indications, usually quantitative ones, there is likely to be nothing for confirmatory data analysis to consider. ([Tukey, 1977], p. 3)

3.6 Bibliography Notes

Most of the graphical methods for exploratory data analysis are discussed in detail by Cleveland [1993], which are implemented in R (the `lattice` package). The connection between statistics and science is also discussed by Box [1976]. A more philosophical examination of the importance of EDA can be found in Lenhard [2006].

3.7 Exercises

1. In a normal Q-Q plot, we expect to see data points line up to form a straight line when they are random samples from a normal distribution. As in almost all statistical rules, this expectation is literally an "expectation" or what we expect to see on average. When the sample size of the data is small, a normal Q-Q plot may not entirely resemble a straight line even when the data points are truly random samples from a normal distribution. Use `my.qqnorm(rnorm(20))` repeatedly to draw several normal Q-Q plots, each using 20 random numbers drawn from the standard normal distribution to see the likely departure from a straight line.

2. In a boxplot (Figure 3.8), the height of the box represents the interquartile range r (often used as a measure of spread, close to the standard deviation), and the upper and lower adjacent values are the farthest data points from the center of the box within $1.5r$ away from the lower and upper quartiles. Can you guess why $1.5r$?

3. Use the water quality monitoring data from Heidelberg University (Chapter 2, Exercise 4) to

   - plot TP (total phosphorus) against sampling dates to visualize the changes of TP over time and compare the temporal pattern of SRP to that of river discharge (flow). Describe the seasonal patterns in both.
   - plot TP against flow, both in the original and log scales to examine the correlation between TP and flow.

4. The Great Lakes Environmental Research Laboratory of the U.S. National Oceanic and Atmospheric Administration (NOAA-GLERL) routinely monitors the western basin of Lake Erie from May to October every year. The data file `LakeErie1.csv` includes all available data up to the end of 2014. Two variables are of particular interest in studying lake eutrophication. They are the total phosphorus (TP) and chlorophyll a (chla) concentrations. These are concentration variables and we often assume that their distributions are approximately normal.

   - Read the data into R and use the graphics to evaluate whether TP and chla are normally distributed.
   - Western Lake Erie's nutrient concentrations are largely associated with Maumee River input, which varies from year to year due to variation in weather conditions. As a result, we expect that TP, as well as chla, concentration distributions vary by year. Use the function `qqmath` (from package `lattice`) to draw normal Q-Q plots of

TP and chla concentrations by year. Are the annual concentration distributions closer to being normal?

5. In addition to NOAA-GLERL, several other institutions also have routine monitoring programs on Lake Erie. Data file `LakeErie2.csv` contains TP and chla concentration data collected by NOAA-GLERL, Ohio Department of Natural Resources (ODNR), and The University of Toledo (Toledo).

   (a) Compare the distribution of the TP concentration data from NOAA to the same from ODNR:
      - Are the two distributions different?
      - If so, is the difference between the two TP concentration distributions more likely to be additive or multiplicative?
      - Are the variances of the two distributions the same?

   (b) Describe the difference in the two distributions in non-technical terms.

   (c) Repeat the previous comparison to compare TP distributions from Toledo and NOAA, and ODNR and Toledo, and summarize the results.

6. The 1971 edition of Dr. Seuss's *The Lorax* included lines describing the dire fate of the "humming fish" after their pond was polluted:

   "They'll walk on their fins and get woefully weary
   in search of some water that isn't so smeary.
   I hear things are just as bad up in Lake Erie."

   The last line was removed in the 1985 edition after Dr. Seuss realized that Lake Erie is "the happy home of smiling fish" again after the input of a key culprit of lake eutrophication, phosphorus, was successfully reduced, particularly in the Maumee River basin. The input of phosphorus from the Maumee River has since stablized. However, since the late 1990s, harmful algal blooms have returned to western Lake Erie. Some suggested that the widespread use of an inorganic form of phosphorus fertilizer is to blame. Because much of the phosphorus in western Lake Erie are from Maumee River, we can use the long-term monitoring data from Heidelberg University to evaluate whether this hypothesis is supported by data.

   - Plot daily SRP (soluble reactive phosphorus) concentration against time. Can you see an increasing trend over time?
   - Input of nutrient to a lake is better measured by the mass loading rate (the product of flow and concentration). Plot the daily SRP loading rate against time. Is there a temporal trend in loading rate?

- If a temporal trend is not obvious, it is often because of the large daily fluctuations. Calculate the annual total of SRP loading rates and plot them against the respective years. Is there a trend?
- If we repeat the above steps for TP, we will see that the annual TP loadings are more or less the same in the last 20 years. Could the return of harmful algal blooms in the last 20 years be caused by increased proportion of SRP in the total phosphorus? Plot the ratio of SRP over TP over time (both at the daily and annual scales).

7. Figure 3.16 explores the dependency of the ozone concentration–solar radiation relationship on wind speed and temperature. Use the same conditional plot to examine the dependency of the ozone concentration–temperature relationship on wind speed and solar radiation, and the ozone concentration–wind speed relationship on solar radiation and temperature.

Chapter 4

Statistical Inference

4.1	Introduction	77
4.2	Estimation of Population Mean and Confidence Interval	78
	4.2.1 Bootstrap Method for Estimating Standard Error	86
4.3	Hypothesis Testing	90
	4.3.1 t-Test	91
	4.3.2 Two-Sided Alternatives	98
	4.3.3 Hypothesis Testing Using the Confidence Interval	99
4.4	A General Procedure	101
4.5	Nonparametric Methods for Hypothesis Testing	102
	4.5.1 Rank Transformation	102
	4.5.2 Wilcoxon Signed Rank Test	103
	4.5.3 Wilcoxon Rank Sum Test	104
	4.5.4 A Comment on Distribution-Free Methods	106
4.6	Significance Level α, Power $1-\beta$, and p-Value	109
4.7	One-Way Analysis of Variance	116
	4.7.1 Analysis of Variance	117
	4.7.2 Statistical Inference	119
	4.7.3 Multiple Comparisons	121
4.8	Examples	127
	4.8.1 The Everglades Example	127
	4.8.2 Kemp's Ridley Turtles	128
	4.8.3 Assessing Water Quality Standard Compliance	134
	4.8.4 Interaction between Red Mangrove and Sponges	137
4.9	Bibliography Notes	142
4.10	Exercises	142

4.1 Introduction

As we discussed earlier, statistics attempts to find the likely underlying probability distribution that produced the data we observed. In almost all applications of statistics, the true underlying probability distribution (or model) is unknown. As a result, the process of finding the correct model is a process of careful sleuthing, which inevitably will include two general steps. One is the initial guess on the model form (what distribution), and the other is the estimation of the unknown model parameters. In this book, I use the term *model* as a generic term to describe the probability distribution model. Inevitably, the first question in any statistical analysis should be about the form of the distribution. How should we decide which model is appropriate for the problem we have? This question, a version of the problem of induction originated from David Hume's *An Enquiry Concerning Human Understanding* first published in 1748 [Hume, 1748, 1777] , is impossible to answer in general. This can be explained on two levels. First, there are many alternative models that are equally likely to have produced the data we observed. Second, even when we find a unique model that can be used to explain the observation made so far, we cannot be sure that the model would still be correct for the future. In Hume's words, our inductive practices have no *rational foundation*, for no form of reason will certify it. Philosophical arguments about the impossibility of causal inference aside, statistical thinking is a form of inductive process that follows a quasi-falsificationism approach. The basis of Fisher's statistical reasoning can be interpreted using Popper's falsification theory, which is an attempt to solve the problem of induction. Popper suggests that there is no positive solution to the problem of induction ("no matter how many instances of white swans we may have observed does not justify the conclusion that all swans are white"). But theories, while they cannot be logically proved by empirical observations, can sometimes be refuted by them (e.g., sighting a black swan). Furthermore, a theory can be "corroborated" if its logical consequences are confirmed by suitable experiments. Statistical inference starts with an assumption or theory, usually in the form of a specific probabilistic distribution (a model). Because statistical assumptions cannot be directly refuted, inference is usually based on the evidence presented in data that is contradictory to the theory. If the evidence is strong, we reject the theory. Once a theory is corroborated, that is, a probability distribution model is established as the likely representation of the true underlying distribution, model parameters are estimated. In most statistical analysis, statistical inference is presented in terms of the estimation of model parameters and hypothesis testing with respect to specific values of the parameter of interest. This is because the theory about probability distribution is inevitably subject-matter specific. As a result, the discussion of statistical inference is largely conditional on the knowledge of the underlying distribution.

4.2 Estimation of Population Mean and Confidence Interval

The Everglades TP reference data set discussed in Section 1.2 is intended to make an inference about the background TP distribution in the Everglades.

This is a typical statistical inference problem about a population distribution. Here, the underlying TP distribution is to be estimated using a limited number of samples, a case of inductive inference from particular to general. Although the underlying TP concentration probability distribution is unknown, many studies suggest that environmental concentration distributions can be approximated by the log-normal distribution (e.g., [Ott, 1995]). As a result, we need only to estimate the log-mean and log-standard deviation of the distribution. A natural way of estimating the mean and standard deviation of the population distribution is to use the sample average and sample standard deviation:

$$\bar{y} = \frac{1}{n}\sum_{i=1}^{n} y_i$$

and

$$\hat{\sigma} = \sqrt{\frac{(y_i - \bar{y})^2}{n-1}}$$

where y_i are the logarithm of the observed TP concentrations. But, if it is possible to repeatedly take samples, each sample will produce a different sample average and sample standard deviation. That is, $\bar{y}$ and $\hat{\sigma}$ are random variables. As a result, the validity of any given estimate $\bar{y}$ is questionable. The question must be related to the variability of $\bar{y}$. If the variance of $\bar{y}$ is large, we expect to see very different $\bar{y}$ from sample to sample, thereby reducing the reliability of any given estimate. If the variance of $\bar{y}$ is small, we don't expect the next sample average would differ from the current one substantially. If we know the distribution of sample averages, we can quantitatively describe the relationship between the estimated sample average and the population mean. This quantitative description should give us information on the reliability of the estimate and whether additional samples are needed. The distribution of statistics such as sample average ($\bar{y}$) and sample variance ($\hat{\sigma}^2$) is known as the *sampling distribution.*

The central limit theorem (CLT) describes the sample average distribution. For any random variable Y, the sample average $\bar{Y}$ distribution is approximated by the normal distribution when the sample size is large enough. A normal distribution has two parameters, mean and standard deviation. CLT states that the mean of the sample average distribution is the same as the population mean and the standard deviation of the sample average distribution is the population standard deviation divided by the square root of the sample size:

$$\bar{Y} \sim N(\mu, \sigma/\sqrt{n})$$

The quantity $\sigma/\sqrt{n}$ is the *standard error* (or *se*) of the sample average, the standard deviation of the sample average distribution. From this result, we can describe the variability of sample average by using the standard error, or we can use the range of the most frequently observed sample means. For example, $\mu \pm 2se$ gives the range of approximately the middle 95% all possible sample means. The number "2" is obtained through a linear transformation of the variable $\bar{Y}$:

$$z = \frac{\bar{Y} - \mu}{\sigma/\sqrt{n}} \tag{4.1}$$

and z follows the standard normal distribution, or $z \sim N(0,1)$. The middle 95% range of z is the 2.5 and 97.5 percentiles of the standard normal distribution, approximately $(-2, 2)$. That is the probability that z takes value between -2 and 2 is 0.95:

$$\Pr(-2 \leq z \leq 2) = \Pr(-2 \leq \frac{\bar{Y} - \mu}{\sigma/\sqrt{n}} \leq 2) = 0.95$$

which is equivalent to $\Pr(\mu - 2\sigma/\sqrt{n} \leq \bar{Y} \leq \mu + 2\sigma/\sqrt{n}) = 0.95$. This relationship is, however, of no practical meaning because it describes the distribution of $\bar{Y}$ using two population parameters in which we are interested. However, if we know σ, this relationship can be further revised to be $\Pr(\bar{Y} - 2\sigma/\sqrt{n} \leq \mu \leq \bar{Y} + 2\sigma/\sqrt{n}) = 0.95$. The interval $(\bar{Y} - 2\sigma/\sqrt{n}, \bar{Y} + 2\sigma/\sqrt{n})$ gives us a measure of uncertainty. The interval is random, and the probability that the interval includes the population mean μ is approximately 0.95. This interval is the 95% *confidence interval.* In general, a $100 \times (1-\alpha)\%$ confidence interval of the estimated sample average is $\bar{Y} \pm z_{\alpha/2}\sigma/\sqrt{n}$, where $z_{\alpha/2}$ is the $\alpha/2$ quantile of the standard normal distribution.

When the population standard deviation is unknown and is replaced by the sample standard deviation $\hat{\sigma}$ in equation 4.1, the transformed variable is no longer a normal random variable. Instead, the linear transformed variable

$$t = \frac{\bar{Y} - \mu}{\hat{\sigma}/\sqrt{n}} \tag{4.2}$$

has a t-distribution with degrees of freedom of $n-1$. Likewise, the confidence interval $\bar{Y} - t_{\alpha/2,n-1}\hat{\sigma}/\sqrt{n}, \bar{Y} + t_{\alpha/2,n-1}\hat{\sigma}/\sqrt{n}$ covers the population mean with the probability $1 - \alpha$.

The multiplier $t_{\alpha/2,n-1}$ reflects the confidence level on the estimated sample mean. The multiplier varies by sample size. For example, it is 2.23 when sample size is 10 and 2.08 when sample size is 20. But for a moderate sample size (20–50), the multiplier of a 95% confidence interval is very close to 2. Therefore, we will often use $\bar{Y} \pm 2se$ as a rough estimate of the 95% confidence interval. The multiplier for a 68% confidence interval is approximately 1, and a 50% confidence interval has a multiplier approximately 2/3. In R, the multiplier is calculated by using the function `qt`. For example, suppose the TP concentration data are named `TP.conc` in R:

```
#### R code ####
    y <- log(TP.conc)
    n <- length (y)
    y.bar <- mean(y)
```

```
se <- sd(y)/sqrt(n)
int.50 <- y.bar + qt(c(0.25, 0.75), n-1)*se
int.95 <- y.bar + qt(c(.025, .975), n-1)*se
```

A 95% confidence interval indicates that the probability of the confidence interval includes the true mean μ is 0.95. The 95% confidence interval is usually about 3 times as wide as the 50% confidence interval. It is important to recognize that the true mean μ is not random, and the confidence interval is random.

Using the Everglades data collected from the three stations on transects labeled as "U" (for unimpacted) in 1994 (see Section 4.8 on page 127 for reasons), the estimated log-mean is 2.048 and log standard deviation is 0.342. With the sample size of 30, the standard error is 0.06244. The 50% confidence interval is then (2.005, 2.090) and the 95% confidence interval is (1.920, 2.176).

The interpretation of a confidence interval is often confusing. This is because when we say "the 95% confidence interval of the mean is (1.9, 2.2)," we are often tempted to interpret the 95% as the probability of the true mean being bounded by the interval. This interpretation is wrong because the true mean is not a random variable. The true value is either inside or outside the interval. The confidence interval itself is random – a different sample will lead to a different confidence interval. Therefore, a probability statement should be applied to the confidence interval. The 95% refers to the probability that a confidence interval includes the true value of the mean, and the probability should be interpreted in terms of a long run frequency. In other words, if it is possible to repeatedly sample the Everglades and calculate the 95% confidence interval each time, we expect that 95% of the time the calculated confidence intervals will include the true mean. To understand this interpretation, we can perform a simulation. A simulation is to use random numbers to summarize a statistical inference. In this case, we assume that the true distribution of the log TP concentration is N(2.05, 0.34) and let the computer mimic the sampling process by taking a sample of 30 random numbers from this true distribution and calculate the confidence interval. When repeating this process many times (e.g., 1000), we expect 95% of the confidence intervals will include 2.05.

```
#### R code ####
    n.sims <- 1000
    n.size <- 30
    inside <- 0
    for (i in 1:n.sims){  ## looping through n.sims iterations
        y <- rnorm(n.size, mean=2.05, sd=0.34)
            ## random samples from N(2.05, 0.34)
        se <- sd(y)/sqrt(n.size)
        int.95 <- mean(y) + qt(c(.025, .975), n.size-1)*se
        inside <- inside + sum(int.95[1]<2.05 & int.95[2]>2.05)
```

```
    }
  inside/n.sims  ## fraction of times true mean inside int.95
```

The result of this simulation will vary from run to run, but close to 0.95. The variability of the result depends on the values of `n.sims` and `n.size`.

The CLT indicates that the sample average distribution is normal regardless of the population distribution. Using simulation can also check what would happen if the data are not from a normal distribution. For example, we can change the distribution in the above simulation from a normal distribution to a uniform distribution (i.e., `y<-runif(n.size,min=1.05,max=3.05)`). Because the central limit theorem describes the asymptotic behavior of the sample average, it is important to know what value of sample size is large enough to ensure an approximately normal sample mean distribution. There are many "rules of thumb" in the literature to suggest a minimum sample size. These rules are usually unreliable. For example, Figure 4.1 shows a simulation result for two population distributions. In the figure, three sample sizes are used ($n = 5, 20, 100$). A simulation of 10,000 samples with the given sample sizes were taken from the two population distributions for calculating sample averages. The resulting sample averages are presented using histograms. According to the central limit theorem, the sample average distribution should approach normal (expecting a symmetric histogram), with the mean equal to the population mean and standard deviation equal to population standard deviation divided by square root of the sample size. In the figure, $\hat{\mu}$ and $\hat{\sigma}$ are the average and standard deviation of the sample averages, and μ and σ are the mean and standard deviation predicted by the central limit theorem. Clearly, the three sample average distributions in the top row are not symmetric, indicating that a sample size of 100 is not large enough for this particular population distribution. The sample average distributions on the bottom row are all approximately symmetric, indicating that a sample size of 5 is large enough for this less skewed distribution. Therefore, specific suggestions on how large a sample size is large enough (often suggested to be 30) are not reliable.

The second part of the inference is the standard deviation. The sample distribution of $\hat{\sigma}$ is more complicated than the sampling distribution of $\bar{x}$. When the data are from a normal distribution, the distribution of $\hat{\sigma}^2$ is proportional to an inverse χ^2 distribution because the sample variance $\hat{\sigma}^2 = \frac{1}{n-1}\sum_{i=1}^{n}(x_i - \bar{x})^2$ formula can be re-expressed as

$$\frac{n-1}{\sigma^2}\hat{\sigma}^2 = \frac{1}{\sigma^2}\sum_{i=1}^{n}(x_i - \bar{x})^2. \tag{4.3}$$

The right-hand side of equation 4.3 is a χ^2 random variable with degrees of freedom of $n-1$ (or $\chi^2(n-1)$). As a result, sample variance $\hat{\sigma}^2$ is a rescaled χ^2 random variable ($\hat{\sigma}^2 = \frac{\sigma^2}{n-1}\chi^2(n-1)$). We can calculate the 95% confidence interval by first calculating the 95% interval of the χ^2 distribution on the

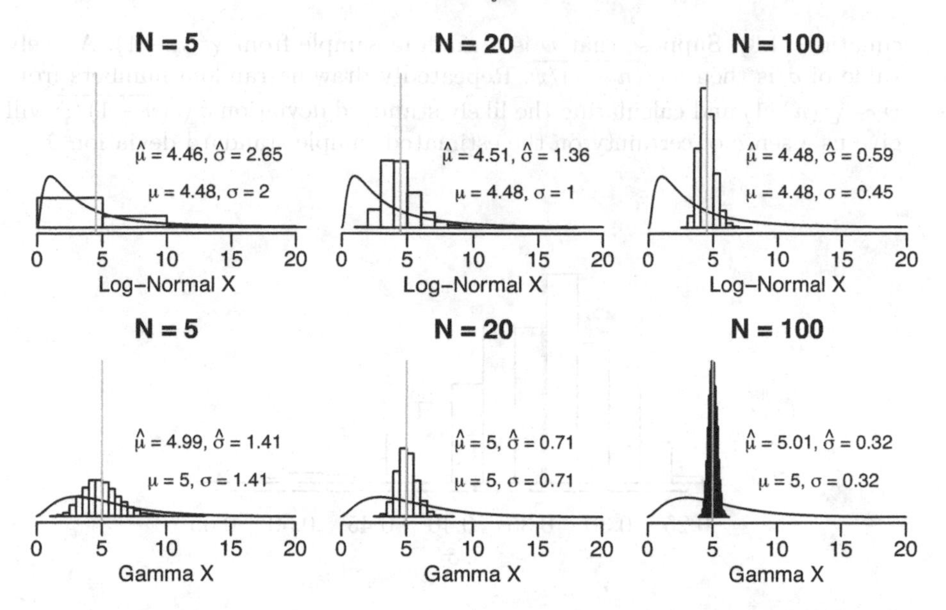

FIGURE 4.1: Simulating the Central Limit Theorem – Simulated sample mean distributions show the rate of convergence of the sample mean distribution depending on not only the sample size, but also the population distribution. Sample mean distributions from the log-normal distribution (upper row) are somewhat skewed when the sample size is 100, while the sample mean distributions from the Gamma distribution (lower row) are close to symmetric when the sample size is 5.

right-hand side of equation (4.3) – $\chi^2_{0.025}, \chi^2_{0.975}$. That is:

$$\Pr(\chi^2_{0.025} \leq \frac{n-1}{\sigma^2}\hat{\sigma}^2 \leq \chi^2_{0.975}) = 0.95$$

Rearranging the inequality inside the probability parentheses with respect to σ we have:

$$\frac{(n-1)\hat{\sigma}^2}{\chi^2_{0.975}} \leq \sigma^2 \leq \frac{(n-1)\hat{\sigma}^2}{\chi^2_{0.025}}$$

That is, the 95% confidence interval of σ^2 is $\left(\frac{(n-1)\hat{\sigma}^2}{\chi^2_{0.975}}, \frac{(n-1)\hat{\sigma}^2}{\chi^2_{0.025}}\right)$.

One way to understand the uncertainty we have on the population standard deviation after observing $\hat{\sigma}^2$ is to use the distribution of the quantity on the left-hand-side of equation (4.3), which is $\chi^2(n-1)$. Let us use the Everglades data (Figure 4.13 on page 129) as an example. We choose the 1994 data because data from that year are approximately normal. We can use the χ^2 distribution to summarize the uncertainty in the estimated $\hat{\sigma}^2$ through a simulation. Random numbers drawn from $\chi^2(n-1)$ can be used to represent the uncertainty we have on the quantity represented by the left-hand side of equation (4.3). Suppose that ψ is a random sample from $\chi^2(n-1)$. A likely value of σ is then $\hat{\sigma}\sqrt{(n-1)/\psi}$. Repeatedly drawing random numbers from $\psi \sim \chi^2(n-1)$ and calculating the likely standard deviation $\hat{\sigma}\sqrt{(n-1)/\psi}$ will give us a sense of certainty on the estimated sample standard deviation $\hat{\sigma}$.

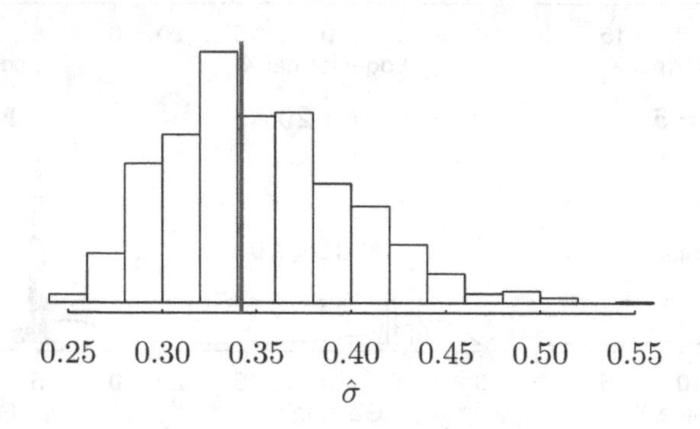

FIGURE 4.2: Distribution of sample standard deviation – Simulated uncertainty about the population distribution standard deviation is proportional to an inverse χ^2 distribution.

With an estimate of the mean and an estimate of the standard deviation, along with their uncertainties summarized in the confidence intervals, the step of model parameter estimation is complete. But the question behind the Everglades study is to set an environmental standard for TP. Because the U.S. EPA recommended that the 75th percentile of the background concentration distribution should be used as a tentative standard, the question now is how to estimate the 0.75 quantile. If we know the population distribution to be normal and we know the true values of the mean and standard deviation, the 0.75 quantile can be directly estimated. Suppose the estimated mean (2.05) and standard deviation (0.34) are the true values:

```
#### R output ####
    qnorm(0.75, mean=2.05, sd=0.34)
    [1] 2.279
```

The 0.75 quantile of the TP concentration distribution is $e^{2.279} = 9.77$ μg/L (or ppb). But we know that the estimated log mean of 2.05 is likely to be different from the true mean, and so is the estimated log standard deviation of 0.34. How can we assess the uncertainty in the estimated 0.75 quantile? A simple and straightforward way for estimating the uncertainty is to perform a simulation. In this case, we can approximate the uncertainty of the sample average by using the sampling distribution (the central limit theorem), and approximate the uncertainty in the standard deviation using the relation expressed in equation 4.3 (described in Figure 4.2). From the distribution of σ, we draw random numbers as realizations of the standard deviation, which is used to form the sample mean distribution to generate a mean. The resulting mean and standard deviation pair is used to estimate one realization of the 0.75 quantile.

```
#### R code ####
    n.sims <- 1000
    n <- 30
    y.bar <- mean(log(y))
    se <- sd(log(y))
    X <- rchisq (n.sims, df=n-1)
    sigma.chi2 <- se * sqrt((n-1)/X)
    sample.mean <- rnorm(n.sims, y.bar, sigma.chi2/sqrt(n))
    q.75 <- qnorm(0.75, sample.mean, sigma.chi2)
    hist(exp(q.75), axes=F, xlab="0.75 Quantile Distribution",
         main="")
    axis(1)
```

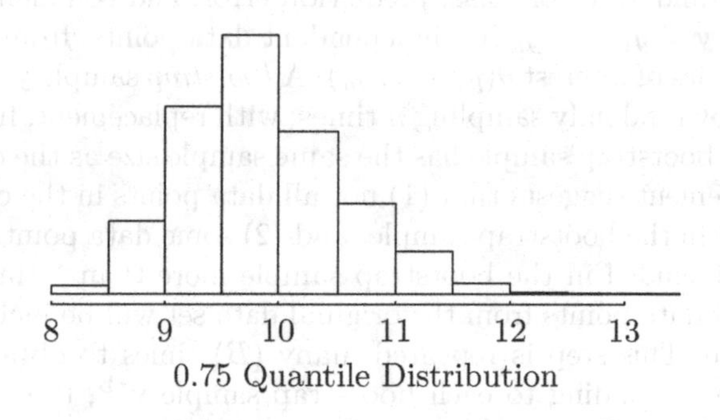

FIGURE 4.3: Distribution of the 75th percentile of Everglades background TP concentration – Simulated uncertainty about the 0.75 quantile of the background distribution of TP.

From the simulated uncertainty, we can present a 95% confidence interval

```
#### R output ####
    quantile(exp(q.75), prob=c(0.025, 0.975))
    2.5% 97.5%
    8.699 11.446
```

Simulation is an alternative to the frequently used bootstrapping method for estimating confidence intervals that are otherwise difficult to estimate.

4.2.1 Bootstrap Method for Estimating Standard Error

Bootstrap is a simulation-based method for assigning measures of accuracy to statistical estimates. The standard error of a sample average $\bar{x}$ is an example of a measure of accuracy. From *se* we learn that an estimator ($\bar{x}$) will be less than one *se* away from its expectation $\sim$ 68% of the time, and less than two *se* away $\sim$ 95% of the time. If *se* is very small, we know that $\bar{x}$ is likely close to the true value and vice versa. Confidence intervals are also measures of accuracy. Both standard error and confidence interval are easy to obtain for a sample average or sample standard deviation because we have statistical theory on their sampling distributions. When the quantity of interest to be estimated is not a statistic with a known sampling distribution such as sample average and sample variance, bootstrap and other simulation methods are often used for estimating measures of accuracy.

The basic idea of the bootstrap method is that the original sample represents the population from which it was drawn. So re-samples from this sample represent approximations of what we would have gotten if we took many samples from the population. The bootstrap distribution of a statistic, based on many re-samples, represents an approximation of the sampling distribution of the statistic. From these re-samples, the following accuracy measures can be estimated: standard error, bias, prediction error, and confidence interval.

Suppose $\mathbf{y} = y_1, \cdots, y_n$ are independent data points, from which we compute a statistics of interest $\theta(y_1, \cdots, y_n)$. A *bootstrap* sample $\mathbf{y}^* = (y_1^*, \cdots, y_n^*)$ is obtained by randomly sampling n times, with replacement, from the original data $\mathbf{y}$. The bootstrap sample has the same sample size as the original sample. With replacement suggests that (1) not all data points in the original data set are included in the bootstrap sample, and (2) some data points in the original sample are included in the bootstrap sample more than 1 time. On average, about 2/3 of data points from the original data set will be included in a bootstrap sample. This step is repeated many (B) times to obtain B bootstrap samples. Corresponding to each bootstrap sample $\mathbf{y^{*b}}$, the statistics $\theta(\mathbf{y^{*b}})$ can be calculated. The bootstrap estimated standard error is:

$$\widehat{se}_{boot} = \sqrt{\frac{\sum_{b=1}^{B}\left(\theta(\mathbf{y^{*b}}) - \bar{\theta}^*\right)^2}{(B-1)}}$$

For example, we have a data set

```
x <- c(94, 38, 23, 197, 99, 16, 141)
```

from which we can estimate the standard error of the sample average to be `se = sd(x)/sqrt(7) = 25.24`. The bootstrap method for estimating the standard error will take the following steps:

1. Taking a bootstrap sample, that is, take a sample of size 7 from the original data with replacement:

```
#### R Code ####
     boot.sample <- sample(x, size=length(x), replace=T)
```

2. Treating each bootstrap sample as a sample from the population, calculate the statistics of interest:

```
#### R Code ####
     boot.mean <- mean(boot.sample)
```

3. Repeating steps 1 and 2 B times:

```
#### R code ####
    boot.mean <- numeric()
    B <- 2000
    for (i in 1:B){
      boot.sample <- sample(x, size=length(x), T)
      boot.mean[i] <- mean(boot.sample)
    }
```

These steps produce $B = 2000$ realizations of the sample average. Statistical theories suggest that the distribution of these bootstrap samples approaches the theoretic sampling distribution of the statistics as the sample size n increases. As a result, we can directly estimate the standard deviation of the 2000 bootstrap sample averages to obtain an estimation:

```
#### R Code and output ####
    boot.se <- sd(boot.mean)}
    boot.se
    [1] 23.36
```

We can write simple R programs to carry out the bootstrap steps because the statistics of interest is simple. For more complicated statistics, these steps can be implemented in the R function `bootstrap`. The same steps can be simplified to be

```
#### R code and output ####
    require(bootstrap)
    boot.mean <- bootstrap(x, 2000, mean)
    sd(boot.mean$thetastar)
    [1] 23.41
```

Obviously we don't need to estimate the sample average standard error using bootstrap. But if the statistics of interest is the median, bootstrap is good to have:

```
#### R output ####
    boot.median <- bootstrap(x, 2000, median)
    sd(boot.median $ thetastar)
    [1] 38.64895
```

Running `bootstrap(x, 2000, median)` returns `nboot=2000` bootstrap estimated medians, from which we can calculate the bias:

```
mean(boot.median$thetastar) - median(x),
```

standard error of the median

```
sd(boot.median$thetastar),
```

and the distribution of sample median

```
hist(boot.median$thetastar).
```

In addition, there are several methods for estimating the confidence interval for the statistics of interest.

Bootstrap-t Confidence Interval: Suppose that the bootstrap distribution of a statistic from a simple random sample of size n is approximately normal and that the bootstrap estimate of bias is small. An approximate $(1-\alpha)\times 100\%$ confidence interval for the parameter that corresponds to this statistic by the plug-in principle is

$$\text{statistic} \pm t^* se_{boot}$$

where t^* is the critical value of the t_{n-1} distribution with area $(1-\alpha)$ between $-t^*$ and t^*. For our sample median example:

```
#### R code and output ####
    boot.median <- bootstrap(x, 2000, median)
    sd(boot.median$thetastar)
    [1] 38.65
    mean(boot.median$thetastar)
    [1] 79.1105

    CI <- mean(boot.median$thetastar) + qt(c(0.025,0.975), 6)
    [1] 76.66359 81.55741
```

Bootstrap Percentile Confidence Interval: The bootstrap-t confidence interval assumes that the bootstrap distribution of the statistics is approximately normal. When the distribution is asymmetric, the t-confidence interval often produces confidence bounds that are meaningless. A percentile-based $(1-\alpha)100\%$ CI is the interval between the $(\alpha/2)100$ and $(1-\alpha/2)100$ percentiles of the estimated bootstrap statistic θ^{*b}:

```
#### R code and output ####
     CI.percent <- quantile(boot.median$thetastar,
                            prob=c(0.025, 0.975))
     CI.percent
       2.5% 97.5%
        23   141
```

A "good" bootstrap confidence interval should closely match an exact confidence interval (when available), and provide accurate coverage probability. The bootstrap-t intervals have good theoretical coverage probabilities, but tend to be erratic in practice. The percentile intervals are less erratic but have less accurate coverage properties. As a result the bootstrap Bias-Corrected accelerated (BCa) interval is often used.

The BCa Method: The bootstrap Bias-Corrected accelerated interval is a modification of the percentile method that adjusts the percentiles to correct for bias and skewness. Details of this method can be found in Efron and Tibshirani [1993]. It is implemented in R function `bcanon`.

Now we use the Everglades example to illustrate the use of bootstrap for estimating the confidence interval of the 0.75 quantile of the background TP concentration distribution.

There are two approaches for estimating the 0.75 quantile of the TP background distribution. One is to directly estimate the 0.75 quantile from the data:

```
TP.75Q <- quantile(y, prob=0.75)
```

That is, the statistics of interest is calculated by the function `quantile`

```
#### R code and output ####
  results <- bootstrap(y, 2000, quantile, prob=0.75)

  ## bootstrap-t CI
  CI.t <- mean(results $ thetastar) + qt(c(0.025,0.975), 29)
  CI.t
  [1] 0.1997 4.2902

  ## percentile CI
   CI.percent <- quantile(results$thetastar,
                          prob=c(0.025, 0.975))
   CI.percent
   2.5% 97.5%
  2.079 2.485

  ## BCa CI
  bca.results <- bcanon(y,2000,theta=quantile, prob=0.75,
          alpha=c(0.025, 0.975))
  bca.results$confpoints
       alpha bca point
  [1,] 0.025     2.079
  [2,] 0.975     2.485
```

The bootstrap-t confidence interval is (0.1997, 4.2902) or in original scale (1.2, 73) μg/L, which is quite unreasonable because observed TP concentration values range from 4 to 15 μg/L. The percentile and BCa confidence intervals are both (2.097, 2.485) or (8.1, 12.0) μg/L, slightly wider than the CI of (8.7 11.4) from our simulation study (Figure 4.3).

The bootstrap is a rich collection of data re-sampling methods. The discussion presented in this section represents a small part of these methods. A more complicated example is presented in Chapter 9.

4.3 Hypothesis Testing

Hypothesis testing is a topic with many terminologies and many confusing concepts. Fisher's original work on this topic was intended to provide tools for inductive inference. In Fisher's hypothesis testing, one hypothesis is put forward and data are used to assess evidence against the hypothesis. The evidence is in the form of a probability – the probability of observing data as contradictory or more to the hypothesis as the observed data. The evidence against the hypothesis is strong if the probability is small. When the evidence is strong, the chance of observing the data we have when the hypothesis is true is small; we can logically conclude that either a small chance event happened or the hypothesis is wrong. If we reject the hypothesis, we have a small chance of making a mistake. The probability in this context is called the p-value. For example, we hypothesize that the mean TP background concentration in the Everglades is equal to or smaller than 10 μg/L and test the hypothesis by taking one sample to measure the TP concentration. If the background concentrations follow a log-normal distribution with log mean $\leq \log(10)$ and a known log standard deviation of 0.34, the probability of observing a concentration value of 20 μg/L or higher is less than `1- pnorm(log(20), log(10), 0.34) = 0.02`. In this simple example, the p-value is the probability of observing a TP concentration $\geq \log(20)$ (the observed value) assuming the hypothesis is true. When the hypothesis is true, the likelihood of observing a TP concentration value closer to the hypothesized mean of log(10) is larger. Therefore, the term "as contradictory or more" means observing a TP concentration equal to or higher than the observed one. The p-value of 0.02 is a measure of evidence against the hypothesis that log mean of TP concentration distribution is $\leq \log(10)$; the chance of observing a concentration value $\geq \log(20)$ is only about 1 in 50. Although a p-value is a probability, it is not the probability of the hypothesis being true. Suppose instead of log(10), we hypothesize that the log mean is log(15) and the p-value is now 0.20 or a 1 in 5 chance of observing a concentration value $\geq \log(20)$ if the hypothesis is true. So, when comparing these two hypotheses, the chance of observing the single observation of 20 (or larger) is 1 in 50 for the first hypothesis and the chance is 1 in 5 for the second hypothesis. That is, the evidence against the first hypothesis is stronger. The natural follow-up question is how strong the evidence must be before we conclude the hypothesis to be unlikely. Fisher suggested that a hypothesis be "disproved" if the data deviate from the hypothesized mean by more than a specific criterion. In Fisher's words, 5% is a convenient standard "level of significance" to disprove or reject the hypothesis. Both the value (5%) and the term (significance) have remained in the statistical lexicon. But when rejecting or accepting the hypothesis, the probability of making a mistake is unknown. Furthermore, many believe that testing a single hypothesis without an alternative is unreasonable.

The Neyman–Pearson hypothesis testing procedure was proposed to "improve" Fisher's method. In the Neyman-Pearson framework, two competing hypotheses, the null hypothesis (H_0) and the alternative hypothesis (H_a), are formulated. In the Neyman–Pearson statistical orthodoxy, hypothesis testing is viewed as a way to guide coherent inductive *behavior*. It is a decision process to choose among the two hypotheses. In this decision process, two types of error were defined and the decision to choose one hypothesis over the other as the true one is based on a predetermined critical region that limits the error of false positive and minimizes error of false negative. For example, when testing the hypothesis that the mean background TP concentration is less than or equal to 10 μg/L, we set the null hypothesis to be $H_0 : \mu \leq \log(10)$ and the alternative hypothesis $H_a : \mu > \log(10)$, where μ is the true log mean of the background TP distribution. Only one of the two competing hypotheses is true. When choosing H_a or believing that $\mu > \log(10)$, we risk making the Type I error (or false positive), that is, rejecting the null hypothesis when the null is true. When choosing H_0 or believing that $\mu \leq \log(10)$, we risk making the Type II error (or false negative), that is, accepting the null hypothesis when the alternative is true. Once the acceptable risk of making a type I error is decided, the test procedure ensures that the risk of type II error is minimized.

The classical statistical inference as it is commonly presented in most introductory texts is essentially a hybrid of these two approaches. This is because the computation involved in the two competing approaches is essentially the same. In the rest of this section, we introduce the hybrid testing procedure using the t-test as an example. We then discuss the general procedure for hypothesis testing and some nonparametric hypothesis testing methods.

4.3.1 t-Test

We start the introduction of the t-test using the same Everglades data set the Florida Department of Environmental Protection used for setting the environmental standard for total phosphorus. In this data set, each monitoring site is classified either as "I" for "impacted" or "R" for "reference" depending on whether or not the site is located in the area that is known to be affected by an anthropogenic source of phosphorus. Naturally, we want to know (1) what is the background TP concentration distribution and (2) what is the difference between the background TP concentration distribution and the impacted TP concentration distribution. Furthermore, once the TP environmental standard is set, the U.S. Clean Water Act (CWA) mandates that states regularly evaluate water quality status, which is a hypothesis testing problem to decide whether a water body is in compliance with water quality standards. All these problems require statistical inference about the *population* distribution from *samples.* In this and many problems, the population mean is the quantity of interest. Consequently, the objective of drawing a random sample from a population is to learn about the population mean. To learn about the population mean from a random sample, we must know the *sampling distribution* of an average. The central limit theorem suggests that the sample average distribution will approach a normal distribution as sample size increases. In the Everglades data set, we have a sample of 30 TP measurements. Because we do not know the true value of population mean, the sample mean distribution specified by the central limit theorem cannot be used directly to infer the true mean. However, if we want to know whether the true mean is equal to a specific value or in a specific range, we can *assume* that the true mean is equal to the value or within the range we are interested in, making it a hypothesis. Suppose that our hypothesis is that the log mean of the background TP concentration is less than or equal to log(10) and we set this hypothesis as the null hypothesis: $H_0 : \mu \leq \log(10)$. The alternative hypothesis is then: $H_a : \mu > \log(10)$. Assuming that the null hypothesis is true, the sample mean has a normal distribution:

$$\bar{y} \sim N(\log(10), \sigma/\sqrt{30}) \tag{4.4}$$

Because we don't know σ, the population standard deviation, we must use the sample standard deviation, $\hat{\sigma}$, as an approximation. To simplify the sample average distribution in equation 4.4, we can introduce a statistic:

$$t = \frac{\bar{y} - \log(10)}{\hat{\sigma}/\sqrt{30}}$$

or more generally:

$$t = \frac{\bar{y} - \mu_0}{\hat{\sigma}/\sqrt{n}} \tag{4.5}$$

If we know σ or the sample size n is large, t follows the unit or standard normal distribution, or $t \sim N(0, 1)$. When the sample size is small and σ is unknown (but approximated by $\hat{\sigma}$), the distribution of t will have noticeable discrepancies from the unit normal distribution. This is because as the sample size decreases, the sample standard deviation is "subject to an increasing error, until judgments reached in this way may become altogether misleading" wrote William S. Gosset (under the pen name "Student") in his 1908 *Biometrika* paper entitled "The Probable Error of a Mean" [Student, 1908]. Because of Gosset's work, the distribution of the quantity t is known as the Student's t-distribution (Figure 4.4). The quantity t is known as the test statistic and its distribution under the null hypothesis is known as the *null distribution.*

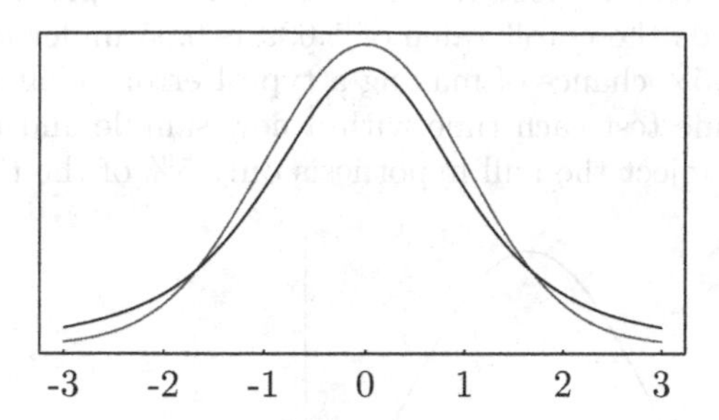

FIGURE 4.4: The t-distribution – Student's t-distribution with degrees of freedom of 3 (the dark line) is compared to a unit normal distribution (the gray line).

Using Fisher's approach, we calculate the p-value, the probability of a test statistic being as contradictory or more than the one observed: $\Pr(t \geq t_{obs}|H_0)$. The null distribution in our example is a t-distribution with degrees of freedom of $n-1$. The observed sample mean of log TP concentrations is 2.05, standard error is 0.062, and the sample size is 30. Hence, observed t value is $\frac{2.05-\log(10)}{0.062} = -4.08$; the p-value is `1 - pt(-4.08, 30-1)=0.9998`. Following Fisher's recommendation of using a level of significance of 0.05, we would conclude that the evidence against the null hypothesis is weak.

Using Neyman–Pearson's approach, we first decide the acceptable risk of making a type I error, that is, the probability of erroneously rejecting H_0. For the Everglades example, a large observed sample average would suggest that the null hypothesis is likely wrong. In other words, we would reject the null hypothesis if the sample average is larger than a critical value. The acceptable risk of a type I error is used to decide what value of the sample average (or the test statistic) is large enough to reject the null hypothesis. The probability of making a type I error is to reject the null hypothesis (i.e., the sample average is larger than a cutoff point) when the null hypothesis is true. Since the sample mean distribution is now defined through the t-distribution, the cutoff point can be defined in terms of t: $\Pr(t \geq t_{cutoff}|H_0) = \alpha$ (Figure 4.5). In the Everglades example, this cutoff point is calculated in R using the function `qt`: `qt(0.95, 30-1)`, or 1.699. That is, the null hypothesis will be rejected if the observed t is larger than 1.699. The cutoff point $t_{cutoff} = 1.699$ divides the t statistic space into an acceptance region ($t < t_{cutoff}$) and a rejection region ($t \geq t_{cutoff}$). The observed t is -4.08, inside the acceptance region. We would accept the null hypothesis. Using the cutoff value for t, we can estimate how large a sample average would result in the rejection of the null hypothesis from equation 4.5: $\mu_0 + t_{cutoff} \cdot se$. In our case, if the standard error is 0.062, we would reject the null hypothesis only when the sample average is larger than $\log(10) + 1.699 \times 0.062 = 2.408$. Because the probability that the test statistic t exceeds the cutoff value of 1.699 is 0.05 under the null hypothesis, we have only a 5% chance of making a type I error. That is, if we repeatedly perform the same test each time with a new sample and the null hypothesis is true, we will reject the null hypothesis only 5% of the time.

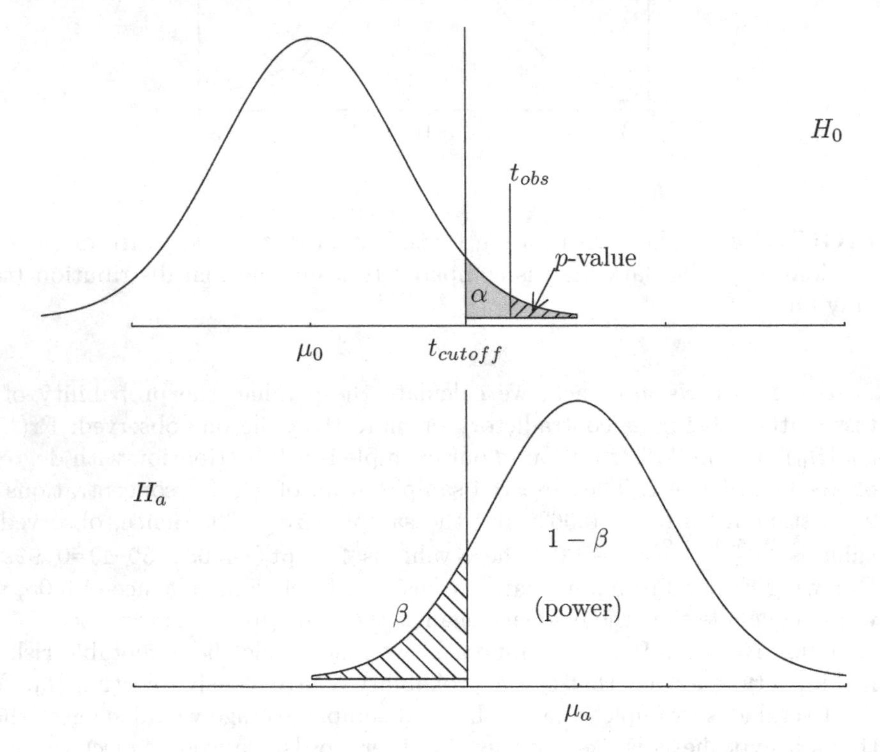

FIGURE 4.5: Relationships between α, β, and p-value – the relations of the three values are shown using a one-sided test with hypothetical sample average distributions under the null (H_0, top panel) and alternative (H_a, bottom panel) hypotheses. On the top panel, the gray-shaded area is α and the 45° line shaded area is the p-value. On the bottom panel, the $-45°$ line-shaded area is β.

The testing process can be carried out in R using the function `t.test`:

```
#### R code and output ####

  t.test(y, mu=log(10), alternative="greater")
        One Sample t-test
  data:  y
  t = -4.0802, df = 29, p-value = 0.9998
```

```
alternative hypothesis: true mean is greater than 2.3026
95 percent confidence interval:
 1.9417   Inf
sample estimates:
mean of x
   2.0478
```

This is a *one sample* t-test. In this test, we have only one sample and we are interested in learning whether the population, from which the sample was drawn, has a mean less than or equal to log(10). The test output reported the p-value, and the observed test statistic t_{obs}, but the t_{cutoff} is not included. This is because p-value is the only information we need in hypothesis testing, whether we follow Fisher's method or the Neyman–Pearson approach. Figure 4.5 illustrates the relationship between the rejection region and the p-value. Both the p-value and t_{cutoff} are calculated conditional on the null hypothesis being true (only the upper distribution in Figure 4.5 is relevant). For the Everglades example, this means that the test statistic distribution for these two approaches is the same t-distribution with degrees of freedom of 29 $(30 - 1)$. In Figure 4.5, the rejection region is the gray-shaded region to the right of t_{cutoff}, and its area (size) under the curve is α. The p-value is the area of the line-shaded region under the curve to the right of t_{obs}. The p-value is less than α only when $t_{obs} > t_{cutoff}$, and vice versa. When we see a reported p-value to be less than α (e.g., 0.05, the conventional type I error probability), we know that the t_{obs} is larger than t_{cutoff} even when t_{cutoff} is not reported.

In the Everglades example, we are also interested in learning the difference between the background TP concentration distribution (free of anthropogenic influence) and the TP concentration distribution from an area known to be affected by human activities. In statistical terms, we now have two distributions, one describes the TP concentration distribution from the reference sites and the other from the impacted sites. The first step of a comparison of the two populations is to examine the possible nature of the difference. Using data from 1994, we compare the two data sets using Q-Q plot, and the difference between the two populations are likely multiplicative. As a result, the difference in log TP concentrations is likely additive. Quantifying the difference in the two population means will be sufficient in describing the difference between the populations. The difference of the two sample averages is a random variable because the sample averages are random variables: $\delta = \bar{y}_1 - \bar{y}_2$. Based on the central limit theorem, the distribution of δ is normal with mean equal to the difference of population means and the variance of δ is the sum of the variances of $\bar{y}_1$ and $\bar{y}_2$:

$$\sigma_\delta^2 = \sigma_{\bar{y}_1}^2 + \sigma_{\bar{y}_2}^2$$

If the two populations have the same standard deviation of σ, then, $\sigma_{\bar{y}_1} = \sigma/\sqrt{n_1}$ and $\sigma_{\bar{y}_2} = \sigma/\sqrt{n_2}$, and the standard deviation of δ is $\sigma\sqrt{\frac{1}{n_1} + \frac{1}{n_2}}$. Because we have reason to believe that the two populations have the same standard deviation, its estimation can be improved if we pool the residuals ($\epsilon_i = y_{1i} - \bar{y}_1$ and $\varepsilon_j = y_{2j} - \bar{y}_2$) from their respective sample averages (see Section 3.5). The resulting estimate of the common standard deviation is called the *pooled standard deviation* (which is the standard deviation of the pooled residuals ($\{\epsilon_i, \varepsilon_j\}$). If the standard deviation of each sample is calculated separately to be $\hat{\sigma}_{\bar{y}_1}$ and $\hat{\sigma}_{\bar{y}_2}$, the pooled standard deviation can be expressed as:

$$\hat{\sigma}_p = \sqrt{\frac{(n_1-1)\hat{\sigma}^2_{\bar{y}_1} + (n_2-1)\hat{\sigma}^2_{\bar{y}_2}}{n_1+n_2-2}}$$

The standard deviation for the difference of the sample means is estimated by:

$$\hat{\sigma}_\delta = \hat{\sigma}_p\sqrt{\frac{1}{n_1} + \frac{1}{n_2}}$$

To test whether the two population mean is different or not, we set the two competing hypotheses to be:

$$\begin{array}{ll} H_0: & \mu_1 - \mu_2 = 0 \\ H_a: & \mu_1 - \mu_2 > 0 \end{array}$$

If the null hypothesis is true, the test statistic

$$t = \frac{\bar{y}_1 - \bar{y}_2}{\hat{\sigma}_\delta}$$

follows a t distribution with degrees of freedom of $n_1 + n_2 - 2$. For the Everglades data, $\bar{y}_1$ is the sample average of the log TP from the impacted sites and $\bar{y}_2$ is the sample average of log TP concentrations from the reference sites. The observed t-statistic is $t_{obs} = 5.40$ and the p-value is 9.61×10^{-7}.

In R, the same function `t.test` is used for a two sample t-test problem:

```
#### R code ####
   t.test(x=x, y=y, alternative="greater", var.equal=T)
       Two Sample t-test

#### R output####
   data:  x and y
   t = 5.4022, df = 49, p-value = 9.61e-07
   alternative hypothesis: true difference in means is
        greater than 0
   95 percent confidence interval:
    0.58144     Inf
   sample estimates:
   mean of x mean of y
      2.8909    2.0478
```

When the two population standard deviations are not the same, the difference between the two distributions are no longer additive. To accurately describe the differences between the two populations, we need to compare both the location (e.g., mean) and spread (standard deviation). If variable transformation can make the transformed variables differ roughly additive, we should perform a t-test on the transformed data, as we did for the TP concentration data. Particularly, when the difference in original scale is close to multiplicative, a logarithm transformation will result in an additive difference in the transformed data. The estimated difference is the logarithm of the proportional constant.

If no transformation can be easily found and the difference in population means is still of interest, the Welch's t-test should be used. The test statistic under the Welch's t-test remains the same, but its standard deviation is directly calculated to be $\hat{\sigma}_\delta = \sqrt{\frac{\hat{\sigma}_1^2}{n_1} + \frac{\hat{\sigma}_2^2}{n_2}}$ and the null distribution can only be approximated by the t-distribution with the degrees of freedom approximated by the Scatterwaite's approximation:

$$df_W = \frac{\hat{\sigma}_\delta^4}{\frac{(\hat{\sigma}_1/\sqrt{n_1})^4}{n_1-1} + \frac{(\hat{\sigma}_2/\sqrt{n_2})^4}{n_2-1}}.$$

Because the Scatterwaite's approximation (df_W) is always smaller than the degrees of freedom under the equal standard deviation ($df = n_1 + n_2 - 2$), the same t_{obs} will result in a larger p-value when using the Scatterwaite's approximation. This is perceived as "conservative" in that we will be less likely to reject the null hypothesis when using the Welch's t-test. For this reason, the R function `t.test` uses the Welch's t-test as the default (`var.equal=FALSE`).

```
#### R code ####
        t.test(x=x, y=y, alternative="greater")
            Welch Two Sample t-test

#### R output ####

    data:  x and y
    t = 4.7943, df = 25.816, p-value = 2.941e-05
    alternative hypothesis: true difference in means is
                greater than 0
    95 percent confidence interval:
     0.54307      Inf
    sample estimates:
    mean of x mean of y
       2.8909    2.0478
```

Be aware that when the population standard deviations are known to be different, the difference in population means describes only one aspect of the difference between the two population distributions.

4.3.2 Two-Sided Alternatives

So far, the tests we discussed are *one-sided* hypothesis testing. Often we are only interested in whether the population mean or the difference of two population means is equal to a particular value μ_0. Consequently, the null hypothesis is either $H_0 : \mu = \mu_0$ or $H_0 : \mu_\delta = 0$. Under this null hypothesis, we say the data evidence contradicts the null if the sample average is too large or too small. In our one-sample t-test example, a two-sided test will change our null hypothesis to $H_0 : \mu = \log(10)$ ($\log(10) = 2.3$) and the alternative hypothesis to $H_a : \mu \neq \log(10)$. If the observed sample log average is 1.7, the level of contradiction is measured by the distance between the observed sample averages and the null hypothesis mean, or $|1.7 - 2.3| = 0.6$. Therefore, 2.9 is "as contradictory" to the null hypothesis as 1.7 is. As a result, the p-value, defined as the probability of observing data as contradictory or more, is the probability of observing the sample mean to be less than or equal to 1.7 or larger than or equal to 2.9. Operationally, we calculate the test statistic t_{obs} and the p-value is $1 - \Pr(-|t_{obs}| \leq t \leq |t_{obs}|)$ (Figure 4.6), the sum of the two tail areas. The two-sided t-test is the default in the R function `t.test`:

```
#### R code ####
    t.test(x, y, var.equal=T)

#### R output ####
        Two Sample t-test

    data:  x and y
    t = 5.4022, df = 49, p-value = 1.922e-06
    alternative hypothesis: true difference in means is
                not equal to 0
    95 percent confidence interval:
     0.52947 1.15672
    sample estimates:
    mean of x mean of y
       2.8909    2.0478
```

and for the one sample t-test:

```
#### R code ####
    t.test(y, mu=log(10))


#### R output ####
        One Sample t-test
    data:  y
    t = -4.0802, df = 29, p-value = 0.0003217
    alternative hypothesis: true mean is not equal
             to 2.3026
    95 percent confidence interval:
```

```
   1.9201 2.1755
sample estimates:
mean of x
   2.0478
```

Since the only difference between a one-sided test and a two-sided test is the measure of the level of contradiction to the null hypothesis, using the same data, a two-sided test will have a *p*-value that equals to 2 times the corresponding one-sided test *p*-value.

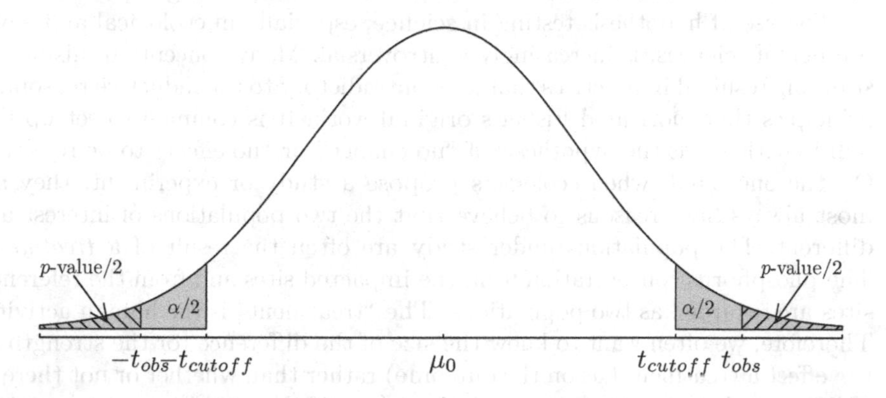

FIGURE 4.6: A two-sided test – In a two-sided test, t_{obs} is as contradictory to the null as $-t_{obs}$ is.

4.3.3 Hypothesis Testing Using the Confidence Interval

The R output includes the 95% confidence interval for the estimated mean (or the difference of the means). The confidence interval and the hypothesis testing are connected, at least in terms of computation, if not conceptually. In calculating a $(1-\alpha)100\%$ confidence interval, we need to calculate the sample average ($\bar{y}$), standard error ($se = \hat{\sigma}/\sqrt{n}$), and the *t*-multiplier $t(1-\alpha/2, df)$. For a two-sided *t*-test with a type I error probability of α, we also need to calculate these three items to determine t_{obs}, t_{cutoff}. The confidence interval is $(\bar{y} \pm t(1-\alpha/2, df) \cdot se)$. The rejection region for a two-sided alternative is defined by the value of $t_{cutoff} = \pm t(1-\alpha/2, df)$, and $t_{obs} = \frac{\bar{y}-\mu_0}{se}$. We reject the null hypothesis when $|t_{obs}| > |t_{cutoff}|$, which is:

$$\frac{|\bar{y}-\mu_0|}{se} > |t(1-\alpha/2, df)|$$

Defining the rejection region in terms of the sample average, we reject the null hypothesis when

$$\bar{y} > \mu_0 + |t(1-\alpha/2, df)| \cdot se$$

or

$$\bar{y} < \mu_0 - |t(1-\alpha/2, df)| \cdot se$$

Comparing the rejection region and the confidence interval of $(\bar{y} - |t(1-\alpha/2, df)| \cdot se, \bar{y} + |t(1-\alpha/2, df)| \cdot se)$, we know that the null hypothesis is rejected (with a type I error probability of α) when the null hypothesis mean μ_0 is outside the confidence interval.

In our one-sided test, R returned a "one-sided" confidence interval, with `Inf` or `-Inf` as the upper or lower bound. This is done so that the confidence interval will lead to the same conclusion as the corresponding hypothesis test.

The use of hypothesis testing in science, especially in ecological and environmental sciences, is increasingly controversial. Many conceptual misunderstanding resulted in practices that are contradictory to the inductive reasoning principles that motivated Fisher's original work. It is common to set up the null hypothesis as the hypothesis of "no change" or "no effect" to be rejected. On the one hand, when ecologists propose a study or experiment, they almost always have reasons to believe that the two populations of interest are different. The populations under study are often the result of a *treatment*. The phosphorus concentration from the impacted sites and from the reference sites are regarded as two populations. The "treatment" is the human activity. Therefore, we often want to know the size of the difference (or the strength of the effect a treatment has on the outcome) rather than whether or not there is a difference between the two populations (or whether an effect exists or not). By using a significance test where the inference is based on the assumption of no difference, we emphasize the type I error rate often at the expense of our capability of detecting the difference. On the other hand, a nonexistent difference can be shown to be statistically significant if one tries often enough [Ioannidis, 2005]. As a result, the estimated mean (or difference in means) along with its confidence interval should be always reported. The estimate gives us a sense of the magnitude of the mean (or the difference), which by itself is informative. The width of the confidence interval provides information on the uncertainty we have on the estimate. If the confidence interval is wide (hence the null is not rejected) but the size of the estimate is large, we may have reasons to explore the likely source of uncertainty and plan a new study accordingly to reduce the uncertainty. If the size of the estimate is small but the null hypothesis is rejected, we should evaluate the difference in terms of practical implications. If the difference in TP concentration means is 1 μg/L between the impacted and the reference sites, whether or not the difference is statistically significant is irrelevant because the difference could be well within the margin of error of the chemical analytical method used to measure the concentration.

4.4 A General Procedure

The general hypothesis testing procedure is based on the Neyman–Pearson's "inductive behavior" approach. Under this framework, a hypothesis testing problem is a decision-making problem of choosing one hypothesis over the other. The objective of the method is to have an acceptable type I error probability (α) and minimize the type II error probability (β). The process can be summarized in the following steps.

1. Set up the competing hypotheses: H_0 and H_a. The null hypothesis should be formulated such that a meaningful test statistic can be derived and its distribution is known when the null is true.

2. Determine the acceptable type I error probability α based on the seriousness of erroneously rejecting H_0. (But we rarely define what is a type I error in scientific terms, let alone its seriousness; a default of 0.05 is almost always used.)

3. Determine the rejection region in terms of the test statistic using the null distribution.

4. Calculate the test statistic using the observed data (the observed test statistic).

5. The null hypothesis is rejected if the observed test statistic is inside the rejection region, and the null hypothesis is accepted otherwise.

The terms "reject" and "accept" are used to define a decision or "behavior," rather than a statement about the hypothesis in question. By carrying out the test under this framework, we do not attempt to learn whether H_0 is true or not. The decision prescribed by the hypothesis testing procedure ensures that the risk of making a type I error is fixed and the risk of making a type II error is minimized.

I find that the value of hypothesis testing as described by this general procedure is rather limited for environmental and ecological studies, and for inductive reasoning in general. This is mostly because of the inductive nature of our field. Even when the hypothesis testing process is interpreted as a inductive reasoning tool as used by Fisher, the process is often incompatible to scientific methods in practice. Although Popper's descriptive account of the scientific method is accurate in that scientists do abandon or revise a theory when it is refuted by experimental evidence, the rejection of the null hypothesis will not lead to a narrowed-down alternative hypothesis sufficient enough to improve our understanding of the problem. In practice, scientists often use a weight of evidence approach to systematically evaluate or rank alternative theories of their worthiness of receiving further consideration. When using a hypothesis testing process, we stop at rejecting a meaningless null hypothesis that we do not believe to be true in the first place. In Chapter 11, I will illustrate this point with an example.

4.5 Nonparametric Methods for Hypothesis Testing

The t-test assumes normality about the data. Although the test is quite robust against the violation of normality, in many situations, data distribution cannot be approximated by the normal distribution. As a result, a series of test procedures were developed for testing the hypothesis when the normality assumption is obviously inadequate. These methods, often referred to as nonparametric or distribution-free methods, follow the same general procedure described in the previous section. The tests developed by these methods assume that data are independent and identically distributed under the null hypothesis, but without requiring the data to be from a normal distribution. In this section, the two nonparametric alternatives to the t-test, the Wilcoxon's sign rank test and rank sum test, are introduced. A nonparametric test based on permutation simulation will be discussed in Section 11.2.

4.5.1 Rank Transformation

Most nonparametric tests are based on the rank transformation of the data. A rank transformation replaces each data value by its rank in the combined sample. By using a rank transformation, the actual value or magnitude of a data point is no longer of any importance. Consequently, probability distribution of a variable is no longer important. For example, when considering a t-test on a variable that is likely to have a log-normal distribution, such as the phosphorus concentration distribution in the Everglades, a log transformation is always recommended before any statistical analysis such that the transformed data will be approximately normal. Because the logarithm transformation (or power transformation in general) is monotonic and does not change the rank of a data point in the sample, the rank transformation will be exactly the same whether it is applied to the original scale or the logarithm scale of the phosphorus data. The rank transformation is attractive to many because it eliminates the impact of outliers. If the highest TP concentration value is mistakenly recorded to be 2000 μg/L instead of the correct value of 20, it will not affect the rank of this data point. In addition, rank transformation is not affected by censored observations. An observation is *censored* if its value is known to be less than (or greater than) a certain number. Censorship occurs when we report that a TP concentration is below the method reporting limit (or MRL, which was 4 μg/L for TP concentration when the Everglades data were collected).

The R function **rank** is one of several functions that can be used for the rank transformation. For example, the vector **x** has 7 values

```
x <- c(17.0, 4.0,  7.0, 11.0, 21.5, 4.0,  24.0)
```

and **rank(x)** will return the ranks for each value:

```
#### R code and output ####
    rank (x)
    [1]  5.0 1.5 3.0 4.0 6.0 1.5 7.0
```

The function **rank** can handle ties (e.g., 4.0 in **x**) in several ways. The default is to use the average rank, and the method can be selected by specifying the **ties.method** when using the function:

```
#### R code and output ####
    rank (x, ties.method = "min")
    [1] 5 1 3 4 6 1 7
```

4.5.2 Wilcoxon Signed Rank Test

The signed rank test is for a one-sample location problem, where we want to test whether the location measure (median) is equal to a specific number.

Suppose we obtain n observations: $z_1, \cdots, z_n$. (If a paired replicate problem, we obtain $2n$ observations $(x_1, y_1), \cdots, (x_n, y_n)$, which is equivalent to observing $z_i = x_i - y_i$.) We assume that (1) z's are mutually independent, and (2) each z comes from a population that is continuous and symmetric about θ, where θ is the "location" (or median) of the distribution.

The null hypothesis of the test is $H_0 : \theta = \theta_0$. The first step of defining the test statistic is to modify $z : z'_i = z_i - \theta_0$ and rank transform $|z'|$. That is, we form a new data set $|z'_1|, \cdots, |z'_n|$, and each is assigned a rank R_i. The second step is to define an indicator variable ψ:

$$\psi_i = \begin{cases} 1 & \text{if } z'_i > 0 \\ 0 & \text{if } z'_i < 0 \end{cases}$$

The test statistic is

$$V = \sum_{i=1}^{n} R_i \psi_i$$

the rank sum of the positive z'_is.

Under the null hypothesis, the distribution of V (the null distribution) is known. However, the null distribution cannot be expressed by an algebraic formula. It is instead tabulated. Using these tables we can find the critical value of V for a given sample size. In R, the function **wilcox.exact** in package **exactRankTests** can be used to carry out the exact test. When sample size is large, the test statistic is approximately defined to be $Z = (V - \mu_V)/\sigma_V$, where $\mu_V = \frac{n(n+1)}{4}$ and $\sigma_V = \sqrt{\frac{n(n+1)(2n+1)}{24}}$ are the mean and variance of V. The null distribution is the unit normal distribution $N(0,1)$. In R, the test is implemented in function `wilcox.test`. The usage of `wilcox.test` is similar to that of `t.test`.

Applied to the Everglades data, the exact test is carried out by calling the function `wilcox.exact`:

```
#### R code

    require(exactRankTests)
    wilcox.exact(y, mu=log(10))

#### R output

        Exact Wilcoxon signed rank test

    data:  y
    V = 49, p-value = 0.0003513
    alternative hypothesis: true mu is not equal to 2.3026
```

The normal approximation (also known as the *continuity correction*) is implemented in function `wilcox.test`:

```
#### R code

    wilcox.test(y, mu=log(10))

#### R output
        Wilcoxon signed rank test with continuity correction

    data:  y
    V = 49, p-value = 0.0007723
    alternative hypothesis: true location is not equal to 2.3026
```

4.5.3 Wilcoxon Rank Sum Test

The rank sum test (also known as the Mann–Whitney two-sample test) is for a two-sample location problem, where we want to test whether the medians of two samples are equal to each other.

The data consist of two variables, $x_1, \cdots, x_m$ and $y_1, \cdots, y_n$. The basic assumptions about the test are as follows:

1. The model is:

$$\begin{array}{rcll} x_i & = & e_i, & i = 1, \cdots, m, \\ y_i & = & e_{m+j} + \Delta, & j = 1, \cdots, n \end{array} \quad (4.6)$$

where $e_1, \cdots, e_{m+n}$ are unobservable random variables and Δ is the parameter of interest (the unknown shift in location or treatment effect).

2. The N e's are mutually independent, where $N = n + m$.

3. Each e comes from the same (but unknown) continuous distribution.

The null hypothesis is $H_0 : \Delta = 0$. The test statistic is defined in two steps:

1. Order the N observations from least to greatest and let R_i be the rank of y in this ordering.

2. Set: $W = \sum_{i=m+1}^{n+m} R_i$ (sum of ranks of y's).

The null distribution of the test statistic is tabulated for selected combinations of m and n. To test the null against alternative, $H_a : \Delta > 0$, we find the critical value of W that has a tail area of approximately α from the respective table and compare it to the observed test statistic. In R, the same functions `wilcox.test` and `wilcox.exact` are used.

To illustrate the use of these functions for a two-sample location problem, we go back to the air quality data set we discussed in Figure 3.11. To compare the ground level ozone concentration in May and in August using the exact test, we use

```
#### R code ####
   require(exactRankTests)
   wilcox.exact(Ozone ~ Month, data = airquality,
       subset = Month==5|Month==8)

#### R output ####

       Exact Wilcoxon rank sum test

data:  Ozone by Month
W = 127.5, p-value = 6.109e-05
alternative hypothesis: true mu is not equal to 0
```

When the sample sizes are large, the test statistic is approximated by $Z = (W - \mu_w)/\sigma_w$, where $\mu_w = \frac{n(n+m+1)}{2}$ and $\sigma_w = \sqrt{\frac{nm(n+m+1)}{12}}$. The null distribution of Z is $N(0, 1)$. The normal approximation is implemented in the function `wilcox.test`. We use

```
#### R code ####


   wilcox.test(Ozone ~ Month, data = airquality,
           subset = Month==5|Month==8)}.
```

```
#### R output ####

        Wilcoxon rank sum test with continuity correction

    data:  Ozone by Month
    W = 127.5, p-value = 0.0001208
    alternative hypothesis: true location shift is
         not equal to 0
```

4.5.4 A Comment on Distribution-Free Methods

In his 1976 article [Box, 1976], George E.P. Box gave us not only the memorable quote "all models are wrong," but also a computer simulation demonstrating that the violation of the normality assumption may be a small problem (a mouse) when compared to the problem of lack of independence (a tiger). The point here is that the normality assumption is unlikely to be the main obstacle to a successful application of statistical hypothesis tests. Experimental design and factors related to data collection are more likely to be the source of error.

In the spirit of Box, a set of R code is produced here such that readers can conduct similar simulations. The simulation evaluates the frequencies of type I error of a two-sample t-test and a Wilcoxon rank sum test. When using a hypothesis test, we want to limit the type I error probability to a small value (α) and minimize the type II error probability. To evaluate a test, we often estimate the type I error probability. This is because the null hypothesis is defined by a specific probability distribution from which we can draw random samples to perform the test repeatedly. The type II error is associated with the alternative hypothesis, which is not represented by a specifically defined probability distribution. As a result, we always use simulation to assess a test's type I error probability. We want the type I error probability to be close to α. If the actual type I error probability is larger than α, the test is more likely to reject H_0 erroneously. If the type I error probability is smaller than α, we will have a higher type II error probability (see Figure 4.5).

The basic design of Box's simulation is as follows:

1. Generate a series of 20 random numbers from a known distribution, $u_1, \cdots, u_{20}$,
2. Produce serially correlated variable $y : y_i = u_i - \theta u_{i-1}$.
3. Split the data y into two groups each with 10 numbers, using two different methods:
   (a) randomly divide $y_1, \cdots, y_{20}$ into two groups,
   (b) take $y_1, \cdots, y_{10}$ as the first group and $y_{11}, \cdots, y_{20}$ as the second group.

4. Conduct a t-test and the Wilcoxon rank sum test on the resulting data and record whether the null is rejected.

5. Repeat steps 1–4 many times, e.g., 5000, and calculate the fraction of time that the null is rejected.

Because the two groups in a given test are from the same random variable, their mean or median should be the same. That is, the null hypothesis is true. We expect that a test rejects the null hypothesis about 5% of the time when we use $\alpha = 0.05$. When a test rejects the null more than 5% of the time, the test has a larger type I error probability than declared. If a test rejects the null far less than 5% of the time, the test has a larger type II error than we would expect (see Figure 4.5).

```
#### R code ####
    hypo.sim <- function(n.sims, rdistF, theta, ...){
      reject.t1<-0; reject.t2<-0; reject.w1<-0; reject.w2<-0
      for (i in 1:n.sims){
        u <- rdistF(20, ...)
        y <- u
        for (j in 2:20)
            y[j] <- u[j] - theta*u[j-1]
        samp1 <- data.frame(x=y, g=sample(1:2, 20, TRUE))
            ### randomized sample
        samp2 <- data.frame(x=y, g=rep(c(1,2), each=10))
            ### correlated sample
        reject.t1 <- reject.t1 +
            (t.test(x~g, data=samp1, var.equal=T)$p.value<0.05)
        reject.t2 <- reject.t2 +
            (t.test(x~g, data=samp2, var.equal=T)$p.value<0.05)
        reject.w1 <- reject.w1 +
            (wilcox.exact(x~g, data=samp1)$p.value<0.05)
        reject.w2 <- reject.w2 +
            (wilcox.exact(x~g, data=samp2)$p.value<0.05)
      }
      return(rbind(c(reject.t2,reject.t1),
                   c(reject.w2,reject.w1))/n.sims)
    }
```

To use this function, we supply the number of simulations (`n.sims`), the population distribution for u (`rdistF`), the θ value (`theta`), and variables required in the distribution function. The function will return a 2×2 matrix. The first row shows the results from t-test, and the second row shows the Wilcoxon rank sum results. The left column is results using data that were not randomized and the right column is results using randomized data. For example:

```
#### R output ####
```

```
hypo.sim(n.sims=5000,rdistF=rnorm,theta=-0.4,mean=2,sd=4)
    ## u from N(2,4)
      [,1]  [,2]
[1,] 0.12 0.049
[2,] 0.10 0.049

hypo.sim(n.sims=5000,rdistF=rpois,theta=-0.4,lambda=3)
    ## u from Poisson(3)
      [,1]  [,2]
[1,] 0.11 0.046
[2,] 0.11 0.053

hypo.sim(n.sims=5000,rdistF=runif,theta=-0.4,max=3,min=-3)
    ## u from uniform(-3,3)
      [,1]  [,2]
[1,] 0.13 0.059
[2,] 0.11 0.051
```

In the three simulations, the distributions for u are normal, Poisson, and uniform, respectively. In all three cases, the right column shows two numbers close to 0.05, while the left column shows number above 0.10. No matter what the population distribution is, if the data are not properly randomized, both the t-test and the Wilcoxon test will reject the null hypothesis more than 10% of the time. When the data are properly randomized, both tests rejected the null hypothesis about 5% of the time. Now we change θ to be 0.4:

```
#### R output ####
  hypo.sim(n.sims=1000,rdistF=rnorm,theta=0.4,mean=2,sd=4)
       [,1]  [,2]
[1,] 0.003 0.055
[2,] 0.002 0.047

hypo.sim(n.sims=1000,rdistF=rpois,theta=0.4,lambda=3)
       [,1]  [,2]
[1,] 0.003 0.060
[2,] 0.004 0.061

hypo.sim(n.sims=1000,rdistF=runif,theta=0.4,max=3,min=-3)
       [,1]  [,2]
[1,] 0.004 0.064
[2,] 0.004 0.062
```

The correlated data now resulted in too few rejections.

In light of these results, I will downplay the importance of distribution-free methods in the rest of the book.

4.6 Significance Level α, Power $1 - \beta$, and p-Value

The three important numbers in a hypothesis testing process are the significance level α (the type I error probability), the statistical power $1 - \beta$ (the probability of rejecting the null hypothesis when the alternative is true), and the p-value. The significance level and power belong to the Neyman–Pearson paradigm and the p-value is distinctly Fisherian. Although these three quantities can be graphically presented in a single plot (Figure 4.5), they represent the two very different approaches in statistical inference. Fisher complained about the use of his p-value in the context of a Neyman–Pearson hypothesis testing, and Neyman never accepted the value of inductive reasoning in statistical analysis. From these two giants' perspective, p-value and α should never be mentioned in the same sentence. When these two quantities are mixed together, the p-value is often misinterpreted. Because α is the probability of committing a type I error, and both α and p-value are calculated as tail areas on the null distribution, p-value is often misinterpreted as the "observed" type I error rate or the probability the null being true. These interpretations often result in overconfidence in the result. Their interpretation is also encouraged by statistical software (including R) output tables that mark a p-value with 1 to 3 asterisks to indicate the range of the p-value. For example, when reporting a regression model result, R puts three asterisks (***) next to the reported p-value indicating that p-value < 0.001, two asterisks (**) indicating that $0.001 <$ p-value < 0.01, and one asterisk (*) indicating $0.01 <$ p-value < 0.05.[1] Obviously, one can interpret the practice as a matter of convenience to a user. But many would equate the p-value and α in the expression $p < \alpha$ used in many scientific papers. On the one hand, when using the p-value as evidence against the null hypothesis, a small p-value (strong evidence against the null) will not automatically translate to strong evidence supporting the alternative hypothesis because the alternative is a composite of an infinite number of hypotheses. On the other hand, because p-value is a random number associated with the specific sample, a small p-value cannot be used as assurance that future p-values will also be small, which is why a p-value cannot be interpreted as the type I error rate. When using α to set up the rejection region, the specific value of the p-value is of no concern.

The concept of statistical power is either neglected, because the power cannot be defined for a composite alternative hypothesis (e.g., $H_a : \mu > \log(10)$), or misused, because of the mix-up of the Fisherian and Neyman–Pearson approaches. A test's power is defined as the probability of accepting the alternative hypothesis when it is true. Figure 4.5 illustrates that power can only be calculated for a specific alternative hypothesis mean value (μ_a). If we do not know the value of μ_a, we cannot calculate the area under the curve to

[1]Fortunately, we can remove these asterisks by changing the default options in R: `options(show.signif.stars=FALSE)`.

the right of t_{cutoff}. The difference between the null hypothesis mean and the mean of a specific alternative hypothesis mean is called the *effect size* (δ). The ability to detect an effect depends on (1) effect size, (2) sample size (n), (3) inherent variability in the data (σ), and (4) the level of the Type I error (α) we are willing to tolerate. The power will increase if the effect size increases, or the sample size increases, or α increases (it is easier to find a difference if you take a bigger chance on a false positive). These factors are illustrated in Figure 4.7, which shows calculated powers for 4 one-sided one-sample t-tests with the same effect size of 2 and four different combinations of n, α, and σ.

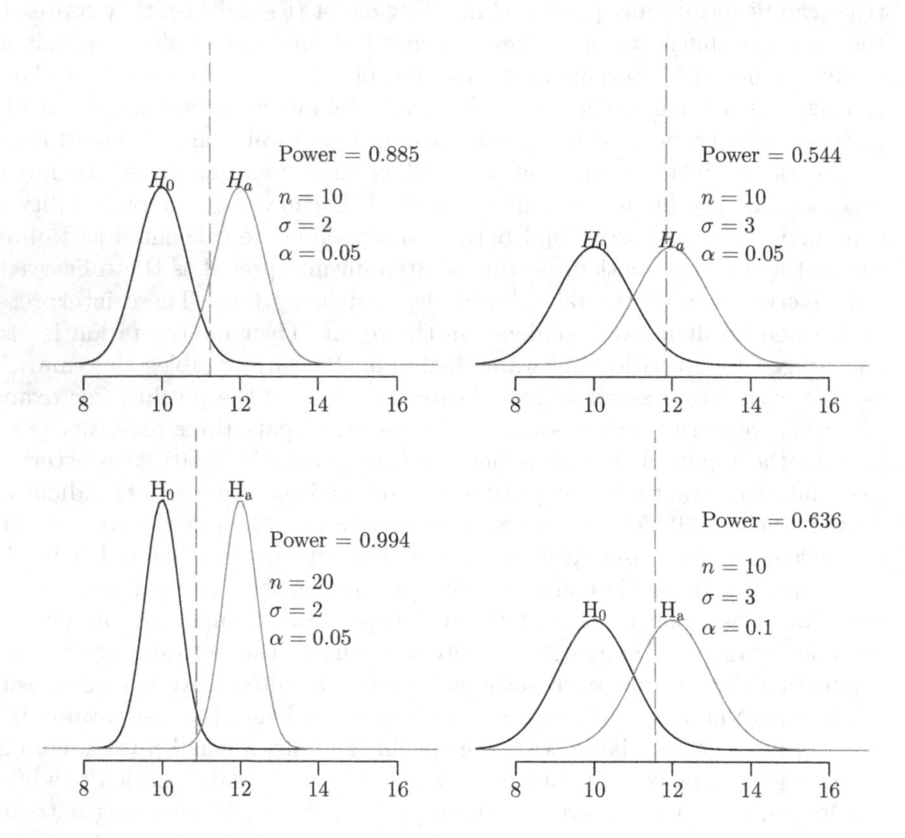

FIGURE 4.7: Factors affecting statistical power – Statistical powers are calculated for four different combinations of n, α, and σ. All tests are one-sample one-sided tests, with the t_{cutoff} shown by the dashed lines.

The power of a test is an important consideration when planning an experiment or designing a sampling activity. When planning a study, we would like the study to be able to detect a difference that is practically important with a relatively high probability. For example, states in the Great Lakes region of the United States have fish consumption guidelines to prevent overdose of PCB. The "safe" level of fish-tissue PCB concentration is below 0.05 mg/kg.

If the concentration is between 0.05 and 0.2 mg/kg, consumption of such fish is recommended to be restricted to no more than 1 meal per week. As a result, if the true concentration is close to 0.2, we want to be able to detect it and warn the angler about the risk of exposing to high level of PCB. The difference between this more "dangerous level" and the "safe level" of 0.05 is the effect size we want to be able to detect with a high probability. If we know, based on previous data, the standard deviation of PCB concentration in fish, we can estimate the minimum sample size needed to achieve this goal (setting the null mean to be 0.05 and the alternative mean to be 0.2). Conversely, if we only have 12 fish-tissue samples for analyzing fish-tissue PCB concentrations, we should estimate the statistical power (or the probability) of detecting a mean concentration at the dangerous level of 0.2 mg/kg. A low power is an indication of an inadequate sample size. The calculation of a sample size (to achieve a given power) or the power of a test with a given sample size can be easily done using the R function `power.t.test`. To use this function, we need to know four of the five quantities we discussed earlier, sample size n, effect size δ, significance level α, power $1-\beta$, and population standard deviation σ. For example, to calculate the power of a sample size of $n = 12$, we need to know δ, σ, and α. Suppose that $\delta = 0.15, \sigma = 0.5$, and $\alpha = 0.05$,

```
#### R code ####
power.t.test(n = 12, sd = 0.5, sig.level = 0.05,
       delta=0.15, type = "one.sample",
       alternative = "one.sided")

#### R output ####
     One-sample t test power calculation

              n = 12
          delta = 0.15
             sd = 0.5
      sig.level = 0.05
          power = 0.25
    alternative = one.sided
```

The power (0.25) seems too low. If we want to achieve a power of 0.85, we use the same function to calculate the necessary sample size:

```
#### R code ####

power.t.test(sd = 0.5, sig.level = 0.05,power=0.85,
       delta=0.15, type = "one.sample",
       alternative = "one.sided")

#### R output ####
```

```
     One-sample t test power calculation

              n = 81
          delta = 0.15
             sd = 0.5
      sig.level = 0.05
          power = 0.85
    alternative = one.sided
```

The sample size should be at least 81.

Although simple and straightforward, the statistical power concept is often misinterpreted. The confusion is largely due to the hybrid nature of the hypothesis testing procedure discussed in this chapter and in most statistics texts. The Neyman–Pearson approach is a decision-making process. When the null hypothesis is rejected at a predetermined α, the alternative hypothesis is accepted. The relevant type of error is the type I error and we know the probability of make an error is α. When the null is not rejected, it should be accepted. The relevant error is the type II error. However, the type II error probability (β) is unknown when the null is not rejected. As a result, we are uneasy about "accepting" the null hypothesis. When an experiment results in a p-value larger than 0.05, the result is referred to as a negative result in the literature. But, often the null hypothesis is associated with the desired (or no change) state of the world in environmental and ecological studies. As a result, discussion in the literature in how to deal with a "negative" result is often focused on the fact that a type II error rate is undefined. Behind this uneasiness is the desire for evidence supporting the null hypothesis, as the null is often indicative of a desired state. Because the type II error probability is undefined, accepting the null may be because the null is true or because the data are quite variable. Thus a small sample size or highly variable data can result in statistics that would favor the acceptance of this null hypothesis.

An influential work by Rotenberry and Wiens [1985] suggested that a power analysis should be used to provide evidence supporting the null when the null hypothesis is not rejected. Their reason for this analysis is that "if a large effect is expected to be present, and we fail to detect it (i.e., do not reject H_0), then we can be reasonably certain (small β) that it is, indeed, not present." This approach is, however, conceptually problematic. First, when performing a hypothesis testing of the form $H_0 : \mu \leq \mu_0$ versus $H_a : \mu > \mu_0$, the alternative hypothesis is a composite hypothesis, including many possible values. Rejecting the null hypothesis does not imply a support of any specific alternative value of $\mu > \mu_0$. In the same token, when calculating β given a specific alternative mean value, β is the probability of rejecting the specific alternative hypothesis when it is true. It conveys no support for any specific values outside the current hypothesis. Therefore, a small β will not provide direct evidence supporting the null hypothesis. The fact that β is a conditional probability is often neglected.

Furthermore, using a power analysis as a support for the null is difficult because the power is calculated for a specific effect size. Rotenberry and Wiens [1985] point out this difficulty of selecting an effect size needed for calculating β because "there is no conventional methodology for estimating a priori the magnitude of an effect size for ecological problems." To resolve this difficulty in selecting a specific effect size for calculating the power (or β), Rotenberry and Wiens suggested that the comparative detectable effect size (CDES) of Cohen [1988] be used. A CDES is the effect size calculated by setting a specific value of β, e.g., 0.05. In other words, the test in question has a power of $1-\beta$ if the effect size is CDES. They stated (citing [Cohen, 1988]) that "it is proper to conclude that the population ES is no larger than CDES, and that this conclusion is offered at the β significance level" and that "if CDES is deemed negligible, trivial, or inconsequential, this conclusion is functionally equivalent to affirming the null hypothesis with a controlled error rate." In other words, if the test has a large power to detect a small effect size and yet the test failed to reject the null hypothesis, the null hypothesis must have a strong support.

This line of thinking implies that CDES can be used as evidence supporting the null hypothesis – the smaller the CDES is, the stronger the evidence is. However, CDES can contradict with p-value. Suppose that we are interested in a one-sided t-test $H_0 : \mu \leq 0$ versus $H_a : \mu > 0$ and we have two experiments; both have the same sample mean of 0.5 and both have the same sample size of $n = 10$. Suppose further the p-value of the first experiment (experiment A) is 0.06, while the p-value for the second experiment (experiment B) is 0.3. If we examine the p-value from a Fisherian perspective, the evidence against the null in experiment A is stronger than the evidence from experiment B. Based on the numbers, we know that $\hat{\sigma}_1$ is 0.93 and $\hat{\sigma}_2 = 2.9$. This result implies that the same sample average of 0.5 from the two experiments resulted in different interpretation in terms of evidence against the null hypothesis. The smaller population standard deviation in the first experiment implies that the probability of observing a sample mean of 0.5 in experiment A is smaller than the same in experiment B under the null hypothesis. The CDES for the two experiments can be calculated using the R function `power.t.test`. For experiment A:

```
#### R code ####
power.t.test(n = 10, sd = 0.93, sig.level = 0.05,
             power = 1-0.05, type = "one.sample",
             alternative = "one.sided")

#### R output ####

     One-sample t test power calculation

              n = 10
          delta = 1.1
             sd = 0.93
      sig.level = 0.05
```

```
        power = 0.95
  alternative = one.sided
```

The estimated effect size (`delta`) for achieving a power of 0.95 is 1.1. That is, the estimated CDES is 1.1. For experiment B:

```
#### R code ####

power.t.test(n = 10, sd = 2.9, sig.level = 0.05,
             power = 1-0.05, type = "one.sample",
             alternative = "one.sided")

#### R output ####

     One-sample t test power calculation

              n = 10
          delta = 3.3
             sd = 2.9
      sig.level = 0.05
          power = 0.95
    alternative = one.sided
```

the estimated CDES is 3.3. So the interpretation is that experiment A has a stronger support for the null hypothesis, while the p-values suggest the opposite.

In yet another variation of posthoc power analysis aimed at providing evidence supporting the null hypothesis, Hayes and Steidl [1997] recommended the use of a "biologically significant" effect size (BSES) for calculating power. Presumably, at a given effect size, the larger the power, the stronger the support for the null hypothesis. If the biologically significant effect size is 1.5 in our previous example, the estimated statistical power for experiment A is almost 1:

```
#### R code ####
power.t.test(n = 10, sd = 0.93, sig.level = 0.05,
             delta = 1.5, type = "one.sample",
             alternative = "one.sided")

#### R output ####

     One-sample t test power calculation

              n = 10
          delta = 1.5
             sd = 0.93
      sig.level = 0.05
```

```
          power = 0.99878
    alternative = one.sided
```

While the estimated power for experiment B is only 0.45:

```
#### R code ####

power.t.test(n = 10, sd = 2.9, sig.level = 0.05,
             delta = 2, type = "one.sample",
             alternative = "one.sided")

#### R output ####

     One-sample t test power calculation

              n = 10
          delta = 1.5
             sd = 2.9
      sig.level = 0.05
          power = 0.44707
    alternative = one.sided
```

This is the same contradictory conclusion as in the CDES case.

Post hoc power analysis is unlikely to provide more information about the null than the p-value has already provided. Interestingly, á priori power analysis for supporting experimental design (selecting the necessary sample size) is rarely reported in ecological literature, although the practice is recommended in almost every statistics text for life sciences, and strongly advocated by Steidl et al. [1997].

The motivation of the use of power to support the action of accepting the null is to provide a measurable support as in the significance level α. When setting $\alpha = 0.05$, the null is rejected only when the evidence against it is strong enough. When accepting the null, the evidence against the alternative is undefined. The setup of a typical hypothesis test puts the burden of proof on rejecting the null. This framework is best illustrated by the relation between a two-sided hypothesis testing and the confidence interval. If the null hypothesis mean is inside the confidence interval, the null hypothesis is not rejected. In other words, the confidence interval defines a set of population mean values that cannot be refuted by the data at hand. If, as recommended by Rotenberry and Wiens [1985], we want to accept the null with a "stringent probabilistic standard" similar to the standard of rejecting the null using a significance level of $\alpha = 0.05$, we can use the concept of biologically significant effect size to shift the burden of proof on demonstrating that the effect is no larger than the BSES. For example, in a one-sided test of $H_0 : \mu \leq \mu_0$ versus $H_a : \mu > \mu_0$, instead of using a power analysis to prove that the effect is 0, we can revise the test to show that the effect is no larger than the "biologically significant effect size" δ: $H_0 : \mu \geq \mu_0 + \delta$ versus $H_a : \mu < \mu_0 + \delta$. This approach is natural because the concepts of both CDES and BSES concedes an acceptable level of departure from the null. Examples of such applications can also be found in water quality standard compliance assessment and in statistical evaluations of mechanistic water quality models. Reckhow et al. [1990] discussed the problem of using hypothesis testing for assessing the performance of a water quality model. The problem is the same as those problems in Rotenberry and Wiens [1985] and Steidl et al. [1997] – an acceptable model would have prediction error close to 0 and a water body that is in compliance with a water quality standard would have a mean concentration of the water quality constituent equal to or less than the standard. In both cases, the null hypothesis would be the desired state of no difference. When assessing a water body's compliance to a specific water quality standard, the default null is $H_0 : \mu \leq \mu_0$, where μ_0 is the water quality standard and μ is the true mean concentration. When verifying a water quality model, we test whether the model's predictive residual mean is 0 (i.e., $H_0 : \mu = 0$). In both cases, the desired outcome (i.e., reporting a water body to be in compliance and recommending the use of a water quality model) is the acceptance of the null hypothesis. Obviously, the problem discussed by Rotenberry and Wiens [1985] and Steidl et al. [1997] is the same here. Reckhow et al. [1990] suggested, in the context of model verification, that an acceptable model error be proposed by the modeler. For example, if 2 mg/L is an acceptable level of error for a model predicting dissolved oxygen levels in a stream, the hypothesis testing procedure for model verification may use a null hypothesis of mean absolute model predictive residual to be at least 2 mg/L, or $H_0 : \mu \geq 2$, and the alternative to be $H_1 : \mu < 2$.

4.7 One-Way Analysis of Variance

When comparing means from more than two populations, we can consider using a t-test to compare them pairwise. We can also consider using a linear modeling approach for estimating the sample means. In the Everglades example, the TP concentration data were collected from five reference sites over six years. The five sites were classified as reference sites based on a study using a method similar to a t-test. However, the yearly differences were not discussed by the Florida Department of Environmental Protection. A natural question would be whether the annual means of the five sites were the same. If they are the same, the use of the combined six-year data is appropriate. If the means are not the same, what is the implication to the process of setting the TP standard using these data? Because the state of Florida uses a hypothesis testing procedure for their 305(b) reporting and 303(d) listing (see Section 4.8.3 for more), will the standard be violated in some of the year even for the reference sites?

The question is now "Is there any difference between the annual means?" To answer this question, the t-test is inefficient, not only because the number of tests required, but also the increased rate of making a type I error. When comparing annual means from six years, there are a total of $6 \times 5/2 = 15$ pairs of annual mean differences. If each t-test uses $\alpha = 0.05$, the chance of making a type I error in at least one of the 15 tests is much higher than 0.05. In a t-test, the test statistic is the ratio of the difference of the two sample averages and the standard error of the sample mean difference. The sample average difference is a measure of the distance between the centers of the two populations, or the *between-population* difference. The standard error is a measure of *within* population variability. In other words, the t-test statistic is the ratio of between and within population variability. A large t-statistic suggests that the between population difference is large compared to the within population variation, and we reject the null hypothesis of no difference when the ratio is larger than a predetermined value. A generalization of the t-test was proposed by Fisher for comparing means of more than two populations. The test is known as the analysis of variance (ANOVA). As in the t-test, a measure of between population difference and a measure of within population variability are used to test the null hypothesis of no difference. In this section, ANOVA is first introduced as a hypothesis test procedure. At the end of the section, a linear model interpretation of ANOVA will be discussed.

4.7.1 Analysis of Variance

Consider now that the question we are interested in is whether or not the six annual means in the reference data set are the same. For now, we are not interested in the specifics of the differences, but in whether differences exist. Following the procedures discussed in Section 4.4, we first introduce the null hypothesis:

$$H_0 : \mu_1 = \cdots = \mu_k$$

The alternative hypothesis is simply that H_0 is not true. Let the data be x_{ij} with $i = 1, \cdots, k$ to indicate the years ($k = 6$ for this example), and $j = 1, \cdots, n_i$, and $N = \sum_{i=1}^k n_i$, where n_i is the sample size of the ith year. If the null hypothesis is true, we would expect that the sample averages for each year ($\hat{\mu}_i = \frac{1}{n_i}\sum_{j=1}^{n_i} x_{ij}$) will be close to the overall average ($\hat{\mu} = \frac{1}{N}\sum_{i=}^{k}\sum_{j=1}^{n_i} x_{ij}$). The overall variance is calculated with respect to the overall average ($\frac{\sum(x_{ij}-\hat{\mu})^2}{N-1}$) and the within group variance is calculated with respect to group averages ($\sum_{i=1}^k \frac{\sum(x_{ij}-\hat{\mu}_i)^2}{n_i-1}$). As a result, if the null hypothesis is true, the between population and within population variances should also be close to each other. If the null hypothesis is not true, we expect that the group averages will be different from the overall average, and the overall variance will be larger than the within population variance. To measure the difference among populations, the between population variance is introduced:

$\frac{\sum_{i=1}^{k} n_i(\hat{\mu}_i - \hat{\mu})^2}{k-1}$. As in the t-test, the test statistic is a ratio of the measure of the difference between populations and the measure of the difference within populations. A variance is proportional to the sum of squares term in the numerator, and the total variance in the data proportional to

$$SST = \sum_i \sum_j (x_{ij} - \hat{\mu})^2,$$

the within population variance is proportional to

$$SSE = \sum_i (x_{ij} - \hat{\mu}_i)^2,$$

and the between population is proportional to

$$SSG = \sum_i n_i(\hat{\mu}_i - \hat{\mu})^2.$$

If the null hypothesis is true, we expect SSG to be very close to 0 and SSG/SSE close to 0. When the null hypothesis is not true, we expect SSG to be larger than 0 and $SSG/SSE > 0$. Fisher showed that the ratio of the "mean" sum of squares $MSG = SSG/(k-1)$ and the mean sum of squares $MSE = SSE/(N-k)$ follows the F-distribution with degrees of freedom of $k-1$ and $N-k$ when the null hypothesis is true:

$$MSG/MSE \sim F(k-1, N-k)$$

This result is contingent of three assumptions. Data are independent random samples (independence assumption), the within population variances are the same for all k populations (equal variance), and the distribution of each population can be approximated by the normal distribution (normality). If the independence and equal variance assumptions hold, the normality assumption can be imposed either on data from each population or on the residuals (differences between the observed data and their respective population means).

The ratio of MSG/MSE is the test statistic, known as the F-statistic. If the null hypothesis is true, we expect to see a small MSG. A large observed MSG contradicts the null hypothesis. As a result, the p-value of the test is calculated as the (right) tail area of the observed F-statistic. The famous ANOVA table is used for tabulating the calculations of the sum of squares (Table 4.1).

The R function aov can be used for ANOVA:

```
#### R code ####
Everg.aov <- aov(log(TP) ~ factor(Year), data=TP.reference)
summary (Everg.aov)

#### R output ####
              Df Sum Sq Mean Sq F value  Pr(>F)
factor(Year)   5    6.5     1.3     6.7 5.1e-06
Residuals    430   83.9     0.2
```

TABLE 4.1: ANOVA table

Source of Variation	Sum of Squares	df	Mean Sum of Square	F-value
Between population	SSG	$k-1$	$MSG = SSG/(k-1)$	MSG/MSE
Within population	SSE	$N-k$	$MSE = SSE/(N-k)$	
Total	SST	$N-1$		

Given the small p-value of 0.0000051, we reject the null hypothesis and conclude that there are differences in the six annual means. The hypothesis test associated with the ANOVA table is known as the F-test.

4.7.2 Statistical Inference

With the ANOVA table and the F-test result, we conclude that the six annual means are not the same. We must, however, ask two questions. The first question is about the three assumptions necessary for carrying out the F-test, and the second is about the nature of the difference among the populations.

To check for normality, we use the normal Q-Q plot either on the residuals or on the observed data. Typically, using residuals is more efficient as the three assumptions apply to residuals as well.

```
#### R code ####
qqmath(~resid(Everg.aov),
       panel = function(x,...) {
          panel.grid()
          panel.qqmath(x,...)
          panel.qqmathline(x,...)
       },
       ylab="Residuals",
       xlab="Unit Normal Quantile")
```

The residual distribution is likely skewed, with more large values than a normal distribution can predict (Figure 4.8).

To evaluate the equal variance assumption, we use the S-L plot, where square root of the absolute values of residuals are plotted against the estimated group means:

```
xyplot(sqrt(abs(resid(Everg.aov)))~fitted(Everg.aov),
    panel=function(x,y,...){
        panel.grid()
        panel.xyplot(x, y,...)
        panel.loess(x, y, span=1, col="grey",...)
    }, ylab="Sqrt. Abs. Residuals", xlab="Fitted")
```

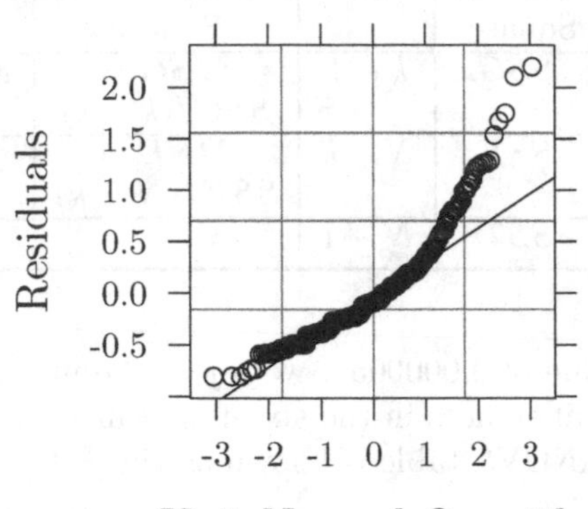

FIGURE 4.8: Residuals from an ANOVA model – A normal Q-Q plot of the ANOVA model residuals indicates a likely nonnormal residual distribution.

The residual variances are approximately constant across all six groups (Figure 4.9).

Independence of the residuals is more difficult to evaluate. One obvious plot is the scatter plot of the residuals against the estimate group averages, so that we can observe whether the residual distribution pattern changes from group to group.

```
#### R Code ####
xyplot(resid(Everg.aov)~fitted(Everg.aov),
    panel=function(x,y,...){
        panel.grid()
        panel.xyplot(x, y,...)
        panel.abline(0, 0)
    }, ylab="Residuals", xlab="Fitted")
```

The residual distribution may vary from year to year (Figure 4.10). This is judged by visual inspection for symmetry about 0. The six years appear to be divided into two separate categories: the year with mean less than 1.9 and the years with means larger than 2.1. When the mean is larger than 2.1, it seems that the larger the mean, the more likely the residual distribution to be skewed. For the year with mean below 1.9, we know from the data the skewness is largely due to the large number of censored data.

Problems revealed in Figures 4.8 and 4.10 are quite common in environmental and ecological data. To resolve these problems, one often uses power transformation of the response variable, y^λ. We can try various values of λ and find the appropriate transformation. However, when a variable is transformed, interpreting the results may be difficult. For example, when using the

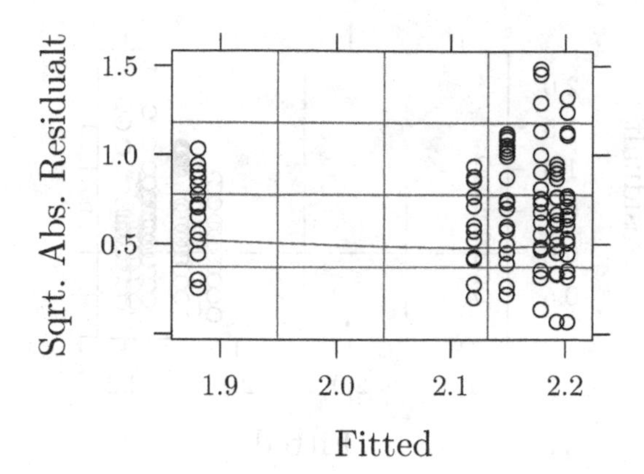

FIGURE 4.9: S-L plot of residuals from an ANOVA model – Residual variances for the 6 years are likely to be the same.

logarithm transformation, the resulting differences between the means of 1994 and subsequent years can be interpreted as the log of multiplicative factors. For example, the estimated difference between 1995 and 1994 is −0.2394. In the concentration scale, this estimate suggests that the mean in 1995 is a fraction ($e^{-0.2394} = 0.79$) of the 1994 mean. If we used $\lambda = -0.75$, the resulting ANOVA model will have a residual distribution much closer to normality (Figure 4.11). But there is no easy interpretation of the estimated difference of −0.1 between 1995 and 1994.

A significant ANOVA result (the null is rejected) should always be treated as the initial exploratory result, because ANOVA does not provide further information on the nature of the difference. With a significant result, the second question is then how the means are different among groups. This is a question addressed by multiple comparisons.

4.7.3 Multiple Comparisons

Before discussing multiple comparisons, let us first look at the alternative way for conducting an ANOVA using the linear regression model function `lm`:

```
#### R code ####
Everg.aov.lm <- lm(log(TP) ~ factor(Year), data=TP.reference))
summary.aov (Everg.aov.lm)

#### R output ####
                 Df Sum Sq Mean Sq F value  Pr(>F)
```

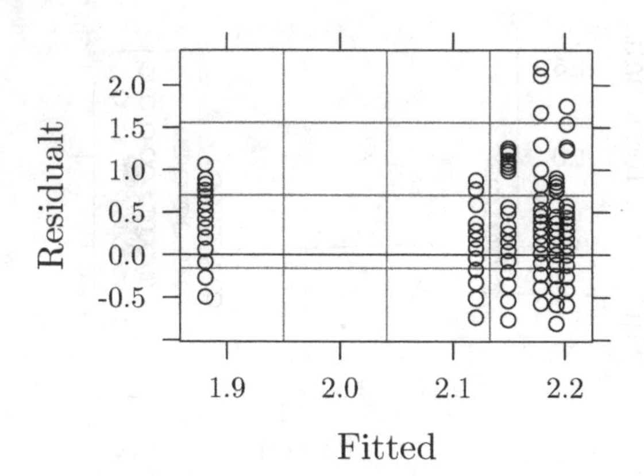

FIGURE 4.10: ANOVA residuals – Residual distributions may vary from year to year.

```
factor(Year)   5    6.5     1.3       6.7 5.1e-06
Residuals    430   83.9     0.2
```

The resulting ANOVA table is exactly the same as in the previous section. Using `lm` we can also summarize the model directly using the function `summary` for estimated model coefficients (population means):

```
#### R code ####
summary (Everg.aov.lm)

#### R output ####
Call:
lm(formula = log(TP) ~ factor(Year), data = TP.reference)

Residuals:
    Min      1Q  Median      3Q     Max
-0.8062 -0.2715 -0.0892  0.1822  2.2036

Coefficients:
                 Estimate Std. Error t value Pr(>|t|)
(Intercept)        2.1204     0.0631   33.59   <2e-16
factor(Year)1995  -0.2394     0.0774   -3.09   0.0021
factor(Year)1996   0.0288     0.0800    0.36   0.7187
factor(Year)1997   0.0814     0.0839    0.97   0.3325
factor(Year)1998   0.0581     0.0779    0.75   0.4560
factor(Year)1999   0.0721     0.0884    0.82   0.4150
```

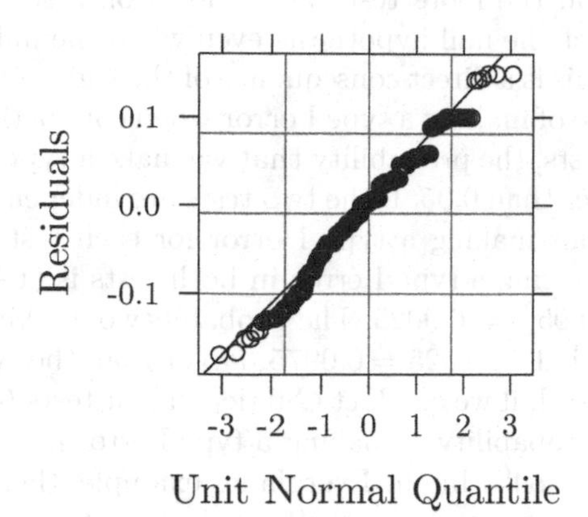

FIGURE 4.11: Normal quantile plot of ANOVA residuals – Residual distribution is much closer to normality when the response variable (TP) is transformed using a power of -0.75.

The estimated model coefficients are, however, not exactly the estimated sample averages for each year. The coefficient labeled as "Intercept" is the sample average for year 1994, the coefficient for year 1995 (-0.2394) is the difference in sample averages between 1995 and 1994, and the same with the subsequent years. This model is focused on the comparisons between the baseline mean (1994) and the means of other years. When originally developed, ANOVA was proposed for analyzing experimental data for causal inference. A typical setup is a randomized experimental design that assigns experimental units with different treatments. An experimental unit can be a field, a treatment can be a type of fertilizer, and the objective of the experiment is often to test whether or not different treatments (e.g., fertilizers) will result in different responses (e.g., yield of a crop). If the experiment was intended to compare several new fertilizers to the conventional one (the control), the objectives are then (1) to study if differences exist in crop yield when using different fertilizers, and (2) if differences exist, which new fertilizer will result in the highest yield. The first objective is achieved by using the F-test of the ANOVA model. If the null hypothesis of no difference is rejected, we will proceed to study the nature of the difference. The default linear model output is designed for comparing several "treatments" to "control" and the output includes t-test results of these comparisons. This is one form of multiple comparisons.

Comparing treatments to the control is one of many possible forms of the alternative hypothesis tested by the F-test. In general, a rejection of the null hypothesis in an F-test may indicate a difference in any pair of two groups. We may want to make pairwise comparisons using the t-test to find out which pairs of means are different. But the problem is that we will have to perform many such tests, and the more tests we perform on a set of data, the more likely we are to reject the null hypothesis even when the null hypothesis of no difference is true. This is a direct consequence of the logic of hypothesis testing: we have a 5% chance of making a type I error when conducting one test. When we perform many tests, the probability that we make a type I error for at least one test will be larger than 0.05. If the two tests are independent of each other, the probability of not making a type I error for each test is $1-\alpha$, and the probability of not making a type I error in both tests is $(1-\alpha)^2$. If $\alpha = 0.05$, this probability is $(0.95^2) = 0.9025$. The probability of making a type I error in any of the two tests is $1-0.9025 = 0.0975$, larger than the type I error rate for a single test. In general, if we conduct C independent tests (comparisons) each with $\alpha = 0.05$, the probability of making a type I error in at least one test can be as large as $1-(1-\alpha)^C$. In our Everglades example, there are $6 \times 5/2 = 15$ possible pairwise comparisons (tests). If all these tests are independents, the probability of a type I error can be as high as $1-(1-0.05)^15 = 0.54$! These tests are not independent, so the actual probability of making at least one type I error is less than 0.54. If the ANOVA null hypothesis is true and at least one pair of comparisons is "significant," the difference between the smallest sample average and the largest sample average is among the significant pairs. In other words, if only the smallest and the largest averages are compared, the result is likely a false positive because the type I error probability can be far larger than the declared 0.05. We often fall into this "multiple comparison trap" when comparing the largest differences (see Chapter 11).

To protect from the inflation of the α level, one strategy is to correct the α level when performing multiple tests. Making the α level smaller will reduce the chance of having errors, but it may also make it harder to detect real effects. The type I error α is the probability of error per test, and the calculated probability of $1-(1-\alpha)^C$ is the probability of error per collection or family of tests. To distinguish the two types of error, we use α_t to denote the per test type I error rate and α_f to denote per family type I error rate. The relationship between the two type I error rates for independent tests is:

$$\alpha_f = 1-(1-\alpha_t)^C \tag{4.7}$$

One way to adjust the per test type I error rate is to set a fixed per family type I error rate and calculate the per test type I error rate by rearranging equation 4.7:

$$\alpha_t = 1-(1-\alpha_f)^{1/C}$$

Historically, because the fractional power in the equation is difficult to compute by hand, several authors derived approximations (the linear term of the Taylor expansion of equation (4.7)). The most well known is the Bonferroni method, which sets $\alpha_t = \alpha_f/C$.

The concern about the inflated α level is often illustrated using computer simulation to show the probability of rejecting the null when it is true. Suppose we are to compare means from six populations all from the same normal distribution. We draw six equal sample size (20) samples from the same normal distribution (i.e., the null hypothesis is true) and run ANOVA and a t-test to compare the group with the smallest mean to the group with the largest means. This process is repeated many times, counting the fraction of times that the ANOVA null is rejected (at $\alpha = 0.05$) and the fraction of times the t-test null hypothesis is rejected.

```
#### R code ####
    anova.p <- t.p <- numeric()
    for (i in 1:1000){
        data.sim <- data.frame(y=rnorm(120),
                               g=rep(1:6, each=20))
        sample.mean <- tapply(data.sim$y, data.sim$g, mean)
        data.sim$g <- ordered(data.sim$g,
                        levels=names(sort(sample.mean)))
        data.sim$g <- as.numeric(data.sim$g)
        anova.p[i] <- summary(aov(y~factor(g),
                                 data=data.sim))[[1]][1,5] < 0.05
        t.p[i] <- t.test(y~g, data=data.sim,
                         subset=g==1|g==6)$p.value < 0.05
    }
    print(c(mean(anova.p), mean(t.p)))

#### R output ####

[1] 0.047 0.346
```

The simulation rejected the ANOVA null hypothesis about 5% of the time (as expected) and rejected the t-test null close to 35% of the time. When the Bonferroni method is used, the t-test should have a significance level $\alpha_t = 0.05/15 = 0.0033$. We conduct the simulation again by using the "corrected" significance level, that is, replacing the line for t-test with:

```
        t.p[i] <- t.test(y~g, data=data.sim,
                      subset=g==1|g==6)$p.value < 0.0033
```

and rejection rate for the t-test is ~ 0.035, somewhat smaller than the expected 0.05. Methods for multiple comparisons based on the basic relationship in equation (4.7) will be inevitably overconservative, and making the detection of the true difference difficult. In addition to the Bonferroni method, other authors proposed different methods for adjusting the per test type I error rate. These multiple comparisons methods are implemented in the R function `glht` (from package `multicomp`). The various α-level adjustments are implemented in the function `p.adjust`, which converts a per test p-value to a per family p-value.

A different approach to the multiple comparison problem is represented by the Tukey's Honestly Significant Difference (HSD). When comparing two populations with the same mean, the test statistics is Student's t. When there are g populations, there are $g(g-1)/2$ pairwise comparisons that can be made. Tukey discovered the distribution of the largest of these t-statistics when there were no underlying differences among the g population means. When using the Tukey's HSD method, we conduct all possible t-tests, but using the null distribution for the largest difference. The Tukey's HSD method is implemented in the R function `TukeyHSD`.

Another approach is the Fisher's least significant distance (or LSD), where all possible t-tests are conducted only when the initial ANOVA F-test null is rejected. The name "least significant difference" refers to the smallest difference between two population means that will result in a rejection of the null hypothesis of no difference. It is obviously developed for easy calculation by hand. The emphasis of the Fisher's LSD is that the t-tests cannot be performed unless the overall F-test null is rejected. There is only a 5% chance that the overall F ratio will reach statistical significance when there are no differences. Therefore, the chance of reporting a significant difference when there are none is held to 5%.

Furthermore, many authors suggested that "planned" comparisons should be allowed without adjustments for multiple comparisons because these tests are planned á priori, not picked after the fact.

Considerations of multiple comparisons are about the two different types of type I errors, the per test type I error, and the per family type I error. I believe that these discussions are important to warn users against overconfidence about their statistically significant results, which can be due entirely to chance. However, the practice of adjusting the α level should be used with caution. Gotelli and Ellison [2004] argued that adjusting the α level based on the number of tests made in a study will lead to the loss of the single standard applied to all hypotheses tests in the scientific literature. They further argued that the adjustment of α should be made in the other direction. For example, if three independent tests are conducted, the Bonferroni adjustment would require that the per test α be $0.05/3$ (or 0.016). But if all three tests obtained the same p-value of 0.11, one would argue that the chance of the null hypothesis (all three means are the same) being true is fairly small.

Two practical considerations often make multiple comparisons less attractive. First, an experiment is proposed often because we have reasons to believe a significant treatment effect. As a result, the objective of the experiment is often to estimate the magnitude of the effect, rather than to test whether or not the effect exists. Adjusting the α level downward will lead to the wider confidence intervals of treatment effects, unnecessarily inflating the level of uncertainty. The emphasis on the per family type I error is therefore counterproductive. Second, multiple comparisons are frequently used for detecting small treatment effects. Because multiple comparisons are achieved through analysis of the relative magnitude of between and within group variances, small effects suggest a relatively large within group variance. As a result, these effects require a large sample size, and the results of such studies are susceptible to random sampling error. We will revisit the issue in Chapter 10.

4.8 Examples

The examples in this section are intended for illustrating exploratory data analysis as a means for ensuring that the resulting statistical inference is scientifically sound.

4.8.1 The Everglades Example

This example illustrates the use of graphical display of distributions for identifying potential anomalies in the data. The data were collected for setting the phosphorus standard for the Everglades wetlands from five sampling locations. These locations are long-term sampling stations. The data used for the study were collected from 1994–1999. The Everglades has a dry season (winter and spring) and a wet season (summer and autumn). Phosphorus concentrations are affected by precipitation and the variation of precipitation. During the sampling period, annual precipitation in the nearby Everglades National Park varied from slightly above the long-term average of $\sim$ 125 cm in 1994 to near 200 cm in 1995 and then dropped to a level slightly above 100 cm in 1996, before moving back to normal and slightly above normal from 1997 to 1999 (Figure 4.12). The high precipitation in 1995 resulted in many low TP concentrations in that year. In fact, most of the concentration values that are below the method detection limit occurred in this year. In the next two years (1996 and 1997), either the total rainfall was low (1996) or the month-to-month variation was low (1997), resulting in more high TP concentrations being observed. Furthermore, throughout the 6 years when TP samples were included in the data set, samples were taken unevenly across the year. More summer and autumn months were sampled in 1996 and 1997 (Table 4.2). As a result of these problems, the data cannot be regarded as independent random samples.

The changes in precipitation and the uneven distribution of monthly sample sizes resulted in clustering of high or low TP concentration values in different years. Figure 4.13 shows the normal Q-Q plot of the log TP concentration data for each year. The distributions in years 1996 to 1998 are clearly non-normal with more large concentration values than a normal distribution can expect.

Based on these observations on the data, only data from 1994 were used earlier in this chapter.

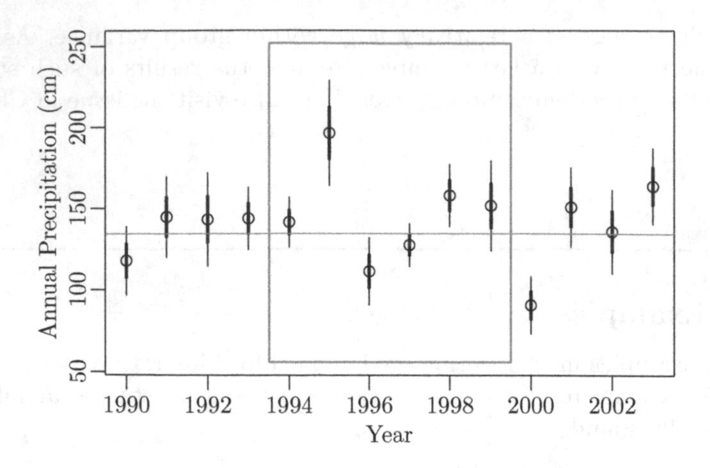

FIGURE 4.12: Annual precipitation in the Everglades National Park – Annual precipitation in the Everglades National Park experienced a few years of ups and downs during the period when TP samples were taken and used for the standard study.

TABLE 4.2: Everglades data sample sizes – Monthly sample sizes in the Everglades data set vary

	Jan	Feb	Mar	Apr	May	Jun	Jul	Aug	Sep	Oct	Nov	Dec
1994	3	1	6	4	0	5	7	0	0	9	7	7
1995	8	0	5	10	10	5	9	10	10	10	10	10
1996	6	3	0	5	5	9	10	5	11	5	15	7
1997	9	0	0	0	5	5	10	5	10	5	10	5
1998	10	5	10	10	5	5	10	5	9	5	10	10
1999	8	0	2	0	0	5	6	5	0	5	15	5

4.8.2 Kemp's Ridley Turtles

This is a data set reported in Ruckdeschel et al. [2005] for studying the sex ratio in Kemp's ridley sea turtle, the most endangered species of sea turtle in the world. The study claimed that the sex ratio shifted between 1989 and 1990, from a slightly male-biased ratio in 1983–1989 to significantly female-biased ratio from 1990 to 2001. Data were collected from stranded (dead) Kemp's ridley sea turtles on Cumberland Island, Georgia, U.S.A, between 1983 and 2001. From 1983 to 1989 there were 16 males and 10 females and from 1990 to 2001 there were 19 males and 56 females. To verify the authors' claim, we can perform a series of hypothesis tests. Although the authors of the study did not mention the possible cause of the shift, it is suggested elsewhere that sea turtle hatcheries established nearby during that time ($\sim$1990) may have contributed to the change.

First, we note that the data we have now are the number of male and female turtles and the quantity we are interested in is the ratio of male as a

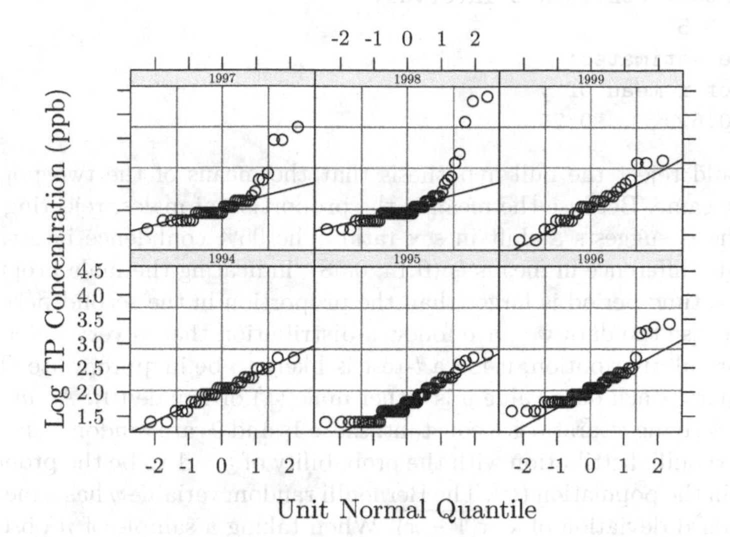

FIGURE 4.13: Yearly variation in Everglades TP concentrations – Normal Q-Q plots of log TP concentration from three reference sites in the Everglades are plotted using the yearly data. The yearly differences in TP concentration distribution are largely due to the different precipitation patterns (Figure 4.12).

fraction of the total number, that is 16/26 from 1983 to 1989, and 19/75 from 1990 to 2001. If we record a male as 1 and a female as 0, the data set we have is sixteen 1s and ten 0s in the first time period and nineteen 1s and fifty-six 0s in the second time period. The quantity of interest is the mean of these 1s and 0s. The central limit theorem suggested that the sample average distribution will approach normality when the sample size is large enough. The simplest test is to test whether the means are equal before and after the apparent shift using a t-test.

Using a two-sample t-test,

```
#### R code and output ####
t.test(x=c(rep(1, 16), rep(0, 10)),
       y=c(rep(1, 19), rep(0, 56)) )

        Welch Two Sample t-test

data:  c(rep(1, 16), rep(0, 10)) and c(rep(1, 19), rep(0, 56))
t = 3.3, df = 39, p-value = 0.00205
alternative hypothesis: true difference in means is not
    equal to 0
```

```
95 percent confidence interval:
 0.14 0.58
sample estimates:
mean of x mean of y
     0.62      0.25
```

we would reject the null hypothesis that the means of the two populations are the same. Because the mean is the proportion of males, rejecting the null hypothesis suggests a shift in sex ratio. The 95% confidence interval of the estimate difference in means is (0.14, 0.58), indicating the male proportion in the first time period is larger than the proportion in the second period.

Because the data we have follow a distribution that is very different from the normal distribution, using a t-test is likely to be inappropriate. The data are binary. Each data value y is either male (1) or female (0). Assuming that the underlying sex ratio is a constant, these 1s and 0s are random samples from the Bernoulli distribution with the probability of $y = 1$ to be the proportion of males in the population (π). The Bernoulli random variable y has a mean π and a standard deviation of $\sqrt{\pi(1-\pi)}$. When taking a sample of n observations $y_1, \cdots, y_n$, the sample total $S = \sum_{i=1}^{n} y_i$ is a count of the number of 1s. The variable S, having the binomial distribution, will take one of the values: $0, 1, \cdots, n$. The probability of $S = k$ is:

$$\Pr(S = k) = \frac{n!}{k!(n-k)!}\pi^k(1-\pi)^{n-k}$$

The mean of S is $n\pi$ and the standard deviation of S is $\sqrt{n\pi(1-\pi)}$. Using the sample distribution of S, we can develop a formal hypothesis testing procedure for testing whether the proportion is equal to a specific value. In the Kemp's ridley turtle example, we may test whether the proportion of male is 0.5. That is, setting the null hypothesis to be $H_0 : \pi = 0.5$ and the alternative to be $H_a : \pi \neq 0.5$. The test statistic is S, which has a binomial distribution when the null hypothesis is true. In this case, the null distribution for the period of 1983 to 1989 is $n = 26, k = 16$. The p-value is $2\Pr(S \geq 16|H_0)$, which is calculated in R as `2*(1-pbinom(16-1, 26, 0.5))`, or 0.3269. For the period of 1990 to 2001, p-value $= 2\Pr(S \leq 19|H_0)$, which is `2*pbinom(19, 75, 0.5)`, or 0.000022. Alternatively, we can directly use the function `binom.test`:

```
#### R code and output ####
binom.test(x=16, n=26, p=0.5)

        Exact binomial test

data:  16 and 26
number of successes = 16, number of trials = 26,
    p-value = 0.3269
```

```
alternative hypothesis: true probability of success is not equal
    to 0.5
95 percent confidence interval:
 0.4057075 0.7977398
sample estimates:
probability of success
             0.6153846

#### R code and output ####
binom.test(x=19, n=75, p=0.5)

        Exact binomial test

data:  19 and 75
number of successes = 19, number of trials = 75,
    p-value = 2.243e-05
alternative hypothesis: true probability of success is not
    equal to 0.5
95 percent confidence interval:
 0.15993 0.36701
sample estimates:
probability of success
               0.25333
```

These two tests provide some information, but do not directly address the question of whether the sex ratio had a shift between 1989 and 1990. A direct test would estimate the difference of the two proportions. The sampling distribution for the difference between two sample proportions is somewhat difficult to obtain directly. But we know that the mean of the sample mean difference

$$\hat{\pi}_1 - \hat{\pi}_2$$

is equal to the difference of population means $\pi_1 - \pi_2$ and the standard deviation of the sample mean difference is

$$\sqrt{\pi_1(1-\pi_1)/n_1 + \pi_2(1-\pi_2)/n_2}$$

If the sample size is large the distribution of $\hat{\pi}_1 - \hat{\pi}_2$ is approximately normal. As a result, to test whether the two proportions (or population means) are the same, we can use the calculated difference as the test statistic and the approximate normal distribution as the null distribution. For this example, $\hat{\pi}_1$ is 0.62 and $\hat{\pi}_2$ is 0.25, $n_1 = 16$, and $n_2 = 75$. The null hypothesis is $H_0 : \pi_1 - \pi_2 = 0$; therefore the null distribution of $\hat{\pi}_1 - \hat{\pi}_2$ is approximately normal with mean 0 and standard deviation

$$\sqrt{\pi_1(1-\pi_1)/n1 + \pi_2(1-\pi_2)/n2} \approx$$
$$\sqrt{0.62(1-0.62)/16 + 0.25(1-0.25)/75} =$$
$$0.1312$$

The p-value is then calculated in R as `2*(1-pnorm(0.62-0.25, 0, 0.1312))`, or 0.0048. The estimated difference of $0.62 - 0.25 = 0.37$ has a confidence interval of $0.37 \pm 2 \times 0.1312$ or (0.11, 0.63).

The exact test for this problem is provided by Karl Pearson in the form of a general method for comparing the observed to the expected (based on the null hypothesis). This class of test is known as the χ^2 test of goodness of fit. For this particular example, the data can be arranged in a 2×2 table:

	Male	Female
1983-1989	16	10
1990-2001	19	56

or, more generally,

	Response 1	Response 2	Totals
Factor 1	n_{11}	n_{12}	R_1
Factor 2	n_{21}	n_{22}	R_2
Totals	C_1	C_2	T

For each of the 4 cells, the numbers $16, 26, 19, 56$ are called the *Observed.* If the null hypothesis is true, the proportion of male turtles in the first row should be the same as the same proportion in the second row. Using the general table, the total male proportion is C_1/T. If the proportion is independent of the rows (i.e., the null is true), we expect the number of males in the first time period to be R_1C_1/T. That is, for each cell, we have an expected number:

	Response 1	Response 2	Totals
Factor 1	R_1C_1/T	R_1C_2/T	R_1
Factor 2	R_1C_1/T	R_2C_2/T	R_2
Totals	C_1	C_2	T

Pearson combined the expected and observed numbers into a single statistic:

$$\chi^2 = \sum \frac{(Expected - Observed)^2}{Expected}$$

which has an approximate sampling distribution; the χ^2 distribution with degrees of freedom equal the number of cells (4) minus the number of parameters estimated (2) minus 1. In R, the test is implemented in function `prop.test`:

```
#### R code and output ####
prop.test(x=c(16, 19), n=c(26, 75))

	2-sample test for equality of proportions with
	        continuity correction

data:  c(16, 19) out of c(26, 75)
```

```
X-squared = 9.6343, df = 1, p-value = 0.001910
alternative hypothesis: two.sided
95 percent confidence interval:
 0.12483 0.59927
sample estimates:
 prop 1  prop 2
0.61538 0.25333
```

Three different methods were used for testing whether or not there is a sex ratio shift between 1989 and 1990. The p-values are 0.00205 from the crude two-sample t-test, 0.0048 from the normal approximation method, and 0.00191 from the χ^2 test. These tests are used to illustrate the common problem often faced in environmental and ecological studies, that is, there is not a single best model that can be directly used in the analysis at hand. Certain level of approximation is inevitable. For this example, the χ^2 test is the one that most would agree to be the appropriate model. However, our conclusion would not change when using the other two approximations. This is another example of the robustness of the many statistical procedures against the normality assumption.

As discussed before, the independence assumption is often a more influential assumption. It is the same for this example. A careful examination of the data gave me reasons to question the conclusion of a sex ratio shift. The data may not be independent samples of the turtle population.

First, the authors reported that only 50% of the stranded turtles provided gender information. Is there a sex bias in those turtles whose sex cannot be identified? Those turtle bodies that cannot be positively identified of their sex were partially decomposed. Some suggest that male turtle bodies (especially the sex organ) decompose faster than female turtle bodies.

Second, there are seasonal differences in sex ratios among the turtles visiting a beach. No information was provided in the paper with regard to seasonal composition of the data. The authors discussed seasonal differences in the sex ratio. In general, there are more females on the beach in spring and summer. Can the observed shift in the sex ratio be attributed to the increased data collection effort in spring and summer in the second period? Before we conclude that there is indeed a sex ratio shift, we must examine the seasonal composition of the data. If the increased number of turtles in the second time period was due to the participation of residents and tourists, who would more likely visit the beach during spring and summer, there may be a sampling bias and we cannot conclude a sex ratio shift had happened. At least, we can be certain that the sex ratio shift, if indeed exists, is not the result of the establishment of the hatchery in 1990. A quick Google search on the live history of Kemp's ridley turtle shows that these turtles spend their first two years of life floating in open ocean and need 10 to 20 years to reach sexual maturity. Any indication of a sex ratio shift due to the hatchery would not be evident until after the data collection period.

4.8.3 Assessing Water Quality Standard Compliance

Under the U.S. Clean Water Act (CWA) Section 305(b), states of the United States are required to assess conditions of their waters for compliance of their respective designated uses. The U.S. EPA collects and utilizes this information to prepare a biennial report, known as the National Water Quality Inventory, commonly referred to as the "305(b) Report," for the Congress. Section 303(d) of the CWA requires U.S. states to prepare lists of "surface waters that do not meet applicable water quality standards," referred to as impaired waters, and to establish Total Maximum Daily Loads (TMDLs) for pollutants causing the impairment of these waters on a prioritized schedule. A TMDL establishes the maximum daily amount of a pollutant that a water body can assimilate from all sources without causing exceedances of water quality standards. As such, the development of TMDLs is an important step toward restoring surface waters to their designated uses.

The U.S. EPA guidelines once required a water body to be listed as impaired when greater than 10% of the measurements of water quality conditions exceed numeric criteria. This rule for declaring a water body to be impaired can be viewed as a hypothesis testing procedure. The null hypothesis is that the water body is in compliance (hence $\mu \leq WQ_c$) and the alternative hypothesis is that the water body is out of compliance (hence $\mu > WQ_c$). Here, we loosely used μ as a measure of water quality condition (which may be the mean or other measure) and WQ_c is the numerical water quality criterion. Under the null hypothesis, we expect most measurements of the water quality to be small. The test statistic is the fraction of measurements that exceeds the criterion. Obviously, the "10% rule" is not a statistical hypothesis testing procedure, so we cannot derive the null distribution. Consequently, we don't know the probabilities of making type I and type II errors. But it is nevertheless a decision process. We must decide whether or not to declare that the water body is impaired, and each decision is associated with error because the decision will be inevitably based on random samples. Using hypothesis testing terminology, the U.S. EPA has decided a test statistic (proportion of measurements exceeding standards) and set a rejection region (>10%).

Smith et al. [2001] interpreted the U.S. EPA's rule as to ensure that when a water body is declared to be in compliance, a standard violation will occur no more than 10% of the time in the future. Based on this interpretation, the rule can be represented by statistical hypothesis testing about the underlying true rate of exceedance. In this test, we assume that each water quality measurement is a random sample from the population representing the water body with an unknown probability (π) of exceeding the water quality criterion. The null hypothesis is then $H_0 : \pi \leq \pi_0$, where π_0 is the acceptable exceedance rate. When the null hypothesis is true, each measurement has a probability of exceeding the criterion of no more than π_0. As a result, the U.S. EPA's 10% rule is now interpreted as $\pi_0 = 0.1$. If we call a measurement that exceeds the criterion a success and record it as 1, and a measurement that is below the criterion a failure and record it as 0, we transform the water quality measurements into a binary variable of 1s and 0s. Under the null hypothesis, the transformed binary variable follows a Bernoulli distribution with probability of success to be π_0. The total number of successes in a sample, our test statistic, follows a binomial distribution. From this distribution, we can calculate the rejection criterion, that is, find x_r such that $\Pr(x \geq x_r | H_0) \leq \alpha$, or the $1 - \alpha$ quantile of the null distribution. That is, we can use the same exact binomial test discussed in the Turtle example. However, the 303(d) listing is a decision process; a state resource manager wants to have a clear rule for listing. In that regard, tabulating the condition for listing is a better way for clearly describing the decision-making process. In R, this can be achieved by using the function `qbinom`:

```
#### R code and output ####
    qbinom(1-0.05, size=12, prob=0.1)
    [1] 3
```

That is, the 0.95 quantile of the distribution is approximately 3. It is approximate because the binomial distribution is discrete. The rejection region is the number of successes greater than 3, that is, 4 or more measurements exceeding the criterion. Based on this procedure, the type I error (listing a water body that is in compliance with the designated use) rate is less than or equal to 5%, depending on the sample size. If the U.S. EPA's rule is used for this test, the null will be rejected if 10% of the measurements exceed the standard, which is 2 or more; the type I error rate is 0.34 (Exercise 2 of Chapter 2), much too large to be acceptable. In fact, the U.S. EPA's 10% rule can have a type I error rate as high as 0.61 (when $n = 20$), and approaching to 0.5 as sample size increases.

Because the binomial distribution is not continuous, the rejection region estimated here is actually associated with a type I error rate of `1-pbinom(3, 12, 0.1) = 0.026`. The rejection criterion of 4 or more is the smallest number that is associated with a p-value less than or equal to 0.05. (If we set the rejection region to be the number of successes greater than 2, the type I error rate will be 0.11.) A table of minimum number of exceedances needed for listing can be generated for a range of sample sizes (e.g., 5 to 20):

```
#### R code and output ####
    qbinom(1-0.05, size=5:20, prob=0.1) + 1
   [1] 3 3 3 3 4 4 4 4 4 4 5 5 5 5 5 5
```

Again, each sample size has a different type I error rate (Figure 4.14, left panel). Because 303(d) listing is a decision-making process, not only the type I error probability should be controlled to be less than α, the type II error probability should also be limited to an acceptable level. When calculating the type II error, we need to know the effect size (the difference between the alternative and null hypothesis proportions). In California, the type II error is based on an effect size of 0.15, or the alternative proportion of 0.25.

For a sample size of 10, the condition for listing is 4 or more measurements exceeding the standard. The statistical power is the probability of rejecting the null when the alternative is true. Equivalently, the probability of observing 4 or more 1s when $\pi = 0.25, n = 10$:

```
#### R code and output ####
   1-pbinom(4-1, size=10, prob=0.25)
   [1] 0.22412
```

The power of the test is about 22%, obviously too small. The power is a function of sample size:

```
#### R code ####
sample.size <- 10:40
reject <- qbinom(1-0.05, size=sample.size, prob=0.1) + 1
decision.table <- data.frame(n=10:40, reject=reject,
                             power=1-pbinom(reject-1,
                               size=sample.size, prob=0.25))
plot(power~n, data=decision.table, type="l",
   xlab="Sample Size", ylab="Power")
```

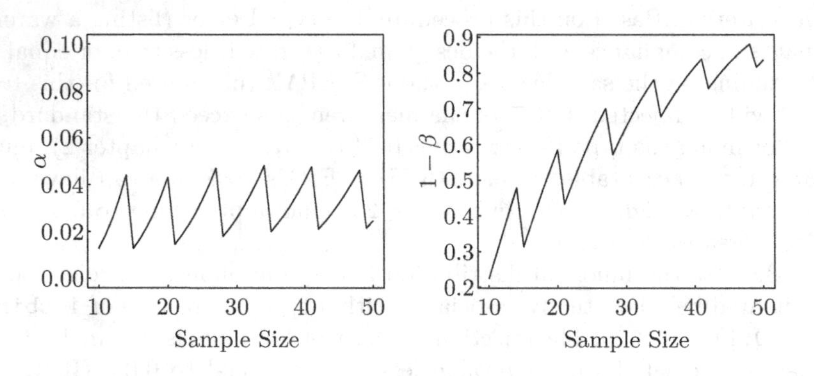

FIGURE 4.14: Statistical power is a function of sample size.

Figure 4.14 (right panel) shows that the statistical power increases in a zigzag pattern. This is because the type I error rate varies as sample size changes. The statistical power is affected by both n and α. But in general, a sample size of 30 or more is necessary to achieve a modest power of 70%.

Once a water body is listed as impaired, a TMDL program will be developed and implemented. When the water quality is improved such that its exceedance rate is below 0.1, the water body will be removed from the 303(d) list, a process called "de-listing." The recommended procedure currently used in many states in the United States is to perform the following test:

$$H_0: \quad \pi \geq \pi_0 \quad \text{impaired}$$
$$H_a: \quad \pi < \pi_0 \quad \text{unimpaired}$$

Therefore, to remove a water body with a sample of 12 measurements, there should be fewer than `qbinom(0.05, 12, 0.1) = 0`. However, the probability of success is a very small number (0.1), it is impossible to test the null hypothesis against the alternative because the probability of observing 0 is 0.28 under the null hypothesis. To properly assess the water quality, the sample size must be increased. For example, if we set the rejection region at ≤ 1, the sample size must be 46, because `pbinom(1, 46, 0.1) = 0.048`.

4.8.4 Interaction between Red Mangrove and Sponges

The next example is a typical experimental study designed for the use of ANOVA. The data were from a study of the interaction between red mangrove (*Rhizophora mangle*) roots and root-fouling sponges [Ellison et al., 1996]. Mangroves are plants and shrubs that grow in saline coastal habitats in the tropics and subtropics. They grow on prop roots, which arch above the water level, giving stands of this tree the characteristic "mangrove" appearance. The roots of red mangroves are colonized by a number of species of sponges, barnacles, algae, and smaller invertebrates and microbes. The question Ellison et al. are interested in is whether mangrove plants benefit from animal assemblages grown on their roots. In the study, the effects of two common sponge species on the root growth of red mangrove were experimentally studied. Stands of red mangrove trees were randomly assigned four different treatments: (1) unmanipulated control, (2) foam (fake sponge) attached to bare mangrove roots, (3) living red fire sponge (*Tedania ignis*) colonies transplanted to bare mangrove roots, and (4) purple sponge (*Haliclona implexiforms*) living colonies transplanted to bare mangrove roots. The measured response variable was mangrove root growth in mm/day.

The data contain only two "outside values" (Figure 4.15), potentially unusual values or outliers. The distribution of root growth data from each treatment can be approximated by the normal distribution (Figure 4.16). The normal Q-Q plots in Figure 4.16 do not suggest a systematic departure between the data points and the reference line. Based on these two plots, we feel that ANOVA is an appropriate method for analyzing the data.

```
#### R code ####
mangrove.lm <- lm(RootGrowthRate ~ Treatment,
    data=mangrove.sponge)
summary.aov(mangrove.lm)

#### R output ####
            Df Sum Sq Mean Sq F value  Pr(>F)
Treatment    3   4.40    1.47    6.87 0.00041 ***
Residuals   68  14.51    0.21
---
Signif. codes:  0 '***' 0.001 '**' 0.01 '*' 0.05 '.' 0.1 ' ' 1
```

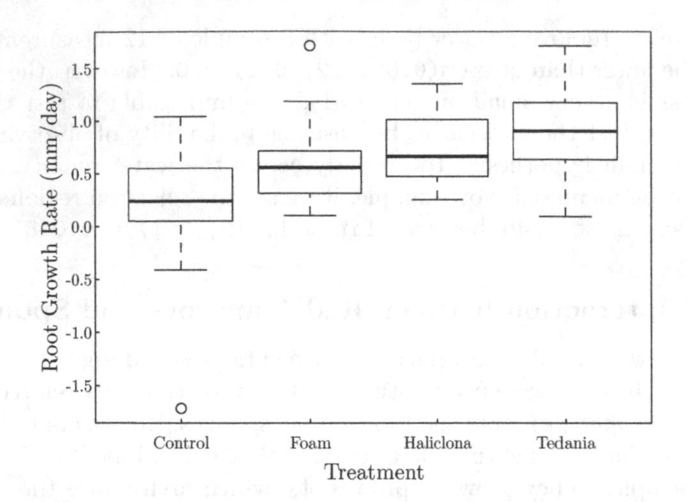

FIGURE 4.15: Boxplots of the mangrove-sponge interaction data – these plots suggest that root growth is potentially higher under the live sponge treatments.

The ANOVA table shows a p-value less than the significance level of 0.05, indicating the existence of some treatment effects. There are many possible comparisons. The most obvious are the comparisons of the control to the three treatments (`Foam`, `Haliclona`, and `Tedania`). Also, it is reasonable to compare the mean of the two live sponge treatments to the mean of the foam treatment, and to the control mean. The R function `TukeyHSD` implements the Tukey's HSD method:

```
#### R code ####
mangrove.aov <- aov(RootGrowthRate ~ Treatment,
    data=mangrove.sponge)
mangrove.HSD <- TukeyHSD(mangrove.aov)

#### R output ####
mangrove.HSD
  Tukey multiple comparisons of means
    95% family-wise confidence level

Fit: aov(formula = RootGrowthRate ~ Treatment,
    data = mangrove.sponge)
$Treatment
                      diff       lwr     upr   p adj
Foam-Control       0.35436 -0.025798 0.73451 0.07650
Haliclona-Control 0.49109  0.094128 0.88806 0.00927
Tedania-Control    0.67643  0.256617 1.09624 0.00039
```

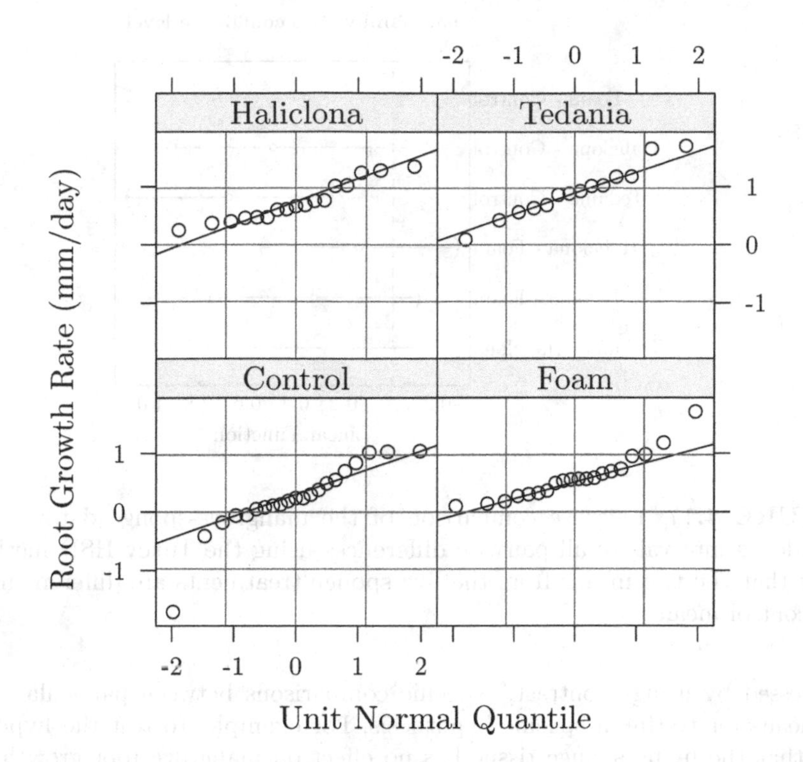

FIGURE 4.16: Normal Q-Q plots of the mangrove-sponge interaction data – root growth rate distributions are approximately normal.

```
Haliclona-Foam     0.13674 -0.264644 0.53812 0.80630
Tedania-Foam       0.32207 -0.101917 0.74606 0.19790
Tedania-Haliclona  0.18534 -0.253787 0.62446 0.68369
```

The function returns confidence intervals for all pairwise differences, in both tabulated and graphical forms (Figure 4.17). Using the generic `plot` function, these confidence intervals are shown in a graph. The Tukey's HSD method is also implemented in the package `multcomp`

```
#### R code ####
library(multcomp)
q2<-glht(mangrove.aov, linfct=mcp(Treatment="Tukey"))
summary(q2)
plot(q2)
```

Specific questions such as whether the mean of live sponge treatments is different from the foam treatment mean or the control mean are traditionally

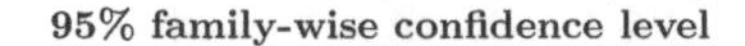

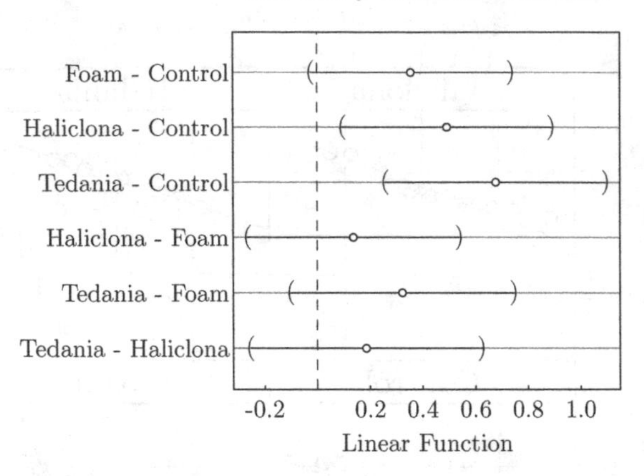

FIGURE 4.17: Pairwise comparison of the mangrove-sponge data – The confidence intervals of all pairwise differences using the Tukey HSD method show that the two means from the live sponge treatments are different from the control mean.

addressed by using "contrast," specific comparisons between particular sets of means for testing a specific hypothesis. For example, to test the hypothesis that the living sponge tissue has no effect on mangrove root growth we would consider the comparison of the control mean and the mean of the two sponge treatment means. The null hypothesis is that the difference between the control mean and the mean of the two sponge treatment means is 0. The difference is $\delta = -\mu_{control} + 1/2(\mu_{Haliclona} + \mu_{Tedania})$. The difference is often expressed as a linear combination of all 4 treatment means:

$$\delta = -1\mu_{control} + 0\mu_{foam} + 1/2\mu_{Haliclona} + 1/2\mu_{Tedania}$$

In general, a contrast is a linear combination of treatment means, usually comparing means of different groups of treatments:

$$\delta = \sum_{i=1}^{k} a_i \mu_i$$

The linear combination coefficients of a contrast must sum to 0: $\sum_{i=1}^{k} a_i = 0$. The standard error of δ is $se_\delta = \sigma\sqrt{\sum_{i=1}^{k} a_i^2/n_i}$, where σ is the residual standard deviation and n_i is the sample size of treatment i. The t-ratio of δ/se_δ has a t-distribution with degrees of freedom $df = \sum_{i=1}^{k} n_i - k$, which can be used to construct a confidence interval of δ or a hypothesis test about δ. In R, contrast can be specified in the function `glht`:

```
#### R code ####
contr <- rbind("F - C" = c(-1, 1, 0, 0),
               "H - C" = c(-1, 0, 1, 0),
               "T - C" = c(-1, 0, 0, 1),
               "S - F" = c(0, -1, 1/2, 1/2),
               "S - C" = c(-1, 0, 1/2, 1/2))
q3 <-  glht(mangrove.aov, linfct = mcp(Treatment = contr))

#### R output ####

summary(q3, test=adjusted(type=c("none")))

         Simultaneous Tests for General Linear Hypotheses

Multiple Comparisons of Means: User-defined Contrasts

Fit: aov(formula = RootGrowthRate ~ Treatment,
    data = mangrove.sponge)

Linear Hypotheses:
           Estimate Std. Error t value p value
F - C == 0    0.354      0.144    2.45  0.0167
H - C == 0    0.491      0.151    3.26  0.0018
T - C == 0    0.676      0.159    4.24 6.8e-05
S - F == 0    0.229      0.133    1.73  0.0885
S - C == 0    0.584      0.131    4.46 3.1e-05
---
(Adjusted p values reported -- none method)
```

The results suggest that (1) compared to the control, the effects of living sponge on mangrove root growth are positive and statistically different from 0, (2) the effects of living sponge on mangrove root growth are virtually the same as the effect of the inert foam, and (3) the effect of biologically inert form on root growth is also statistically significant when compared to the control.

Many texts recommended that a priori (or planned) comparisons be used to guide against the inflation of α level. But I find this recommendation is vague and cannot be expected to achieve the intended objective. In the original article of this example, the à priori comparisons were the three comparisons between the control and the three treatments. While I was reading the paper, I thought that the two additional comparisons (living sponge versus control, and living sponge versus foam) would be informative. In a reanalysis of the data, Gotelli and Ellison [2004] compared living sponge against foam, and the mean of all three treatment against the control:

```
#### R code ####
  contr2 <- rbind("F - C" = c(-1, 1/3, 1/3, 1/3))
```

```
q4 <-  glht(mangrove.aov, linfct = mcp(Treatment = contr2))
summary(q4, p.adjust.methods="none")

#### R output ####

Linear Hypotheses:
           Estimate Std. Error t value p value
F - C == 0    0.507      0.120    4.22 7.4e-05
---
(Adjusted p values reported)
```

In this particular example, these different comparisons may not change the conclusion. But with different researchers, different à priori comparisons are to be expected. I find that the interpretation of the estimated differences is more important, as long as the test results are not blown out of proportion. A multilevel modeling approach (Chapter 10) is more natural for a multiple comparison problem.

4.9 Bibliography Notes

The problem of induction was first discussed by David Hume [Hume, 1777]. Popper [Popper, 1959] restated the problem to be a method that "passes *singular statements*, such as accounts of the results of observations or experiments, to *universal statements*, such as hypothesis or theory." Fisher's discussion on hypothesis testing can be found in Fisher [1955]. George Box [Box, 1976] also discussed the role of statistics in science. Discussions on the abuse of statistical power analysis can be found in Hoenig and Heisey [2001]. Statistical issues related to water quality assessment are discussed by Smith et al. [2001]. The use of the χ^2 distribution for assessing the uncertainty in the estimated standard deviation (Figure 4.2 on page 84) is unconventional. See Gelman and Hill [2007] for more detail.

4.10 Exercises

1. In a study of water quality trends, EPA compiled stream biological monitoring data in the Mid-Atlantic region before and after 1997. They are interested in whether there was a shift in biological conditions in streams in the area. The indicator they used is EPT taxa richness (number of taxa belong to three genera of flies commonly known as mayfly, stonefly,

and caddisfly). As distributions of count variables are typically skewed, log transformation was used. The log mean and standard deviation of EPT taxa richness before 1997 are 2.2 and 6.9 ($n = 355$), and are 1.8 and 5.4 after 1997 ($n = 280$). Is the difference in log mean statistically significant? How do you report the difference in terms of EPT taxa richness?

2. The Student's t-test

   The famous t-test was initially illustrated by "Student" in a paper published in 1908 (available at: `http://biomet.oxfordjournals.org/cgi/reprint/6/1/1`). In the paper, Student used several examples to illustrate the process. One example (Illustration 3) discussed the yield of barley of seeding plots with two different kinds of seed. Each type of seed (kiln-dried and not kiln-dried) was planted in adjacent plots in two different years, accounting for 11 pairs of "split" plots. The data listed below are from the table on page 24 of the 1908 paper.

N.K.D	K.D.	Diff.
1903	2009	106
1935	1915	-20
1910	2011	101
2496	2463	-33
2108	2180	72
1961	1925	-36
2060	2122	62
1444	1482	38
1612	1542	-70
1316	1443	127
1511	1535	24

   The statistical method used in Student (1908) is very different from the ones we use now. On page 24, Student concluded that the odds that kiln-dried seeds have a higher yield is 14:1. Conduct the t-test using the "head corn" yield data shown above. Can you guess where the 14:1 odds come from?

3. PCB in Fish.

   In the PCB in fish example, we learned that lake trout switch diet when they are about 60 cm long. Large trout (> 60 cm) tend to have higher PCB concentrations. Assuming that PCB concentration distribution can be approximated by the log-normal distribution,

   (a) use a statistical test to compare the mean concentrations of large and small trout, and

   (b) discuss any potential problems of using this test based on relevant graphic methods

4. The Everglades wetland ecosystems are phosphorus limited. After the Everglades Agriculture Areas (EAA) were established (enabled by a series of federal government constructed water diversion systems for draining part of the Everglades wetland), phosphorus-rich agriculture runoff reached the Everglades wetland and resulted in dramatic changes in parts of the Everglades wetlands. To better protect the Everglades, many studies were conducted in the late 1980s and the 1990s to learn about the effects of phosphorus enrichment in the Everglades. One study focused on estimating the background level of phosphorus concentration. To identify which site is not affected by the agriculture runoff, researchers measured phosphatase activity (APA) in sites known to be affected (TP > 30 μg/L) and sites that are unaffected by agriculture runoff. Phosphatase is an enzyme produced by organisms in low P environment. Because producing this enzyme costs energy, organisms do not produce them when bio-available phosphorus are present. As a result, high APA is an indicator of P limitation. The data file `apa.s` contains both APA and TP concentrations. It can be imported into R using function `source`.

    (a) Compare the distributions of APA from sites with TP > 30 μg/L and APA from sites with TP < 30 μg/L using graphical tools we learned in Chapter 3.

    (b) What is the nature of difference between the two populations of APA?

    (c) Use an appropriate test to determine whether the difference is statistically significant and describe the result in non-technical terms.

5. In the Kemp's ridley turtle example, we tested whether the sex ratio has shifted. The data were observational, that is, turtles used in the data set were not randomly selected from the population, rather, these were turtles stranded on the beach. One particular feature of the data is the large difference in the number of turtles observed during the two time periods. Suppose that the data from the first time period (1983–1989) were collected by regularly scheduled beach patrols throughout the year, while the data in the second time period (1990–2001) also include observations reported by tourists. Discuss the possible causes of the observed sex ratio differences.

6. Problems of a significance test with a low power.

   Reproducible research findings are a cornerstone of the scientific method. However, studies have shown that results of published research can be difficult to replicate when only statistically significant results are published, particularly when the statistical significance tests are conducted with small sample sizes (hence with low statistical power). Let's consider a one sample t-test with $H_0 : \mu \leq 0$ versus $H_a : \mu > 0$.

(a) If the population standard deviation is 0.5, what is the power of the test when the effect size is $\delta = 0.1$ and sample size is $n_1 = 10$?

(b) In order to reject the null hypothesis, how large must the sample mean $\bar{x}$ be?

(c) If you rejected the null using data from one experiment, how likely can the statistically significant result be verified if you repeat the same experiment?

(d) Suppose that you realized that the test has a low power and you decided to repeat the experiment with a much larger sample size (e.g., $n_2 = 100$). What is the power of the new test?

(e) Assuming $\delta = 0.1$, how likely is it that you will obtain a sample mean as large as the statistically significant result from the experiment with $n_1 = 10$?

7. Eutrophication due to increased input of nitrogen in the Neuse River Estuary in eastern North Carolina, USA, was considered the primary cause of large scale fishkills in the late 1990s. The North Carolina General Assembly established laws to protect the estuary, including a requirement of reducing nutrient (particularly, nitrogen) input to the estuary. Because eutrophication in North Carolina is measured by the concentration of chlorophyll *a*, assessing the success of the nutrient reduction program relies on the demonstration of a reduction in chlorophyll *a* concentration in the estuary. Three institutions have water quality monitoring programs including chlorophyll *a* in the variables they measure: NC Division of Water Quality (DWQ), University of North Carolina Institute of Marine Sciences (IMS), and Weyerhaeuser Corp. (WEY). Because the estuary is large and chlorophyll *a* concentrations vary spatially, methods used in sampling and measuring can affect the reported chlorophyll *a* concentrations. To demonstrate the success of the nitrogen reduction program, we need to compare the chlorophyll *a* concentrations measured before and after the implementation of the program to the concentrations before. But which series of data should we use? To answer this question, we need to compare the reported chlorophyll *a* concentrations from the three institutions and determine whether they are different. If they are the same, we may want to combine the three sources of data to increase the power of any statistical test we will use. If they are different, we need to describe the nature of the difference and decide how to best use them to describe the effect of the nitrogen reduction program.

   For this problem, we are to compare the chlorophyll *a* concentrations from the three institutions and discuss the differences among them:

   - Exploratory data analysis – a summary of data distributions and potential problems with the data.
   - A decision on whether a transformation is necessary. In general, we use log-transformation for environmental concentration variables.

- ANOVA to test whether the mean (or median, if log-transformed) varies by institution.
- Present the estimated differences (and interpret the differences in the original concentration scale)
- A short discussion on other factors that may affect the result of this comparison.

8. Harmel et al. [2006] compiled a cross-system data set to study the effects of agriculture activities on water quality. The data included in the study were mostly field scale experiments that measured nutrients (P, N) loading leaving a field. The data set (`agWQdata.csv`) includes the measured TP loading (`TPLoad`, in kg/ha), land use (`LU`), tillage method (`Tillage`), and fertilizer application methods (`FAppMethd`). You are to determine whether tillage methods affect TP loading.

   (a) Estimate the mean TP loading for each tillage method (an easy way to do this in R is to use the function `tapply`):

   ```
   > tapply(agWQdata$TPLoad, agWQdata$Tillage, mean)
   ```

   (b) Discuss briefly whether logarithm transformation is necessary.

   (c) Use statistical test to study whether different tillage methods resulted in different TP loading (state the null and alternative hypothesis, conduct the test, report the result).

   (d) Discuss briefly how useful is the test result.

9. Using the same data as in the last question, fit two ANOVA models using log TP loading (`log(TPLoad)`) and the square root of the square root of TP loading (`TPLoad ^ 0.25`) as the response and tillage method as the predictor.

   (a) Plot residual normal Q-Q plots of both models; discuss which transformation is better.

   (b) Suppose that we can treat the residuals from both models as approximately normal. Try to explain the results from both models in plain English.

Part II

Statistical Modeling

Chapter 5

Linear Models

5.1 Introduction 149
5.2 From t-test to Linear Models 152
5.3 Simple and Multiple Linear Regression Models 154
 5.3.1 The Least Squares 154
 5.3.2 Regression with One Predictor 156
 5.3.3 Multiple Regression 158
 5.3.4 Interaction 160
 5.3.5 Residuals and Model Assessment 162
 5.3.6 Categorical Predictors 170
 5.3.7 Collinearity and the Finnish Lakes Example 174
5.4 General Considerations in Building a Predictive Model 185
5.5 Uncertainty in Model Predictions 189
 5.5.1 Example: Uncertainty in Water Quality Measurements 191
5.6 Two-Way ANOVA 193
 5.6.1 ANOVA as a Linear Model 193
 5.6.2 More Than One Categorical Predictor 195
 5.6.3 Interaction 198
5.7 Bibliography Notes 200
5.8 Exercises 200

5.1 Introduction

In Chapter 4, we defined a model as a probability distribution model. Once a model is proposed, we make inference about the unknown model parameters based on data. In a one-sample t-test problem, we are interested in learning about the mean of a normal distribution.

$$y_i \sim N(\mu, \sigma^2) \tag{5.1}$$

It is often convenient to think of the data y_i in terms of the mean and a remainder:

$$y_i = \mu + \varepsilon_i \tag{5.2}$$

That is, we can split an observed value into two parts, the mean (μ) and the remainder (ε_i). Mathematically the above two expressions are equivalent. The remainder is the difference between the observed and the mean, often known as residuals, has a normal distribution with mean 0 and standard deviation σ ($\varepsilon_i \sim N(0, \sigma^2)$). In a two sample t-test problem, we are interested in the difference between the means of two populations or groups. We present the problem as follows:

$$\begin{array}{rcl} y_{1i} & \sim & N(\mu_1, \sigma^2) \\ y_{2j} & \sim & N(\mu_2, \sigma^2) \end{array} \tag{5.3}$$

and we are interested in the difference between the two means $\delta = \mu_2 - \mu_1$. We can present the problem in the format of equation (5.2) by combining the data from the two groups together into a data frame with a second column to indicate the group association (or "treatment"). A mathematically convenient construction of the treatment column is to use a column of 0's (for y_{1i}) and 1's (for y_{2j}). The data frame consists of two columns, the data column (y) and the treatment (or more generally, group) column (g). Each row represents an observed data point and its group association (0 for group 1 and 1 for group 2). The two-sample t-test problem in equation (5.3) can be expressed in the form of equation (5.4):

$$y_j = \mu_1 + \delta g_j + \varepsilon_j \tag{5.4}$$

where j is the index for the combined data, g_j is the group association of the jth observation. For data from group 1 ($g_j = 0$), equation (5.4) reduces to $y_j = \mu_1 + \varepsilon_j$ and for data from group 2 ($g_j = 1$), the model is $y_j = \mu_1 + \delta + \varepsilon_j$.

The group indicator g is often known as a "dummy variable." A dummy variable takes value 0 or 1. When we have data from more than two groups, we will use $p - 1$ dummy variables to represent the p groups. For example, if we have three groups in an ANOVA problem (e.g., Exercise 7 in Chapter 4), we combine observed data from all three groups into one column. The first dummy variable g_1 takes value 1 if the observation is from group 2 and 0 otherwise. The second dummy variable g_2 takes value 1 if the observation is from group 3 and 0 otherwise. The ANOVA problem can now be expressed as a linear model problem:

$$y_j = \mu_1 + \delta_1 g_{1j} + \delta_2 g_{2j} + \varepsilon_j \tag{5.5}$$

For data from group 1, the model is reduced to $y_i = \mu_1 + \varepsilon_i$. For data from group 2, the model is $y_i = \mu_1 + \delta_1 + \varepsilon_i$, and for group 3, $y_i = \mu_1 + \delta_2 + \varepsilon_i$.

By represent the t-test and ANOVA problems in terms of a "statistical model," I want to convey two main messages. First, we use different models for different problems. Second, statistical inference is mostly about the relationship among variables. Likewise, a main goal in science is the understanding of the relationship among important variables. The relationship, either described qualitatively or quantitatively, is a *model*. In a statistical problem, we define a model as the probability distribution of the variable of interest (the response variable). A probability distribution has a mean (or location) parameter and a parameter representing spread (e.g., standard deviation). When a distribution model is specified, we want to understand how the mean of the distribution varies as a function of other variables (predictor variables). In equation (5.2), the mean is a constant (no predictor variable). In equations (5.4) and (5.5), the mean varies by groups (g is the predictor variable). For a response variable with a normal distribution, the standard deviation can be estimated from the residuals. As a result, we can often express a statistical model as $y_i = f(x, \theta) + \varepsilon_i$, where x represents predictor variable(s), θ represents unknown parameter(s) to be estimated, and ε_i is a normal random variable with mean 0 and an unknown standard deviation (σ). In equation (5.5), x represents both g_1 and g_2, and θ includes μ_1, δ_1, and δ_2. The function $f(x, \theta)$ is an example of a mean function of a statistical model – a function defines the relationship between the mean parameter of the response variable distribution and a number of predictors. Using equation (5.5), we can define a statistical modeling problem as follows:

- Model formulation – response variable is a normal random variable with different group means and a constant standard deviation (e.g., equation (5.5)).
- Parameter estimation – how to estimate unknown parameters (e.g., $\mu_1, \delta_1, \delta_2, \sigma$ in equation (5.5)).
- Model evaluation – is the proposed model appropriate based on the data and are the statistical assumptions necessary for parameter estimation met? In this case, the assumptions can be summarized in terms of the residual distribution – ε_i is independently and identically distributed (iid) $N(0, \sigma^2)$.
- Statistical inference – can the estimated differences (e.g., δ_1, δ_2) be attributed to the randomness of the data?

A difficulty in developing a statistical model is that the exact model form is unknown, and there can be numerous model forms that can produce the same data. Statistical modeling does not provide a means to determine the proper functional form of $f(x, \theta)$. Instead, statistical modeling provides methods for assessing whether a chosen model form is likely appropriate. Once a specific model form is chosen, statistical analysis can provide uncertainty assessment of the model's fit to the data and the model's predictive capability. While choosing the correct model form is the important first step of an application, we do not necessarily have a specific application in mind when learning statistics. As a result, the learning process is typically a process of learning available statistical models one at a time, without a specific application in mind. In the learning process, the importance of selecting the appropriate model is usually not emphasized. This part of the book explores a number of models. These models fall into the three types of response variables most relevant to biological science – continuous variables (normal distribution), binary variables (binomial distribution), and count variables (Poisson distribution). In a typical application, we can easily distinguish these three types of variables and a model selection process will be focused on the selection of a proper mean function.

The class of linear models is of particular interest, mostly because of its simplicity, but also, to some extent, due to the desirable statistical properties such as unbiasedness and efficiency. We start with the simple and multiple regression models in Chapter 5 with an emphasis on the use of residuals for model assessment, followed by nonlinear models in Chapter 6. Both linear and nonlinear models assume that the response variable can be approximated by the normal distribution. As a result, model residuals are normal, independent, and homogeneous. Subsequent chapters will discuss models for situations where the response variable is binary or count variable. A binary response variable is often modeled using the binomial distribution and a count variable is modeled using the Poisson distribution.

5.2 From *t*-test to Linear Models

The linear model is a class of statistical models with normal response variable and linear mean function $f(x, \theta) = \beta_0 + \beta_1 x_1 + \cdots + \beta_p x_p$. Among this most frequently used class of models are simple and multiple regression models for continuous predictors and analysis of variance models (ANOVA) for categorical predictors. In this chapter, we will focus on the process of building a linear model. I will use the PCB in fish example to illustrate the process of building a model by incrementally increasing the complexity of the model until we are satisfied. The model building process is a process of iterative model fitting, evaluation, and updating. In the PCB in fish example, we have two goals in supporting the need of managing Lake Michigan fishery: (1) understanding the risk of exposure to PCB posed to the public when consuming fish caught in the lake and (2) learning about the rate of PCB dissipation over time. We will use the lake trout data.

Madenjian et al. [1998] reported that lake trout in Lake Michigan make dietary shift at a size of about 60 cm. Large lake trout (longer than 60 cm) eat large alewife with a much higher average PCB concentration than the average concentration in small alewife and rainbow smelt consumed by small lake trout. A natural first step of risk analysis would be to compare the average PCB concentrations between small ($<$ 60 cm) and large ($\geq$ 60 cm) lake trout. This comparison will naturally be unsatisfactory as we know that PCB concentration in fish reduces over time as there have been no new sources of PCB since 1970. Time since the PCB production ban should be a factor, in addition to fish size.

Let us first look at a *t*-test for comparing two populations: PCB concentrations in small lake trout versus PCB concentrations in large lake trout. I added a column to the data frame to indicate whether a fish is large or small.

The column is named as "large." It takes value 1 if the PCB measurement was from a large fish and 0 otherwise. Before applying a t-test, we first need to explore the nature of the difference between the two populations. We use the Q-Q plots of PCB and log PCB (Figure 5.1). Using a Q-Q plot we are interested in determining whether the difference between the two populations is predominantly additive or multiplicative. The Q-Q plot in the concentration scale seems to suggest a multiplicative difference. The Q-Q plot in the log scale suggests an approximately additive difference, but the line formed by data points shows a small angle pointing away from the reference line, suggesting that there may be a systematic departure from an additive shift in the logarithmic scale. For now, I will assume that the difference is additive in the logarithmic scale.

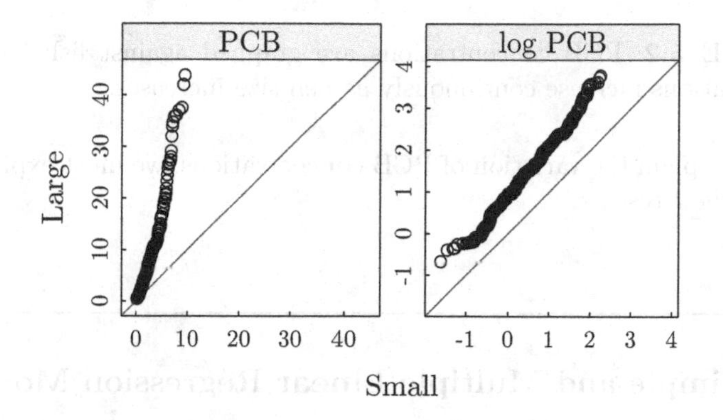

FIGURE 5.1: Q-Q plots compare PCB concentrations in large and small fish. The left panel shows the comparinson in PCB concentration scale and the right panel shows the comparison in log scale.

A t-test of log PCB concentrations calculates sample averages of small and large fish; they are 0.296 and 1.254, respectively. The difference is associated with a t-statistic of -14.5 (with a p-value of $< 2.2 \times 10^{-16}$). The result suggests that PCB concentrations in large fish are significantly higher. The log difference of -0.958 is translated to a multiplicative factor of $e^{-0.958} = 0.384$. In other words, average concentration in small fish is slightly more than one third of the average concentration in large fish.

Although the t-test is informative, the information from the test is sketchy. That is, we are limited to two categories of fish, ignoring systematic variations within each category. Mathematically, a t-test simplifies the relationship between PCB concentration and fish length to a step function: one mean for small fish and a second mean for large fish. The data, however, show that PCB concentrations continuously increase as fish size increases (Figure 5.2).

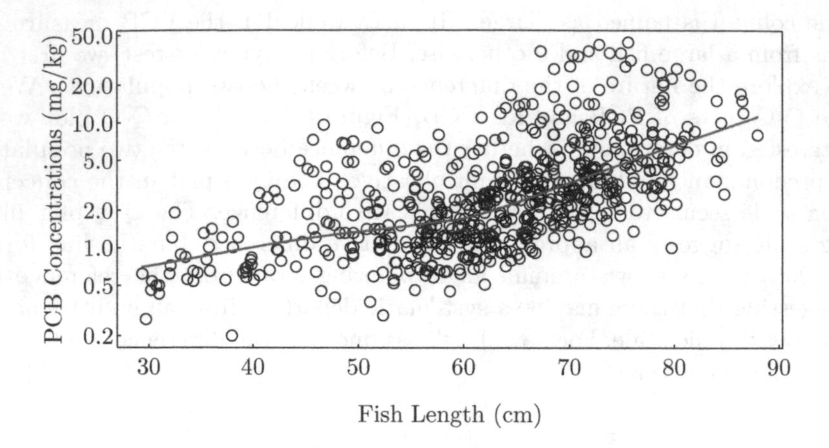

FIGURE 5.2: PCB concentrations are graphed against fish length. PCB concentrations increase continuously as fish size increases.

To fully explain the variation of PCB concentrations, we must explore models beyond the t-test.

5.3 Simple and Multiple Linear Regression Models

The linear regression is the simplest and most studied model. In a linear regression model, $f(x)$ is parameterized by a linear function of $x : f(x) = \beta_0 + \beta_1 x$, where β_0 and β_1 are unknown model coefficients to be estimated from data. Under the usual statistical assumptions, the residuals are independent random variates from a normal distribution with mean 0 and a constant (but unknown) standard deviation σ. To specify a simple linear model with one *predictor* x, we need to estimate three quantities: β_0, β_1, and σ. Another way of expressing the model is through the use of probability distribution.

$$
\begin{array}{rcl}
y_i & \sim & N(\mu_i, \sigma^2) \\
\mu_i & = & \beta_0 + \beta_1 x_i
\end{array}
$$

An important aspect of statistical modeling is to develop a method for estimating unknown model coefficients. The least squares method and the maximum likelihood method are the two most frequently used ones.

5.3.1 The Least Squares

The least squares estimator (LSE) yields a set of estimates of model coefficients that minimizes the sum of squared residuals. Residuals are the difference between model predicted ($\beta_0+\beta_1 x_i$) and observed (y_i). Defining the residuals in terms of a function of model coefficient,

$$\epsilon_i = y_i - \beta_0 - \beta_1 x_i$$

we have the residual sum of squares

$$RSS = \sum_{i=1}^{n}(y_i - \beta_0 - \beta_1 x_i)^2$$

RSS is a function of β_0 and β_1. To minimize RSS, we set its partial derivatives with respect to β_0 and β_1 to 0:

$$\begin{aligned}\frac{\partial RSS}{\partial \beta_0} &= -2\sum_{i=1}^{n}(y_i - \beta_0 - \beta_1 x_i) = 0\\ \frac{\partial RSS}{\partial \beta_1} &= 2\sum_{i=1}^{n} x_i(y_i - \beta_0 - \beta_1 x_i) = 0\end{aligned}$$

The least square estimates are given as follows, where $\bar{y}$ and $\bar{x}$ are the mean values of y_i and x_i:

$$\begin{aligned}\hat{\beta}_1 &= \frac{\sum_{i=1}^{n}(y_i-\bar{y})(x_i-\bar{x})}{\sum_{i=1}^{2}(x_i-\bar{x})^2}\\ \hat{\beta}_0 &= \bar{y} - \hat{\beta}_1\bar{x}\end{aligned}$$

We note that these well-known estimates require no distributional assumption about the model residuals. In addition, the least squares method does not apply to σ, which needs to be estimated separately.

Although it is difficult to justify the use of the least squares method for parameter estimation beyond the usual "intuitive plausibility," the least squares estimator coincides with the maximum likelihood estimator when the residuals are independent random variates from a normal distribution with mean 0 and a constant standard deviation. The maximum likelihood estimator is based on the distribution assumption on the residuals. For a given data point, the residual has a normal distribution:

$$\epsilon_i \sim N(0, \sigma)$$

The likelihood of ϵ_i is the normal density of $\epsilon_i = y_i - \beta_0 - \beta_1 x_i$, or

$$\frac{1}{\sqrt{2\pi}\sigma}e^{-\frac{(y_i-\beta_0-\beta_1 x_i)^2}{2\sigma^2}}$$

and the likelihood of observing all data points is:

$$L(\beta_0, \beta_1, \sigma) = \prod_{i=1}^{n}\frac{1}{\sqrt{2\pi}\sigma}e^{-\frac{(y_i-\beta_0-\beta_1 x_i)^2}{2\sigma^2}}$$

The estimator maximizing $L(\beta_0, \beta_1, \sigma)$ is the maximum likelihood estimator (MLE). Again we can set the partial derivatives of the likelihood function with respect to β_0, β_1, σ to 0. But the derivatives are much easier for the logarithm of the likelihood function:

$$\log(L) = -\frac{n}{2}\log(2\pi\sigma^2) - \frac{\sum_{i=1}^{n}(y_i - \beta_0 - \beta_1 x_i)^2}{2\sigma^2} \tag{5.6}$$

By setting the partial derivatives of the log-likelihood function to 0, we obtain the same formulas for β_0 and β_1 as obtained from the least squares and the MLE of σ is $\hat{\sigma} = \sqrt{\frac{\sum_{i=1}^{2}\hat{\epsilon}_i^2}{n}}$. Note that the log likelihood function in equation (5.6) is proportional to RSS. If σ is known, $-2\log(L) \propto \sum_{i=1}^{n}(y_i - \beta_0 - \beta_1 x_i)^2$. In general (i.e., for normal and other probability distributions), negative 2 times log likelihood is known as the *deviance*.

Once a linear model form is chosen, the process of model fitting includes estimating model coefficients and evaluating the fitted model. The objectives of analyzing PCB concentration in fish are (1) assessing the temporal trends of PCB in fish to determine whether or not meaningful reductions are still occurring, and (2) providing a basis for fish consumption advisories to caution the public of possible risks associated with eating contaminated fish.

For both objectives, linear (or log-linear) regression models were used. In assessing the temporal trend, PCB declining in fish is often assumed to follow an exponential model [Stow et al., 2004]. An exponential model suggests that the logarithm of PCB concentration declines linearly over time. In assessing the risk of PCB exposure through fish consumption, regression models were developed to predict PCB concentrations using fish size [Stow and Qian, 1998]. Most consumption advisories are based on fish tissue PCB concentrations. For example, the state of Wisconsin recommended that fish be divided into "no limit" (PCB below 0.05 mg/kg), "one meal per week" (0.05–0.20 mg/kg), "one meal per month" (0.20–1.00 mg/kg), "six meals per year" (1.00–1.90 mg/kg), and "no consumption" (>1.90 mg/kg). Since anglers cannot easily know the PCB concentration of their catch, the advisory translates these concentration-based consumption categories into fish size ranges for the important recreational species.

Data used here are PCB concentrations in lake trout collected by the Wisconsin Department of Natural Resources from 1974 to 2003 (Figure 5.3). The PCB concentration–fish size relationship (Figure 5.2) represents the biological accumulation of PCB over time, as a larger fish is likely to be older.

5.3.2 Regression with One Predictor

The simple linear regression used for assessing a temporal trend is a log linear model:

$$\log(\text{PCB}) = \beta_0 + \beta_1 \text{Year} + \varepsilon \tag{5.7}$$

The model coefficients β_0, and β_1 are estimated using the least squares method (Section 5.3.1), implemented in R function `lm()`:

```
#### R code ####
```

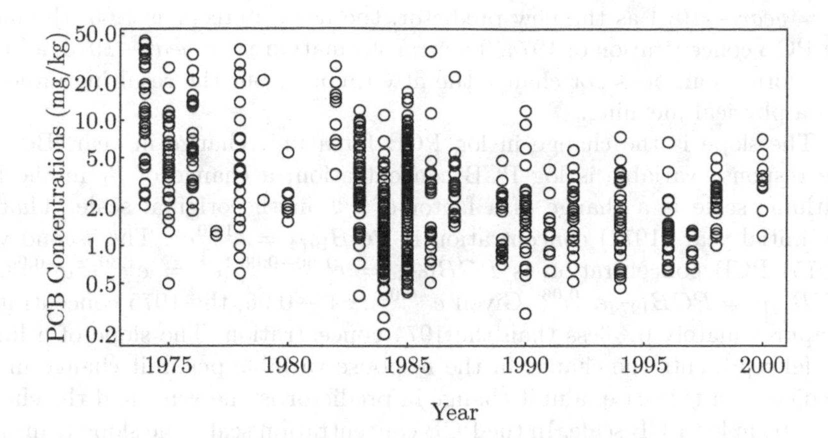

FIGURE 5.3: Temporal trend of fish tissue PCB concentrations – PCB concentrations in lake trout from Lake Michigan decline over time, but have shown a stabilizing trend in the last few years.

```
lake.lm1 <- lm(log(pcb) ~ year, data=laketrout)
display(lake.lm1, 3)

#### R output ####

lm(formula = log(pcb) ~ year, data = laketrout)
(Intercept) 119.8467  10.9689
year         -0.0599   0.0055
---
n = 631, k = 2
residual sd = 0.8784, R-Squared = 0.16
```

The estimated β_0 (the intercept) is 119.85 and the estimated β_1 (the slope) is -0.06. The fitted model has two parts, the deterministic part and the random part. The deterministic part is $\beta_0 + \beta_1 year$, the expectation or average of log PCB for a given year. The random part ε describes the variability or uncertainty. When putting the two parts together, the fitted model can be seen as a conditional normal distribution describing the probability distribution of log PCB concentrations. The mean of the log PCB distribution is the deterministic part of the model, and the standard deviation is the same as the standard deviation of the residuals (the random part). For example, the estimated log PCB distribution for year 1974 is $N(\beta_0 + \beta_1 \times 1974, 0.88)$ or $N(1.60, 0.88)$.

The intercept of a simple regression model is the expected value of the response variable when the predictor is 0. For this model, we don't believe that the model can be extrapolated to year 0. Consequently, the intercept does not have any physical meaning. However, if the model is refit with using

$yr = year - 1974$ as the new predictor, the new intercept is 1.66, the mean log PCB concentration of 1974. The transformation $yr = year - 1974$, a linear transformation, does not change the fitted model, but the resulting intercept has a physical meaning.

The slope is the change in log PCB for a unit change in year. Because the response variable is log PCB concentration, a change of β_1 in the logarithmic scale is a change of a factor of e^{β_1} in the original scale. That is, the initial year (1974) concentration is $PCB_{1974} = e^{1.60}e^{\varepsilon}$. The second year (1975) PCB concentration is $PCB_{1975} = e^{1.60-0.06\cdot 1}e^{\varepsilon} = e^{1.60}e^{\varepsilon}e^{-0.06}$, or $PCB_{1975} = PCB_{1974}e^{-0.06}$. Given $e^{-0.06} \simeq 1 - 0.06$, the 1975 concentration is approximately 6% less than the 1974 concentration. The slope of a linear model represents the change in the response variable per unit change in the predictor. In this case, a unit change in predictor is one year, and the change is -0.06 in log PCB scale. In the PCB concentration scale, the slope translates into a 6% annual rate of decreasing in PCB concentration.

The residual or model error term ε describes the variability of individual log concentrations. For this model, the estimated residual standard deviation is 0.87. When interpreting the fitted model in the original scale of PCB concentration, the predicted PCB concentration has a log normal distribution with log mean $1.6 - 0.06yr$ and log standard deviation 0.88. This model suggests that the middle 50% of the PCB concentrations in 1974 will be bounded between `qlnorm(c(0.25,0.75), 1.60, 0.88)` or (2.74, 8.97) mg/kg, and the middle 95% of the concentration values are bounded by (0.88, 27.79) mg/kg. The estimated mean concentration in 1974 is $e^{1.6+0.88^2/2} = 7.3$ mg/kg, and the estimated standard deviation is $e^{1.6+0.88^2/2}\sqrt{e^{0.88^2}-1} = 7.89$, or $\sqrt{e^{0.88^2}-1} = 1.081$ times of the mean (i.e., the coefficient of variation $cv = 1.081$). The model can be summarized graphically as in Figure 5.4.

5.3.3 Multiple Regression

The regression model with *year* or *yr* as the single predictor has a fairly large residual standard deviation, and the predicted PCB distribution has a standard deviation that is as large as the mean. As indicated in Figure 5.2, fish length is a good predictor of log PCB concentration. Including length as a second predictor will improve the model's predictive accuracy.

```
#### R code ####

lake.lm2 <- lm(log(pcb) ~ I(year-1974)+length, data=laketrout)
display(lake.lm2, 3)

#### R output ####
lm(formula = log(pcb) ~ I(year - 1974)+length, data = laketrout)
               coef.est coef.se
(Intercept)    -1.834    0.120
```

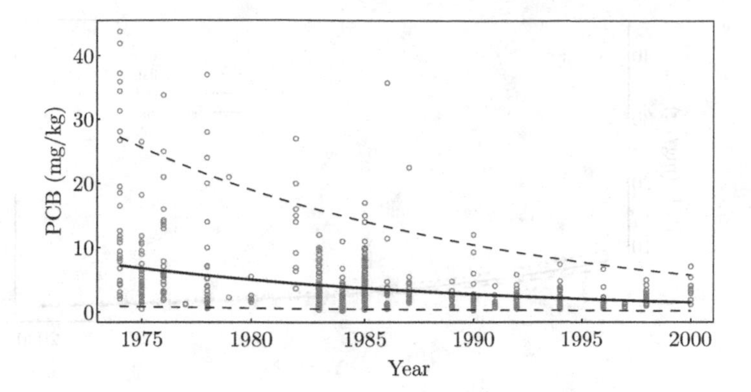

FIGURE 5.4: Simple linear regression of the PCB example – PCB concentration data are plotted against year. The simple linear regression resulted in highly uncertain predictions. The solid line is the predicted mean PCB concentration and the dashed lines are the middle 95% intervals.

```
I(year - 1974) -0.086    0.004
length          0.060    0.002
---
n = 631, k = 3
residual sd = 0.555, R-Squared = 0.66
```

The fitted model is

$$\log(\text{PCB}) = -1.834 + 0.060 Length - 0.086 yr + \varepsilon$$

The intercept (-1.834) is the expected log PCB concentration when both predictors are 0. $yr = 0$ indicates 1974, but length 0 is meaningless. As a result, the intercept is again meaningless. A commonly used linear transformation for this situation is `x-mean(x)`, or centering the predictor around its mean. When the centered predictor is used, the intercept is the expected response variable value when the predictor is at its mean:

```
#### R code ####
laketrout$len.c <- laketrout$length - mean(laketrout$length)
lake.lm3 <- lm(log(pcb) ~ I(year-1974)+len.c, data=laketrout)
display(lake.lm3, 3)

#### R output ####
lm(formula = log(pcb) ~ I(year - 1974)+len.c, data = laketrout)
               coef.est coef.se
(Intercept)     1.899    0.047
I(year - 1974) -0.086    0.004
len.c           0.060    0.002
```

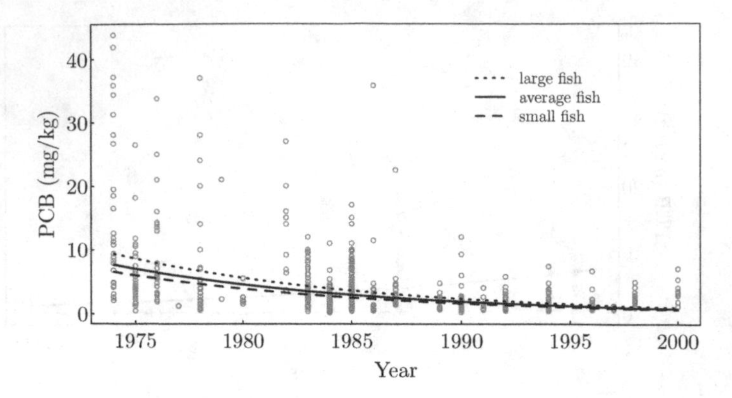

FIGURE 5.5: Multiple linear regression of the PCB example – PCB concentration data are plotted against year. The multiple regression predictions are for specific-sized fish. The solid line is the predicted mean PCB concentration for an average-sized fish (62.48 cm), the dashed line is for a small fish (56 cm), and the dotted line is for a large fish (71 cm).

```
---
n = 631, k = 3
residual sd = 0.555, R-Squared = 0.66
```

As a result, the new intercept is the average log PCB concentration for an average-sized fish in 1974. The slope (0.060) is the change in log PCB per unit change in fish length. Each 1 cm increase in fish length will result in a factor of $e^{0.060} = 1.062$ increase in PCB concentration (or about 6%) *in a given year.* The slope of yr is now -0.086 or an annual reduction rate of 8.6% *for a given sized fish.*

When fish size is included as a second predictor, the prediction is for a specific sized fish in a given year. Much of the variation not explained by the simple linear model with the year as the only predictor can be attributed to the variation in fish size. For a fish with an average length (62.48 cm), its average PCB concentration in 1974 has a log normal distribution with log mean 1.942 and log standard deviation 0.543. The predicted mean is $e^{1.899+0.555^2/2} = 7.79$ mg/kg, and the cv is $\sqrt{e^{0.555^2} - 1} = 0.60$. Figure 5.5 shows the model predicted mean PCB concentrations in fish with three different sizes.

5.3.4 Interaction

When fitting the multiple regression model with yr and `len.c` as the predictors, an important assumption is that the effect of year (the slope of year) is not affected by the size of the fish and the effect of fish size (the slope of length) is the same throughout the study period. This is the additive-effect assumption imposed on a multiple regression model. This assumption suggests that the annual rate of dissipation of PCB is the same for fish of all sizes. Likewise, the rate of increase in PCB concentration as a function of fish size is the same for all years. Is this assumption reasonable? Madenjian et al. [1998] reported that small lake trout (< 40 cm) eat small alewives (*Alosa pseudoharengus*, which have an average PCB concentration of 0.2 mg/kg), intermediate-size lake trout (40 ∼ 60 cm) eat alewives and rainbow smelt (*Osmerus mordax*, whose PCB concentrations ranged from 0.2 to 0.45 mg/kg), and large lake trout (≥ 60 cm) eat large alewives (with an average PCB concentration of 0.6 mg/kg). On the one hand, because larger fish tend to consume food with higher concentrations of PCB, its reduction over time should be slower than the rate of reduction of small fish. On the other hand, because PCB was banned in the 1970s, the natural reduction of PCB through microbial metabolism resulted in the overall reduction of PCB concentration in the environment and in fish. We expect that the PCB–length relationship will change over time. In other words, the slope of year in the multiple regression model is expected to change with the size of a fish and the slope of length is expected to change over time. To model this "interaction" effect, we add a third predictor, the product of *yr* and `len.c` in the model:

```
#### R code ####
lake.lm4 <- lm(log(pcb) ~ I(year-1974)*len.c, data=laketrout)
display(lake.lm4, 4)

#### R output ####
lm(formula = log(pcb) ~ I(year - 1974)*len.c, data = laketrout)
                      coef.est coef.se
(Intercept)            1.8967   0.0465
I(year - 1974)        -0.0873   0.0036
len.c                  0.0510   0.0038
len.c:I(year - 1974)   0.0008   0.0003
---
n = 631, k = 4
residual sd = 0.5520, R-Squared = 0.67
```

When the interaction term (`len.c:I(year - 1974)`) is included, the model is expressed as:

$$\log(PCB) = 1.89 - 0.087yr + 0.051Len.c + 0.00085yr \cdot Len.c + \varepsilon \quad (5.8)$$

Because of the product term, the model is no longer a linear model with respect to the two predictors. The slopes of centered length (`len.c`) and year (yr) are no longer constant. We can rearrange the model to understand the interaction effect. First, the interaction term is grouped with yr:

$$\log(PCB) = 1.89 + (-0.087 + 0.00085Len.c)yr + 0.051Len.c + \varepsilon$$

That is, the effect (or slope) of yr is now a linear function of $Len.c$. The estimated slope of yr (-0.087) is the slope when $Len.c = 0$ or the year effect for an average-sized fish. When the fish size is 10 cm above average, the yr effect is $-0.087 + 0.00085 \cdot 10 = -0.0785$. In other words, not only a larger fish has a higher PCB concentration on average, PCB in a larger fish tend to dissipate at a lower rate. This interpretation is true only when we are comparing same-sized fish over time. So, when comparing fish of the average length ($Len.c = 0$), the annual rate of dissipation is 8.7%. The annual dissipation rate is 7.6% for fish with a size 10 cm above average.

When examining the $\log(PCB)$ fish length relationship, the model can be rearranged to be:

$$\log(PCB) = 1.89 + (0.051 + 0.00085yr)Len.c - 0.087yr + \varepsilon$$

The relationship is still linear for any given year. But the slope changes over time. Initially, ($yr = 0$ or 1974), the size effect is 0.051. Each unit (1 cm) increase in size will result in a 5.1% increase in PCB concentration. Ten years later (1984), the slope was $0.051 + 0.00085 \cdot 10 = 0.0595$. The size effect is stronger. This is reasonable because the rate of concentration decreasing for a large fish is smaller than the rate for a small fish. Consequently, the difference in log concentration between the same two fish increases over time.

The interaction effect is small (albeit statistically different from 0). Can this small interaction effect be practically meaningful? Because the response variable is in logarithmic scale, we need to be careful in interpreting a small effect. For the slope of yr, the slope value for a small fish (-6.7 cm below average, or the first quartile) is $0.09 - 0.00085 \times (-6.7) = 0.095$ and the slope is $0.09 - 0.0008 \times (8.5) = 0.083$ for a large fish (8.5 cm above average, the third quartile). PCB concentration reduction is at a lower rate ($\sim 8\%$) for a large fish and a higher rate ($\sim 10\%$) for a small fish. The slope of `len.c` increases from 0.05 in 1974 to 0.074 in 2004, indicating a much larger relative difference in PCB concentration between a large and a small fish.

5.3.5 Residuals and Model Assessment

The fitted multiple regression model with an interaction term in the previous section can be easily interpreted. The interaction effect can be explained by the diet shifts of lake trout. Although interpretable model results is a necessary feature of a good model, the problem of model evaluation is a quantitative one. We must ask questions about model form (e.g., is a linear model adequate?). Analyzing residuals, the discrepancy between model predictions and observations, is the most effective means for answering questions about a model's fit to the data. We will use the full model (equation 5.8 on page 161) as an example to illustrate the necessary steps for model assessment and evaluation. We will use both graphs and summary statistics.

The model in equation 5.8 was fitted after initial examination of the data and considerations of the factors that may influence PCB concentration in fish. When using a log-linear model we assume an exponential model for PCB reduction over time, and assume that PCB concentration in a lake trout increases in a fixed proportion to a unit increase in its size. These assumptions are made with little theoretical support. How can these assumptions be tested based on the data? To answer this question, we must first clarify the objective of a model. In general, a model is developed with one of the two general objectives – causal inference and prediction.

A predictive model is developed for predicting the outcome using predictor variable values outside of the data set used for model fitting. A good predictive model should be simple and adequately accurate. A causal inference model is aimed at establishing a causal relationship, which requires a higher standard than establishing a correlation. In both cases, we need to justify the model based on statistical inference. In this section, we describe the necessary steps for assessing a predictive model. These include the summary statistics of a fitted model, methods for evaluation of model assumptions, and prediction and validation.

Summary Statistics

When a model is fitted using the R function `lm()`, all necessary model summaries and diagnostic information is included in the resulting R object. For example, the PCB in the fish model we discussed in Section 5.3.4 is stored in the model object `lake.lm4`. For an overall assessment of the model, we often use the coefficient of determination or the R^2 value and a hypothesis test (the F-test) to compare the fitted model with a model with no predictor variable ($y = \beta_0 + \varepsilon$). For assessing whether an individual predictor is necessary, a t-test is used to test whether or not the slope of the variable is different from 0. The test result is often used to determine whether or not the effect of a predictor variable is statistically different from 0. These summary statistics and test results can be presented by using the R function `summary`:

```
#### R output ####
summary(lake.lm4)

Call:
lm(formula = log(pcb) ~ I(year - 1974)*len.c ,
    data = laketrout)

Residuals:
    Min      1Q  Median      3Q     Max
-2.4796 -0.3411  0.0197  0.3387  1.9711

Coefficients:
                    Estimate Std. Error t value Pr(>|t|)
(Intercept)         1.890718   0.046465   40.69   <2e-16
I(year-1974)       -0.087393   0.003604  -24.25   <2e-16
len.c               0.051037   0.003841   13.29   <2e-16
len.c:I(year-1974) 0.000848   0.000329    2.58    0.010
---
```

TABLE 5.1: ANOVA table of a linear model

Source of Variation	Sum of Squares	df	Mean Sum of Square	F-value
Model	$SSreg$	p	$MSreg = SSreg/p$	$MSreg/MSE$
Residual	SSE	$n-p$	$MSE = SSE/(n-p)$	
Total	SST	$n-1$		

```
Residual standard error: 0.55 on 627 degrees of freedom
  (15 observations deleted due to missingness)
Multiple R-Squared: 0.668,      Adjusted R-squared: 0.667
F-statistic:  421 on 3 and 627 DF,  p-value: <2e-16
```

The R^2 and the F-test results are displayed near the bottom of the output. The adjusted R^2 is defined by $R^2_{adj} = 1 - \frac{n-1}{n-p}(1-R^2)$, where n is the sample size and p is the number of predictors. The R^2 value is the proportion of the total variance in the data ($SST = \sum_{i=1}^{n}(y_i - \bar{y})^2$) explained by the model. It is calculated as 1 minus the ratio of residual variance ($SSE = \sum_{i=1}^{n}\epsilon_i^2$) over the total sum of squares, or $R^2 = 1 - SSE/SST$. The adjusted R^2 value (R^2_{adj}) is a modified version of R^2 adjusted for the number of predictors in the model. R^2 will always increase when additional predictors are added to a regression model. By adjusting to the number of predictors in the regression model, R^2_{adj} may not always increase as more predictors are added to the model. The R^2_{adj} is a redundant statistic and less informative than the analysis of variance of individual predictors.

The R^2 of this model is 0.668, or the model explains about 66.8% of the total variability in log PCB concentration data. Compared to the null model (a model with no predictor) which predicts log PCB using the average of the data, the amount of variability explained (66.8%) is unlikely due to random noise, as shown in the large F-statistic or the small p-value. The F-test is based on the same analysis of variance concept discussed in Section 4.7.1 (page 117). In a linear regression model, the analysis of variance is comparing the full model

$$y = \beta_0 + \beta_1 x_1 + \cdots \beta_p x_p + \varepsilon$$

to the model with no predictor

$$y = \beta_0 + \varepsilon$$

or the null model. For the null model, $\hat{\beta}_0 = \bar{y}$ and the residual sum of square is SST (see Section 4.7.1). For the full model, the residual sum of squares (SSE) is less than or equal to SST. The difference between SST and SSE (call it $SSreg$) is the sum of squares explained by including the predictors. An ANOVA table for a linear model summarizes these results (Table 5.1).

ANOVA is a very rich technique for dividing the total variance for model comparison. In general, it can be used to compare a model with fewer predictors (the reduced model) to a model with more predictors (the full model). This comparison of different models can be used to infer whether a predictor variable should be included in a model. For example, we can decide whether to add `length` as the second predictor based on statistics only. That is to compare the models

```
log(pcb) ~ I(year-1974)
```

and

```
log(pcb) ~ I(year-1974)+len.c
```

The analysis of variance of the simple linear model is shown below.

```
#### R output ####
summary.aov(lake.lm1)

                Df Sum Sq Mean Sq F value Pr(>F)
I(year - 1974)   1     91      91     118 <2e-16
Residuals      629    485       1

15 observations deleted due to missingness
```

The model has a residual sum of squares of 485, part of the total variance not explained by the predictor yr. This amount of unexplained variability is further used to evaluate whether a second predictor is necessary.

```
#### R output ####
summary.aov(lake.lm2)
                Df Sum Sq Mean Sq F value Pr(>F)
I(year - 1974)   1     91      91     295 <2e-16
len.c            1    292     292     950 <2e-16
Residuals      628    193     0.3

15 observations deleted due to missingness
```

The second predictor (centered length) explains 292 of the 485 unexplained sum of squares, which resulted in a large F-statistic and a very small p-value, suggesting that including `len.c` will improve the model significantly.

In general, if the predictors in the reduced model is a subset of the predictors in the full model, the comparison can be used to infer whether the increased variability explained by the full model is a statistically significant improvement over the reduced model. When the full model has only one more predictor than the reduced model, the F-test is equivalent to a t-test on whether or not the slope of the additional predictor is different from 0. These t-tests are summarized in the model coefficient summary table:

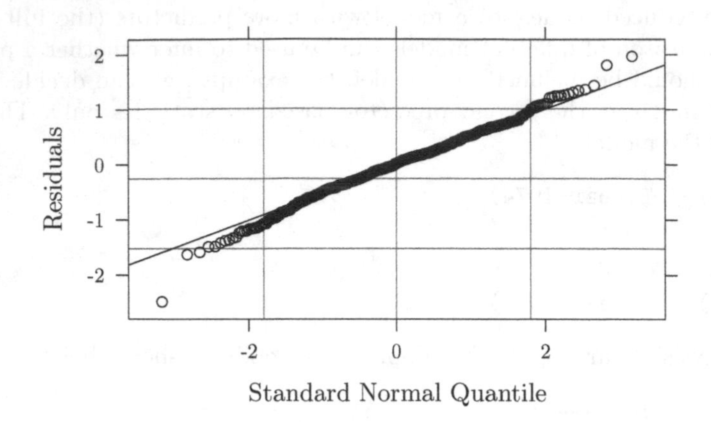

FIGURE 5.6: Normal Q-Q plot of the PCB model residuals – Q-Q plot of residuals of the model (equation 5.8) indicates a symmetric residual distribution, but with more extreme values than a normal distribution can predict.

```
#### R output ####

summary(lake.lm4)$coef
                      Estimate Std. Error t value Pr(>|t|)
(Intercept)            1.89072    0.04646    40.7 4.3e-178
I(year - 1974)        -0.08739    0.00360   -24.2  4.0e-92
len.c                  0.05104    0.00384    13.3  1.1e-35
len.c:I(year - 1974)   0.00085    0.00033     2.6  1.0e-02
```

The small p-value for the slope of `len.c:I(year - 1974)` suggests that the slope is significantly different from 0 after the effects of yr and $len.c$ have been accounted for. The R default summary output has too much information that may never be needed. As a result, the function `display` from package `arm` is preferred. All frequently used summary information is included. The display output does not include any hypothesis testing result, but we can easily use the standard errors of the estimated coefficients to determine whether a slope is statistically different from 0 based on the approximate confidence interval (estimated value $\pm$ 2 standard errors). If the interval includes 0, the slope is not different from 0 (or the effect of this predictor is not significant).

These summary statistics suggest that both predictors and their interaction should be included in the model. But these summary statistics do not provide enough information for us to judge whether the fitted model is adequate.

Graphical Analysis of the Residuals

For the summary statistic (especially the hypothesis testing results) to be

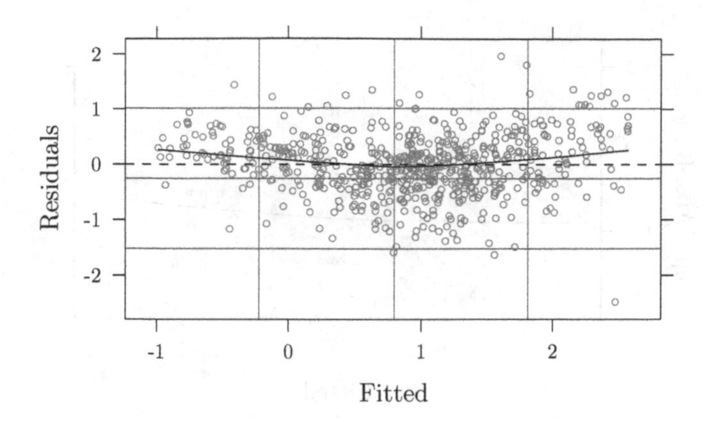

FIGURE 5.7: PCB model residuals vs. fitted – PCB model (equation 5.8) residuals are plotted against the estimated mean log PCB concentrations. The plot suggests that the model tends to underpredict when predicting low or high concentrations.

valid, the residuals should be independent and close to a normal distribution with mean 0 and a constant standard deviation: $\varepsilon_i \sim N(0, \sigma)$. A graphical evaluation of residuals will inevitably include the assessment of the three assumptions (normality, independence, and constant standard deviation). The normal Q-Q plot of the residuals is the obvious device for checking normality (Figure 5.6). Figure 5.6 shows a typical residual distribution that is symmetric, but with slightly more extreme values than a normal distribution can expect. The independence is typically assessed by plotting the residuals against the fitted values (Figure 5.7), which shows a systematic pattern: the model tends to underpredict the log PCB concentrations when the predicted concentrations are small or large. The loess curve suggests that the residuals can be predicted to some extent. The constant standard deviation assumption is assessed using the S-L plot, a scatter plot of the square root of the absolute value of the residuals against the fitted values (Figure 5.8). The residual standard deviation may be larger when the predicted log PCB is larger.

Figures 5.6 to 5.8 suggest that the fitted model may not be adequate. A closer look at Figure 5.2 (page 154) (and adding the loess line) suggested that the linear relationship between log PCB and fish length may not be appropriate. To test for nonlinearity, we can add the square of the length as a third predictor:

```
#### R code ####
lake.lm5 <- lm(log(pcb) ~ I(year-1974)*len.c +
                    I(len.c^2), data=laketrout)
```

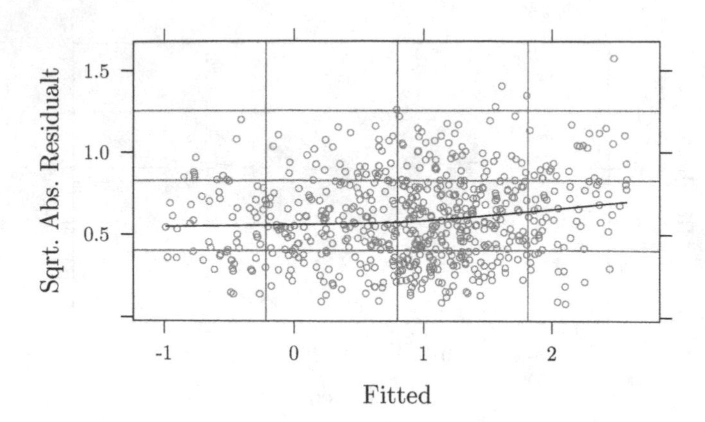

FIGURE 5.8: S-L plot of PCB model residuals – The plot suggests that residual standard deviation increases as the predicted log PCB concentration increases.

```
display(lake.lm5, 4)

#### R output ####
lm(formula = log(pcb) ~ I(year - 1974) * len.c +
    I(len.c^2), data = laketrout)
                     coef.est coef.se
(Intercept)           1.8133   0.0496
I(year - 1974)       -0.0863   0.0036
len.c                 0.0590   0.0043
I(len.c^2)            0.0005   0.0001
I(year - 1974):len.c  0.0004   0.0003
---
n = 631, k = 5
residual sd = 0.5452, R-Squared = 0.68
```

The estimated coefficient for the squared length is 0.0005 with a standard error of 0.0001. It seems that the model may be nonlinear with respect to length. A loess line, fit to the scatter plot of log PCB against fish length, hints at a piecewise linear model. That is, the slope of fish length may have two distinct values, one for fish shorter than 60 cm and another for fish longer than 60 cm. We will revisit this model again in Chapter 6.

The fitted model should also be evaluated for potential influential or leverage data points. A data point is influential if model coefficients estimated with or without the data point are different. A data point's influence is measured by the Cook's distance, a metric measuring the effect of a particular point on regression coefficient estimates. Its distribution is known ($F(2, n-2)$, where

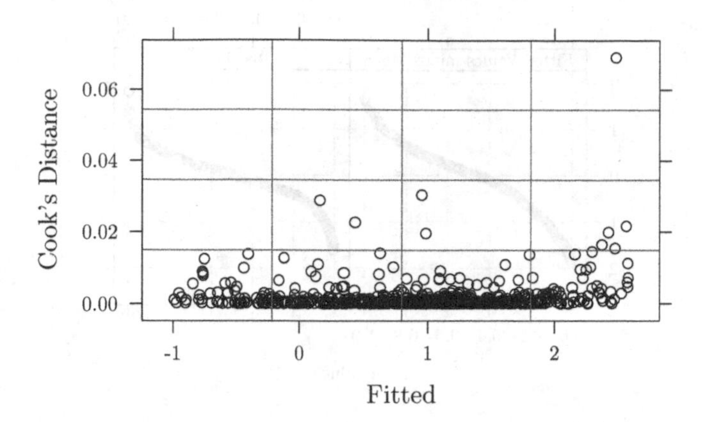

FIGURE 5.9: The Cook's distances of the PCB model – The Cook's distances for data points are plotted against the fitted log PCB concentrations. The Cook's distance for all data points are less than 1. But the lone data point with a Cook's distance above 0.8 is curious.

n is the number of observations). Consequently, just how large the Cook's distance of an observation is can be expressed in terms of quantiles of the F distribution. A heuristic argument also shows that Cook's distance may be considered "large" if it is substantially larger than 1. If observations with "large" Cook's distance exist, we should examine the data to make sure that these data points are obtained without obvious errors. For the PCB in fish data, the Cook's distances (Figure 5.9) for all data points are less than 1, suggesting no obvious influential data point.

Finally, although the R^2 value was never meant to be a statistic for variable selection, in many applications the R^2 value is used as the sole criterion for model assessment. To discourage this practice, the residual-fitted-spread (or rfs) plot (Figure 5.10) should be used instead of the R^2. A rfs plot displays the quantile plot of the fitted values and the quantile plot of the residuals. The quantile plot of the fitted values is centered around the predicted overall mean. Thus, the point 0 on the y-axis is the predicted overall mean log PCB concentration. Because the fitted and the residuals are measured in the same unit (log PCB), by placing them side by side, we can easily visualize the relative ranges covered by the model (the fitted) and by random error (the residuals). The R^2 measures the variance, and the rfs plot shows the relative spread explained by the model. The figure shows that the fitted values cover about the same range as the residuals do, although the R^2 suggests that the model explains 2/3 of the total variability.

In the GitHub page of the book (`https://github.com/songsqian/eesR`), I have included a function to quickly generate the diagnostic plots discussed

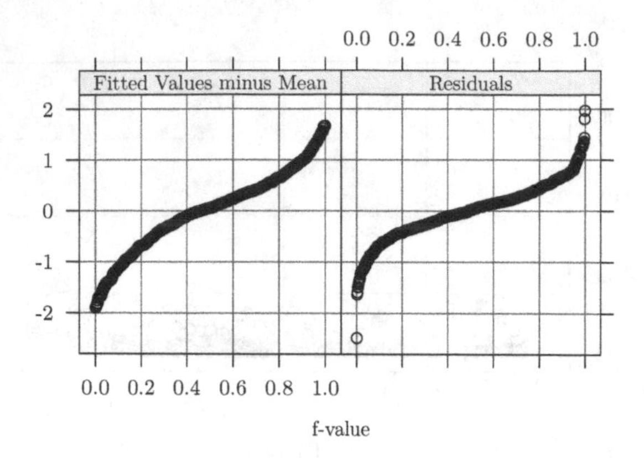

FIGURE 5.10: The rfs plot of the PCB model – The plot compares the fitted log PCB value range and the residual range. The rfs shows that the model accounts for about the same range as the residuals range.

in this section. The function (`lm.plots`) takes the fitted linear model object as the input and produces six diagnostic plots.

5.3.6 Categorical Predictors

The change of diet at a size about 60 cm is important information for model building. Although the study also suggested a diet shift around a size of 40 cm, we do not have many fish smaller than 40 cm. Thus, the shift around 60 cm is more relevant to this dataset. If smaller lake trout eat other small fish with a lower PCB concentration than the PCB concentration in food consumed by larger lake trout, we would expect the slope of the variable `len.c` to be different for the two size categories. To model this effect, we create a categorical predictor `size`:

```
#### R code ####

laketrout$size<- "small"
laketrout$size[laketrout$length>60] <- "large"
```

In our last model, we justified the inclusion of the interaction term. As a result, we also expect that the slope of *yr* changes as a function of length. One possible explanation of the interaction effect is that small fish and large fish should not be pooled together for developing a single model. To fit two separate models for small and large fish, we allow both the intercept and the slopes to vary between the two categories:

```
#### R code ####
```

```
lake.lm6 <- lm(log(pcb) ~ I(year-1974)*factor(size) +
                             len.c * factor(size),
                             data=laketrout)
display(lake.lm6, 4)

#### R output ####
lm(formula = log(pcb) ~ I(year - 1974) * factor(size) +
    len.c * factor(size), data = laketrout)
                                coef.est coef.se
(Intercept)                       1.7394  0.0667
I(year - 1974)                   -0.0846  0.0044
factor(size)small                -0.0647  0.1197
len.c                             0.0776  0.0044
I(year - 1974):factor(size)small  0.0001  0.0074
factor(size)small:len.c          -0.0345  0.0063
---
n = 631, k = 6
residual sd = 0.5426, R-Squared = 0.68
```

When a factor (or categorical) predictor is included in the model, the factor variable is converted into *dummy* variable(s) taking value 0 or 1. For example, the categorical variable `size` has two levels: `large` and `small`. R creates a dummy variable (named `factor(size)small`) taking value 0 if `size` is `large` and 1 if `size` is `small`. The fitted model is now (ignoring the interaction effect between year and length)

$$\begin{aligned} \log(PCB) &= 1.74 - 0.0846yr - 0.0647Dummy + 0.0776Len.c \\ &\quad +0.0001yr \cdot Dummy \\ &\quad -0.0345Dummy \cdot Len.c + \varepsilon \end{aligned} \tag{5.9}$$

Equation (5.9) combines two separate models for small and large fish into one equation. For a large fish, the dummy variable ($Dummy$) takes value 0. The model is then:

$$\log(PCB) = 1.74 - 0.0846yr + 0.0776Len.c + \varepsilon.$$

For a small fish, the dummy variable takes value 1, and the model becomes:

$$\begin{aligned} \log(PCB) &= (1.74 - 0.0647) + (-0.0646 + 0.001)yr+ \\ &\quad (0.0776 - 0.0345)Len.c + \varepsilon \end{aligned}$$

The intercept for small fish is the sum of the large fish intercept (1.74) plus the slope of the dummy variable (−0.0647). That is, the reported slope for the term `factor(size)small` is the difference in intercept between the small fish model and the large fish model. In the same way, the slope of 0.001 for the term `I(year - 1974):factor(size)small` is the difference in slope of *yr* between the model for small fish and the model for large fish. When creating a dummy variable, R's default is to set large fish as the baseline and fit separate models (in alphabetical order). But the model output compares the model for small fish to the baseline model. If the categorical predictor has more than two levels (e.g., we can divide fish into small, medium, and large categories) R will create several dummy variables (number of levels minus 1) and set a baseline level (the first one in alphabetical order if the order is not manually defined). Computer output is organized to compare nonbaseline models to the baseline model.

The output for our model includes the estimated intercept and slopes for large fish (`(Intercept)`, `I(year-1974)` and `len.c`), and the difference between small and large models in intercept and slopes (`factor(size)small`, `I(year - 1974):factor(size)small` and `factor(size)small:len.c`). The estimated difference in intercepts is -0.0647 with a standard error of 0.1197, suggesting that the intercepts for small and large fish are statistically not different. The difference between the slopes of `I(year-1974)` is 0.0001 and statistically not different from 0, but the difference of the slopes for length is -0.0345 ($se = 0.0063$) and statistically different from 0. As a result, we may consider further simplifying the model to allow a common slope for *yr*:

```
#### R code ####
lake.lm7 <- lm(log(pcb) ~ I(year-1974) +
        len.c * factor(size), data=laketrout)
display(lake.lm7, 4)

#### R output ####
lm(formula = log(pcb) ~ I(year - 1974) +
    len.c * factor(size), data = laketrout)
                        coef.est coef.se
(Intercept)              1.7389   0.0588
I(year - 1974)          -0.0846   0.0035
len.c                    0.0776   0.0044
factor(size)small       -0.0631   0.0779
len.c:factor(size)small -0.0345   0.0062
---
n = 631, k = 5
residual sd = 0.5422, R-Squared = 0.68
```

The estimated differences in intercepts and slopes of `len.c` did not change. To directly report the intercepts and slopes of `len.c` for the two size categories, we change the R formula by adding `-1-len.c`:

```
#### R code ####
lake.lm8 <- lm(log(pcb) ~ I(year-1974) +
    len.c * factor(size)-1-len.c, data=laketrout)
display(lake.lm8, 4)
```

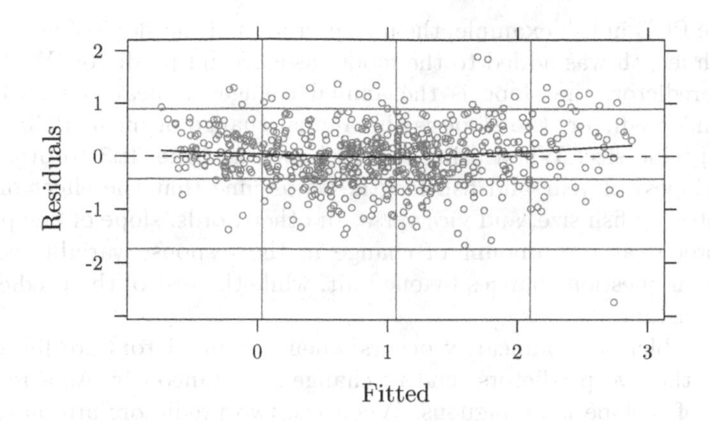

FIGURE 5.11: Modified PCB model residuals vs. fitted – Residuals of PCB model fitted to two size categories are plotted against the estimated mean log PCB concentrations. The plot suggests that the problem shown in Figure 5.7 still exists.

```
#### R output ####
lm(formula = log(pcb) ~ I(year - 1974) +
    len.c * factor(size) - 1 - len.c,
    data = laketrout)
                        coef.est coef.se
I(year - 1974)          -0.0846   0.0035
factor(size)large        1.7389   0.0588
factor(size)small        1.6758   0.0795
len.c:factor(size)large  0.0776   0.0044
len.c:factor(size)small  0.0431   0.0045
---
n = 631, k = 5
residual sd = 0.5422, R-Squared = 0.83
```

Using the categorical predictor `size` allows us to further refine the model using scientific information, although the problem of potential biased prediction still remains to a lesser extent (Figure 5.11).

The model `lake.lm7` is exactly the same as the model `lake.lm8` in all respects except the R^2 values. The R^2 is a relative measure of the sum of squares. It is the sum of squares explained by the full model as a fraction of the sum of squares not explained by the null model. When the intercept is estimated, the null model is the mean of the response variable data. When using `-1` in the model formula, the null model is $y = 0 + \varepsilon$. As a result, the unexplained sum of squares is estimated to be $\sum y^2$. The resulting R^2 value is no longer meaningful.

5.3.7 Collinearity and the Finnish Lakes Example

In the PCB in fish example, the interpretation of the slope of year changed when fish length was added to the model as a second predictor. Without the second predictor, the slope is the annual change of mean log PCB. With the second predictor, the slope is the annual change of mean PCB of a fish with a specific size. When using a model such as `lake.lm3` to describe the effects (slopes) of year and fish length, we assume that the effect of year is not affected by fish size, and vice versa. In other words, slope of one predictor is interpreted as the amount of change in the response variable when the predictor in question changes by one unit, while the rest of the predictors are held in constant.

The problem of collinearity occurs when two predictors are linearly correlated – the two predictors tend to change simultaneously. As a result, the meaning of a slope is ambiguous. When the two predictors are independent of each other (correlation coefficient, ρ, is 0), changes in one predictor are not related to changes in the other. As a result, the effects of the two predictors can be separated using regression. If the two predictors are perfectly collinear (i.e., correlation coefficient is $|\rho| = 1$, or one predictor is a linear function of the other), only one of the two predictors is needed. This is because the two predictors change simultaneously and their effects cannot be separated. The problem of collinearity describes the situation when the two predictors are correlated ($|\rho| < 1$). A strong correlation indicates that the two predictors tend to change together; as a result, it is difficult to separate their effects on the response variable. The stronger the correlation, the harder it is to separate the effects of the two predictors. A strong correlation between two predictors is often reflected in the large standard errors of the estimated slopes of the two predictors. Many texts recommended that we select one of the two predictors to be included in the regression model. I find that correlation between two relevant predictors is often an indicator of an interaction effect. The Finnish lakes example provides an illustration for the need to explore interaction before deciding to drop one of the two predictors.

Linear regression analysis is the most commonly used method in quantifying the relationship between nutrient input to a lake or a group of lakes and the growth of phytoplankton in lake water. Because nitrogen and phosphorus are the two nutrients algae need in large quantities, they are often the limiting factor of algal growth. To understand the relationship, algal growth is often represented by the in-lake chlorophyll *a* concentration. Malve and Qian [2006] developed linear regression models for Finnish lakes using data from the Finnish national water quality monitoring network, which started monitoring most lakes in Finland in 1965 after the passage of the Water Act in 1962. In January 2000, 253 lake sites from the national monitoring network were integrated into the European Environment Agency's Eurowaternet to produce statistically reliable information that allow the European Commission, member states, and the general public to judge the effectiveness of the environmental policy. These 253 lakes were classified into 9 categories or types according to expert assessments on lake morphological and chemical metrics such as depth, surface area, and color. More data from similar lakes may be combined when lakes are grouped. As a result, we may have a better understanding of the natural ecological status of the lakes under different conditions. One important relationship required for assessing lake water quality status is the in-lake chlorophyll a concentration and nutrient input. Frequently, both nitrogen and phosphorus are used in developing statistical models often justified by data plots such as the one in Figure 5.12.

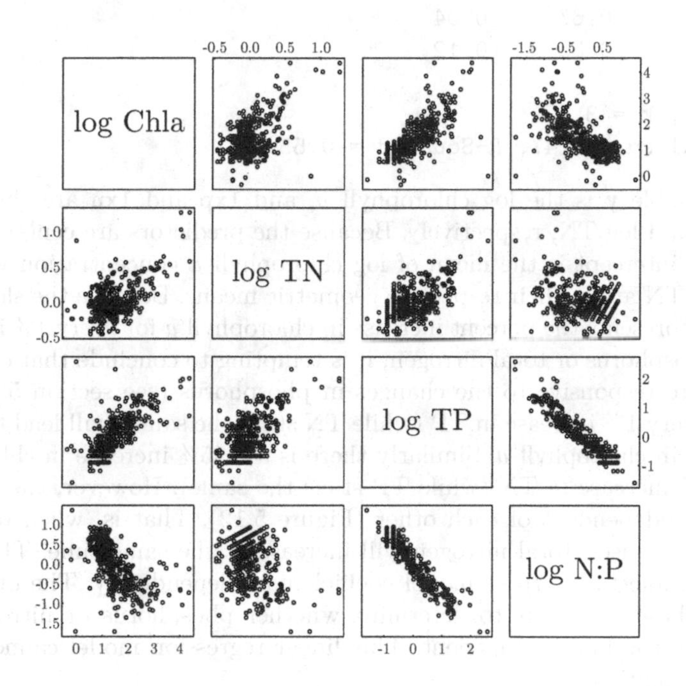

FIGURE 5.12: Finnish lakes example: bivariate scatter plots – scatter plot matrix shows strong linear relationships between log chlorophyll a and log TP, log chlorophyll a and log TN, and the log N:P ratio. The data were from the lake with the largest sample size.

This relationship is most frequently expressed as a log-log linear regression model:

$$\log(Chla) = \beta_0 + \beta_1 \log(TP) + \beta_2 \log(TN) + \varepsilon \qquad (5.10)$$

where TP and TN are total phosphorus and total nitrogen concentrations, respectively, and $Chla$ is chlorophyll a concentrations. I use data from three lakes to illustrate the connection between collinearity and interaction. These lakes represent three different types of interactions between the two correlated predictors.

Let us examine the first lake in detail. Both log TP and log TN by themselves are linearly correlated with log chlorophyll a (Figure 5.12). Using both TP and TN as predictors, the resulting model is somewhat satisfactory:

```
#### R Output ####
> display(Finn.lm2)
lm(formula = y ~ lxp + lxn, data = lake2)
            coef.est coef.se
(Intercept) 1.43     0.02
lxp         0.67     0.04
lxn         0.55     0.12
---
n = 441, k = 3
residual sd = 0.47, R-Squared = 0.55
```

The variable `y` is the log chlorophyll *a*, and `lxp` and `lxn` are the centered log TP and log TN, respectively. Because the predictors are centered, the regression intercept is the mean of log chlorophyll *a* concentration when both TP and TN are at their respective geometric means. Because the slope in this model represents the percent increase in chlorophyll *a* for every 1% increase in total phosphorus or total nitrogen, it is tempting to conclude that chlorophyll *a* is more responsive to the changes in phosphorus (see section 5.4 on page 185). Every 1% increase in TP (while TN stays the same) will lead to a 0.67% increase in chlorophyll *a*. Similarly there is a 0.55% increase in chlorophyll *a* for a 1% increase in TN (while TP stays the same). However, the predictors are not independent of each other (Figure 5.12). That is, when total phosphorus increases, total nitrogen will increase at the same time. Therefore, it is impossible to interpret model coefficients independently. The objective of fitting these models is to determine whether phosphorus or nitrogen is, or both are, the limiting nutrient. This linear regression model cannot provide this information.

Furthermore, when two predictors are strongly correlated, regression model coefficients tend to have higher estimation uncertainty. The estimated coefficients are sensitive to small changes in input data, which also contributes to the difficulty in interpretation. In many cases, when predictor variables are correlated, their interaction should be examined. In a lake eutrophication problem, the limiting nutrient is the one of the smaller quantity. The cellular nitrogen and phosphorus ratio of a given species of algae is relatively stable. In theory, algal growth uses the same ratio of nitrogen and phosphorus from water to build their cells. Suppose the optimal nitrogen to phosphorus ratio is 16 for a community of phytoplankton, the supply of nitrogen exceeds the demand when the actual N:P ratio in lake water is above 16, and vice versa. Consequently, the interaction effect of TP and TN on chlorophyll *a* is expected.

When collinearity is a potential problem, we can either ignore the problem by treating the fitted model as a predictive model and not interpret the fitted model coefficients (which is not an option for this example), or include the interaction of the two correlated predictors. The commonly used interaction term is the product of the two predictors. In a lake eutrophication problem, the nitrogen to phosphorus ratio is likely the key factor in determining the relative importance of TP and TN in the model. Some reported the use of the N:P ratio as the third predictor.

In general, a good starting point of assessing the problem of collinearity is plotting the data, specifically, the conditional plots. A conditional plot is a series of bivariate scatter plots for examining the conditional relationship among three or more variables. In this case, we are interested in the relationship between chlorophyll *a*, TP, and TN. Because of the strong correlation between TP and TN, the linear relationship between chlorophyll *a* and TP (or TN) is hard to interpret, because of the additive assumption of the multiple regression. We cannot explain the slope of TP as the change in chlorophyll a concentration as a function of TP when TN is held in constant. A strong correlation implies that TP and TN will change simultaneously. A conditional plot is to examine the, say, chlorophyll *a*–TP relationship at a relatively constant TN. To achieve a relatively constant TN, we can divide the range of TN into multiple intervals and divide the data accordingly. The chlorophyll *a*–TP relationship is then examined for each group along the gradient of TN. The conditional plot is implemented in R function `coplot`:

```
#### R Code ####
 given.tn <- co.intervals(lake2$lxn, number=4,
                          overlap=.1)
 coplot(y ~ lxp | lxn, data = lake2,
        given.v=given.tn, rows=1,
        panel=panel.smooth)

 given.tp <- co.intervals(lake2$lxp, number=4,
                          overlap=.1)
 coplot(y ~ lxn | lxp, data = lake2,
        given.v=given.tn, rows=1,
        panel=panel.smooth)
```

The function `co.intervals` splits the *conditioning* variable into `number=4` groups, each with roughly the same number of data points and about 10% (`overlap=0.1`) overlap between neighboring intervals. The option `panel = panel.smooth` adds a loess line to the plots (Figures 5.13 and 5.14).

The chlorophyll *a*–TP relationship is relatively stable (Figure 5.13). The smooth line in each of the 4 panels show a similar slope, which suggests that the effect of phosphorus on chlorophyll *a* is consistent regardless of the level of TN – an indication of a phosphorus limited lake. The chlorophyll *a*–TN relationship is less pronounced until the right-most panel where TP is the

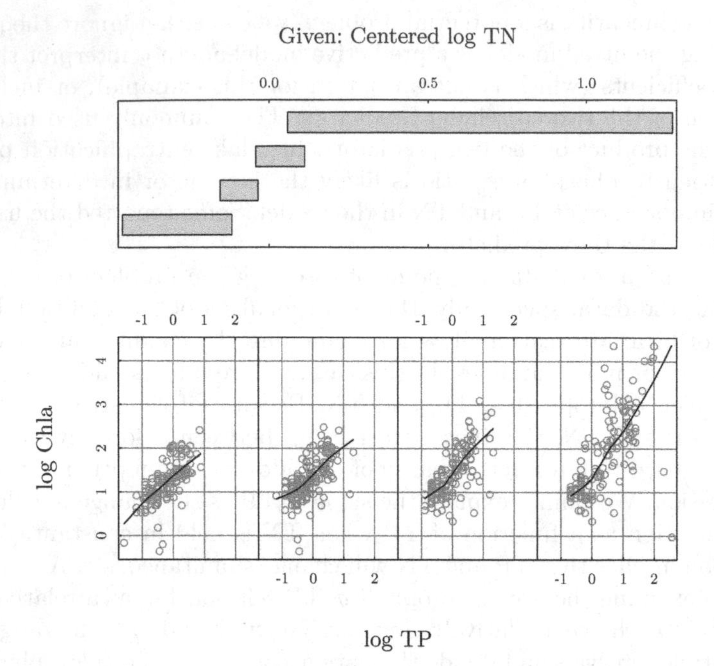

FIGURE 5.13: Conditional plot: chlorophyll *a* against TP conditional on TN – The relationship between log chlorophyll *a* and log TP is plotted conditional on the range of TN. From left to right, each panel represents a different range of log TN shown in the top panel (from left to right and bottom to top). This plot shows the case of no interaction between TP and TN.

highest (Figure 5.14). We note that the TN intervals were selected to have similar data points. As a result of the skewed distribution, the last interval has the widest range (more than half of the entire range). Again, these plots are intended as tools for exploratory analysis.

These conditional plots suggest that the lake is phosphorus limited. As a result, chlorophyll *a* concentration responds to the changes of phosphorus rapidly. In-lake nitrogen concentration reflects an overall level of nutrient enrichment. But, changes in nitrogen alone are unlikely to cause changes in chlorophyll *a* concentrations. This suggests that the interaction effect of the two predictors should be weak. We fit the following model with a product interaction term:

$$\log(Chla) = \beta_0 + \beta_1 \log(TP) + \beta_2 \log(TN) + \beta_3 \log(TP)\log(TN) + \varepsilon \quad (5.11)$$

```
#### R Output ####
> display(Finn.lm4)
lm(formula = y ~ lxp * lxn, data = lake2)
            coef.est coef.se
```

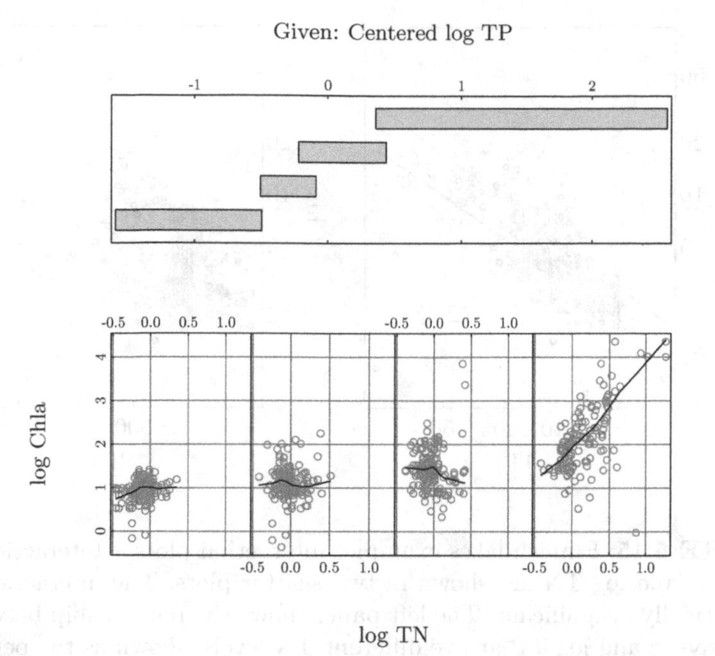

FIGURE 5.14: Conditional plot: chlorophyll *a* against TN conditional on TP – The relationship between log chlorophyll *a* and log TN is plotted conditional on the range of log TP. From left to right, each panel represents a different range of log TP shown in the top panel (from left to right and bottom to top). This plot shows the case of no interaction between TP and TN.

```
(Intercept) 1.43      0.02
lxp         0.66      0.04
lxn         0.52      0.13
lxp:lxn     0.05      0.10
---
n = 441, k = 4
residual sd = 0.47, R-Squared = 0.55
```

The relatively large standard error of the interaction coefficient suggests that the interaction effect is statistically not different from 0.

The conclusion is that this lake is likely to be phosphorus limited. As a result, the model should be interpreted in terms of a phosphorus effect (0.66% change in chlorophyll *a* for every 1% change in phosphorus).

As discussed in Section 5.3.4, including an interaction effect changes the model from linear to nonlinear. Specifically, the effect of one predictor (its slope) is a function of the other. To present this change in slopes, we can make two separate plots. In the first plot, the response variable is plotted against one

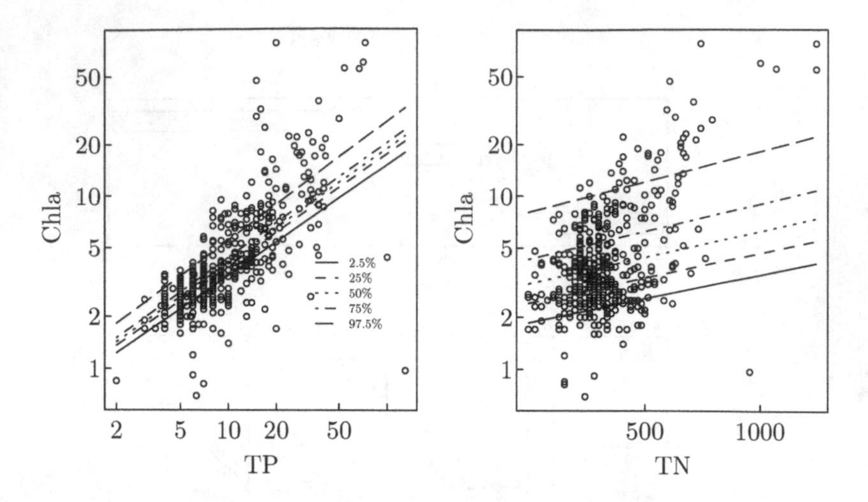

FIGURE 5.15: Finnish lakes example: interaction plots – Interaction effect of log TP and log TN are shown in two scatter plots. The interaction effect is statistically insignificant. The left panel shows the relationship between log chlorophyll *a* and log TP at five different TN levels, shown as the percentiles in the legend. The right panel shows the relationship between log chlorophyll *a* and log TN at five TP levels (the same 5 percentiles in the left panel).

predictor and the regression model is superimposed on the scatter plot using selected values of the other predictor values. For example, the left panel of Figure 5.15 shows log chlorophyll *a* against log TP, and the regression model is evaluated at five values of log TN (the 2.5, 25, 50, 75, and 97.5 percentiles). In the second plot, the same setup is used with the response variable plotted against the other predictor (right panel of Figure 5.15). As expected of a zero interaction between TP and TN, the relationship between log chlorophyll *a* and log TP vary only slightly at different values of total nitrogen, while the other relationship shows notable changes as a function of TP.

Each panel of Figure 5.15 shows five lines that are almost parallel. These lines suggest that the effect (slope) of one predictor is not affected by the other variable. No interaction effect between TP and TN means that no matter what the value of TN, each 1% increase in phosphorus (nitrogen) concentration will lead to a 0.66% (0.52%) increase in chlorophyll *a* concentration in this lake. But this interpretation is unlikely to be appropriate because of the correlation between the two predictors. Including the interaction term in this case did not directly tell us whether the lake is likely limited by phosphorus or nitrogen. But the conditional plots suggest that phosphorus is likely the limiting nutrient. For this lake, dropping TN as a predictor is a sensible choice.

While the first lake shows no obvious interaction effect between the two correlated predictors (i.e., a 0 interaction effect), the second lake shows a strong positive interaction effect:

```
#### R Output ####
lm(formula = y ~ lxp * lxn, data = lake3)
            coef.est coef.se
(Intercept) 1.59     0.03
lxp         0.57     0.07
lxn         0.75     0.14
lxp:lxn     0.31     0.12
---
n = 236, k = 4
residual sd = 0.33, R-Squared = 0.74
```

This positive interaction effect suggests that the effect of phosphorus increases as the nitrogen concentration increases, and vice versa. In a conditional plot (Figures 5.16 and 5.17), this feature is demonstrated by the increasingly steeper slopes of one predictor, as the values of the other predictor represented by these panels increase.

Representing the fitted model with a positive interaction effect, we use the same interaction plot as in Figure 5.15. A positive interaction is expressed by the increasing slopes of one predictor when the value of the other predictor increases (Figure 5.18).

The third lake has a negative interaction:

```
#### R Output ####
lm(formula = y ~ lxp * lxn, data = lake4)
            coef.est coef.se
(Intercept)  2.84     0.05
lxp          0.35     0.18
lxn          0.73     0.29
lxp:lxn     -0.31     0.18
---
n = 105, k = 4
residual sd = 0.44, R-Squared = 0.31
```

We leave the conditional plots to Section 10.4.2 (page 453) where data from multiple lakes in the same type are plotted to reduce noise. A negative interaction suggests that the effect of phosphorus reduces as nitrogen level increases (Figure 5.19).

The problem of collinearity is mostly in interpretation. Mathematically, when two predictors are perfectly co-linear, the multiple regression is *singular* and only one of these two predictors should be included in the model. When the correlation between two predictors is strong, removing one of them from the model is often recommended. Although choosing one over the other often makes no difference mathematically, the interpretation of the resulting model may be very different because the predictor is often viewed as the cause. It is,

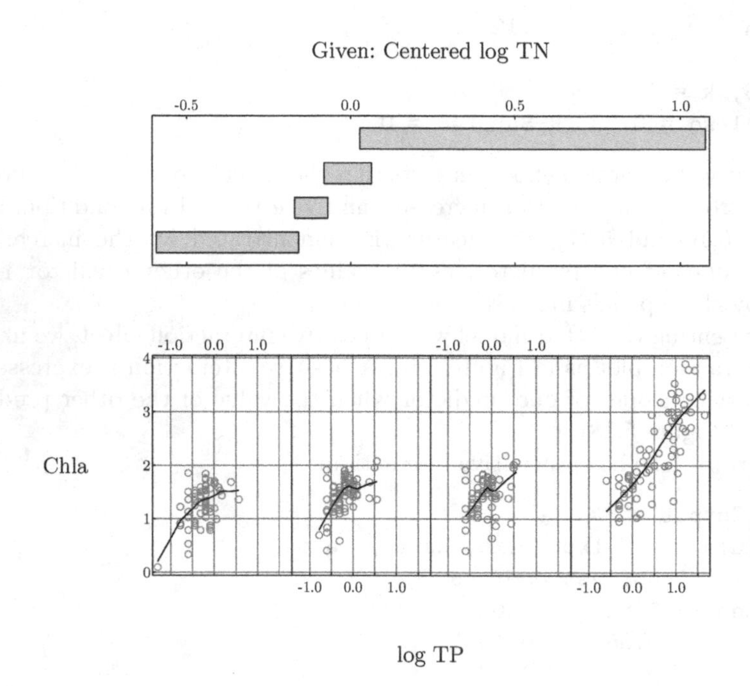

FIGURE 5.16: Conditional plot: chlorophyll *a* against TP conditional on TN – The relationship between log chlorophyll *a* and total phosphorus is plotted conditional on the range of total nitrogen. From left to right, each panel represents a different range of TN shown in the top panel (from left to right and bottom to top). This plot shows the case of a positive interaction between TP and TN.

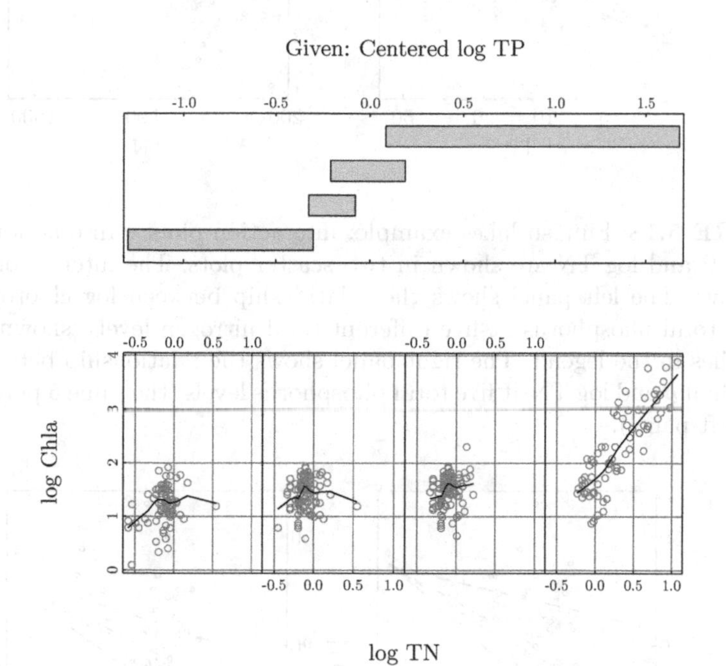

FIGURE 5.17: Conditional plot: chlorophyll a against TN conditional on TP – The relationship between log chlorophyll a and total nitrogen is plotted conditional on the range of total phosphorus. From left to right, each panel represents a different range of TP shown in the top panel (from left to right and bottom to top). This plot shows the case of a positive interaction between TP and TN.

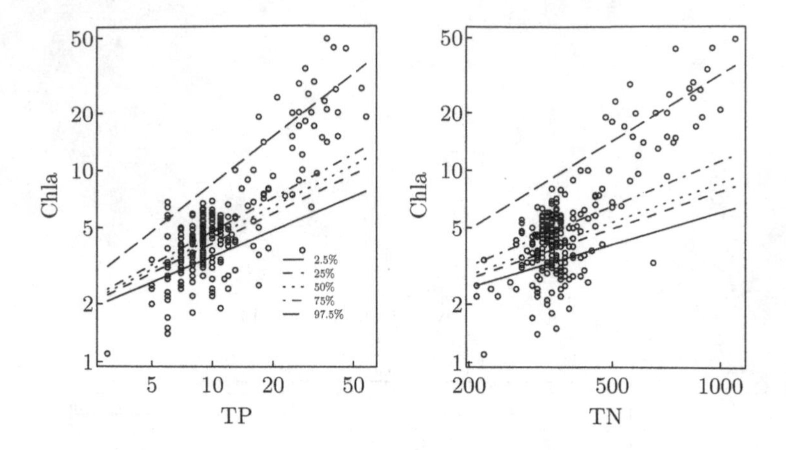

FIGURE 5.18: Finnish lakes example: interaction plots – Interaction effect of log TP and log TN are shown in two scatter plots. The interaction effect is positive. The left panel shows the relationship between log chlorophyll *a* and log total phosphorus at five different total nitrogen levels, shown as the percentiles in the legend. The right panel shows the relationship between log chlorophyll *a* and log TN at five total phosphorus levels (the same 5 percentiles in the left panel).

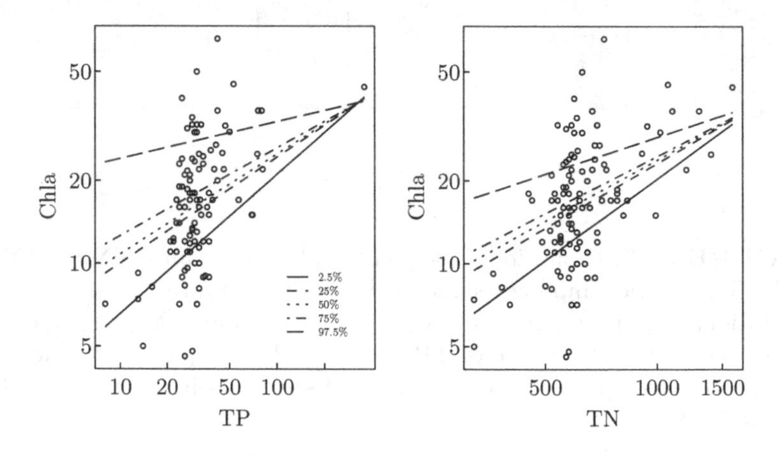

FIGURE 5.19: Finnish lakes example: interaction plots – Interaction effect of log TP and log TN are shown in two scatter plots. The interaction effect is negative. The left panel shows the relationship between log chlorophyll *a* and log total phosphorus at five different total nitrogen levels, shown as the percentiles in the legend. The right panel shows the relationship between log chlorophyll *a* and log TN at five total phosphorus levels (the same 5 percentiles in the left panel).

therefore, prudent to use the conditional plots as a guide for interpretation or choosing the predictor to be removed.

For a lake eutrophication problem, the estimated model coefficients of the model represented in equation 5.11 (page 178) can be used, along with conditional plots, to determine whether phosphorus or nitrogen is the limiting nutrient. If the interaction effect is close to 0, it is likely that the lake is limited by only one nutrient. Conditional plots should help us determine whether it is nitrogen or phosphorus. If the interaction is strong and positive, both nitrogen and phosphorus are likely limiting. A strong negative interaction effect suggests that the lake may be limited by neither nitrogen nor phosphorus (see Section 10.4.2 on page 453 for more discussion).

5.4 General Considerations in Building a Predictive Model

A linear model is simple and straightforward, both to fit and to interpret. However, the use of a linear model is usually not simple and straightforward. The PCB in the fish example represents a relatively simple problem. But finding the appropriate model is difficult. This is typical in environmental and ecological studies. The difficulty is also the result of the inductive nature of statistical modeling. Suppose that we are only interested in building a predictive model; we should follow some general guidelines to make the resulting model more interpretable.

- Centering and Standardization of predictors.
  Linear transformations such as centering around the mean (or a conventional/convenient value, such as 60 cm of fish length in the PCB in fish example) will not change the fitted model but will make the estimated model coefficients easier to interpret. For example, the PCB in fish example models (e.g., model `lake.lm8`) can be fitted by centering the length at 60 cm. The current fitted model is centered at the average length of 62.6. The intercept for large fish (1.7389) is the mean log PCB concentration for a fish of size 62.6 cm in 1974. The intercept for small fish (1.6758) cannot be directly interpreted because it is the predicted mean log PCB for a fish of size 62.6 (a large fish, but the model is fitted for small fish). By using a slightly different linear transformation:

  ```
  laketrout$len.c2 <- laketrout$length-60,
  ```

  the fitted intercept for small fish will be the mean log PCB for the largest possible "small" fish, and the intercept for large will be the mean log PCB concentration for a smallest fish in the "large" category. Centering is especially useful for the ease of interpretation when interaction terms are included in the model. For example, in the model in equation 5.8 (page 161), the length is centered around the mean and the year is centered at 1974. The slope of the centered length is the slope when year is 1974, and slope of the centered year is the slope of year for an average-sized fish. Were the two predictors not centered, the slope of length would be the slope when year is 0.

In general, a linear transformation implies $x^T = a + bx$. The multiplicative factor $b \neq 1$ can be a factor for converting the predictor variable from one unit to the other. With $b \neq 1$, the meaning of the slope is changed from the change in response variable mean per unit change of the predictor, to per b units change of the predictor. For example, if the unit of fish length is converted from centimeter to millimeter, that is $b = 10$, the resulting model slope would be the increase of log PCB concentration per 1 mm increase in size. The slope for small fish would change from 0.0431 to 0.00431. This is fine, but we really don't think changes of fish size in terms of millimeter. Likewise, if the length is measured in meters ($b = 0.1$), the slope for small fish would be 0.431, which would be very unnatural to explain. But for people who live around the lake (Lake Michigan), a more familiar length measure would be inches ($b = 1/2.54$).

Frequently, we "standardize" a predictor, or $x^{st} = \frac{x-\bar{x}}{\hat{\sigma}_x}$. The resulting slope will be the change of the response variable per standard deviation change of the predictor.

- Logarithm transformation.

In environmental and ecological studies, most variables take positive values only. These variables are often skewed. The logarithm transformation is the most commonly used transformation to achieve approximate normality. Furthermore, the additive assumption is often not reasonable. Log transforming the response variable will lead to a multiplicative model in the original scale. That is, the linear model in the logarithmic scale

$$\log(y) = \beta_0 + \beta_1 x_1 + \cdots + \beta_p x_p + \varepsilon$$

becomes

$$\begin{aligned} y &= e^{\beta_0 + \beta_1 x_1 + \cdots + \beta_p x_p + \varepsilon} \\ &= B_0 B_1^{x_1} \cdots B_p^{x_p} E \end{aligned}$$

in the original scale, where $B_0 = e^{\beta_0}, B_1 = e^{\beta_1}$, and $E = e^{\varepsilon}$. Each unit increase of a predictor, say x_i, will result in a multiplicative factor increase in y. Suppose we increase x_1 from its current value to $x_1 + 1$ and all other predictors are unchanged, the change in y will be from the current value

$$y = e^{\beta_0 + \beta_1 x_1 + \cdots + \beta_p x_p + \varepsilon} = y_0$$

to

$$y = e^{\beta_0+\beta_1(x_1+1)+\cdots+\beta_p x_p+\varepsilon} = y_0 e^{\beta_1}$$

If β_1 is a small number, for example, $0.0k$, where k is an integer, the quantity $e^{0.0k}$ is approximately $1 + 0.0k$, or a change of $k\%$. The slope of length for small fish is 0.04; thus, a unit (1 cm) change in fish size will result in a 4% change in mean PCB concentration in a small fish.

In some cases, both the response variable and the predictors are log-transformed. Such models are the commonly used power functions in engineering literature. The log-log linear model

$$\log(y) = \beta_0 + \beta_1 \log(x_1) + \cdots + \beta_p \log(x_p) + \varepsilon$$

is

$$y = e^{\beta_0} x_1^{\beta_1} \cdots x_p^{\beta_p} e^{\varepsilon}$$

The slope of each predictor can be interpreted as the percent change in y per 1% change in the respective predictor. To see this approximation, we hold all other predictors in constant, and change, say x_1, by 1% from x_1 to $x_1(1 + 0.01)$. This would result in the response variable changing from the baseline of $y_0 = e^{\beta_0} x_1^{\beta_1} \cdots x_p^{\beta_p} e^{\varepsilon}$ to $y_1 = e^{\beta_0}(x_1 \times (1 + 0.01))^{\beta_1} \cdots x_p^{\beta_p} e^{\varepsilon}$ or $y_0(1 + 0.01)^{\beta_1}$. The multiplier $(1 + 0.01)^{\beta_1} \simeq (1 + 0.01\beta_1)$.

- Other transformations.

  In general, transformation of the response variable is aimed at achieving approximate normality in model residuals. The logarithmic transformation is used in most cases because (1) most environmental and ecological variables take only positive values and are likely to have a log-normal distribution, the logarithm of these variables are likely to be normal, and (2) interpretation of the resulting model is easy. When the logarithmic transformation is unable to achieve this goal, a general power transformation y^λ can be used to achieve normality in the residuals. A proper λ value exists in most cases such that the resulting linear model residuals are approximately normal. The procedure for finding the proper value of λ is described by Box and Cox [1964], implemented in R function `boxcox`. To use this function, a linear model without transformation is first fitted and saved into a linear model object. For example:

```
#### R Code ####
lake.lm0 <- lm(pcb ~ I(year-1974) + len.c, data=laketrout)
PCBboxcox <- boxcox(lake.lm0)
```

  The function `boxcox` produces a figure showing the likelihood profile of the estimated λ. The λ value corresponding to the highest likelihood value (the y-axis in Figure 5.20) is the best estimate of λ. In this case, the best estimate is -0.18:

```
> PCBboxcox$x[PCBboxcox$y==max(PCBboxcox$y)]
[1] -0.18
```

very close to 0 (the log transformation). If the estimated $\lambda = -0.18$ is used, the resulting model would be very difficult to interpret. In general, we choose the transformation that is a compromise between the best estimate and interpretability of the transformation.

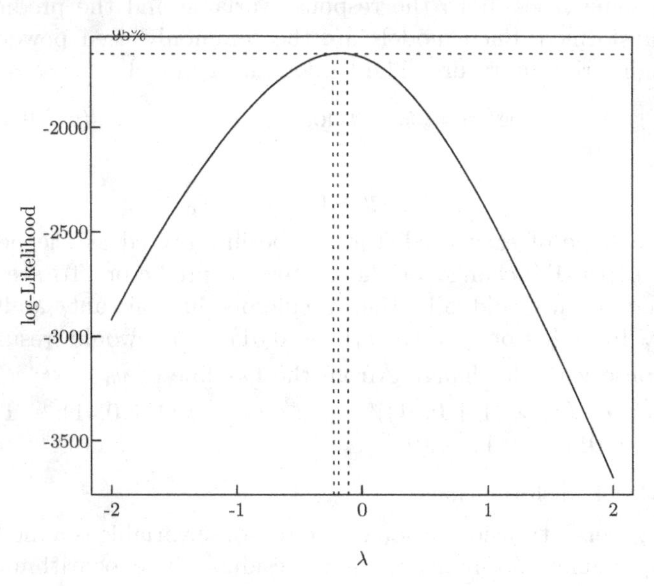

FIGURE 5.20: Box–Cox likelihood plot for response variable transformation – The likelihood profile of λ shows the maximum likelihood is close to 0; hence, a log-transformation is a good choice for a response variable transformation.

- Strategies of building a predictive model.

  The PCB in the fish example is atypical in that there are a limited number of predictors. Frequently, we will have more than two predictors and the decision on which predictors to be included in the model is often a difficult one. In general, the following strategies should be considered.

  1. Include all predictors that are substantively relevant in the model. Sometimes it is necessary to combine several predictors into a "compound variable," e.g., creating a variable petal width plus petal length to represent the size of a petal in Figure 3.17 would be helpful.
  2. Using known mechanics to guide the selection of variables as well as model form. The U.S. Geologic Survey (USGS) watershed water quality model SPARROW [Smith et al., 1997] is a good example. In

that model, regression model forms are determined by the processes of nutrient generation and transportation in the landscape.

3. When little or no knowledge about the relationship under study is available, we should try to use the tree-based model to find relevant predictors (Chapter 7).
4. For those predictors that have strong effects, consider including interaction terms.
5. When a predictor has a slope that is statistically not different from 0, include the predictor in the model if the slope is of the "correct" sign, or the expected sign based on our understanding of the subject. Inclusion of a non-significant predictor will provide no help in predicting the response variable, but will help model interpretation.
6. When a predictor is statistically not significant and the slope is of the "wrong" sign, the predictor should be excluded from the model. But we must think about the unexpected sign to understand whether or not the wrong sign is because of a "lurking" variable.

5.5 Uncertainty in Model Predictions

Once a model is fit, we can use the model to make predictions. An important aspect of statistics is the quantification of uncertainty in any estimation. The response variable of a linear regression model is a normal random variable with mean predicted by the linear model and the standard deviation estimated from the residuals. One way to obtain uncertainty information on a prediction is to use this definition and the estimated residual standard deviation $\hat{\sigma}$. For example, the simple linear regression in Section 5.3.2 (log PCB concentration in fish predicted by year) was used for predicting future (2007) mean PCB concentrations by Stow et al. [2004]. The predicted log PCB concentration is a normal distribution: $N(\hat{\beta}_0 + \hat{\beta}_1 \times 2007, \hat{\sigma})$ or $N(-0.3792, 0.88)$. This distribution is called a *predictive distribution*. However, this distribution does not account for the uncertainty on the estimated model coefficients.

Typically, two different prediction standard errors are suggested in regression texts. One is the *fitted* standard error ($sefit$), the standard error of the estimated response variable average value at a given predictor variable value included in the data used for fitting the model. For example, we may want to estimate the average log PCB concentration in 1980 using the fitted model (as an estimate of the "true" average). For a simple regression model $y = \beta_0 + \beta_1 x + \varepsilon$, the *fitted value* is $\hat{y} = \hat{\beta}_0 + \hat{\beta}_1 x$, and its standard error is

$$sefit(\hat{y}|x) = \hat{\sigma}\left[\frac{1}{n} + \frac{(x-\bar{x})^2}{SXX}\right]^{1/2} \tag{5.12}$$

where $SXX = \sum(x_i - \bar{x})^2$.

The fitted model is more frequently used for making predictions. By prediction, we mean estimating the response variable values given new predictor values not used to estimate model coefficients. Estimating the log PCB concentration in 2007 is an example of prediction. In general, we denote a new predictor value as $\tilde{x}$. The predicted mean response variable value is then $\tilde{y} = \hat{\beta}_0 + \hat{\beta}_1\tilde{x}$, with standard error

$$sepred(\tilde{y}|\tilde{x}) = \hat{\sigma}\left[1 + \frac{1}{n} + \frac{(\tilde{x} - \bar{x})^2}{SXX}\right]^{1/2} \tag{5.13}$$

In R, linear regression prediction can be obtained by using the generic function `predict`. For example, to predict the average log PCB concentration in 2007:

```
#### R Code ####
predict(lake.lm1, new=data.frame(year=2007), se.fit=T)

#### R Output ####
$fit
[1] -0.3792
$se.fit
[1] 0.124
$df
[1] 629
$residual.scale
[1] 0.8784
```

which shows that the fitted mean log PCB is −0.3792, the fitted standard error (calculated using equation 5.12) is 0.124, degrees of freedom of the model, and the residual standard deviation is 0.8784. The prediction standard error can be manually calculated to be $\sqrt{\hat{\sigma}^2 + sefit^2}$ or $\sqrt{0.8784^2 + 0.124^2} = 0.8871$. The R function `predict` has an option for calculating the predictive confidence interval using the prediction standard error (*sepred* in equation 5.13):

```
#### R Output ####
predict(lake.lm1, new=data.frame(year=2007), se.fit=T,
        interval="prediction")$fit
          fit    lwr   upr
[1,] -0.3792 -2.121 1.363
```

This interval is calculated as the fitted mean (-0.3792) ± the prediction standard error (0.8871) times the multiplier ($t(0.975, 629)$ or 1.964).

The predicted mean concentration of −0.3792 is in the logarithmic scale. When interested in the mean concentration, we transform the prediction back to its original scale (re-transformation) using the exponential of the predicted logarithmic mean. The retransformation of a log transformed variable can be, however, problematic, because the exponential of the log-mean is not the same as the mean concentration. The error term ε cannot be ignored when transforming a log-linear regression model ($\log(y_i) = X\beta + \varepsilon$) back to its original scale ($y_i = e^{X\beta}$). Although the error term has mean 0, the transformed model should be $y_i = e^{X\beta}e^{\varepsilon}$, and the mean of e^{ε} is larger than 1. This is because the error term ε follows a normal distribution $N(0, \sigma^2)$, and its exponential follows a log-normal distribution with log mean 0 and log variance σ^2. The (arithmetic) mean of e^{ε} is $e^{0+\sigma^2/2}$, a value that is always greater than 1. As a result, if $e^{X\beta}$ is used as the estimate mean concentration, it is biased by a fixed multiplicative factor. Obviously, we can use the estimated residual standard error as an approximate estimate of σ and calculate the arithmetic mean concentration as $\tilde{y} = e^{\tilde{X}\hat{\beta}}e^{\hat{\sigma}^2/2}$. The multiplicative factor $e^{\hat{\sigma}^2/2}$ is often known as the log-transformation bias correction factor [Sprugel, 1983]. With estimation uncertainty in both $\hat{\beta}$ and $\hat{\sigma}^2$, it is difficult, if not impossible, to come up with a formula for the standard error of the estimated mean of the response variable $\tilde{y}$. A simulation-based method is discussed in Section 9.2.

Predictive error is often an ignored part of regression analysis in many applied fields. For example, when measuring water quality parameters such as total phosphorus or total nitrogen, we develop a regression model using a number of samples with known concentration values and measure the values of an indicator (typically changes in color). The resulting concentration–indicator regression model is often known as a "standard curve." With the estimated standard curve, we measure the indicator values of water samples with unknown concentrations. These indicator values are then used as new predictor values for making predictions. These predictions are reported as the "measured" concentration values. The predictive uncertainty of these "measured concentrations" is rarely reported. When concentrations of water quality variables are used to make important public safety decisions, uncertainty associated with these concentrations should be, but rarely are, communicated.

5.5.1 Example: Uncertainty in Water Quality Measurements

On August 1, 2014, the Water Department of the City of Toledo (Ohio, USA) detected a high concentration of microcystin, a group of toxin associated with harmful algal bloom in the source water, in one finished drinking water sample. The standard measurement method at the time was an enzyme-linked immunosorbent assay (ELISA). The ELISA method quantifies microcystin toxin concentrations through a competitive binding process between the toxin and enzyme-labeled microcystin in the inside wall of test wells. The toxin concentration is visualized with a color development process and is inversely proportional to color development (i.e., the darker the sample color, the less toxin present). During a typical test, a number of (typically, six) samples with known toxin concentrations were used to develop a "standard curve" – a regression model of optical density (OD) as a function of toxin concentrations. For example, the ELISA test kit used by Toledo Water Department (manufactured by Abraxis, Warminster, USA) has six standard concentrations (0, 0.167, 0.444, 1.110, 2.220, and 5.550 μg/L). Water samples with unknown microcystin concentrations are put in the remaining wells of the 96-well plate in the same test. Microcystin concentrations from these water samples are predicted by the fitted standard curve.

One recommended standard curve model is a log-linear regression model. The response variable is the log-concentration and the predictor is the relative optical density (rOD), which is the ratio of the measured optical density (OD) divided by the mean OD from the standard solution with a concentration of 0. In the Toledo case, the five rOD values are 0.784, 0.588, 0.373, 0.270, and 0.202, and the respective microcystin concentrations from the standard solutions are 0.167, 0.444, 01.110, 2.220, and 5.550 μg/L. The water sample in question had a relative OD of 0.261.

```
#### R code ####
mc <- c(0.167, 0.444, 01.110, 2.220, 5.550)
rOD <- c(0.784, 0.588, 0.373, 0.270,0.202)
stdcrv <- lm(log(mc) ~ rOD)

### predict:
aug01 <- predict(stdcrv, newdata=data.frame(rOD=0.261),
                 se.fit=T, interval="prediction")
```

The predictive 95% confidence interval is:

```
#### R output ####
> aug01$fit
       fit         lwr      upr
1 1.021967 -0.06211885 2.106053
```

The standard curve is a log-linear regression model. When converting the predicted log-concentration back to the concentration scale, a correction factor is necessary. The correction is typically not considered in an ELISA kit. As a result, the predicted mean concentration for this water sample is 2.78 μg/L and the ad hoc prediction interval in the concentration scale is (0.94, 8.22). The measured concentration of 2.78 is the basis for a "Do Not Drink" advisory which resulted in nearly five hundred thousand people in the Toledo area without drinking water for three days.

The wide confidence intervals of the fitted and predicted concentration in this case are largely due to the small sample sized used in fitting the model (Figure 5.21). The model has a degrees of freedom of 3. The multiplier for calculating the confidence interval is `qt(0.975, 3)` or 3.18.

We will revisit this example in Chapter 9 for uncertainty analysis of a nonlinear regression model used by the Toledo Water Department, with a focus on the predictive uncertainty.

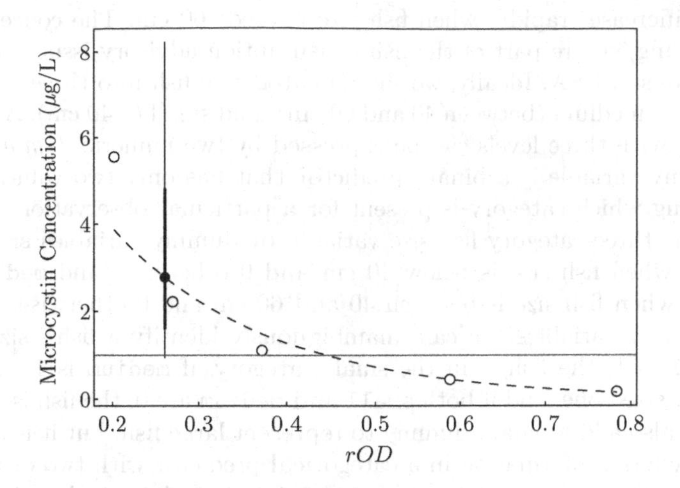

FIGURE 5.21: ELISA standard curve of a log-linear regression model (dashed curve) is fit to five data points (open circles). The predicted microcystin concentration for the water sample in question (black dot) is well above 1, the drinking water standard in Ohio, USA (the shaded horizontal line). The fitted uncertainty (thick vertical line segment) is far smaller than the predictive uncertainty (thin vertical line segment).

5.6 Two-Way ANOVA

5.6.1 ANOVA as a Linear Model

In the PCB in fish example, we initially used a two-sample t-test to compare the log PCB concentrations in large and small fish after the Q-Q plot (Figure 5.1 on page 153) showing a largely multiplicative difference. In carrying out the t-test, we created a variable to separate fish by their size:

```
#### R code ####
aketrout$large<- 0
laketrout$large[laketrout$length>60] <- 1
```

The variable `large` takes values 0 (for small fish) or 1 (large fish). Mathematically, a numeric vector of 0/1 has the same information as a character vector of `small/large` has. As mentioned in the beginning of this chapter, study on lake trout in Lake Michigan also suggested that the initial change in lake trout diet occurs when the fish reaches a size of about 40 cm. In a study of the same data, Qian et al. [2000b] showed that the probability of fish tissue PCB concentration exceeding 1 mg/kg increases rapidly when fish length exceeds 40 cm and the probability of PCB concentration exceeding 1.9

mg/kg increases rapidly when fish size exceeds 60 cm. The concentrations 1 and 1.9 mg/kg are part of the fish consumption advisory issued by the state of Wisconsin, USA. Ideally, we should categorize fish into three groups, large (>60 cm), medium (between 40 and 60 cm), and small (<40 cm). A categorical variable with three levels can be expressed by two numeric *dummy variables.* A dummy variable is a binary predictor that has only two values (1 and 0) indicating which category is present for a particular observation. We can replace the three-category fish size variable by dummy variables `small`, taking value 1 when fish size is below 40 cm and 0 otherwise, and `medium`, taking value 1 when fish size is between 40 and 60 cm and 0 otherwise. With these two dummy variables, we can unambiguously identify a fish's size category. If `small` is 1, the fish is in the small category, if `medium` is 1, the fish is a medium-sized one, and if both `small` and `medium` are 0, the fish is a large one. We can also add a binary dummy to represent large fish, but it is redundant.

In general, information in a categorical predictor with two or more levels can be fully represented by dummy variables and be included in a linear regression model. The number of dummy variables needed is the number of levels of the categorical predictor minus 1. For example, the factor variable `Treatment` in the red mangrove example in section 4.8.4 has four levels – `Control`, `Foam`, `Haliclona`, `Tedania`. This factor variable can be converted into four dummy variables:

```
#### R code ####
attach(mangrove.sponge)
mangrove.sponge$Control <- as.numeric(Treatment=="Control")
mangrove.sponge$Foam <- as.numeric(Treatment=="Foam")
mangrove.sponge$PurpleS <- as.numeric(Treatment=="Haliclona")
mangrove.sponge$RedS <- as.numeric(Treatment=="Tedania")
detach()
```

To compare control to the other three treatments, we fit the following model:

```
#### R Code ####
mangrove.lmDM <- lm(RootGrowthRate ~ Foam+PurpleS+RedS,
                    data=mangrove.sponge)
```

This model is

$$y = \beta_0 + \beta_1 x_{foam} + \beta_2 x_{purple} + \beta_3 x_{red} + \varepsilon$$

The intercept β_0 is the expected value of y when all three predictors are 0 (i.e., not foam, not red fire sponge, and not purple sponge), or when the treatment level is `Control`. When the observation is from the foam treatment, $x_{foam} = 1$ and $x_{purple} = x_{red} = 0$, the expected value of y is $\beta_0 + \beta_1$. The difference between foam treatment mean and control mean is then β_1. Similarly, β_2 and β_3 are the differences in means between purple sponge treatment and control, and between red fire sponge treatment and control, respectively.

```
#### R output ####
display(mangrove.lmDM, 4)

lm(formula = RootGrowthRate ~ Foam + PurpleS + RedS,
              data = mangrove.sponge)
             coef.est coef.se
(Intercept) 0.2371   0.1008
Foam        0.3544   0.1443
PurpleS     0.4911   0.1507
RedS        0.6764   0.1594
---
n = 72, k = 4
residual sd = 0.4620, R-Squared = 0.23
```

This process of converting a factor predictor into dummy variables is exactly what R did when a factor variable is used as a predictor in the linear model function:

```
#### R output ####

display(mangrove.lm, 4)

lm(formula = RootGrowthRate ~ Treatment, data = mangrove.sponge)
                    coef.est coef.se
(Intercept)         0.2371   0.1008
TreatmentFoam       0.3544   0.1443
TreatmentHaliclona 0.4911   0.1507
TreatmentTedania    0.6764   0.1594
---
n = 72, k = 4
residual sd = 0.4620, R-Squared = 0.23
```

5.6.2 More Than One Categorical Predictor

When there is a second factor predictor which will potentially influence the response variable, this second predictor can be transformed into dummy variables as well. In the mangrove example, this second predictor is the locations where the experiments were carried out. These locations are labeled as `bbs`, `etb`, `lcn`, and `lcs`. Different locations may have different conditions that may also affect root growth rate. One way to think about the problem is through multiple regression, with location multiple dummy variables as additional predictors:

```
#### R code ####
attach(mangrove.sponge)
mangrove.sponge$bbs <- as.numeric(Location=="bbs")
```

```
mangrove.sponge$etb <- as.numeric(Location=="etb")
mangrove.sponge$lcn <- as.numeric(Location=="lcn")
mangrove.sponge$lcs <- as.numeric(Location=="lcs")
detach()
mangrove.lmDM2 <- lm(RootGrowthRate ~ Foam+PurpleS+RedS +
                                        etb+lcn+lcs ,
                                        data=mangrove.sponge)
```

and the resulting model uses the controls conducted at location "bbs" as the basis for comparison:

```
#### R output ####
display(mangrove.lmDM2, 4)

lm(formula = RootGrowthRate ~ Foam + PurpleS + RedS +
    etb + lcn + lcs, data = mangrove.sponge)
            coef.est coef.se
(Intercept)  0.1959   0.1378
Foam         0.3508   0.1436
PurpleS      0.4793   0.1503
RedS         0.5968   0.1650
etb          0.2426   0.1507
lcn         -0.0289   0.1575
lcs          0.0116   0.1520
---
n = 72, k = 7
residual sd = 0.4592, R-Squared = 0.28
```

This model is of the form:

$$\begin{aligned} y &= \beta_0 + \beta_1 x_{foam} + \beta_2 x_{purple} + \beta_3 x_{red} + \\ &\quad \beta_4 x_{etb} + \beta_5 x_{lcn} + \beta_6 x_{lcs} + \varepsilon \end{aligned} \tag{5.14}$$

or

$$\begin{aligned} y &= 0.1959 + 0.3508 x_{foam} + 0.4793 x_{purple} + 0.5968 x_{red} + \\ &\quad 0.2426 x_{etb} - 0.0289 x_{lcn} + 0.0116 x_{lcs} + \varepsilon \end{aligned}$$

We are now predicting the root growth using two pieces of information: treatment and location. The expected root growth for control ($x_{foam} = x_{purple} = x_{red} = 0$) at location bbs ($x_{etb} = x_{lcn} = x_{lcs} = 0$) is the intercept ($\beta_0$), the expected root growth for control at location etb ($x_{etb} = 1, x_{lcn} = x_{lcs} = 0$) is $\beta_0 + \beta_4$, and so on. In other words, when there are two factor predictors, to fit a linear regression model is to estimate the expected value of the response at each combination of the two factor predictors. The interpretation of the model coefficients are listed in Table 5.2.

The model assumes that the effects of treatment and location are additive. That is, the difference between foam treatment and control is the same (β_1) no matter which location, and the difference between two locations is the same

TABLE 5.2: Linear model coefficients with two categorical predictors

	bbs	etb	lcn	lcs
Control	β_0	$\beta_0+\beta_4$	$\beta_0+\beta_5$	$\beta_0+\beta_6$
Foam	$\beta_0+\beta_1$	$\beta_0+\beta_1+\beta_4$	$\beta_0+\beta_1+\beta_5$	$\beta_0+\beta_1+\beta_6$
Haliclona	$\beta_0+\beta_2$	$\beta_0+\beta_2+\beta_4$	$\beta_0+\beta_2+\beta_5$	$\beta_0+\beta_2+\beta_6$
Tedania	$\beta_0+\beta_3$	$\beta_0+\beta_3+\beta_4$	$\beta_0+\beta_3+\beta_5$	$\beta_0+\beta_3+\beta_6$

regardless of treatment. This model of equation 5.14 is often referred to as a two-way analysis of variance model and expressed mathematically as

$$y_{ijk} = \beta_0 + \beta_i + \beta_j + \varepsilon_{ijk}$$

where i and j are index of factor predictors and k is the index of individual observations. When β_0 is set to be the expected value of the first particular levels of each of the predictors (the baseline, e.g., `Control, bbs`), β_i is the difference between the means of other levels and the mean of the first level of one predictor and β_j is the difference for the other predictor. Other times, we set β_0 to be the overall mean, and β_i, β_j are known as effects, differences between group means and the overall mean. The additive two-way ANOVA model can be directly fitted using the two-factor predictors:

```
#### R code ####
mangrove.lm2 <- lm(RootGrowthRate ~ Treatment+Location,
    data=mangrove.sponge)
display(mangrove.lm2, 4)

#### R output ####
lm(formula = RootGrowthRate ~ Treatment + Location,
    data = mangrove.sponge)
                     coef.est coef.se
(Intercept)           0.1959   0.1378
TreatmentFoam         0.3508   0.1436
TreatmentHaliclona    0.4793   0.1503
TreatmentTedania      0.5968   0.1650
Locationetb           0.2426   0.1507
Locationlcn          -0.0289   0.1575
Locationlcs           0.0116   0.1520
---
n = 72, k = 7
residual sd = 0.4592, R-Squared = 0.28
```

which is exactly the same as the model using dummy variables. The estimated coefficients for locations have high standard errors. The 95% confidence intervals of the three location differences (mean ± 2 times standard error) include 0, suggesting that location is not an important factor affecting root growth in this study. The formal test for the location effect is the two-way ANOVA, where the total variation in the response is partitioned into the variance due to treatment, variance due to location, and within treatment-location "cell" variance (residual variance).

```
#### R output ####
summary.aov(mangrove.lm2)

          Df Sum Sq Mean Sq F value  Pr(>F)
Treatment    3   4.40    1.47    6.96 0.00039
Location     3   0.81    0.27    1.27 0.29066
Residuals   65  13.71    0.21
```

The two-way ANOVA model can also be fitted using R functions aov:

```
#### R code ####
mangrove.aov2 <- aov(RootGrowthRate ~ Treatment+Location,
    data=mangrove.sponge)
summary(mangrove.aov2 )

#### R output ####
            Df Sum Sq Mean Sq F value  Pr(>F)
Treatment    3   4.40    1.47    6.96 0.00039
Location     3   0.81    0.27    1.27 0.29066
Residuals   65  13.71    0.21
```

The R function model.tables can be used to extract the estimated effects:

```
#### R code and output ####

model.tables(mangrove.aov2)

Tables of effects

 Treatment
    Control      Foam Haliclona Tedania
    -0.3459  0.008444    0.1452  0.3305
rep 21.0000 20.000000   17.0000 14.0000

 Location
        bbs      etb       lcn      lcs
    -0.06204  0.1688 -0.07918 -0.04233
rep 19.00000 19.0000 16.00000 18.00000
```

5.6.3 Interaction

The additive assumption is not always appropriate. That is, the effect of a treatment (e.g., foam) may vary from location to location, or the location difference varies from treatment to treatment. In a two-way ANOVA setting, the interaction is estimated as an adjustment for each treatment-location cell to the additive model estimate. A model with interaction can be expressed in terms of dummy variables. We create dummy variables for each treatment-location combination. Or, we can simplify the notation by using either the product of two factors or using the ":" operator. The R expression:

```
lm(RootGrowthRate ~ Treatment*Location,
                    data=mangrove.sponge)
```

is the same as

```
lm(RootGrowthRate ~ Treatment+Location+
                    Treatment:Location,
                    data=mangrove.sponge)
```

The easiest way to explain the interaction effects is to show the tables of effects:

```
#### R output ####
Tables of effects

 Treatment
    Control      Foam Haliclona Tedania
    -0.3459  0.008444    0.1452  0.3305

 Location
        bbs      etb      lcn      lcs
    -0.06204  0.1688 -0.07918 -0.04233

 Treatment:Location
           Location
Treatment   bbs    etb    lcn    lcs
  Control   -0.074  0.189 -0.052 -0.012
  Foam       0.037 -0.229  0.200 -0.045
  Haliclona  0.129 -0.083 -0.196  0.118
  Tedania   -0.115  0.070  0.096 -0.064
```

The last table suggests how the additive model estimates should be adjusted. For example, the additive model predicts that the mean growth rate for control at location bbs is $(-0.3459 - 0.062 = -0.3512)$ or 0.3512 below overall average. The interaction table suggests that this estimate should be adjusted downwards by 0.074. We can interpret the interaction effects as the differences between the additive model and the treatment-location cell mean. Obviously, we need to use the F-test to see if these differences can be attributed to random noise:

```
#### R output ####
```

```
                    Df Sum Sq Mean Sq F value  Pr(>F)
Treatment            3   4.40    1.47    6.49 0.00076
Location             3   0.81    0.27    1.19 0.32260
Treatment:Location   9   1.04    0.12    0.51 0.85945
Residuals           56  12.66    0.23
```

The interaction effect, just like the location effect, is likely a result of random noise.

5.7 Bibliography Notes

This chapter omitted much of the statistical theories of linear regression models. A good reference for applied regression theories and practices is Weisberg [2005]. The PCB in fish example is based on many published papers, particularly, Stow [1995] and Stow et al. [1995]. These articles documented the details of data collection and analysis. Gelman and Hill [2007] is another good reference on the application of regression analysis in social science. Detailed analysis of the ELISA method for measuring microcystin is presented in Qian et al. [2015a]

5.8 Exercises

1. Huey et al. [2000] studied the development of a fly (*Drosophila subobscura*) that had accidentally been introduced from Europe (EU) into North America (N.A.) around 1980. In Europe, characteristics of the flies' wings follow a "cline" – a steady change with latitude. One decade after introduction, the N.A. population had spread throughout the continent, but no such cline could be found. After two decades, Huey and his team collected flies from 11 locations in western N.A. and native flies from 10 locations in EU at latitudes ranging from 35 to 55 degrees N. They maintained all samples in uniform conditions through several generations to isolate genetic differences from environmental differences. Then they measured about 20 adults from each group. The data set `flies.txt` shows average wing size in millimeters on a logarithmic scale.

   (a) In their paper, Huey et al. used four separate regression models to suggest that female flies from both EU and N.A. have the same wing length – latitude relationship (identical slopes), while the same relationships for male flies from the two continent are close but they were unable to say whether the slopes are the same.

We know that we can create a categorical variable to identify a fly's origin and sex. This variable can be created by pasting the columns `Continent` and `Sex`:

```
Flydata$FlyID <- paste(Flydata$Sex, Flydata$Continent,
                       sep=".")
```

The resulting variable `FlyID` has four levels: `Female.EU`, `Female.N.A.`, `Male.EU`, `Male.N.A`. When fitting a linear regression using:

```
fly.lm <- lm(Wing ~ Latitude * factor(FlyID),
             data=Flydata)
```

we obtain a model with four intercepts and four slopes, and the intercept and slope for the first level of `FlyID` (sorted alphabetically) is estimated and presented as the baseline.

Fit the linear model and interpret the results. Compare your results to the results presented in Huey et al. [2000]. Comment on any differences and why you feel you should use the approach we used here.

(b) The model we fitted here has its limitation. Only the slope and intercept of the first level are presented in the results explicitly. In this case, we will only see the intercept and slope for `Female.EU`, the baseline. Intercepts and slopes for the other three levels are presented in terms of their differences from the baseline. This is set up for hypothesis testing. That is, we can compare whether the slopes for `Female.N.A.`, `Male.EU`, `Male.N.A.` are different from the slope for `Female.EU`. For this particular model, we can directly test whether the difference in slope between `Female.EU` and the slope of `Female.N.A.` is different from 0, but we cannot directly compare the slopes and intercepts for `Male.EU` and `Male.N.A.` To make this comparison, we must set `Male.EU` as the baseline first:

```
Flydata$FlyID <- as.numeric(ordered(Flydata$FlyID,
    levels=c("Male.EU","Male.N.A.",
             "Female.EU","Female.N.A.")))
```

which will change `FlyID` into a numeric variable with integers 1 to 4, and 1 is `"Male.EU"`, 2 is `"Male.N.A."`, 3 is `"Female.EU"`, and 4 is `"Female.N.A."`. Now refit the same model as in (a). Using results from both (a) and (b) to compare whether the slope for male flies from N.A. differs from the slope for male flies from EU, and whether the slope for female flies from N.A. differs from the slope for female flies from EU.

(c) In their paper, the linear regression models have very low R^2 values, and the model we fit has a very high R^2 value. Why? Is our model that much better?

2. Many of the ideas of regression first appeared in the work of Sir Francis Galton on the inheritance of characteristics from one generation to the next. In a paper on "Typical Laws of Heredity," delivered to the Royal Institution on February 9, 1877, Galton discussed some experiments on sweet peas. By comparing the sweet peas produced by parent plants to those produced by offspring plants, he could observe inheritance from one generation to the next. Galton categorized parent plants according to the typical diameter of the peas they produced. For seven size classes from 0.15 to 0.21 inches, he arranged for each of nine of his friends to grow 10 plants from seed in each size class; however, two of the crops were total failures. A summary of Galton's data was published by Karl Pearson (see table 5.3 and the data file `galtonpeas.txt`). Only average diameters and standard deviation of the offspring peas are given by Pearson; sample sizes are unknown.

   (a) Draw the scatter plot of *Progeny* versus *Parent*.

   (b) Assuming that the standard deviations given are population values, compute the regression of *Progeny* on *Parent* and draw the fitted mean function on the scatter plot.

   (c) Galton wanted to know if characteristics of the parent plant such as size were passed on to the offspring plants. In fitting the regression, a parameter value of $\beta_1 = 1$ would correspond to perfect inheritance, while $\beta_1 < 1$ would suggest that the offspring are "reverting" towards "what may be roughly and perhaps fairly described as the average ancestral type" (the substitution of "regression" for "reversion" was probably due to Galton in 1885). Test the hypothesis that $\beta_1 = 1$ versus the alternative that $\beta_1 < 1$.

   (d) In his experiments, Galton took the average size of all peas produced by a plant to determine the size class of the parent plant. Yet for seeds to represent that plant and produce offspring, Galton chose seeds that were as close to the overall average size as possible. Thus, for a small plant, exceptionally large seed was chosen as a representative, while larger, more robust plants were represented by relatively smaller seeds. What effects would you expect these experimental biases to have on (1) estimation of the intercept and slope and (2) estimates of error?

3. Regression analysis is often used as a tool for causal inference. A typical application of regression analysis for casual inference will fit a model using the outcome as the response variable and the potential cause(s) as the predictor(s). Because of the inevitable confounding factors in a

TABLE 5.3: Galton's peas data

Parent Diameter (.01 in)	*Progeny* Diameter (.01 in)	SD
21	17.26	1.988
20	17.07	1.938
19	16.37	1.896
18	16.40	2.037
17	16.13	1.654
16	16.17	1.594
15	15.98	1.763

typical social science study, the regression model will inevitably include other predictors to account for the variability associated with different conditions. Including confounding factors in a regression model is often called controlling in social science. It is this controlling that often leads to the misuse of regression analysis. For example, Kanazawa and Vandermassen [2005] suggested that parent's occupation can predict the likelihood of having boys or girls. Particularly, if the parent's occupation is "systematizing" (e.g., engineering), she/he tends to have more boys, and if the parent's occupation is "empathizing" (e.g., nursing), he/she tends to have more girls. The conclusion was reached by using a regression analysis to the University of Chicago's General Social Survey data. When studying a parent's likelihood of having boys, the article used a regression model of the form:

```
n.boys ~ engineer + n.girls
```

That is, number of boys is predicted by parent's occupation after controlling the number of girls (opposite sex children), plus other predictors (such as income) (Table 1 of Kanazawa and Vandermassen [2005]). The theory was illustrated using the model because the slope for `engineer` is positive and statistically different from 0. In a letter to editor, Gelman [2007] pointed out that this result may be a statistical artifact, and proposed a simulation.

The simulation creates two groups of families (nurses and engineers) of families, each having one or two children. Collectively, child sex ratios of the two groups of families are both one boy to one girl. The difference between a nurse family and an engineer family is how they decide the number of children: nurses will stop at having one child if the first born is a boy, and two children otherwise; engineers will stop at one child with probability 30% and continue on to a second child with probability 70%, regardless of the sex of the first child. In this simulated data, the probability of a boy is exactly 50% for all births; thus the true effect, the difference in sex ratios between engineer and nurse families, is actually zero. Under this simulated model, nurses will have the following distribution of family types: 50% boy, 25% girl-boy, 25% girl-girl. Engineers will have the distribution: 15% boy, 15% girl, 17.5% boy-boy, 17.5% boy-girl, 17.5% girl-boy, 17.5% girl-girl. Use the following scripts to generate 800 families of engineers and 800 families of nurses and fit the regression model:

```
boys.nur <- c(1, 1, 0) ##  # of boys in a nurse family
girls.nur <- c(0, 1, 2)##  # of girls in a nurse family
boys.eng <- c(1, 0, 2, 1, 1, 0)
girls.eng <- c(0, 1, 0, 1, 1, 2)

nur <- sample(1:3, size=800, replace=T,
              prob=c(0.5, 0.25, 0.25))
eng <- sample(1:6, size=800, replace=T,
            prob=c(0.15, 0.15, 0.175, 0.175, 0.175, 0.175))
n.boy.nur <- boys.nur[nur]
n.girl.nur <- girls.nur[nur]
n.boy.eng <- boys.eng[eng]
n.girl.eng <- girls.eng[eng]

### fit a regression
sim.data <- data.frame(n.boys=c(n.boy.eng, n.boy.nur),
                       n.girls=c(n.girl.eng, n.girl.nur),
                       engineer=c(rep(1, 800), rep(0, 800)))
sim.lm <- lm(n.boys ~ engineer + n.girls, data=sim.data)
summary(sim.lm)
```

Is the model result in conflict with the data? Any thoughts on why this would happen (hint: think about the meaning of the slope of `engineer`)?

4. Logarithmic transformations: data set `pollution.csv` (variable definitions are in file `pollution.txt`) contains mortality rates and various environmental factors from 60 U.S. metropolitan areas [McDonald and Schwing, 1973]. For this exercise we shall model mortality rate given nitric oxides, sulfur dioxide, and hydrocarbons as inputs. This model is an extreme oversimplification as it combines all sources of mortality and does not adjust for crucial factors such as age and smoking. We use it to illustrate log transformations in regression.

   (a) Create a scatter plot of mortality rate versus level of nitric oxides. Do you think a linear model will fit these data well? Fit the regression and evaluate a residual plot from the regression.

(b) Find an appropriate transformation that will result in data more appropriate for linear regression. Fit a regression to the transformed data and evaluate the new residual plot.

(c) Interpret the slope coefficient from the model you chose in the previous step.

(d) Now fit a model predicting mortality rate using levels of nitric oxides, sulfur dioxide, and hydrocarbons as inputs. Use appropriate transformations when appropriate. Plot the fitted regression model and interpret the coefficients.

(e) Cross-validate: split the data into two halves and refit the model you chose from the last step to the first half. Use the resulting model to predict the mortality rate using data from the second half. Discuss the result. (A "real" cross-validation often split the data into more, e.g., 20, subsets, and fit the model by leaving one subset out, and make predictions for the set-aside subset.)

(f) Interaction: use conditional plot to investigate potential interaction effects among the three predictors. If you have reason to believe that interaction effects are important, refit the model with these interactions and interpret the fitted model coefficients.

These steps are common for a statistical analysis of observational data. The first four steps are considered exploratory; step 5 verifies a model's predictive capability. Step 6 is often ignored in many studies. In many cases, interaction is more interesting and more informative. Logarithmic transformation is frequently used, but its interpretation is rarely explained clearly in the literature. When explaining the models, you should interpret each model coefficients in plain English.

Write a short report on your findings.

5. The data file `birds.csv` contains measures on breeding pairs of land-bird species collected from 16 islands around Britain over the course of several decades reported in Pimm et al. [1988]. For each species, the data set contains an average time of extinction on those islands where it appeared; the average number of nesting pairs (the average over all islands where the birds appeared, of the nesting pairs per year); the size of the species (categorized as large or small); and migratory status of the species (migrant or resident). It is expected that species with larger numbers of nesting pairs will tend to remain longer before becoming extinct. Pimm et al. were interested in whether, after accounting for number of nesting pairs, size or migratory status has any effect. Furthermore, they were also interested in whether the effect of bird size differs depending on the number of nesting pairs. If any species have unusually small or large extinction times compared to other species with similar values of the predictor variables, it would be useful to point them out. Develop

a regression model to predict the time to extinction using the number of nesting pairs and the two categorical variables as predictors; follow the guidelines outlined in section 5.4. Compare your model to the model reported in Pimm et al. [1988].

6. Increased nitrogen loading to rivers due to human activities was blamed for the rise in abundance of algae in coastal waters. For example, the Neuse River Estuary fishkills in the 1990s were attributed to the increased algae due to increased nitrogen loading from the Neuse River basin, especially from concentrated animal farming in Eastern North Carolina. Conclusions like this are often made from analyzing cross-sectional or metadata. Cole et al. [1993] assembled a data set including many large river systems in the world to study the effect of human activities on nitrogen loadings to rivers (data file `nitrogen.csv`). They included nine variables in the data set: (1) discharge (`DISCHARGE`), the estimated annual average discharge of the river into ocean (in m^3/sec); (2) runoff (`RUNOFF`), the estimated annual average runoff from watershed (in liters/(sec×m^2)); (3) precipitation (in cm/yr) (`PREC`); (4) area of watershed (in km^2) (`AREA`); (5) population density (in people/km^2) (`DENSITY`); (6) nitrate concentration (in μ mol/L) (`NO3`); (7) nitrate export (`EXPORT`), the product of runoff and nitrate concentration; (8) deposition (`DEP`), nitrate loading from precipitation – product of precipitation and precipitation nitrate concentration; and (9) nitrate precipitation (`NPREC`), the concentration of nitrate in wet precipitation at sites located near the watersheds (in μ mol/(sec×km^2)). In the paper, the authors used nitrate concentration (`NO3`) and nitrate export (`EXPORT`) as measures of human impact on rivers. The authors looked at these two response variables separately to determine whether the impact of human activities in river nitrogen level is a result of direct pollutant discharge input to the river or discharged indirectly through atmospheric pollution. They suggested that population density in the watershed can be used as a measure of direct discharge and nitrate precipitation can be served as a measure of indirect discharge.

   Fit a regression model for each of the two response variables and discuss whether anthropogenic impact of river nitrogen is more through direct discharge or through indirect atmospheric deposition. Note that the two factors are correlated and their effects are unlikely to be additive.

7. The data file `co2data.csv` contains monthly mean atmospheric CO_2 concentrations measured at Mauna Loa, Hawaii, from January 1959 to December 2003. Atmospheric CO_2 concentrations show a distinct seasonal pattern, reflecting the annual cycle of plant activities. The data set has four columns: `CO2` (monthly CO_2 concentrations in ppm), `mon` (calendar month), `year`, and `months` (months since January 1959). The data plot (Figure 6.30) showed an unmistakable increasing temporal trend in CO_2 concentrations. In this question you are asked to quantify

the magnitude of this increase. A frequently used statistical method for estimating temporal trend is to fit a linear regression model of the CO_2 concentration against a time variable (e.g., number of months since a starting point). In this case, the column `months` in the data set is such a time variable.

(a) Fit a simple regression model using `CO2` as the response variable and `months` as the predictor variable. Quantify the temporal trend (monthly or annual rate of increase in CO_2 concentration) and discuss the potential problems of the model. (Hint: plot the residuals against `months`.)

(b) Refit the model by using `mon` as a second (factor) predictor and explain the temporal trend in CO_2 concentrations.

In both models, the residuals versus fitted plot shows a systematic pattern. What may be the cause of such pattern?

8. Cigarette smoking is believed to cause lung cancer. In a study, Fraumeni [1968] showed that cigarette smoking is also associated with cancers of the urinary tract. The study collected per capita numbers of cigarettes smoked (*actually, sold*) in 43 states and the District of Columbia in 1960 together with death rates per thousand population from various forms of cancer (data in file `smoking.txt`).

   (a) Is a simple linear model using per capita cigarette consumption (`CIG`) to predict the mortality rate of lung cancer (`LUNG`) appropriate? What about a simple linear model for bladder cancer mortality rate (`BLAD`)?

   (b) Identify (name the state) the two potential outliers in cigarette consumption data.

   (c) Are the two outliers influential to the models developed?

   (d) Should the two outliers be deleted (why)?

9. Mercury contamination of edible freshwater fish poses a direct threat to our health. Large mouth bass were studied in 53 Florida lakes to examine the factors that influence the level of mercury contamination. Water samples were collected from the surface of the middle of each lake in August 1990 and then again in March 1991. The pH level, the amount of chlorophyll, calcium, and alkalinity were measured in each sample. The average of the August and March values were used in the analysis. Next, a sample of fish was taken from each lake with sample sizes ranging from 4 to 44 fish, each used to measure a mercury concentration and the average concentration is reported. The authors of the study [Lange et al., 1993] indicated that alkalinity is best predictor of the average mercury concentration and a linear model was suggested (data in `HgBass.txt`).

(a) Fit a simple linear regression model using the average mercury concentration as the response variable and alkalinity as the predictor variable. Discuss the model fit.

(b) Use log transformation on one or both variables to see if the model in (a) can be improved. (You should try all three and select the one that you think is the best.)

(c) The smallest level of mercury concentration that the measuring instrument can detect is 0.04 (ppm). A data point with value known to be below a certain number is "censored." Any level below the detection limit of 0.04 ppm was set to 0.02 ppm. This, of course, makes the average mercury concentration less accurate. To account for the inaccuracy, the authors also reported the minimum and maximum average Hg concentrations. The minimum (column `min`) is obtained by replacing all censored values with 0 and the maximum (column `max`) is calculated by replacing censored values with the detection limit of 0.04 ppm. Discuss using a simple drawing on what impact of this treatment of "censored value" would have on the estimated slope and intercept. Note that each observed average mercury concentration value is an average of concentrations from 4 to 44 fish samples and we have no way of telling which average mercury concentration values were affected by how many censored data points.

Chapter 6

Nonlinear Models

6.1	Nonlinear Regression		209
	6.1.1	Piecewise Linear Models	220
	6.1.2	Example: U.S. Lilac First Bloom Dates	226
	6.1.3	Selecting Starting Values	229
6.2	Smoothing		240
	6.2.1	Scatter Plot Smoothing	240
	6.2.2	Fitting a Local Regression Model	243
6.3	Smoothing and Additive Models		245
	6.3.1	Additive Models	245
	6.3.2	Fitting an Additive Model	248
	6.3.3	Example: The North American Wetlands Database	250
	6.3.4	Discussion: The Role of Nonparametric Regression Models in Science	254
	6.3.5	Seasonal Decomposition of Time Series	259
		6.3.5.1 The Neuse River Example	261
6.4	Bibliographic Notes		267
6.5	Exercises		269

6.1 Nonlinear Regression

One important assumption in a linear regression model is that the relationship between the response variable and predictor variables is linear. When the relationship is not linear, we often must transform either the response variable or the predictor variable or both. In many situations we have subject matter knowledge that provides the basis for selecting a specific model form, which may be impossible to transform to linear.

For example, in Section 5.3.2 (on page 156) we fitted a log linear model of the form:

$$\log(PCB) = \beta_0 + \beta_1 Yr + \epsilon$$

This model was based on a commonly used mechanistic model describing changes in chemical concentrations. The model assumes that the change in concentration over time is proportional to the concentration:

$$\frac{dPCB}{dt} = -kPCB$$

Assuming the initial PCB concentration to be PCB_0, the PCB concentration at time t is given by

$$PCB_t = PCB_0 e^{-kt}$$

Taking logarithm on both sides of the model we obtain the linear model `lake.lm1` in Section 5.3.2, where the intercept β_0 is equivalent to $\log(PCB_0)$ and the slope β_1 is $-k$ and t is years since 1974. The implication of this model is that PCBs are continually declining, at an ever-slowing rate, toward a concentration of zero.

Given the same data set, we can directly fit this exponential model. We can use the same least squares method for estimating model coefficients. The least squares method estimates model coefficients that minimize the residual sum of squares:

$$RSS = \sum \left([PCB_{ti} - \widehat{PCB}_{ti}]^2 \right)$$

where $\widehat{PCB}_{ti} = \widehat{PCB}_0 e^{-\hat{k}t_i}$. To minimize RSS, we set the partial derivatives of RSS with respect to PCB_0 and k to 0 and solve for $\widehat{PCB}_0$ and k.

In general, we have a specific function describing the relationship between a response variable y and a set of predictors x parameterized using coefficients θ:

$$y = f(x, \theta) + \epsilon$$

The least squares method is to estimate the coefficients θ such that the sum of residual squares

$$SS = \sum([y_i - f(x_i, \theta)]^2)$$

is minimized.

In R, the commonly used function for a nonlinear regression is `nls`, which takes the formula y $\sim$ `f(x,` θ`)` as the first argument:

```
nls.obj <- nls (formula, data, start, control, algorithm,
    trace, subset, weights, na.action, model,
    lower, upper, ...)
```

For example, we can fit the exponential model of PCB in fish:

```
#### R code ####
pcb.exp <- nls(pcb ~ pcb0*exp(-k*(year-1974)),
    data=laketrout, start=list(pcb0=10, k=0.08))
```

A set of initial starting values of model coefficients is supplied as input. These initial starting values are selected based on the data plot and the log linear models we studied in the previous chapter. The model results are presented in a similar form as the linear regression model results:

```
#### R output ###
summary(pcb.exp)
```

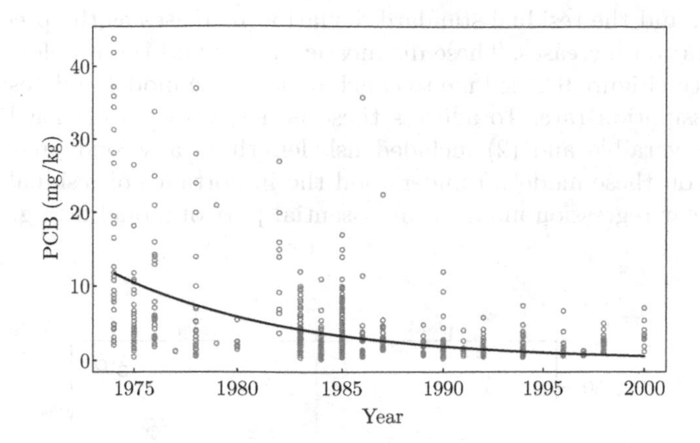

FIGURE 6.1: Nonlinear PCB model – A nonlinear exponential model is fitted to the PCB in the fish data.

```
Formula: pcb ~ pcb0 * exp(-k * (year - 1974))

Parameters:
     Estimate Std. Error t value Pr(>|t|)
pcb0 11.76215    0.64432   18.25   <2e-16
k     0.11487    0.00885   12.98   <2e-16

Residual standard error: 5.14 on 629 degrees of freedom

Number of iterations to convergence: 6
```

This model is similar to the initial log-linear regression model (`lake.lm1`) in Section 5.3.2. The intercept of model `lake.lm1` is comparable to $\log(PCB_0)$, and the slope is comparable to $-k$. The difference is in the probabilistic assumption. The model `lake.lm1` assumes that PCB has a log-normal distribution, while the current nonlinear model `pcb.exp` assumes that PCB has a normal distribution. The fitted model is shown in Figure 6.1.

The standard errors of the estimated model coefficients are based on the assumption that model residuals are independent random variables from a normal distribution with mean 0 and a constant standard deviation. We should evaluate the model fit by using the same graphical methods described in Chapter 5. Because the aim of fitting a nonlinear regression model is to find the coefficients of a mechanistic model, many don't emphasize the statistical side of the problem. However, as we have seen in the PCB in the fish example, statistical analysis is often the key to uncover potential problems associated with data. Diagnostic plots (Figures 6.2 to 6.4) suggest that the residuals are unlikely to follow a normal distribution, the residuals are unlikely to be independent, and the residual standard deviation increases as the predicted PCB concentration increases. These diagnostic figures and the problem of fish size imbalance (Figure 9.2) led me to conclude that the model underestimates the PCB dissipation rate. To address these issues, we used (1) log PCB as the response varaible and (2) included fish length as a second predictor. From working on these models, I understood the importance of residual analysis of a nonlinear regression model as an essential part of model fitting.

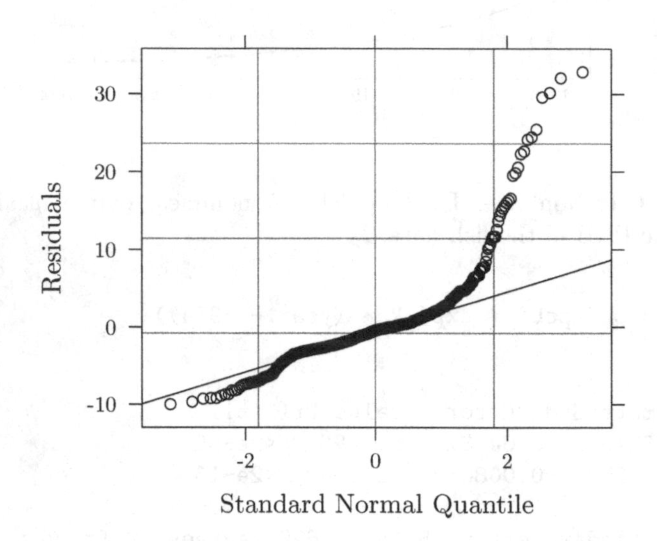

FIGURE 6.2: Nonlinear PCB model residuals normal Q-Q plot – The residual normal Q-Q plot suggests that the residuals are unlikely to have a normal distribution.

When using log-transformed PCB concentration as the response, the exponential model becomes a simple linear model. To be consistent, I will refit the model `pcb.exp` using function `nls`:

```
pcb.exp <- nls(log(pcb) ~ log(pcb0) -k * (year-1974),
               data=laketrout,
               start=list(pcb0=10, k=0.08))
```

The estimated coefficients are very different from the same estimated without log transformation.

```
#### R output ####
> summary(pcb.exp)

Formula: log(pcb) ~ log(pcb0) - k * (year - 1974)
```

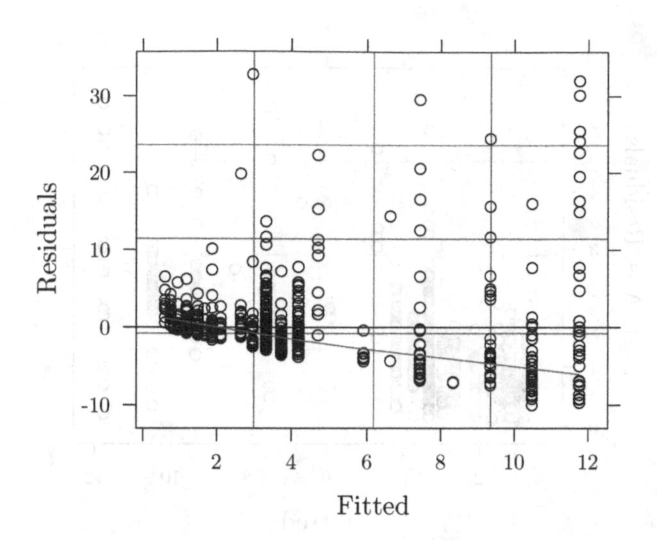

FIGURE 6.3: Nonlinear PCB model residuals vs. fitted PCB – The nonlinear model residuals are plotted against the fitted PCB.

```
Parameters:
     Estimate Std. Error t value Pr(>|t|)
pcb0 4.941199   0.357559   13.82   <2e-16 ***
k    0.059903   0.005525   10.84   <2e-16 ***
---

Residual standard error: 0.8784 on 629 degrees of freedom

Number of iterations to convergence: 4
Achieved convergence tolerance: 5.434e-07
  (15 observations deleted due to missingness)
```

In addition to the exponential model, Stow et al. [2004] used three additional nonlinear models to account for the apparent leveling off of the PCB concentration over time. The first alternative model is an exponential decay model with a nonzero asymptote:

$$PCB_t = PCB_0 e^{-kt} + PCB_a + \epsilon$$

where PCB_a is the asymptotic PCB concentration (mg/kg), a concentration that will persist perpetually. This model also implies a continual decline of PCB concentrations, at a slowing rate, but toward a positive asymptotic concentration. Implicit in this model is the idea that there are two effective PCB

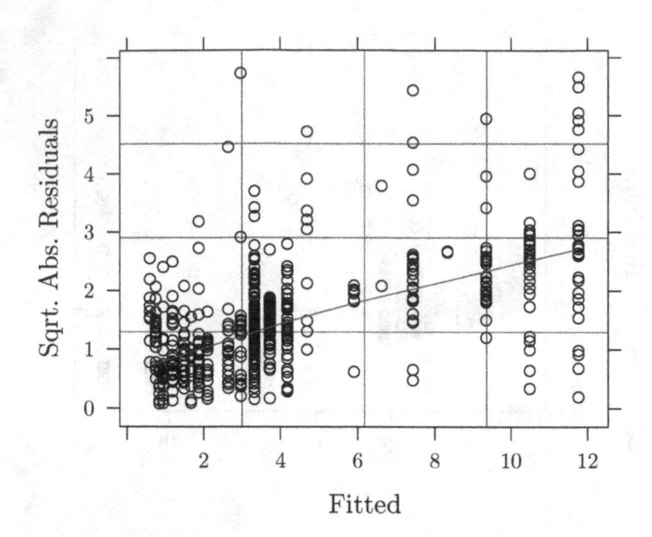

FIGURE 6.4: Nonlinear PCB model residuals S-L plot – The residual S-L plot suggests that the residual standard deviation increases as the predicted PCB increases.

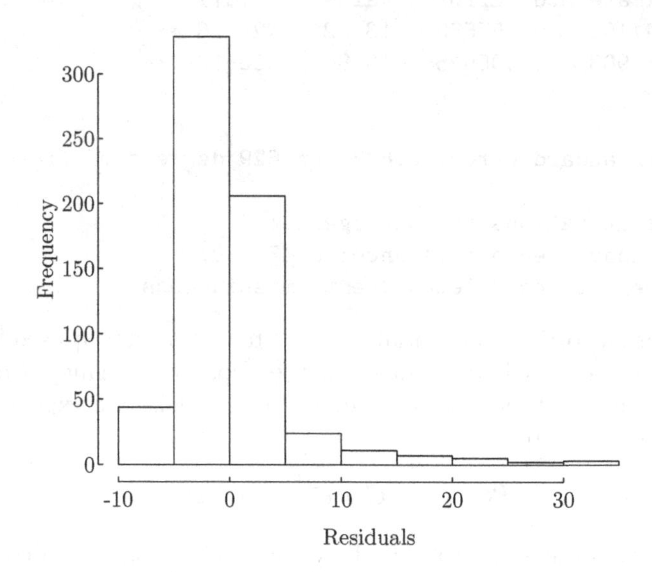

FIGURE 6.5: Nonlinear PCB model residuals distribution – The residual histogram suggests that the residual distribution is highly skewed.

sources supplying the food web, one of which declines rapidly through time and one of which is relatively stable. In R, this model can be fitted as:

```
#### R code ####
pcb.exp2 <- nls(log(pcb) ~ log(pcb0*exp(-k*(year-1974))+pcba),
                data=laketrout,
                start=list(pcb0=10, k=0.08, pcba=1))

#### R output ####
> summary(pcb.exp2)

Formula: log(pcb) ~ log(pcb0 * exp(-k * (year - 1974)) + pcba)

Parameters:
      Estimate Std. Error t value Pr(>|t|)
pcb0    6.2264     0.8386    7.42  3.6e-13
k       0.2479     0.0401    6.18  1.1e-09
pcba    1.6941     0.1369   12.38  < 2e-16

Residual standard error: 0.862 on 645 degrees of freedom
```

The two components of this model can be interpreted as a rapidly decaying component (with a decay rate of 0.25 1/year, or about 22% annual reduction) and a constant component (pcba). The model implies that the part of PCB that can be naturally removed from the lake ecosystem has long disappeared, and what is remaining will be there forever.

The second alternative is a double exponential decay model:

$$PCB_t = PCB_{01}e^{-k_1 t} + PCB_{02}e^{-k_2 t} + \epsilon$$

where k_1 and k_2 represent the coefficients of two distinct decay processes.

```
#### R code ####
pcb.exp3 <- nls(log(pcb) ~ log(pcb01*exp(-k1*(year-1974))+
                           pcb02*exp(-k2*(year-1974))),
                data=laketrout,
                start=list(pcb01=10, pcb02=2, k1=0.24,
                           k2=0.00002))


#### R output ####
summary(pcb.exp3)

Formula: log(pcb) ~ log(pcb01 * exp(-k1 * (year - 1974)) +
    pcb02 * exp(-k2 * (year - 1974)))

Parameters:
       Estimate Std. Error t value Pr(>|t|)
```

```
pcb01    6.7750     0.8577     7.90  1.2e-14
pcb02    0.7339     0.7644     0.96   0.3374
k1       0.1741     0.0528     3.29   0.0010
k2      -0.0359     0.0434    -0.83   0.4086

Residual standard error: 0.862 on 644 degrees of freedom
```

The two components of this model can be interpreted as a rapidly decaying component (with a decay rate of 0.17 1/year, or about 16% annual reduction) and a slowly increasing component (with an annual rate of $\sim$ 3.5%). The second component is somewhat difficult to interpret. If we limit the lower bounds of all coefficients to be 0:

```
#### R code ####
pcb.exp3 <- nls(log(pcb) ~ log(pcb01*exp(-k1*(year-1974))+
                          pcb02*exp(-k2*(year-1974))),
                data=laketrout, algorithm="port",
                lower=rep(0,4),
                start=list(pcb01=10, pcb02=2, k1=0.24,
                          k2=0.00002))

#### R output ####
summary(pcb.exp3)

Formula: log(pcb) ~ log(pcb01 * exp(-k1 * (year - 1974)) +
    pcb02 * exp(-k2 * (year - 1974)))

Parameters:
      Estimate Std. Error t value Pr(>|t|)
pcb01   6.2264     0.9869    6.31  5.2e-10
pcb02   1.6941     0.9338    1.81   0.0701
k1      0.2479     0.0956    2.59   0.0097
k2      0.0000     0.0251    0.00   1.0000

Residual standard error: 0.862 on 644 degrees of freedom
```

The second rate coefficient k_2 becomes 0, reducing to the first alternative model (model `pcb.exp2` on page 215).

A third alternative is a mixed-order model:

$$PCB_t = PCB_0^{1-\phi} - kt(1-\phi)^{(1/1-\phi)} + \epsilon$$

where PCB_0 is the initial concentration, k is the reaction coefficient, and ϕ is the order of the reaction, and all are treated as unknowns. This model is a generalization of the exponential decay model – it assumes the rate of PCB concentration change over time is proportional to the concentration raised to a power ϕ:

$$\frac{dPCB}{dt} = -kPCB^{\phi}$$

Because the exponential model is a special case of the third model (when $\phi = 1$), a function that includes both the general equation and the special case is necessary to avoid dividing by 0:

```
#### R code ####
mixedorder <- function(x, b0, k, theta){
    LP1 <- LP2 <- 0
    if(theta==1){
        LP1 <- log(b0) - k*x
    } else {
        LP2 <- log(b0^(1-theta) - k*x*(1-theta))/(1-theta)
    }
    return( LP1 + LP2)
}

pcb.exp4 <- nls(log(pcb) ~
    mixedorder(x=year-1974, pcb0, k, phi),
    data=laketrout, start=list(pcb0=10, k=0.0024, phi=3.5))
```

This is an example of constructing a more complicated nonlinear model by writing a function.

```
#### R output ####
summary(pcb.exp4)

Formula: log(pcb) ~
    mixedorder(x = year - 1974, pcb0, k, phi)

Parameters:
     Estimate Std. Error t value Pr(>|t|)
pcb0 10.66409    1.72227    6.19  1.1e-09
k     0.00642    0.00271    2.37    0.018
phi   3.28579    0.35091    9.36  < 2e-16

Residual standard error: 0.861 on 645 degrees of freedom
```

These four models were used in Stow et al. [2004] to evaluate the percent reduction of PCB from 2000 to 2007. The three alternative models seem to perform better than the simple exponential model (Figure 6.6). To evaluate the predicted percent reduction from 2000 to 2007, we can compare the predicted concentrations for years 2000 and 2007. Just as in the linear model, we have both fitted and predictive uncertainties in a nonlinear regression. However, there is no analytical solution for the predictive uncertainty. In Chapter 9, we will explore the use of simulation to quantify the predictive uncertainty of nonlinear regression models. Here, we estimate the expected percent reduction using the four alternative models and approximate their uncertainty using residual standard errors.

In R, we use the function `predict` to obtain the predicted PCB concentrations for years 2000 and 2007. Analytical solution of the predictive uncertainty for a nonlinear regression model is unavailable. As a result, when using `predict` with a nonlinear regression object, the arguments `se.fit` and `interval` are ignored.

```
exp1.pred <- predict(pcb.exp,
                     new=data.frame(year=c(2000, 2007)))
exp2.pred <- predict(pcb.exp2,
                     new=data.frame(year=c(2000, 2007)))
exp3.pred <- predict(pcb.exp3,
                     new=data.frame(year=c(2000, 2007)))
exp4.pred <- predict(pcb.exp4,
                     new=data.frame(year=c(2000, 2007)))
```

Because all four models were fit to the log-transformed PCB concentration, we can approximate the uncertainty of the predicted concentrations for the year 2000 and 2007 by ignoring the uncertainty of the estimated model coefficients. However, the uncertainty of the percent change is harder to derive. Although the log difference of PCB concentrations $\log(PCB_{2007})-\log(PCB_{2000})$ is equal to the log ratio $\log(PCB_{2007}/PCB_{2000})$, the predicted log concentrations for the two years are strongly correlated and the standard deviation of the log ratio cannot be easily calculated. The results using simulation (Chapter 9) are in Figure 6.7. These four models provide vastly different estimates of the level of PCB reduction from 2000 to 2007. The first alternative model (exponential decay model with a nonzero asymptote) and the third alternative model (mixed order) predict a reduction close to 0 with very high confidence, while the second alternative model (the double exponential model) cannot predict the level of reduction with certainty. The simple exponential model predicts a level reduction well above the 25% strategic goal. These models, however, failed to account for the PCB–fish length relationship. Because of the imbalance of fish size (Figure 9.2 on page 398), the three alternative models are unlikely to be appropriate. Consequently, fish length must be considered as a predictor.

We discussed that the log PCB and fish length relationship is unlikely to be linear in the previous chapter. We discussed how to fit a linear regression model using the categorical variable `size` to fit, essentially, two linear models, one for large fish (length larger than 60 cm) and one for small fish. The fitted model `lake.lm7` in Section 5.3.6 has five coefficients:

```
#### R code ####
lake.lm7 <- lm(log(pcb) ~ I(year-1974) +
```

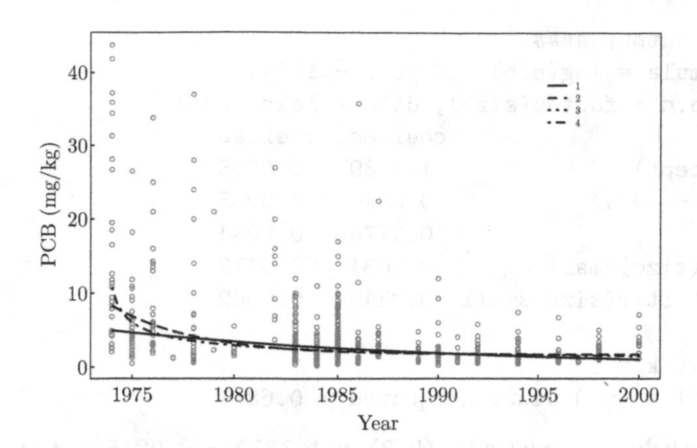

FIGURE 6.6: Four nonlinear PCB models – Four competing models are fitted to the lake trout PCB data.

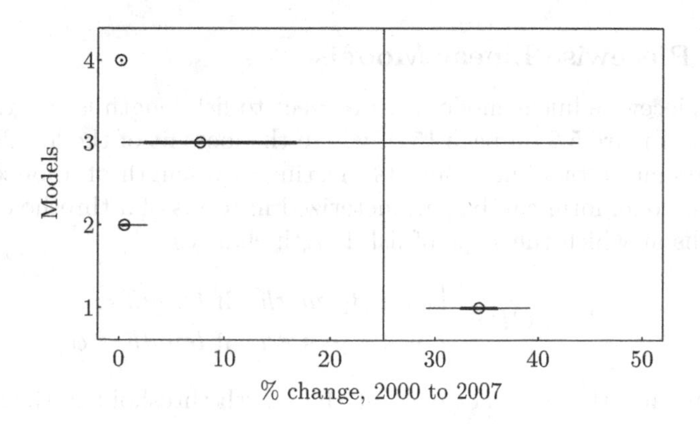

FIGURE 6.7: Simulated % PCB reduction from 2000 to 2007 – The four competing models predict very different reduction between 2000 and 2007. The thin lines are the 95% interval and the thick lines are the 50% interval. The vertical line shows the EPA's 2002 strategic goal of a 25% reduction.

```
        len.c * factor(size), data=laketrout)
display(lake.lm7, 4)

#### R output ####
lm(formula = log(pcb) ~ I(year - 1974) +
    len.c * factor(size), data = laketrout)
                         coef.est coef.se
(Intercept)               1.7389   0.0588
I(year - 1974)           -0.0846   0.0035
len.c                     0.0776   0.0044
factor(size)small        -0.0631   0.0779
len.c:factor(size)small -0.0345    0.0062
---
n = 631, k = 5
residual sd = 0.5422, R-Squared = 0.68
```

For large fish, the model is `log(PCB) = 1.7389 - 0.0846yr + 0.0776Len.c`. For small fish the model is `log(PCB) = (1.7389-0.0631) - 0.0846yr + (0.0776-0.0345)Len.c`. Because these two models are of the same form, when predicting PCB concentration for a fish with length just below the threshold of 60 cm (e.g., 59.999), we use the model for small fish. But for a fish with size just above the threshold (e.g., 60.001), we use the model for the large fish. For year 1974, the predicted log concentrations are 1.53 and 1.56 for small and large fish, respectively. The discontinuity is undesirable. To fix this problem, we will use the piecewise linear regression model.

6.1.1 Piecewise Linear Models

The piecewise linear model with respect to fish length is suggested by the data plot (Figure 5.2 on page 154), where the loess fit of the log PCB against length resembles two line segments meeting at a length of approximately 60 cm. This model form can be parameterized in terms of a threshold parameter on lengths at which the slope of fish length changes:

$$\log PCB = \begin{cases} \alpha_1 + \beta_1 length & \text{if } length < \phi \\ \alpha_2 + \beta_2 length & \text{if } length \geq \phi \end{cases} \tag{6.1}$$

To ensure that the two lines meet at the length threshold ϕ, the model coefficients must satisfy the following condition:

$$\alpha_1 + \beta_1\phi = \alpha_2 + \beta_2\phi \tag{6.2}$$

In addition to the usual intercepts and slopes, we need also to estimate the length threshold ϕ. In general, a piecewise linear model can be parameterized parsimoniously as

$$y = \beta_0 + [\beta_1 + \delta \cdot I(x - \phi)])(x - \phi) + \epsilon \tag{6.3}$$

where

$$I(z) = \begin{cases} 0 & \text{if } z \leq 0 \\ 1 & \text{if } z > 0 \end{cases}$$

and δ is the difference in slope between the two line segments. The model defined by equation 6.3 is nonlinear with four parameters $\beta_0, \beta_1, \delta, \phi$. To simplify the model expression, we define the piecewise regression model as

$$f_{hockey}(x|\beta_0, \beta_1, \delta, \phi) = \beta_0 + [\beta_1 + \delta \cdot I(x - \phi)](x - \phi)$$

Because the first order derivative of a piecewise linear model is not continuous, this model can cause problems in many commonly used numerical optimization programs. To avoid this problem, the piecewise linear model is slightly modified by adding a small quadratic curve at the threshold point to make the first order derivative continuous. The quadratic line (Figure 6.8) is estimated by setting its slopes at two ends to be the same as the slopes of the two lines. An R function `hockey` is then written as:

```
#### R code ####
hockey <-
function(x,alpha1,beta1,beta2,brk,eps=diff(range(x))/100,
    delta=T) {
    ## alpha1 is the intercept of the left line segment
    ## beta1 is the slope of the left line segment
    ## beta2 is the slope of the right line segment
    ## brk is location of the break point
    ## 2*eps is the length of the connecting quadratic piece
        x <- x-brk
        if (delta) beta2 <- beta1+beta2
        x1 <- -eps
        x2 <- +eps
        b <- (x2*beta1-x1*beta2)/(x2-x1)
        cc <- (beta2-b)/(2*x2)
        a <- alpha1+beta1*x1-b*x1-cc*x1^2
        alpha2 <- - beta2*x2 +(a + b*x2 + cc*x2^2)
        lebrk <- (x <= -eps)
        gebrk <- (x >= eps)
        eqbrk <- (x > -eps & x < eps)
        result <- rep(0,length(x))
        result[lebrk] <- alpha1 + beta1*x[lebrk]
        result[eqbrk] <- a + b*x[eqbrk] + cc*x[eqbrk]^2
        result[gebrk] <- alpha2 + beta2*x[gebrk]
        result
}
```

As a result, the piecewise linear model in equation 6.3 can be written in R formula as

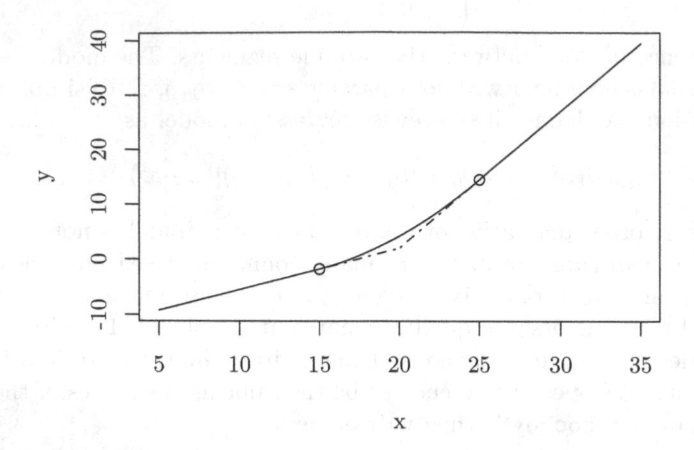

FIGURE 6.8: The hockey stick model – The piecewise regression (or hockey stick) model is reparameterized to create continuous first order partial derivatives. The two straight lines are connected by a small piece of quadratic line.

```
log(PCB) ~ hockey(length, beta0, beta1, delta, theta)
```

and a nonlinear regression model can be fit:

```
#### R code ####
lake.nlm1<-nls(log(pcb)~hockey(length, beta0, beta1, delta, phi),
    start=list(beta0=.6, beta1=0.07, delta=0.03, phi=60),
    data=laketrout, na.action=na.omit)
```

The estimated model coefficients are summarized by the function `summary`:

```
#### R Output ####
> summary(lake.nlm1)

Formula: log(pcb) ~ hockey(length, beta0, beta1, delta, phi)

Parameters:
      Estimate Std. Error t value Pr(>|t|)
beta0   0.5506     0.1316    4.18  3.3e-05
beta1   0.0253     0.0062    4.08  5.2e-05
delta   0.0470     0.0086    5.47  6.4e-08
phi    59.9896     2.3241   25.81  < 2e-16

Residual standard error: 0.751 on 627 degrees of freedom
```

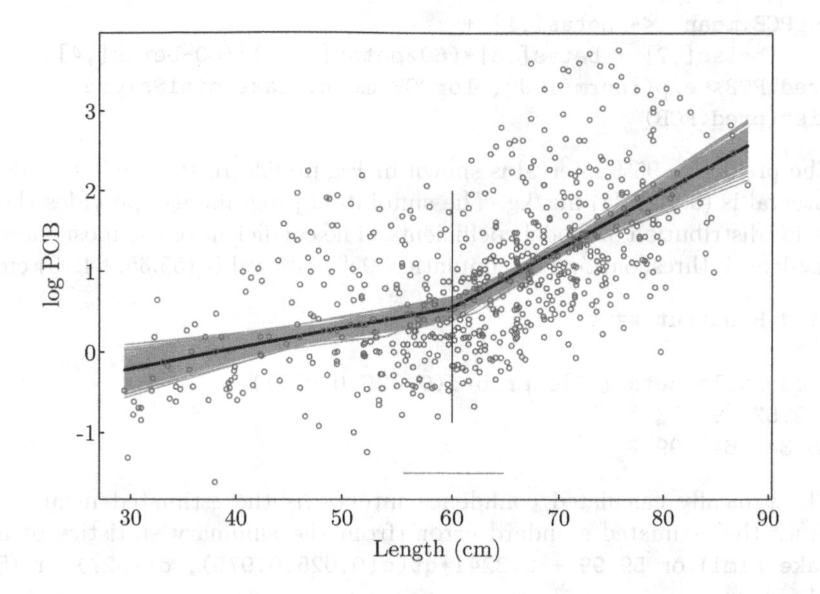

FIGURE 6.9: The piecewise linear regression model – The log PCB concentration and fish length relationship is modeled by a piecewise linear regression model (the black line). Uncertainty in the estimated model coefficients is summarized by using a simulation program to generate possible variations from the fitted mean model (the gray lines). The vertical line is the 95% prediction interval for fish with lengths of 60 cm. The short horizontal line is the 95% interval of the estimated threshold.

The fitted model is shown in Figure 6.9. In addition to the fitted model, a simulation program is written for nonlinear regression models (Chapter 9) such that the uncertainty we have on model coefficients is represented by model coefficient values generated from their respective sampling distributions. Each randomly generated set of model coefficients is used to draw a gray line in Figure 6.9 to represent the propagation of uncertainty in the estimated model coefficients to the fitted values.

The simulation program is similar to the function `sim` (package `arm`) discussed in Section 9.2 (page 390). The gray lines in Figure 6.9 reflect only the uncertainty in the fitted mean model. The model's predictive standard deviation can also be estimated easily using simulation. For example, to predict the PCB concentration distribution of fish with the size of 60 cm, we can first use simulation to generate multiple sets of model coefficients and model error standard deviations, and then generate individual random values of log PCB:

```
#### R Code ####
lake.sim1 <- sim.nls (lake.nlm1, 1000)
betas <- lake.sim1$beta
```

```
logPCB.mean  <- betas[,1] +
    (betas[,2] + betas[,3]*(60>betas[,4]))*(60-betas[,4])
pred.PCB<-exp(rnorm(1000, logPCB.mean, lake.sim1$sigma))
hist(pred.PCB)
```

The predictive 95% interval is shown in Figure 6.9. In the original scale, this interval is (0.41 7.34) mg/kg. The simulation program also provides the posterior distribution of model coefficients. The coefficient of the most interest is the length threshold ϕ. The simulation 95% interval is (55.35, 64.70) cm:

```
#### R Output ####

> quantile(betas[,4], prob=c(0.025,0.975))
  2.5%  97.5%
55.349 64.699
```

The typically calculated confidence interval is the estimated mean $\pm \sim 2$ times the estimated standard error (from the summary statistics of model `lake.nlm1`) or `59.99 + 2.3241*qt(c(0.025,0.975), df=627)` or (55.43, 64.55) cm.

To model changes in log PCB over the year, we can add an additive year effect term to the piecewise linear model:

```
#### R code ####
lake.nlm2 <- nls(log(pcb) ~  beta1*(year-1974) +
        hockey(length, beta0, beta2, delta, phi),
        start=list(beta0=.6, beta1= - 0.08,
                beta2=0.07, delta=0.03, phi=60),
        data=laketrout, na.action=na.omit)

#### R output
> summary(lake.nlm2)

Formula: log(pcb) ~ beta1 * (year - 1974) +
    hockey(length, beta0, beta2, delta, phi)

Parameters:
      Estimate Std. Error t value Pr(>|t|)
beta0  1.59857    0.15338   10.42  < 2e-16
beta1 -0.08459    0.00353  -23.98  < 2e-16
beta2  0.04309    0.00436    9.88  < 2e-16
delta  0.03457    0.00622    5.55  4.1e-08
phi   60.71681    2.26282   26.83  < 2e-16

Residual standard error: 0.542 on 626 degrees of freedom
```

The model is very similar to model `lake.lm7`, which sets the threshold at 60

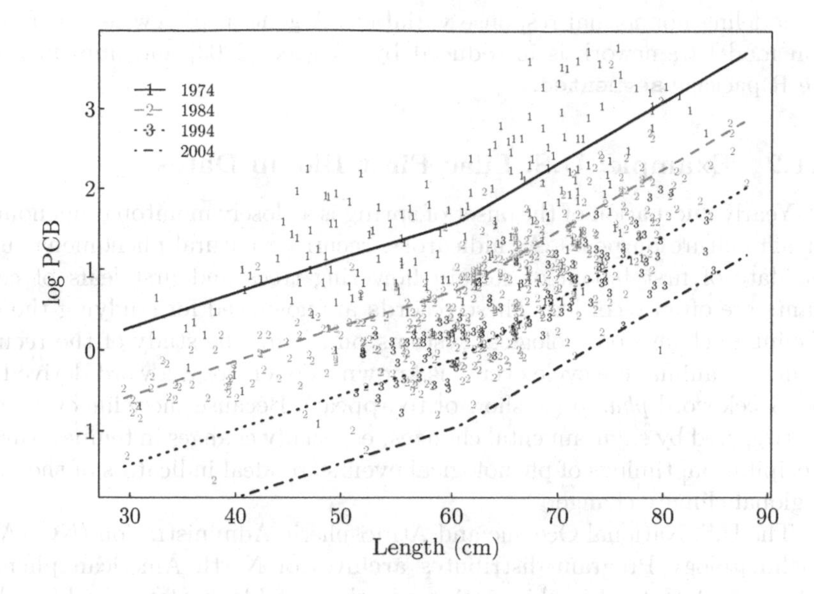

FIGURE 6.10: The estimated piecewise linear regression model for selected years – The log PCB concentration and fish length relationship is estimated for four selected years.

cm based on a study of the lake trout diet. The difference between the two models is that the piecewise linear model is continuous at the threshold, while the linear model `lake.lm7` is not. The nonlinear regression model allows us to estimate the threshold and its standard deviation. From the model output, we expect a lake trout to shift diet at a size about 61 cm, with a 95% confidence interval between 56 and 65 cm. With the added term to model the changes over time, the relationship between PCB and fish length must be presented for a given year. Figure 6.10 shows the estimated log PCB–length relationship for years 1974, 1984, and 1994. The model estimated relationship for 2004 is a prediction. The data points are plotted using three different numbers. The data points labeled as "1" are those measured between 1974 and 1983, those labeled as "2" are from 1984 to 1993, and "3" are from 1994 to 2000.

The piecewise linear model is frequently used for assessing a threshold effect. In environmental management, many attempted to use this model to detect changes in the ecosystem response to environmental changes. The estimated threshold is often used as the basis for setting an environmental criterion. Many such applications were carried out using complicated numerical methods. For example, Qian and Richardson [1997] used a Gibbs sampler for estimating coefficients of a simple piecewise linear regression model. The same computation can be easily carried out using the hockey stick model introduced in this section. The advantage of using the Bayesian approach is the flexibility of modeling nonnormal response variables. A general piecewise linear regression model framework is introduced by Muggeo [2003] and implemented in the R package `segmented`.

6.1.2 Example: U.S. Lilac First Bloom Dates

Yearly fluctuation of the onset of spring is a closely monitored phenomenon for agriculture purposes. As indicators, recurring natural phenomena such as the date of first bloom of certain flowering trees and first leafs of certain plants are often recorded. These records are now used for studying the effect of climate change on biological systems on earth. The study of the recurring plant and animal life cycle events is known as phenology, a word derived from the Greek word *phaino* (to show or to appear). Because these life cycle events are triggered by environmental changes, especially changes in temperature and precipitation, timings of phenological events are ideal indicators of the impact of global climate change.

The U.S. National Oceanic and Atmospheric Administration (NOAA) Paleoclimatology Program distributes archives of North American phenology data. The data used in this section are the first bloom dates for lilac shrubs (*Syringa chinensis* and *Syringa vulgaris*)[Schwartz and Caprio, 2003]. Monitoring of growing season changes may provide insight into ecosystem responses to global climate change. The data set includes first bloom dates collected from more than 1100 stations from the 1950s to 2000s. The longest single station record is about 40 years. This data set was used by many to document the changes in the onset of spring in the Northern Hemisphere. Schwartz et al. [2006] used a simple regression model, where the first bloom date is the response variable and year is the predictor variable. The resulting linear regression model has a negative slope, suggesting that the first bloom starts earlier every year.

Because global climate change is largely due to anthropogenic emission of greenhouse gases through the consumption of fossil fuel, a likely model for the change of the first bloom date over time is a piecewise linear model, consisting two line segments to represent the two temporal periods in the data: a line with a 0 slope before the effect of global climate change can be reflected in the data and a line with a negative slope after the effect has changed the plant behavior. The threshold can be seen as the time when climate change impact started. This threshold pattern can be shown in the time series plots of the first bloom dates for individual monitoring stations (Figure 6.11). If the threshold pattern is the likely pattern of ecosystem response to climate change, the simple regression slopes as used by Schwartz et al. [2006] are likely to underestimate the impact. Furthermore, the estimated thresholds can provide additional information for understanding how plants respond to the change and how this response varies spatially when comparing threshold values from different sites.

Of the more than 1100 stations, those with more than 30 years of data were

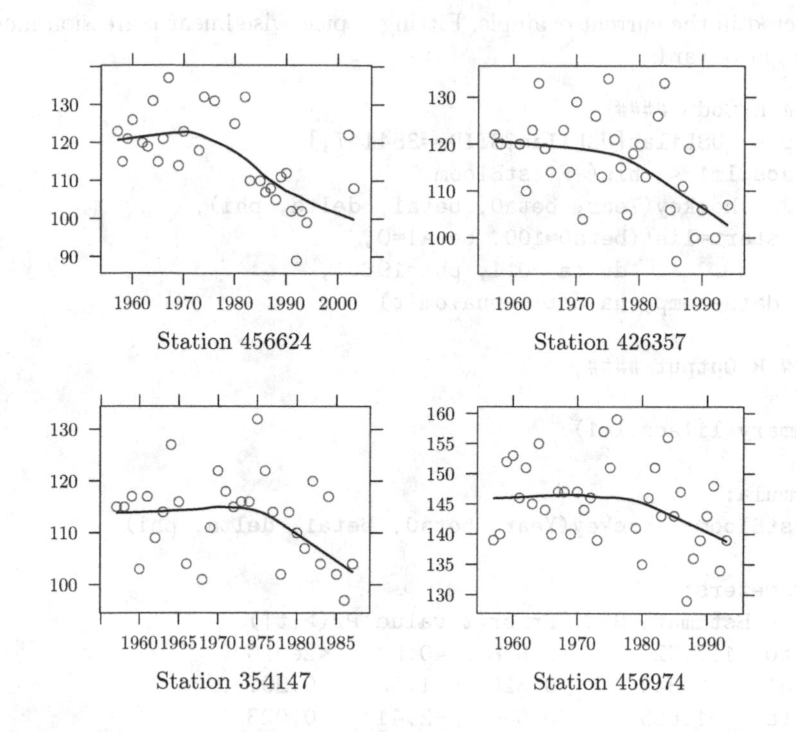

FIGURE 6.11: First bloom dates of lilacs in North America – First bloom dates reported from 4 stations are plotted against year. The loess line in each panel suggests that a threshold model is likely.

TABLE 6.1: Estimated piecewise linear model coefficients (and their standard error) for the data used in Figure 6.11

	Stations			
coefficients	354147	456974	456624	426357
β_0	118(2.9)	148(2.5)	123(5)	117(4.4)
β_1	0.34(0.32)	0.18(0.27)	0.13(0.53)	-0.14(0.27)
δ	-1.7(0.7)	-0.78(0.45)	-0.95(0.6)	-1.48(0.98)
ϕ	1975(3.5))	1976(6.7)	1974(8)	1983(4.9)

selected in the current example. Fitting a piecewise linear regression model is straightforward:

```
#### R Code ####
temp <- USLilac[USLilac$STID==354147,]
lilacs.lm1 <- nls( FirstBloom ~
      hockey(Year, beta0, beta1, delta, phi),
   start=list(beta0=100, beta1=0,
              delta=-0.1, phi=1980),
   data=temp, na.action=na.omit)

#### R Output ####

summary(lilacs.lm1)

Formula:
FirstBloom ~ hockey(Year, beta0, beta1, delta, phi)

Parameters:
      Estimate Std. Error t value Pr(>|t|)
beta0  117.920      2.878   40.97   <2e-16
beta1    0.344      0.320    1.08    0.291
delta   -1.655      0.686   -2.41    0.023
phi   1975.185      3.482  567.31   <2e-16
---
```

The estimated slope before the threshold point is 0.344 with a standard error of 0.32, suggesting that the average first bloom dates did not change over time before the threshold. The estimated slope difference is -1.655 suggesting a negative slope after the threshold, representing that the first bloom has come earlier every year (more than one day annually). The threshold occurred around 1975. For the other three sites represented in Figure 6.11, the estimated coefficients are in Table 6.1

The estimated model coefficients from these four sites located in the west (Utah, site 426357) and northwest (Washington, sites 456974 and 456624;

Oregon, site 354147) of the United States (among those with at least 30 years of observations) show some issues of analyzing such data.

1. Models should be fit to data from individual monitoring sites separately. The difference in first bloom dates can be attributed, in part, to geographical factors. Sites at a higher altitude or latitude tend to have late blooms than sites at a lower altitude or latitude. When combining data together (Figure 6.12), the distinct pattern shown in Figure 6.11 is obscured. This data set is an example of *longitudinal data*, in which individual units are measured repeatedly over time.

2. It is likely that the slope before the threshold (β_1) is always not statistically different from 0, a reasonable result indicating relatively stable timing of the onset of spring before the impact of global climate change were felt by the ecosystem.

3. The response to the climate change varies reflected in the variations in δ and ϕ. Can this variation be explained by local (geographical and climatic) conditions? As an indicator of change, the threshold (ϕ) variation can be used to study timing of the impact of climate change. The variation in δ can be used to study the magnitude of the impact. It appears that δ is negatively correlated with altitude, the higher the altitude, the smaller the absolute value of δ, indicating that the magnitude of the impact decreases as we move higher in altitude.

4. Although different locations may have different temporal patterns, observations from these locations should be analyzed together using either the traditional longitudinal data analysis tools or the multilevel regression model (Chapter 10).

6.1.3 Selecting Starting Values

In fitting a nonlinear regression model, selecting a set starting values can be difficult. Inappropriately selected starting values can lead to unintended results. In our previous examples, we relied on the trial-and-error method. In many cases, we can examine the data and pick values by "eye-balling." For example, in the piecewise linear model of PCB in fish example, the PCB–fish length scatter plot shows that the change point is very close to 60 cm. We can also use the scatter plot to perform a rough calculation of the two slopes. In other cases with more complicated model forms, guessing is often not feasible. I will use the ELISA example in Chapter 5 (Section 5.5.1 on page 191) to illustrate the thought process of deriving appropriate starting values.

In the ELISA example (Chapter 5), we used the log-linear regression model. Another recommended model, used by the Toledo Water Department,

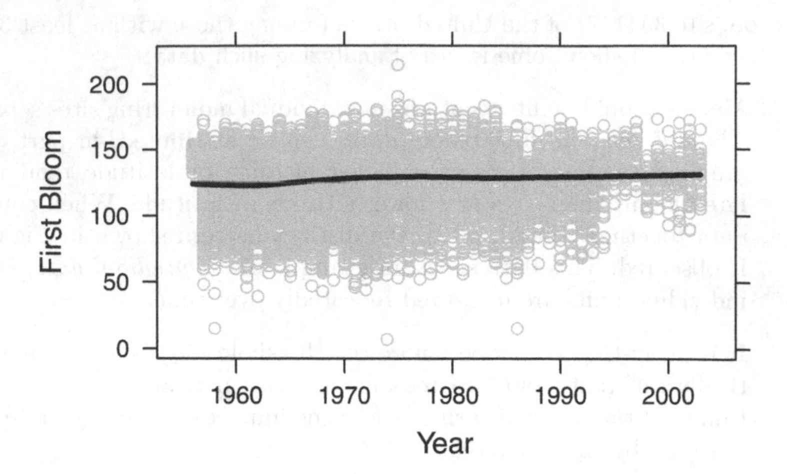

FIGURE 6.12: All first bloom dates of lilacs in North America – First bloom dates from all available stations are plotted against year. The threshold pattern in Figure 6.11 is no longer obvious.

is a four-parameter logistic (FPL) regression model:

$$y = \alpha_4 + \frac{\alpha_1 - \alpha_4}{1 + \left(\frac{x}{\alpha_3}\right)^{\alpha_2}} + \epsilon \tag{6.4}$$

where y is the observed optical density and x is the standard solution concentration. This is a sigmoid function (often known as the Richards function) with the left and right bounds of α_1 and α_4, respectively, and the shape of the curve controlled by the other two parameters. There are many examples of FPL in the literature [Richards, 1959, Ritz and Streibig, 2005]. The Toledo Water Department uses this FPL standard curve to measure microcystin concentrations. The data used to fit the FPL are measured optical densities from six standard solutions. The data from the test that led to the "Do Not Drink" advisory on August 1, 2014 are shown in Figure 6.13.

```
#### R Code ####
## standard solution MC concentrations
stdConc8.1<- rep(c(0,0.167,0.444,1.11,2.22,5.55), each=2)
## measured OD
Abs8.1.0<-c(1.082,1.052,0.834,0.840,0.625,0.630,
            0.379,0.416,0.28,0.296,0.214,0.218)
plot(Abs8.1.0 ~ stdConc8.1, xlab="MC Concentration",
     ylab="Optical Density")
```

The data show the range of the optical density is between 0.2 and 1.08,

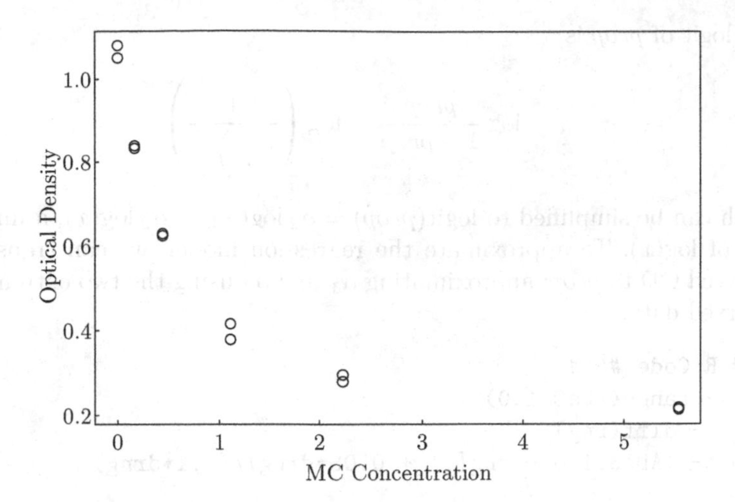

FIGURE 6.13: Data used to fit the standard curve in an ELISA test performed on August 1, 2015.

and it is reasonable to set initial value of α_1 to a value below 0.2 and the initial value of α_4 to be above 1.1. But the initial values of the other two parameters are not easy to derive from the figure. For example, the initial values 1.1, 1.1, 0.5, and 0.2 worked:

```
#### R Output ####
> TM1<-nls(Abs8.1.0~(al1-al4)/(1+(stdConc8.1/al3)^al2)+al4,
           start=list(al1=1.1, al2=1.1, al3=0.5, al4=0.2))
```

But initial values 0.1, 1.2, 0.15, and 1.2 resulted in an error message:

```
#### R Output ####
> TM1<-nls(Abs8.1.0~(al1-al4)/(1+(stdConc8.1/al3)^al2)+al4,
           start=list(al1=.1, al2=1.1, al3=0.15, al4=1.2))
Error in numericDeriv(form[[3L]], names(ind), env) :
  Missing value or an infinity produced when evaluating the model
```

To avoid aimless search of appropriate initial values, we often explore a simpler function or a linear approximation of the nonlinear function and use linear regression to derive the initial values. In this case, we can rearrange equation (6.4) in the following steps:

First, define a fraction term *prop* as

$$prop = \frac{y - \alpha_4}{\alpha_1 - \alpha 4} = \frac{1}{1 + \left(\frac{x}{\alpha_3}\right)^{\alpha_2}}$$

The logit of *prop* is

$$\log \frac{prop}{1 - prop} = \log \left(\frac{1}{\left(\frac{x}{\alpha_3}\right)^{\alpha_2}} \right)$$

which can be simplified to $\text{logit}(prop) = \alpha_2 \log(\alpha_3) - \alpha_2 \log(x)$, a linear function of $\log(x)$. To approximate the regression model, we can transform the observed OD to *prop*: approximating α_1 and α_4 using the two extremes of the observed data:

```
#### R Code ####
rng <- range(Abs8.1.0)
drng <- diff(rng)
prop <- (Abs8.1.0 - rng[1] + 0.05*drng)/(1.1*drng)
```

Fitting a linear regression of `lm(I(log(prop/(1-prop)))~log(stdConc8.1+eps)` we have the estimated intercept and slope. The initial value of α_2 is the negative slope and $\log(\alpha_3)$ is negative intercept divided by slope. Note that one standard solution has a concentration of 0. We can either delete the two 0 concentration replicates, or add a small positive number (e.g., `eps<-0.001`) to x to avoid taking log of 0. Remember that the purpose of this regression model is to obtain starting values for model coefficients, not for accurately estimating these coefficients.

In this case, the intercept and slope are -1.17 and -0.64. Our initial values for α_2 and α_3 are 0.64 and 0.16. Now we are ready to fit the model:

```
TM1 <- nls(Abs8.1.0 ~ (al1-al4)/(1+(stdConc8.1/al3)^al2)+al4,
           start=list(al1=1.5, al2=0.64, al3=0.16, al4=0.15))

> summary(TM1)

Formula: Abs8.1.0~(al1-al4)/(1+(stdConc8.1/al3)^al2)+al4

Parameters:
    Estimate Std. Error t value Pr(>|t|)
al1  1.06556    0.01011 105.363 7.36e-14 ***
al2  1.12384    0.06056  18.557 7.33e-08 ***
al3  0.45203    0.02461  18.371 7.93e-08 ***
al4  0.16150    0.01753   9.212 1.56e-05 ***
---

Residual standard error: 0.01439 on 8 degrees of freedom

Number of iterations to convergence: 6
Achieved convergence tolerance: 3.99e-07
```

This process is tedious and can often be automated by using a *self-starter* function. Before introducing the self-starter function, we introduce the `plinear` algorithm for nonlinear models with conditionally linear parameters. Often a nonlinear regression model has coefficients that are conditionally linear. For example, the four-parameter logistic function of equation (6.4) can be expressed as:

$$y = \alpha_4 + (\alpha_1 - \alpha_4)\frac{1}{1 + \left(\frac{x}{\alpha_3}\right)^{\alpha_2}}$$

The nonlinear term of the model is $1/(1 + (x/\alpha_3)^{\alpha_2})$ and α_4 and $\alpha_1 - \alpha_4$ are the two conditionally linear parameters. That is, conditional on fixed parameters α_2, α_3, the model is linear function of $z = 1/(1 + (x/\alpha_3)^{\alpha_2})$ $(y = \alpha_4 + (\alpha_1 - \alpha_4)z)$. The `plinear` algorithm separates the nonlinear parameters (α_2, α_3) from the conditionally linear parameters. For fixed values of nonlinear parameters, the conditional linear parameters can be estimated using linear regression. As a result, only starter values for the nonlinear parameters are needed. To specify a conditional linear model, the model formula leaves the conditionally linear parameters out. For example, for a model with one conditional linear slope parameter (i.e., $\theta f(x, \beta)$), the model is specified as `nls(y~f(x,beta),algorithm='plinear')`. The estimated linear parameter is denoted `.lin`. When an intercept term is also present, that is, $\theta_0 + \theta_1 f(x, \beta)$, the model is specified using `cbind()`: `nls(y~cbind(1,f(x,beta)),algorithm='plinear')`. For the ELISA data, we have:

```
#### R Code ####
> TM2 <- nls(Abs8.1.0 ~ cbind(1, 1/(1+(stdConc8.1/al3)^al2)),
             start=list(al2=0.64, al3=0.16),
             algorithm="plinear")

#### R Output ####
> summary(TM2)
Formula: Abs8.1.0 ~ cbind(1, 1/(1 + (stdConc8.1/al3)^al2))

Parameters:
      Estimate Std. Error t value Pr(>|t|)
al2    1.12384    0.06056  18.557 7.33e-08 ***
al3    0.45203    0.02461  18.371 7.93e-08 ***
.lin1  0.16150    0.01753   9.212 1.56e-05 ***
.lin2  0.90406    0.02137  42.301 1.08e-10 ***
---

Residual standard error: 0.01439 on 8 degrees of freedom

Number of iterations to convergence: 5
Achieved convergence tolerance: 3.444e-06
```

The coefficient `.lin1` is the estimated α_1 and `.lin2` is the estimated $\alpha_1 - \alpha_4$. By using the `plinear` algorithm in this case, we avoided using data range or scatter plot to determine initial values of two parameters. We now automate the process by constructing a self-starter function.

A self-starter function consists of two parts, a mean function and a function for calculating initial values. The mean function specifies the regression model and its derivatives in an R function. The derivative functions of the nonlinear model are used in typical nonlinear regression fitting algorithms. The first order partial derivatives of the FPL are:

$$
\begin{aligned}
\frac{\partial y}{\partial \alpha_1} &= \frac{1}{1+\left(\frac{x}{\alpha_3}\right)^{\alpha_2}} \\
\frac{\partial y}{\partial \alpha_2} &= -\frac{\alpha_1-\alpha_4}{\left(1+\left(\frac{x}{\alpha_3}\right)^{\alpha_2}\right)^2} \cdot \left(\frac{x}{\alpha_3}\right)^{\alpha_2} \cdot \log\left(\frac{x}{\alpha_3}\right) \\
\frac{\partial y}{\partial \alpha_3} &= \frac{\alpha_1-\alpha_4}{\left(1+\left(\frac{x}{\alpha_3}\right)^{\alpha_2}\right)^2} \cdot \left(\frac{x}{\alpha_3}\right)^{\alpha_2} \cdot \frac{\alpha_2}{\alpha_3} \\
\frac{\partial y}{\partial \alpha_4} &= \frac{\left(\frac{x}{\alpha_3}\right)^{\alpha_2}}{1+\left(\frac{x}{\alpha_3}\right)^{\alpha_2}}
\end{aligned}
$$

The mean function returns a calculated function value, with the derivatives as attributes:

```
## the mean function
#### R Code ####
 fplModel <- function(input, al1, al2, al3, al4){
    .x <- input+0.0001
    .expr1 <- (.x/al3)^al2
    .expr2 <- al1-al4
    .expr3 <- 1 + .expr1
    .expr4 <- .x/al3
    .value <- al4 + .expr2/.expr3
    .grad <- array(0, c(length(.value), 4L),
                   list(NULL, c("al1","al2","al3","al4")))
    .grad[,"al1"] <- 1/.expr3
    .grad[,"al2"] <- -.expr2*.expr1*log(.expr4)/.expr3^2
    .grad[,"al3"] <- .expr1*.expr2*(al2/al3)/.expr3^2
    .grad[,"al4"] <- .expr1/(1+.expr1)
    attr(.value, "gradient") <- .grad
    .value
}
```

The initial values function implements the process of fitting a simple linear regression model for initial values of α_2 and α_3 and using the `plinear` algorithm for initial values of α_1 and α_4:

```
#### R Code ####
```

```
## initial values
fplModelInit <- function(mCall, LHS, data){
    xy <- sortedXyData(mCall[["input"]], LHS, data)
   if (nrow(xy) < 5) {
        stop("too few distinct input values to
              fit a four-parameter logistic")
    }
    rng <- range(xy$y)
    drng <- diff(rng)
    xy$prop <- (xy$y-rng[1]+0.05*drng)/(1.1*drng)
    xy$logx <- log(xy$x+0.0001)
    ir <- as.vector(coef(lm(I(log(prop/(1-prop)))~logx,
                                  data=xy)))
    pars <- as.vector(coef(nls(y~cbind(1,
                                   1/(1+(x/exp(lal3))^al2)),
                                   data=xy,
                                   start=list(al2=-ir[2],
                                              lal3=-ir[1]/ir[2]),
                                   algorithm="plinear")))
    value <- c(pars[4]+pars[3], pars[1], exp(pars[2]),
               pars[3])
    names(value) <- mCall[c("al1","al2","al3","al4")]
    value
}
```

These two functions are assembled into a self-starter function:

```
SSfpl2 <- selfStart(fplModel, fplModelInit,
                    c("al1", "al2", "al3", "al4"))
```

Using SSfpl2, the same model is fit as follows:

```
#### R Code ####
> TM1 <- nls(Abs8.1.0 ~ SSfpl2(stdConc8.1,al1, al2, al3, al4))

#### R Output ####
> summary(TM1)

Formula: Abs8.1.0 ~ SSfpl2(stdConc8.1, al1, al2, al3, al4)

Parameters:
    Estimate Std. Error t value Pr(>|t|)
al1  1.06563    0.01012 105.250 7.42e-14 ***
al2  1.12409    0.06062  18.542 7.38e-08 ***
al3  0.45205    0.02458  18.390 7.87e-08 ***
al4  0.16153    0.01753   9.214 1.56e-05 ***
---
```

```
Residual standard error: 0.01439 on 8 degrees of freedom

Number of iterations to convergence: 2
Achieved convergence tolerance: 1.058e-06
```

The model in equation (6.4) is limited to non-negative predictor values, adequate for fitting ELISA standard curves. A more general form can be derived for log-transformed concentration variable $z = \log(x)$. Substituting $x = e^z$ into equation (6.4), we have a different specification of the four-parameter logistic function:

$$y = A + \frac{B - A}{1 + e^{\frac{z_{mid} - z}{scal}}} \tag{6.5}$$

A self-starter function for regression model based on equation (6.5) is implemented in function `SSfpl` (from package `stats`). To fit the FPL in equation (6.5), we will omit the observations with 0 concentration:

```
#### R Code ####
tmp <- stdConc8.1!=0
TM2 <- nls(Abs8.1.0 ~ SSfpl(log(stdConc8.1),
                            A, B, xmid, scal),
                            subset=tmp)

#### R Output ####
Formula: Abs8.1.0 ~ SSfpl(log(stdConc8.1), A, B, xmid, scal)

Parameters:
     Estimate Std. Error t value Pr(>|t|)
A     0.97787    0.04355  22.452 5.11e-07 ***
B     0.18361    0.01673  10.974 3.40e-05 ***
xmid -0.63690    0.08692  -7.327  0.00033 ***
scal  0.75067    0.07940   9.455 7.97e-05 ***
---
Signif. codes:  0 '***' 0.001 '**' 0.01 '*' 0.05 '.' 0.1 ' ' 1

Residual standard error: 0.01202 on 6 degrees of freedom

Number of iterations to convergence: 0
Achieved convergence tolerance: 1.255e-07
```

When using equation (6.4) we assume that the logistic curve is defined in the concentration scale, while the logistic curve is defined in the log concentration scale when using equation (6.5). It is, therefore, natural to ask which one is more appropriate. In Toledo's report issued after the water crisis, the fitted model is the same as the model `TM1`, suggesting that equation (6.4) was used. However, the fitted model is presented graphically in log concentration scale.

Without further explanation from the ELISA kit manufacturer, I assume that there is a confusion on which model is more appropriate. As in a linear regression problem, we investigate the question based on whether the residuals conform to model assumptions. These assumptions are: normality, variance homogeneity, and independence. Both models fit the data well as shown in the fitted versus observed scatter plots (Figures 6.14 and 6.15). Because of the small sample size, checking normality is difficult, although the normal Q-Q plots show approximately normal residuals for both models. The difference between the two models shows in the residual variances. The model `TM1` shows an increasing residual variance as predicted concentration increases, while the model `TM2` does not. In balance, these diagnostic plots seem to suggest that the FPL of equation (6.5) is more appropriate for data from this test. The conclusion is tentative because of the small sample size used in fitting the model. As an exercise, readers are asked to compare the two model forms using data from the other five tests carried out during the Toledo Water Crisis.

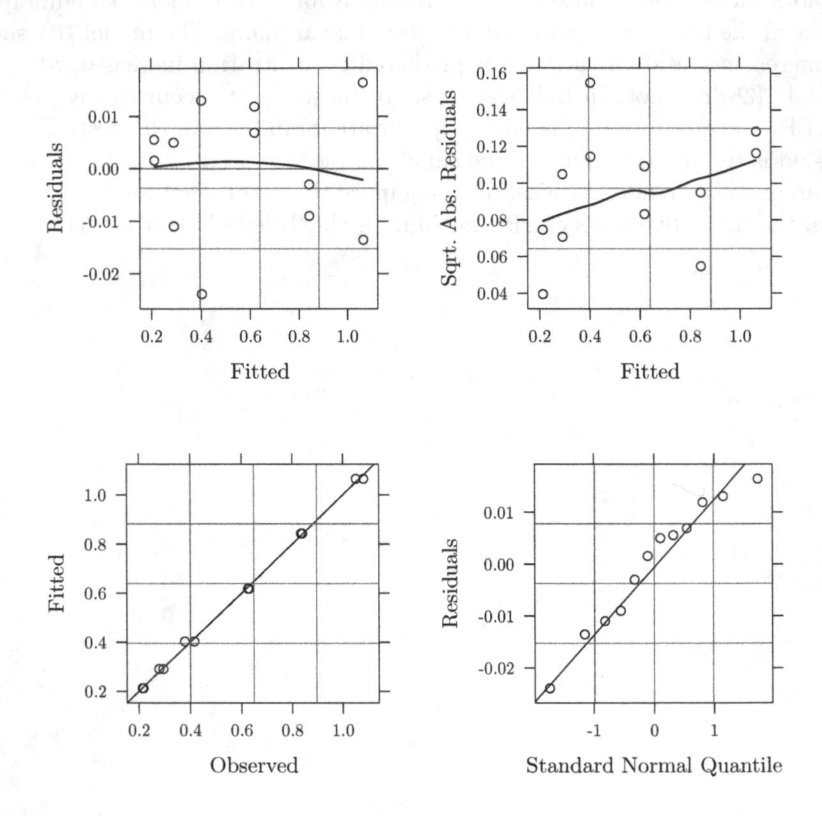

FIGURE 6.14: Diagnostic plots of model based on equation (6.4) consists of a scatter plot of fitted against observed (lower left), residual normal Q-Q plot (lower right), scatter plot of residuals against fitted (upper left), and residual S-L plot (upper right).

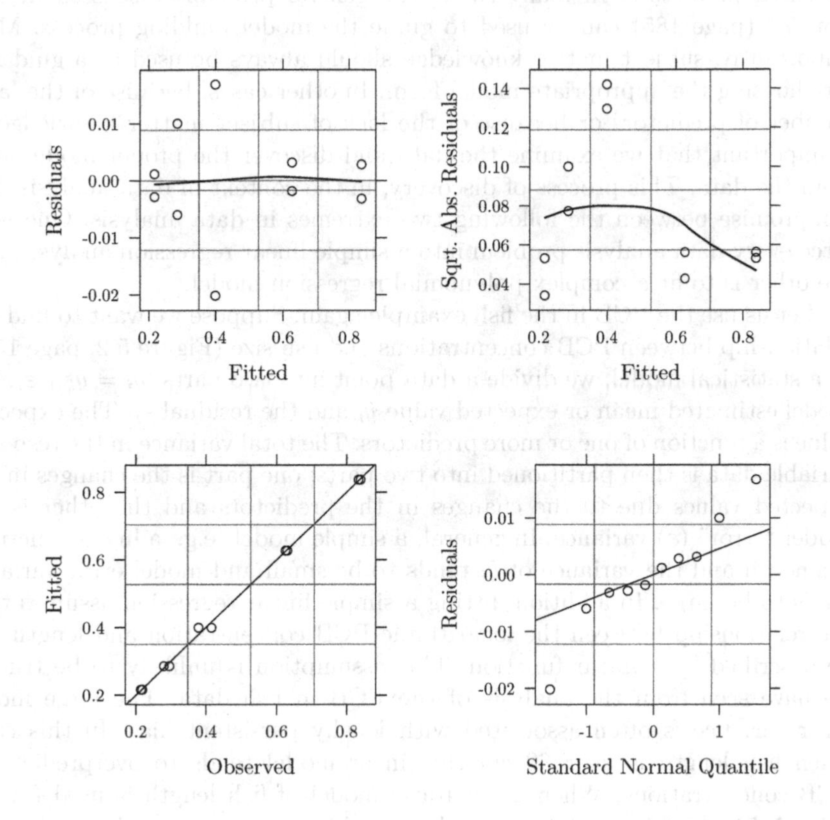

FIGURE 6.15: Diagnostic plots for model based on equation (6.5).

6.2 Smoothing

6.2.1 Scatter Plot Smoothing

In many exploratory studies, the exact model form of a response variable is unknown. The objective of a modeling study is to find the likely functional form to describe the relationship between a response variable and one or more potential predictors. In many cases, the general principle discussed in Section 5.4 (page 185) can be used to guide the model building process. More importantly, subject matter knowledge should always be used as a guidance for choosing the appropriate model form. In other cases, because of the large number of predictors or because of the lack of subject matter knowledge, it is important that we examine the data and discover the proper model form from the data. This process of discovery, in the context of data analysis, is a compromise between the following two extremes in data analysis. One is to force every data analysis problem into a simple linear regression analysis, and the other is to fit a complex polynomial regression model.

Let us use the PCB in the fish example again. Suppose we want to find the relationship between PCB concentrations and fish size (Figure 5.2, page 154). In a statistical model, we divide a data point into two parts: $y_i = \hat{y}_i + \varepsilon_i$, the model estimated mean or expected value $\hat{y}_i$ and the residual ε_i. The expected value is a function of one or more predictors. The total variance in the response variable data is then partitioned into two parts: one part is the changes in the expected values due to the changes in the predictors and the other is the model "error" (ε) variance. In general, a simple model (e.g., a linear function) is smooth and the variance of $\hat{y}_i$ tends to be small and model error variance tends to be large. In addition, fitting a simple linear regression assumes that the relationship between the logarithmic PCB concentration and length can be described by a linear function. This assumption is unlikely to be true as we have seen from the analysis of the PCB in fish data. The large model error variance is often associated with locally persistent bias. In this case, when fish length is near 60 cm the linear model tends to overpredict the PCB concentrations. When a quadratic model of fish length is used (model `lake.lm5`), this bias is reduced, and the residual variance is reduced. We can further reduce the residual variance by adding higher order polynomials of the predictor. In theory, we can always fit a perfect model ($\varepsilon_i = 0$) using a high enough order of polynomials of the predictor. However, such a model offers no additional information than the raw data plot with a line drawn to connect all data points from left to right. It has all the roughness of the data. In fitting a linear model, all data points are used and contribute equally in determining the location of the fitted line at a given predictor variable value. In drawing the line connecting all data points, only one data point is used to determine the location of the line at a given data point. Mathematically, in fitting a linear regression model, we assume that the relationship between y

and x is linear. In drawing the line connecting all data points, we don't make any assumption about the nature of the relationship. If the objective of data analysis is to learn about the relationship between y and x, neither of the two extremes is effective. A compromise between the two extremes would be to estimate the expected response variable value using a number of neighboring observations such that the estimated value is less variable than the data points themselves and yet not determined by the entire data set. This compromise is the essence of smoothing.

Smoothing is an exploratory data analysis tool for uncovering functional forms from data. The goal of using smoothing is to produce a graphical presentation of the underlying relationship that is less variable (or smoother) than the data themselves. By removing random noise in the data, the resulting graph is likely to be easier to understand and a new hypothesis about the relationship can be generated. To construct a smooth line through the data cloud, we need to find a set of plotting points. That is, for a set of given x values, we need to know where to locate the line or what are the expected values of y. The simplest form of smoothing is the moving average. In the PCB in the fish example, a moving average is constructed by selecting a set of fish length values and for each length value we calculate a mean value of the log PCB using a number of neighboring data points. The neighbor can be decided, for example, by a fixed interval in fish length. We can imagine that a fixed width window moves from left to right. At each stop, the window captures a number of data points in the scatter plot and their mean $\bar{y}_j$ is calculated. Connecting these "local" averages results in a smoother line than raw data themselves (Figure 6.16). Obviously, the smoothness of the resulting line depends on the width of the moving window. The wider the window, the more data points it includes and the smaller the variance of the means, hence the smoother the resulting line.

Because the window width determines the smoothness of the fitted line, selecting a proper window width is an important decision in constructing a smoother. If the window width is too wide, a certain locally persistent feature of the relationship may be averaged out, resulting in a line that is too smooth. A line that is too smooth is potentially biased. If the window width is too small, the resulting line may be too jumpy and overstates the variability in the mean function. In selecting the window width, we will make a trade-off between the bias and variance of a fitted line. Another decision in constructing a smoothing line is the selection of a smoother. Figure 6.16 is constructed using a moving average. Other methods include weighted moving average and local regression smoothing. The rationale for using a weighted moving average is that a smoothing line is to reveal locally persistent features of a bivariate relationship; even within a small window data points closer to the point being evaluated should be more relevant than data point far away. So, instead of treating all data points inside a window equally, when calculating the y-axis location of the smoothing line at fish length of 35 cm, a weighted average can be used. Data points farther away from 35 are given lower weights than data

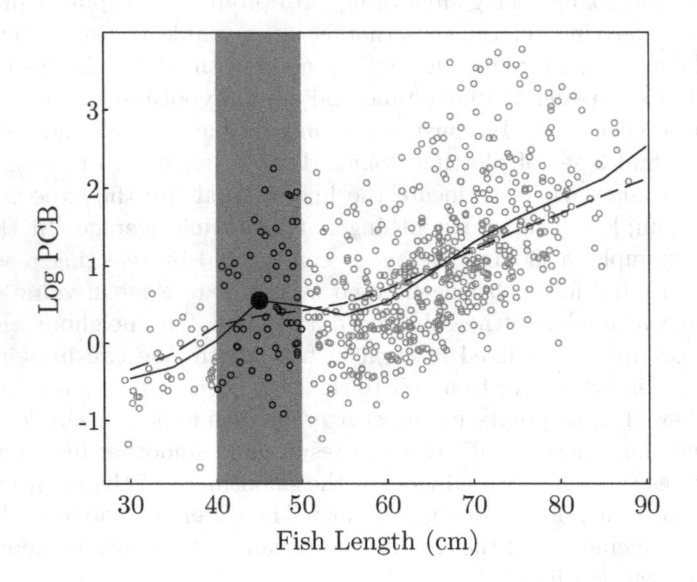

FIGURE 6.16: A moving average smoother – The log PCB concentration and fish length relationship is estimated using a moving average smoother. The solid line is estimated using a window width of 10 cm (shown by the shaded band) and the dashed line is estimated using a window width of 20 cm. When evaluating the expected value of log(PCB) at a length of 35, only those data points bounded inside the shaded window are used for estimating $\hat{y}$.

points closer to 35. Many alternative methods are available for computing the weights. The local regression method constructs a smoothing line by fitting a regression model within a window and estimates the y-axis plotting point using the fitted regression model.

6.2.2 Fitting a Local Regression Model

The most commonly used smoothing method is the *loess*, a method popularized by William Cleveland [Cleveland, 1993] and its implementation in the S language. Although nonparametric smoothing is an active research area in statistics, and many have argued for using one form of smoother over the other, practically all smoothers are more or less equally effective in revealing the underlying relationship between two variables. Furthermore, the effectiveness of a smoother is more related to the selection of the smoothness parameter (e.g., the window width) than the selection of a particular form of smoother. When using scatter plot smoothing as an exploratory tool, *loess* is often the best choice.

As in the moving average shown in Figure 6.16, a loess line is fitted by estimating the expected y-axis variable values ($\hat{y}_i$) at a set of given x-axis variable values (x_i). For each given x-axis value x_i, a weighted regression model (the local regression model) is fit to a subset of data points around x_i, and $\hat{y}_i$ is the fitted value for x_i. To fit the local regression model, we need to choose two parameters: λ and α. In the current implementation in R, the parameter λ can be 1 or 2, representing a linear or quadratic regression, respectively. The parameter α takes a value between 0 and 1 representing the fraction of data points to be used for fitting the local regression model. In Figure 6.17, the loess line was fit using $\lambda = 1$ and $\alpha = 0.5$. The loess model fitting is done by estimating the y values at a set of x values. Figure 6.17 shows the fitting process for $x = 60$ cm, including the data points used for estimating the expected log PCB concentration at length of 60 cm, and the fitted local linear model (a weighted linear regression model). In R, the loess model is fitted by the function `loess`:

```
#### R Code ####
pcb.loess <- loess (log(pcb)~ length,
    data=laketrout, degree=1, span=0.5)
```

A scatter plot smoothing is a nonparametric regression model, in that the resulting model is not defined through an algebraic formula parameterized by using one or more parameters. A scatter plot smoothing is defined graphically. The main objective in using a scatter plot smoothing model is to explore a possible functional relationship between the response and the predictor. As a result, the fitted scatter plot smoothing model should always be treated as the intermediate result that can help us to hypothesize the nature of the relationship. Because of the trade-off between model variability and bias, a smoothing can vary substantially as a function of the smoothing parameter.

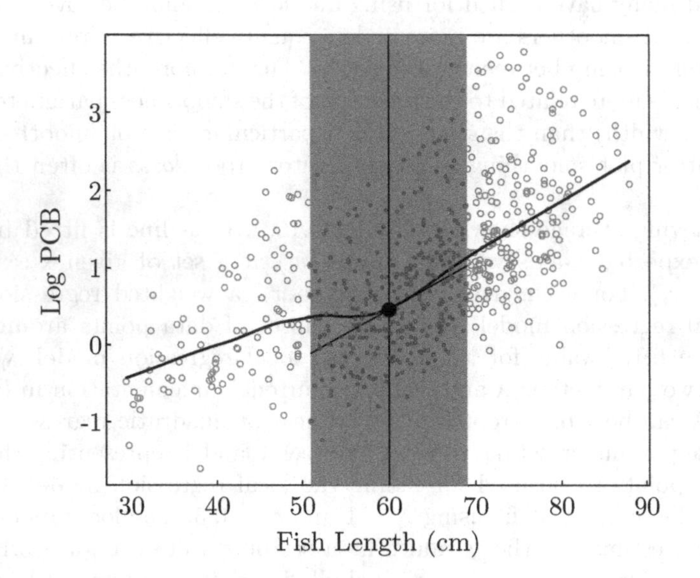

FIGURE 6.17: A loess smoother – The log PCB concentration and fish length relationship is estimated using a loess smoother. The thick solid line is the fitted loess line using parameters $\lambda = 1, \alpha = 0.5$. When evaluating the expected value of log(PCB) at a length of 60, only those data points bounded inside the shaded window are used.

Different levels of smoothness may lead to different interpretations of the underlying relationship. Which interpretation is reasonable should be answered largely by substantive knowledge rather than by statistics. A scatter plot smoothing model is a means to an end, and should not be used as an end by itself.

A common misconception about a nonparametric regression model is that these models do not impose distributional assumptions. Because of the exploratory nature of a nonparametric regression model, we typically do not emphasize the distributional assumption. But when we want to use a smoothing model as a predictive model, distributional assumption is necessary in order to make uncertainty assessments about the model and model prediction.

6.3 Smoothing and Additive Models

If a scatter plot smoothing is a nonparametric generalization of a simple linear regression model, an additive model is then a nonparametric generalization of a multiple regression model. That is, a multiple linear regression model assumes the functional relationship between the response variable and multiple predictors are linear and additive. An additive model assumes additivity, but does not make any specific functional form assumptions about the relationship. An additive model can be expressed as:

$$y_i = \beta_0 + \sum_j f_j(x_{ij}) + \epsilon_i$$

where $j = 1, \cdots, k$, f_j is an unspecified function to be estimated nonparametrically. When $k = 1$, the additive model reduces to scatter plot smoothing. The term "nonparametric" suggests that the function f_j is specified without using parameters. However, the model residual term ϵ is assumed to follow a normal distribution.

6.3.1 Additive Models

Additive models are flexible tools often used to explore possible functional forms between the response variable and several predictors. A fitted additive model is presented graphically. As a transition from a multiple linear regression model to an additive model, I present a multiple linear regression model with two predictors (Figure 6.18). Obviously such graphical presentation is unnecessary in most situations. Figure 6.18 is nevertheless useful for understanding the concept of graphical presentation of a model. With Figure 6.18, a user can make an approximate prediction of the response variable value

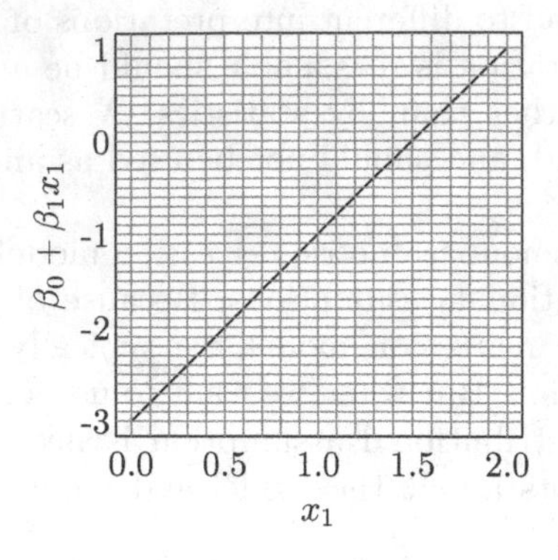

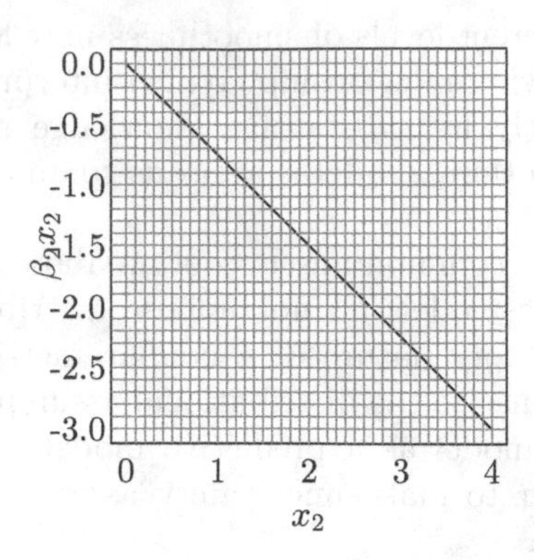

FIGURE 6.18: Graphical presentation of a multiple linear regression model – A multiple linear regression model with two predictors ($y_i = \beta_0 + \beta_1 x_{i1} + \beta_2 x_{i2} + \varepsilon_i$) is presented graphically. The left panel shows the conditional relationship between y and x_1 and the right panel shows the same between y and x_2. A conditional relationship between y and x_1 is the relationship between the two variables when x_2 is held constant.

when given the values of the two predictors. For example, when $x_1 = 1$ and $x_2 = 2$, the corresponding y-axis value from the left and right panels are -1 and -1.5, respectively. Therefore, the predicted response variable value is -2.5. Furthermore, the left panel shows that when x_1 increases (and x_2 is held constant) the response variable will increase, and when x_2 increases (and x_1 is held constant) the response variable will decrease. In other words, a graphical presentation gives us essentially the same information as numerical summary of the fitted linear regression model. The objective of fitting a multiple linear regression model is to make a statistical inference about the dependency of y on predictors. A linear model summarizes this dependency through slopes.

When a transformation is used, for example a logarithmic transformation on x_2, the resulting relationship is no longer linear. Engineers used to use a log-paper for plotting such models for easy prediction (Figure 6.19).

Alternatively, Figure 6.19 can be presented in the original scale of x_2 (Figure 6.20). This presentation tells a user directly that when x_1 is held constant the response variable y increases rapidly when x_2 increases from 0. The rate of increase in y slows down as x_2 increases.

Figures 6.18 to 6.20 show models with known functional forms. When the functional form of the relationship cannot be simplified by simple mathematical functions such as the linear and log-linear functions, the additive model uses scatter plot smoothing to estimate the function numerically and present the result graphically. In other words, an additive model will allow the data to tell us the proper model form. Although an additive model will not produce a formula, the graphs allow us to understand the dependency of y on x_j.

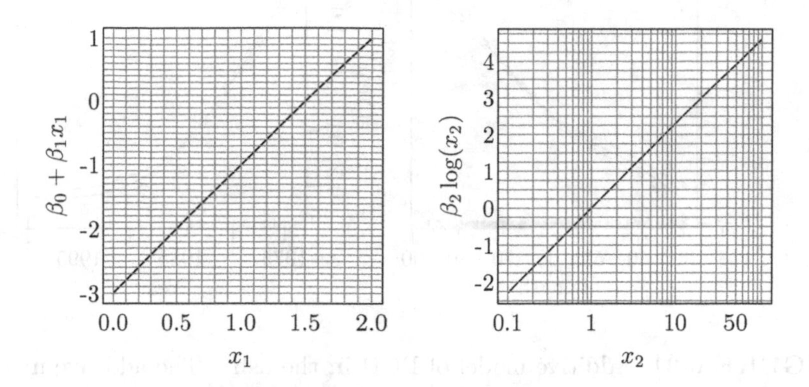

FIGURE 6.19: Graphical presentation of a multiple linear regression model with log-transformation – A multiple linear regression model with two predictors ($y_i = \beta_0 + \beta_1 x_{i1} + \beta_2 \log(x_{i2}) + \varepsilon_i$) is presented graphically. The left panel shows the conditional relationship between y and x_1 and the right panel shows the same between y and x_2. The right panel x-axis is presented in a logarithmic scale.

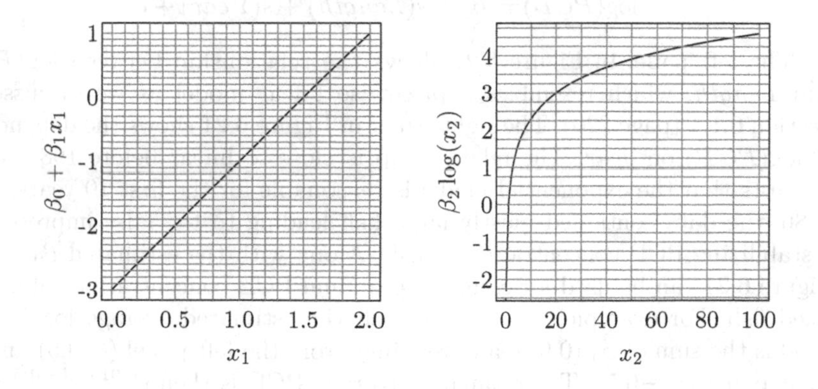

FIGURE 6.20: Graphical presentation of a multiple linear regression model with log-transformation – The model in Figure 6.19 is presented in the original scale of x_2.

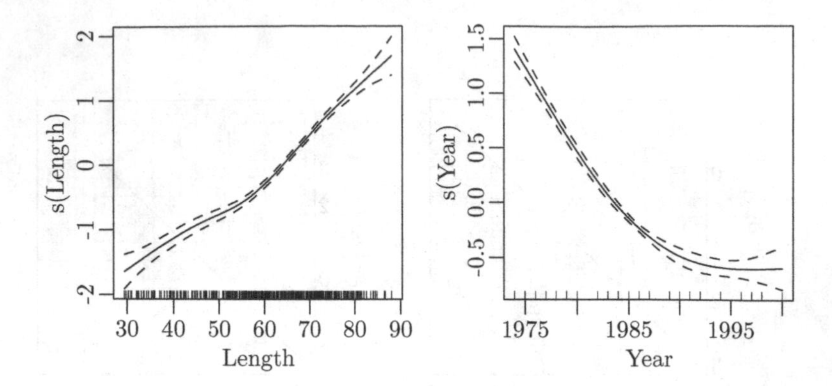

FIGURE 6.21: Additive model of PCB in the fish – The additive model fit of `Length` and `Year` is shown in the left panel and the right panel, respectively. The left panel resembles a piecewise linear model, and the right panel suggests a declining rate of PCB dissipation (a false impression discussed in Section 5.3.4).

From these plots, hypotheses about the functional forms can be generated. For example, when fitting the PCB in fish data, we know that fish size and year since 1974 are two important predictors. Instead of using the exponential model to help decide the model form, we can use the additive model as an initial step to explore possible model forms. The fitted additive model shown in Figure 6.21 can be expressed as:

$$\log(PCB) = \beta_0 + s(Length) + s(Year) + \epsilon \tag{6.6}$$

The left panel in Figure 6.21 shows the relationship between $\log(PCB)$ and *Length*, which resembles a piecewise linear model as we discussed in Section 6.1.1 (page 220). The right panel in Figure 6.21 shows the dependency of $\log(PCB)$ on *year*. The relationship is close to linear before 1985, which suggests that the exponential model is reasonable for the first 10 years. After 1986, the data contained mostly large fish leading to the false impression of a stabilizing PCB concentration in fish. Along with the estimated $\hat{\beta}_0 = 0.91$, Figure 6.21 can be used for estimating annual PCB concentrations of a given sized fish. For example, for a 70 cm fish the estimated average log PCB in 1990 is the sum of $\hat{\beta}_0$ (0.91), and readings from the left panel (~ 0.5) and the right panel (~ -0.5). The estimated average PCB is then $e^{0.91+0.5-0.5} = 2.5$ ppb.

6.3.2 Fitting an Additive Model

An additive model is fit using an iterative procedure, which repeatedly fit scatter plot smoothing until convergence is reached. This procedure is often referred to as the backfitting algorithm. Suppose that we are interested in fitting an additive model with two predictors, $y = \beta_0 + s(x_1) + s(x_2) + \epsilon$, as in the PCB in the fish example. The underlying assumption of this model is that the effect of x_1 on y is independent of x_2 and the effect of x_2 on y is independent on x_1, the additive assumption. The backfitting algorithm for a two predictor model has the following steps:

1. Fit a scatter plot smoothing using only one predictor: $y_i = s(x_{1i}) + \epsilon_i$
2. Calculate the residuals: $y_{r1i} = y_i - s(x_{1i})$
3. Fit a second scatter plot smoothing $y_{r1i} = s(x_{2i}) + \epsilon_i$
4. We have a first estimate of $s(x_1)$ and $s(x_2)$
5. Now calculate residuals $y_{r2i} = y_i - s(x_{2i})$ and fit smoothing $y_{r2i} = s(x_{1i}) + \epsilon_i$
6. Repeat these steps until the estimated $s(x_1)$ and $s(x_2)$ do not change from their respective estimate in the previous iteration.

The backfitting algorithm is implemented in R function `gam` from the package also named `gam`. The name `gam` stands for the *generalized additive models*. As in fitting a scatter plot smoothing, we need to select a type of smoother for each predictor variable (loess, moving average, etc.) and the smoothing parameter. For fitting the model shown in Figure 6.21, the following script can be used:

```
#### R code ####
PCB.gam <- gam(log(pcb)~s(length)+s(year), data=laketrout)
```

This line calls the function `gam` to fit an additive model, with response variable `log(pcb)` and two predictors `length` and `year`. The function `s()` is to specify the smoother to be a smoothing spline. By default, the function `s()` uses a smoothing parameter `df` to target equivalent degrees of freedom. When `df=1` the fitted model is linear and the higher the `df` the more "wiggly" the fitted line will be. For example, Figure 6.22 shows three plots of `s(year)` against `year`: one was fit using `df=2`, the second with `df=4` (the default), and the third with `df=8`. The choice of smoothing parameter value affects the outcome and sometimes different outputs may lead to different interpretations. For example, one may interpret the result from using `df=2` as evidence that the declining trend of the average PCB concentration in fish continues. But the interpretation of the model using `df=4` may be that the average PCB concentration in fish has reached a stable state, and the model with `df=8` may suggest that there is a rebound of PCB in fish. There is only one correct interpretation. Users of nonparametric models such as the additive model should be cautious in interpreting model output. In general, the nonparametric model should be used as an exploratory tool for generating a hypothesis about

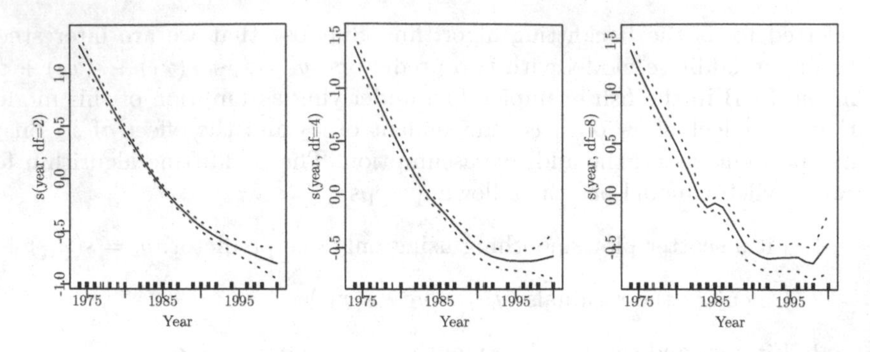

FIGURE 6.22: Effects of smoothing parameter – The effect of the smoothing parameter on the fitted year effect using the same additive model as in Equation 6.6 is illustrated using three different `df` values (from left to right: `df = 2, 4`, and `8`).

the underlying relationship. As a result, a fitted model should be interpreted in words and contradictions to the current understanding of the subject matter should be investigated. As we already know, the apparent slowing down of the rate of PCB dissipation is probably an artifact of the imbalance of the fish size data (Figure 9.2). All three graphs in Figure 6.22 seem to contradict the assumption of a constant rate of PCB dissipation.

An alternative smoother implemented in the function `gam` is loess `lo()`. The same model can be fitted by:

```
#### R code ####
PCB.gam <- gam(log(pcb)~lo(length)+lo(year), data=laketrout)
```

Like fitting a loess line, we can specify `span` and `degree`:

```
#### R code ####
PCB.gam <- gam(log(pcb)~lo(length, span=0.75, degree=1)+
    lo(year), data=laketrout)
```

The default is `span=0.5` and `degree=1`.

The additive model is also implemented in R package `mgcv`, and its use is discussed in the next section.

6.3.3 Example: The North American Wetlands Database

Reckhow and Qian [1994] used additive models to study the effectiveness of using constructed or natural wetlands as a tertiary treatment facility to remove low level pollutants such as phosphorus and nitrogen. In their work, a cross-sectional data set collected by the U.S. EPA was used. The data set includes wetlands performance information from wetlands in the United States and Canada used for wastewater treatment. Variables included are input and output total phosphorus (TP) concentrations (in mg/L), hydraulic loading rate (in mm/day), and input and output TP mass loading rate (in g P m^{-2} yr^{-1}). To evaluate the effectiveness of a treatment wetland, we want to show under what condition a treatment wetland can maintain a low output TP concentration. Naturally, the effluent TP concentration is the response variable. Initial graphical presentation of the data shows no obvious relationship between effluent concentration and variables representing input information (Figure 6.23). As we have already seen in Figure 3.12 (page 64), there is a strong relationship between `TPOut` and `PLI`, visible only when the TP mass loading rate is plotted in a logarithmic scale. Furthermore, when both `TPOut` and `PLI` are plotted in logarithmic scale (Figure 6.24) we note that the effluent TP concentrations are all below 0.1 mg/L when the TP mass loading rate is below a value about 1 g P m^{-2} yr^{-1}.

These graphs suggest a nonlinear relationship between the response and the three input variables. An additive model was fit. Their model is reproduced here by using the `gam` function from package `mgcv`:

```
#### R Code ####
require{mgcv}
nadbGam1 <- gam(logTPOut ~ s(logPLI)+s(logTPIn)+s(logHLR),
    data=nadb)
```

The resulting model is presented graphically by using the generic plotting function `plot`:

```
#### R Code ####
par(mfrow=c(1,3), mar=c(3,3,0.5,0.25),
    mgp=c(1.5,0.5,0))
plot(nadbGam1, select=1, se=T, rug=T, resid=T,
    scale=0, pch=16, cex=0.25)
plot(nadbGam1, select=2, se=T, rug=T, resid=T,
    scale=0, pch=16, cex=0.25)
plot(nadbGam1, select=3, se=T, rug=T, resid=T,
    scale=0, pch=16, cex=0.25)
```

The fitted model (Figure 6.25) is, however, very different from the results in Reckhow and Qian [1994] (see Figure 6.29). This difference is not because a wrong model was fit, rather it is because the model displayed in Figure 6.25 used the default smoothness parameters of the `gam` function in the R package `mgcv` and the model in Reckhow and Qian [1994] used the default smoothness parameters in the S-Plus function `gam` (which is similar to the `gam` function in R package `gam`). The difference raises questions about the value of a nonparametric regression model in scientific data analysis.

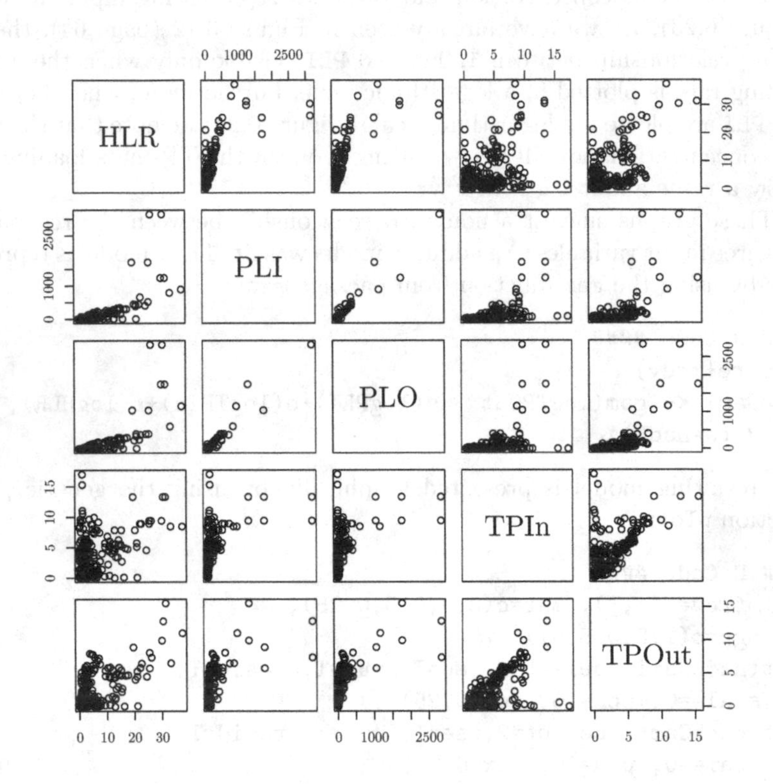

FIGURE 6.23: The North American Wetlands Database – A scatter plot matrix displaying all variables included in the North American Wetlands Database reveals little useful information for hypothesizing the nature of the relationship between the effluent TP concentration (`TPOut`) and input variables (hydraulic loading rate `HRL`, input TP concentration `TPIn`, and input TP mass loading rate `PLI`).

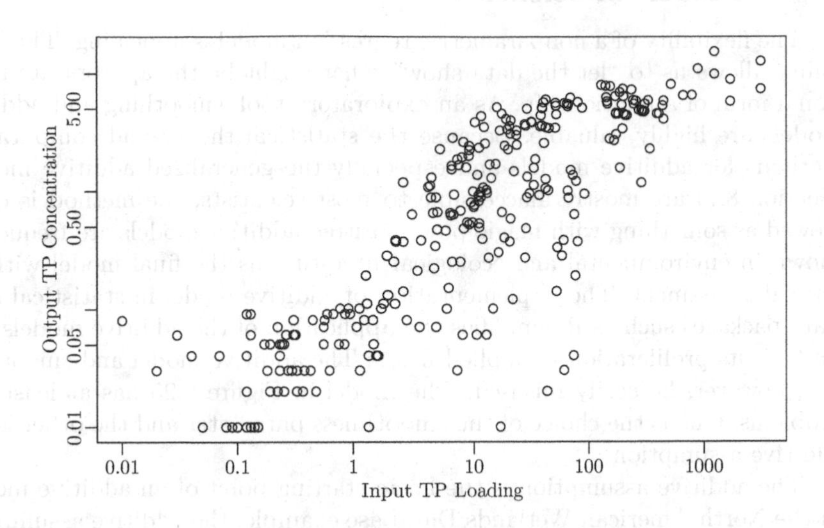

FIGURE 6.24: The effluent concentration–loading rate relationship – When plotted on a logarithmic scale, the TP effluent concentration is seen to be below 0.1 mg/L when the TP mass loading rate is below 1 g/m^2-yr.

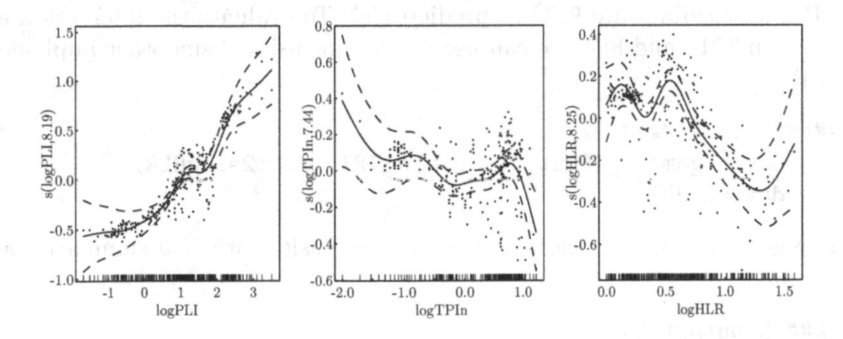

FIGURE 6.25: Fitted additive model using `mgcv` default – The fitted additive model of log effluent TP concentration predicted by the TP mass loading rate (left panel), log TP input concentration (middle panel), and the log hydraulic loading rate (right panel). The model is fit using the `gam` function from package `mgcv` with default smoothness parameter values.

6.3.4 Discussion: The Role of Nonparametric Regression Models in Science

The flexibility of a nonparametric regression model is appealing. This flexibility allows us to "let the data show" what might be the appropriate functional form of a relationship. As an exploratory tool, smoothing and additive models are highly valuable. Because the statistical theory and computation methods for additive models and especially the generalized additive models (Section 8.7) are mostly inaccessible to most scientists, the method is often viewed as something with magic power. Fitted additive models are frequently shown in environmental and ecological literature as the final model without critical assessment. The implementation of additive model in statistical software packages such as R simplifies the application of the additive models and leads to its proliferation in applied fields. The additive model and smoothing can, however, be easily misused. The model in Figure 6.25 has at least two problems. One is the choice of the smoothness parameter and the other is the additive assumption.

The additive assumption is the basic starting point of an additive model. In the North American Wetlands Database example, the additive assumption does not hold. The TP mass loading rate is the product of input TP concentration and flow rate. In other words, `PLI` is proportional to the product of `TPIn` and `HLR`. Consequently, each of the three plots in Figure 6.25 is meaningless. We cannot examine the effect of `log(PLI)` while holding the other two predictors constant because `PLI` will be a constant if `TPIn` and `HLR` are fixed. When the additive assumption is in question, the fitted model should be viewed with skepticism. In this case, `TPIn` and `HLR` are the two independent predictors. We should either use these two variables or their product (the TP mass loading rate `PLI`) as predictor(s). To evaluate the interaction effect between `TPIn` and `HLR`, we can use a two-dimensional smoother implemented in R:

```
#### R Code ####
nadbGam3<-gam(logTPOut~s(var1=logTPIn, var2=logHLR),
    data=nadb)
```

The generic function `summary` will provide basic statistical summaries about the model:

```
#### R output ####
summary(nadbGam3)

Family: gaussian
Link function: identity

Formula:
logTPOut ~ s(var1 = logTPIn, var2 = logHLR)
```

```
Parametric coefficients:
            Estimate Std. Error t value Pr(>|t|)
(Intercept)  0.35403    0.00888    39.9   <2e-16

Approximate significance of smooth terms:
                   edf Est.rank     F p-value
s(logTPIn,logHLR) 25.5        29 38.3  <2e-16

R-sq.(adj) =  0.796   Deviance explained = 81.4%
GCV score = 0.02429   Scale est. = 0.021986   n = 279
```

The resulting model can be plotted either by a contour plot or a three-dimensional perspective plot (Figures 6.26 and 6.27):

```
#### R Code ####
par(mar=c(3,3,1,1), mgp=c(1.5,0.5,0), mfrow=c(1,2),
    pty="s")
plot(nadbGam3, select=1, se=T, rug=T, resid=T, pch=1)
plot(nadbGam3, select=1, se=T, rug=T, resid=T, pers=T)
```

Both Figures 6.26 and 6.27 show a relatively flat area where both `TPIn` and `HLR` are small and steep slopes when both predictors are large. The fitted model led to a long discussion among researchers at Duke University Wetland Center and the concept of wetland assimilative capacity was proposed and discussed by Richardson and Qian [1999], and the TP mass loading rate is used as the predictor to represent the changes in both influent TP concentration and hydraulic loading rate. The resulting single predictor smoothing model (Figure 6.28) shows a clear piecewise linear relationship between effluent TP concentration and mass TP loading rate (both in the logarithmic scale). When the log mass loading rate is less than ~ 0 (or approximately 1 g P $m^{-2}yr^{-1}$), TP effluent concentrations do not change as a function of the loading rate. When the loading rate exceeds 1 g P $m^{-2}yr^{-1}$, the effluent TP concentration increases as a linear function of the loading rate. The TP mass loading threshold of 1 g P $m^{-2}yr^{-1}$ is seen as a critical value when a constructed wetland is designed for removing phosphorus. This threshold is expected to change from wetland to wetland. TP retention capacity of this example illustrates the importance of treating an additive model as an exploratory tool. Results from an additive model should be carefully scrutinized and interpreted using scientific knowledge. The hypothesis about the underlying model should then be proposed and tested. Based on these exploratory analyses, a TP retention model was proposed to quantify a wetland-specific loading threshold [Qian and Richardson, 1997]. Using existing monitoring data from WCA2A at the time, the estimated TP retention capacity was about 1.15 g P $m^{-2}yr^{-1}$ with a 90% interval of (0.61, 1.47). During the late 1990s and early 2000s, a number of constructed wetlands (known as stormwater treatment areas, STAs) were constructed as part of the effort of restoring the Everglades. TP retention ca-

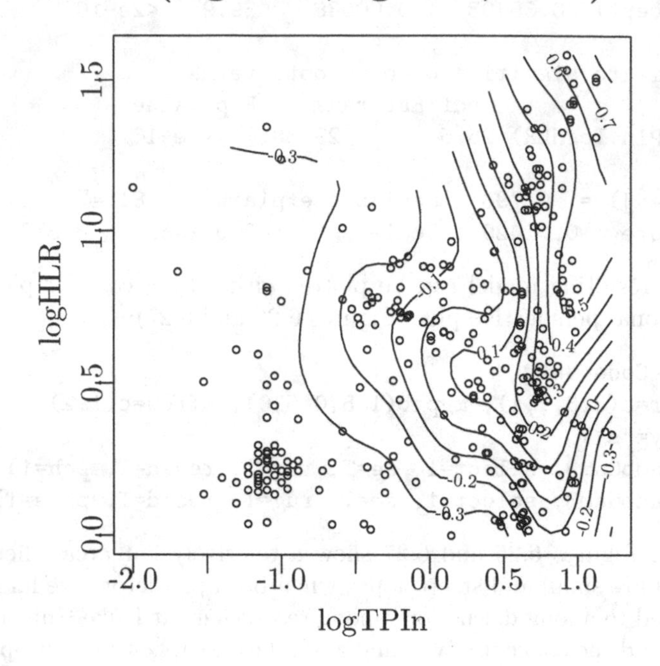

FIGURE 6.26: Contour plot of a two-variable smoother fitted using `gam` – The contour plot shows the fitted two-variable smoother predicting log effluent TP concentration predicted using log TP input concentration, and the log hydraulic loading rate. The model is fit using the `gam` function from package `mgcv`.

pacity of these treatment wetlands was reported to be 1.1 ± 0.5 g P $m^{-2}yr^{-1}$ [Chen et al., 2015].

The second consideration in fitting a GAM is the selection of the smoothness parameter, which we have discussed earlier (Figure 6.22). The fitted additive model in Figure 6.25 used the default smoothness parameter of the `gam` function in package `mgcv`, which is decided by a cross-validation simulation for optimal predictive features. When a different smoothness parameter value is used:

```
#### R Code ####
nadbGam1.5 <- gam(logTPOut ~ s(logPLI, fx=T, k=4)+
                             s(logHLR, fx=T, k=4), data=nadb)
```

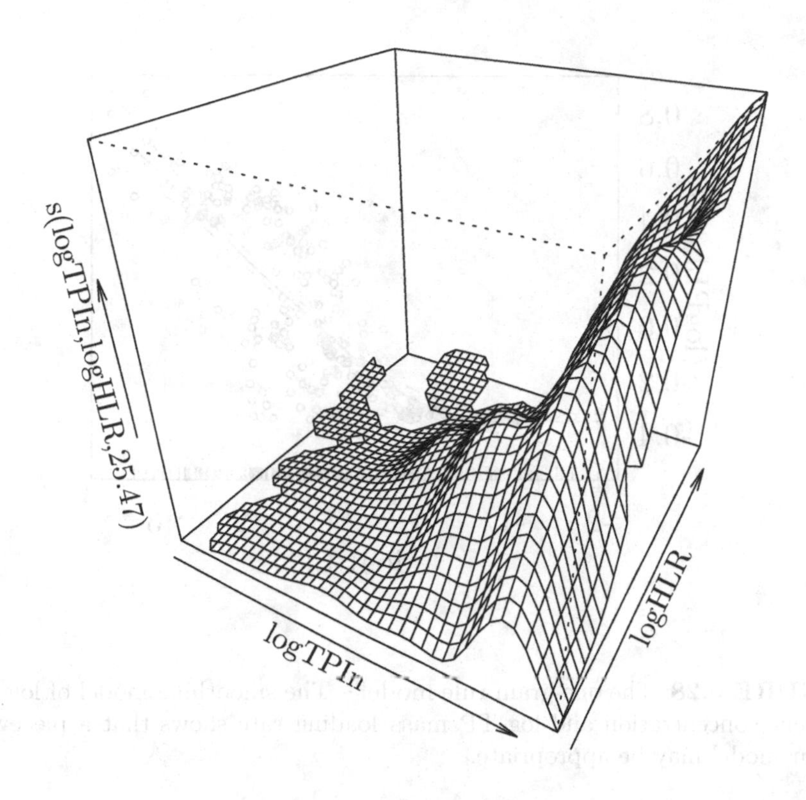

FIGURE 6.27: Three-dimensional perspective plot of a two variable smoother fitted using `gam` – The three-dimensional perspective plot shows the same two-variable smoother in Figure 6.26.

the resulting model (Figure 6.29) is quite different from the default results in Figure 6.25.

Coincidentally, Figure 6.29 is very similar to the result using the `gam` function from the package `gam` with default smoothing parameter values. The differences between the two packages are mainly mathematical methods used for fitting the smoothing model. But the question for an application is how to select the most appropriate smoothness parameter value. Again, if an additive model is used as an exploratory tool rather than an automatic model-generating tool, this question is moot. We should always explore different possibilities and interpret the results using scientific knowledge. If the statistical software and data are allowed to fully control the model-fitting process, models with conflicting interpretations may arise. Scientific knowledge and common sense should be used to guide the process of model selection. In the end, a parametric model should be proposed to reflect both the scientific knowledge and evidence reflected by the data.

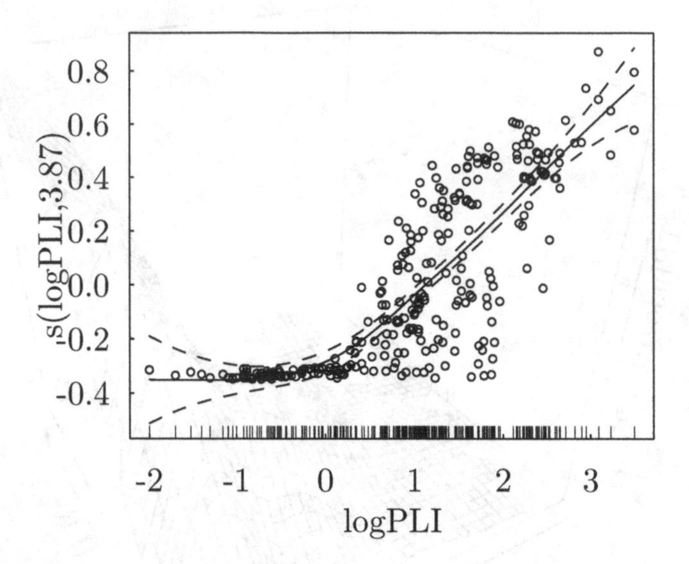

FIGURE 6.28: The one-gram rule model – The smoothing model of log TP effluent concentration and log TP mass loading rate shows that a piecewise linear model may be appropriate.

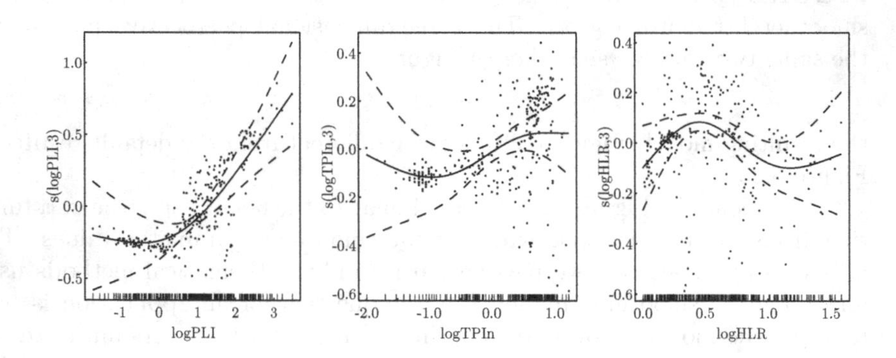

FIGURE 6.29: Fitted additive model using user-selected smoothness parameter value – The fitted additive model of log effluent TP concentration predicted by the TP mass loading rate (left panel), log TP input concentration (middle panel), and the log hydraulic loading rate (right panel). The model is fit using the `gam` function from package `mgcv` with a smoothness parameter value of `k=4`.

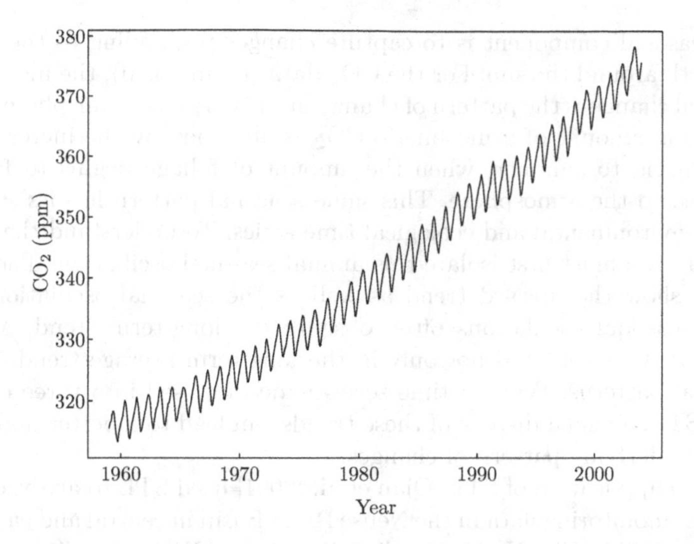

FIGURE 6.30: CO_2 time series from Mauna Loa, Hawaii – Time series of atmospheric CO_2 concentrations measured at the Mauna Loa observatory in Hawaii.

6.3.5 Seasonal Decomposition of Time Series

Detecting and analyzing temporal trends in historical data are important aspects of environmental data analysis. The famous atmospheric carbon dioxide (CO_2) concentrations data measured at the Mauna Loa observatory in Hawaii demonstrated the global impact of anthropogenic emissions of CO_2 (Figure 6.30).

Clearly documented historical trends reveal the response of a natural system to either the unintentional consequences of human activity or the deliberate results of management. The time series plot of the CO_2 concentration data clearly shows an increasing trend and a cyclical seasonal pattern. Although a simple linear regression model is often used for estimating the average temporal rate of increase, the cyclical seasonal pattern will remain in the model residuals. In addition, temporal changes may be affected by seasons. To properly explore the long-term and seasonal trends, the additive model was used for decomposing seasonal and long-term trends. The method is known as the seasonal decomposition of time series using loess or STL (Cleveland et al. [1990]). The technique is a tool for analyzing time series data where observations are made regularly over time. It represents a special case of the additive model with time as the single predictor. The basic setup of the method is to decompose a datum into three parts, representing the long-term trend, seasonal fluctuation, and the remainder:

$$Y_{year,month} = T_{year,month} + S_{year,month} + R_{year,month} \tag{6.7}$$

The seasonal component is to capture changes responding to the rotation of the earth around the sun. For the CO_2 data (Figure 6.30), the mechanism of a seasonal change is the pattern of change in the northern hemisphere foliage. An increasing amount of atmospheric CO_2 is absorbed by the increasing foliage from spring to summer; when the amount of foliage begins to fall, CO_2 is returned to the atmosphere. This same seasonal pattern has its signature on many environmental and ecological time series. To understand the changes in the data, we must first isolate this annual seasonal oscillation. The CO_2 data clearly show the upward trend as well as the seasonal oscillation. In other cases, seasonal oscillations often obscure the long-term trend. More often, trends may be reflected not only in the long-term average trend, but also in seasonal patterns. When a time series is decomposed into three components using STL, a visual display of these trends can lead to a better understanding of the underlying pattern of changes.

In an application of STL, Qian et al. [2001] used STL to analyze long-term nutrient monitoring data in the Neuse River Basin in central and eastern North Carolina, USA. The North Carolina Division of Water Quality maintains 16 mid-channel ambient monitoring stations on the Neuse River and Estuary. Periodic nutrient sampling began at some of these stations in the late 1960s, and regular monitoring, on approximately a monthly basis, began in the late 1970s. The Neuse estuary received considerable public attention in the 10 years since 1995 when symptoms of excessive eutrophication, including algal blooms, low dissolved oxygen, massive fish kills, and outbreaks of toxic microorganisms, were first widely reported in the media. The general perception is that these problems are due to a recent increase in watershed nutrient inputs. The results in Qian et al. [2001] were surprising. Both nitrogen and phosphorus concentrations throughout the river basin have exhibited a decreasing temporal trend, especially phosphorus, due, in large part, to the elimination of phosphate detergents in 1987. The study hypothesized that the changes in the nitrogen to phosphorus ratio (which shows drastic increases in the same time) may be the primary cause of the eutrophication symptoms in the estuary because the balance of the macronutrient is an important consideration of phytoplankton dynamics.

In this section, a longer time series from a Neuse River monitoring station near Clayton, North Carolina, is used to illustrate the STL method. As with most environmental monitoring data, the Neuse River water quality monitoring data have missing months. The current implementation of the STL method requires an evenly spaced time series with no internal missing values. When there are missing values, the model in equation (6.7) is used to derive a systematic method for imputing missing values. This method is the median polish method proposed by John Tukey [Tukey, 1977]. In a median polish method, T is estimated by the median of a given year and S is estimated by the median of a given month. When a value of $Y_{year,month}$ is missing but the other two terms on the right-hand side of equation (6.7) are available, their sum provides a "fitted value" for the missing concentration in the year and month. Median polish is implemented in the R function `medpolish`, which fits T and S iteratively. First, the seasonal component (median of each month) is estimated. Then the long-term component (median for each year) is fit to the residuals. This completes the first iteration. In the subsequent iterations, each component is fit to the residuals from the other component. The procedure stops when the estimates of one iteration do not change significantly from the estimates of previous ones. Imputation of missing values using median polish has been applied previously in the study of trends in urban ozone levels in the Chicago area [Bloomfield et al., 1993].

Seasonal trend decomposition using loess (STL) is a graphics based statistical method for time series analysis [Cleveland, 1993], implemented in the R function `stl`. It is an iterative nonparametric regression procedure using repeated loess fitting. As in equation (6.7), the time series is decomposed into trend, seasonal, and remainder (or residual) components. However, while the median polish process uses median values for the trend and seasonal components, STL uses one continuous loess line for the long-term trend component and 12 month-specific loess lines for the seasonal component. As with median polishing, fitting is done on each component iteratively until the resulting trend and seasonal components are no longer different from the estimates of the previous iterations. Generally three iterations are sufficient [Cleveland, 1993]. The nonparametric nature of STL makes it flexible in revealing nonlinear patterns in seasonal data. Since each season (month) is a subseries in the fitted loess model, seasonal interactions can be captured.

6.3.5.1 The Neuse River Example

Because of the R functions `medpolish` and `stl`, implementation of STL is a straightforward exercise. The difficulty is often in data processing (from state or EPA database to R time series), and graphical presentation of STL results.

The Neuse River water quality monitoring is conducted by both local (North Carolina Division of Water Quality) and federal (U.S. EPA/USGS) agencies. These data are typically stored electronically on U.S. EPA's STORET site.[1] For this example, I used the Neuse River water quality monitoring station maintained by the North Carolina Department of Natural Resources Division of Water Quality (station ID J417000) near Clayton, North Carolina. To obtain data from this station, I followed the STORET site links to search "Regular Results by Station." Total phosphorus, Kjeldahl nitrogen, and fecal coliform were selected as representative water quality variables. Once the data file is downloaded, some preprocessing is necessary because the unit of fecal coliform is "#/100ml." The "#" sign comments out all values behind it and it should be replaced by something like "No." The downloaded file is a text file delimited by ~. After reading the data file into R, a `Date` column is created by converting the STORET time stamp which shows both the date

[1] `http://www.epa.gov/storet/`.

and time when the sample was collected (e.g., "1968-07-14 14:40:00") into a R date object:

```
#### R Code ####
require (survival)
J417$Date <- as.Date(J417$Activity.Start,
                       format="%Y-%m-%d %H:%M:%S")
```

Once a date variable is created, monthly mean concentration values are calculated for the time period of interest (1971–2007):

```
#### R Code ####
FecalColiform <- rep(NA, 12*(2007-1970))
k <- 0
for (i in 1971:2007){ ## year
  for (j in 1:12){ ## month
    k <- k+1
    temp <- as.numeric(format(J417$Date, "%m"))==j &
            as.numeric(format(J417$Date, "%Y"))==i
    if (sum(temp)>0)
      FecalColiform[k] <-
          mean(J417.FecalColiform$Value[temp], na.rm=T)
    }
  }

FecalColiform.ts <- ts(FecalColiform, start=c(1971,1),
    end=c(2007,12), freq=12)
```

The last line creates a time series of monthly average fecal coliform concentrations using function `ts`:

```
#### R Output ####
        Jan   Feb    Mar    Apr    May    Jun    Jul     Aug     Sep   Oct     Nov   Dec
1971     NA    NA     NA 2200.0     NA    0.0     NA    30.0      NA    NA      NA    NA
1972     NA    NA     NA     NA     NA  800.0     NA   253.3   390.0    NA      NA    NA
1973     NA    NA     NA     NA     NA     NA 7000.0      NA    93.3  35.0   216.0  4430
1974 1265.0 276.7 3505.0  310.0 1130.0 1033.3  562.5  8877.5   576.0 173.3   260.0    NA
1975 1117.5 847.5  155.0   85.0 2500.0  120.0   56.7    45.0  1930.0 375.0    30.0    50
1976   10.0   0.0   40.0   70.0   90.0  230.0  110.0     0.0  1800.0 190.0   660.0   100
1977     NA  20.0  990.0   40.0   40.0   50.0   20.0   100.0   330.0  40.0 14000.0     0
1978  650.0  60.0     NA   10.0   70.0 1100.0    0.0   960.0   170.0 560.0   510.0  5600
1979   80.0 240.0  240.0 2366.7  196.7 2983.3 4093.3    55.0 27050.0 280.0  2200.0   810
1980  390.0 490.0  780.0  100.0   50.0   70.0   20.0    10.0    50.0 160.0   150.0   100
1981   20.0  50.0   80.0   20.0   50.0     NA    0.0  2200.0    40.0  10.0    90.0   320
1982    0.0 250.0   70.0   60.0   20.0  100.0  190.0 14000.0   150.0  60.0   390.0    60
1983    0.0 170.0  510.0   50.0   40.0  210.0   70.0    60.0  5200.0  40.0   120.0  9700
1984   30.0  30.0   40.0   40.0  190.0  250.0  160.0    70.0   110.0 110.0   130.0 10000
1985   90.0 150.0   50.0   50.0  500.0   30.0   60.0   690.0    30.0  10.0    70.0    NA
1986    0.0  20.0   50.0   20.0   30.0    0.0     NA      NA      NA 190.0      NA    NA
1987     NA    NA     NA     NA     NA     NA     NA      NA      NA    NA      NA    NA
1988     NA    NA     NA     NA     NA     NA     NA      NA      NA    NA      NA    NA
1989     NA 710.0     NA     NA     NA     NA     NA      NA      NA    NA      NA    NA
1990     NA    NA     NA     NA     NA     NA     NA      NA      NA    NA      NA    NA
1991     NA    NA     NA     NA     NA     NA     NA      NA      NA    NA      NA    NA
1992     NA    NA     NA     NA     NA     NA     NA      NA      NA    NA      NA    NA
1993     NA    NA     NA     NA     NA     NA     NA      NA      NA    NA      NA    NA
```

```
1994     NA     NA     NA     NA     NA     NA     NA     NA  120.0 2500.0  500.0  160
1995  100.0   60.0  110.0  500.0  310.0  700.0  110.0  180.0  700.0  230.0     NA   NA
1996   73.0  170.0 1400.0   97.0   87.0  750.0  271.5  130.0   40.5   91.7   54.5   18
1997     NA   52.5   22.5     NA  127.7   18.0  180.0   45.0   82.0     NA   27.0   27
1998  340.0   82.0   14.0  690.0   81.0   40.0   54.0   73.0   67.0   91.0  230.0   62
1999   33.0   75.0   36.0  100.0   70.0   86.0   20.0   50.0     NA  100.0   80.0  240
2000 1000.0   45.0     NA  170.0   80.0 6000.0 2300.0   36.0  140.0   45.0   90.0   64
2001   71.0  120.0     NA   41.0   55.0  680.0   34.0 2000.0     NA   85.5  230.0   55
2002  820.0   30.0   13.0     NA     NA   93.0  130.0 1700.0     NA 1425.5   56.0  320
2003   64.0  230.0     NA   32.0   36.0   82.0   73.0 2000.0  160.0   74.0  110.0  170
2004   68.0   53.0  120.0  970.0   66.0  200.0  260.0 6800.0  190.0  150.0   83.0  130
2005   66.0  430.0   70.0   54.0     NA  550.0     NA  180.0  300.0   97.0     NA  800
2006   86.0   38.0   78.0   77.0   56.0  190.0 1100.0  100.0     NA   84.0  120.0   45
2007  280.0   45.0     NA   75.0     NA 1041.5     NA   57.0     NA  120.0     NA   NA
```

Months with missing values are common in environmental monitoring data. Using the median polish method, missing values can be "imputed" as long as there are nonmissing values in the same year (row) and the same month (column):

```
#### R Code ####
temp.2w <- medpolish(matrix(data.ts, ncol=12, byrow=T),
    eps=0.001, na.rm=T)
year.temp <- rep(seq(start(data.ts)[1], end(data.ts)[1]),
    each=12)
month.temp <- rep(1:12,
    length(seq(start(data.ts)[1], end(data.ts)[1])))
data.ts[is.na(data.ts)]<-temp.2w$overall +
    temp.2w$row[year.temp[is.na(data.ts)]-start(data.ts)[1]+1]+
    temp.2w$col[month.temp[is.na(data.ts)]]
```

When an entire row (year) or column (month) is missing, imputation using this method is impossible. In this data set, we have no record for 1987, 1988, and 1990–1993. Because the current R implementation of STL in function `stl` does not allow "internal" missing values, these internal missing values will be replaced by the median of all nonmissing months. The resulting data plot shows two distinct groups, before and after the data gap, with visible differences in standard deviations (Figure 6.31).

Because STL is an additive model of seasonal and trend components, two smoothness parameters need to be specified. The smoothness parameter is specified in terms of the span (number of months) of the loess window. The seasonal component smoothness parameter (`s.window`) must be supplied by the user. There is no general guideline on how to specify this parameter. Because STL is intended for exploratory analysis, we should try a few values to visualize the results. The resulting STL model can be plotted in many different ways. Figure 6.32 shows an example.

In Figure 6.32, the top row compares the magnitude of the three components of a time series decomposition. The long-term trend component (top left panel) is centered around its mean to facilitate the visual comparison of the three components. The seasonal component (top middle panel) shows a pattern of change that is the result of replacing the internal missing values

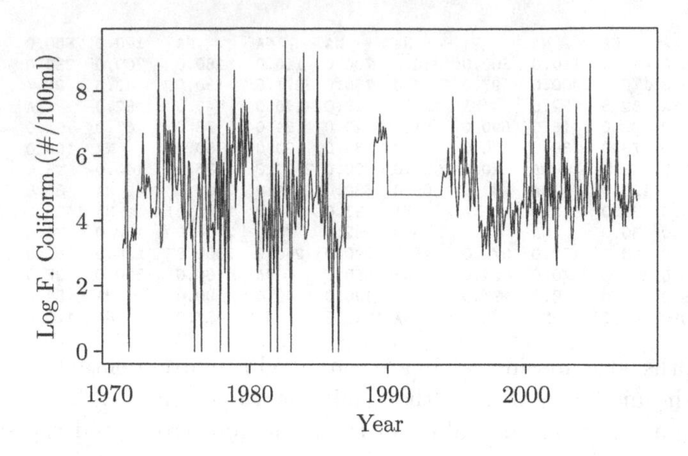

FIGURE 6.31: Fecal coliform time series from the Neuse River – Fecal coliform data (in logarithm) from the NC DWQ monitoring station on Neuse River near Clayton, NC. Internal missing values are replaced by the median of observed monthly values.

with a fixed value that artificially sets seasonal variation to 0. The remainder component shows a change in the residual spread before and after the data gap. (A new lab method was used after the data gap.) The bottom row shows the seasonal trend component. These are the trends of each month over the sampling period. In each plot, the mean of the values is shown by a horizontal line segment. These horizontal line segments show an overall seasonal pattern. In this case, the seasonal pattern is not very obvious, reflecting the fact that this section of the Neuse River receives urban runoff from Raleigh and a generally evenly distributed precipitation in this region. A clear pattern of the seasonal components is that there is either a hump or a valley before or around 1990. This is most likely because of the data gap that was filled by a constant, resulting in a disruption in the cyclical pattern fitted by the model. This feature indicates the importance of maintaining a long-term monitoring station for trend assessment.

The seasonal pattern of phosphorus is very clear (Figure 6.33): the horizontal line segments show that total phosphorus in this section of the Neuse River are generally low in early spring and high in late summer and early fall. In 1987, North Carolina banned the use of phosphate detergent and its effect is clearly shown as a rapid drop in the long-term trend. For each individual month, after the effect of the long-term trend is removed, we see an interesting pattern: early spring to early summer (months with low phosphorus) the decreasing trend before 1985 is reverted to an increasing trend after 1985, while in fall (months with higher phosphorus) the pattern is the opposite, i.e., an increasing trend before 1985 is reverted to an decreasing trend after that. This change is also reflected in the changes in the seasonal amplitude

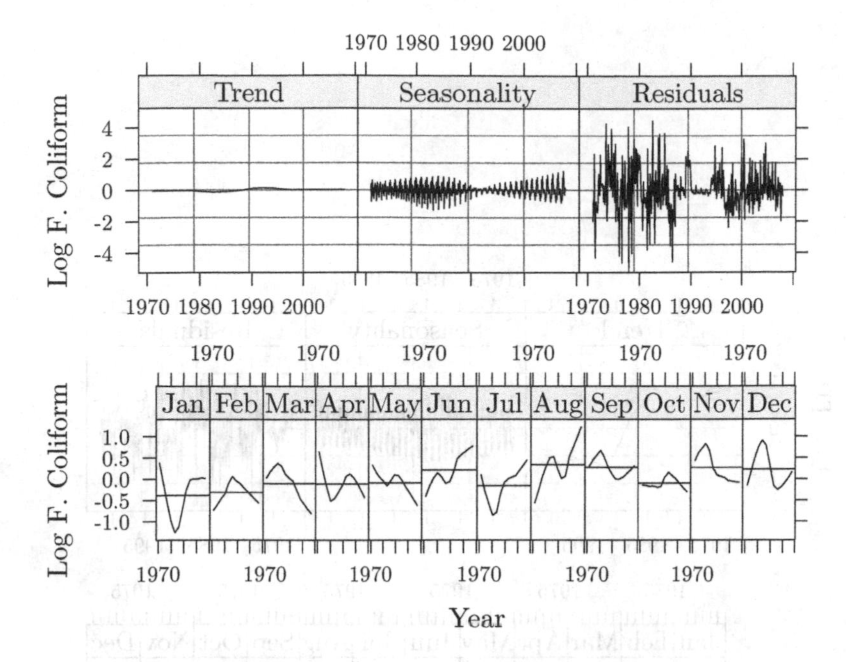

FIGURE 6.32: STL model of fecal coliform time series from the Neuse River – The fitted fecal coliform STL model is shown in two groups of plots. In the first group (top row, from left to right), the fitted trend (centered at its overall mean), seasonality, and remainder are compared. The second group (bottom row) compares seasonal trends of individual months. Each tick mark on the x-axis represents a 10-year increment.

in the seasonal component plot (Figure 6.33, top row middle panel): the amplitude increased from the early 1970s to just before 1985 and then gradually decreased. What might be the explanation of these changes?

STL, like other nonparametric regression methods introduced in this chapter, is an exploratory data analysis tool. Results are to be interpreted with caution. The seasonal pattern shift noted in Figure 6.33 is intriguing. But we have no explanation on why such a shift should happen. Because of the nonparametric nature, graphical results are to be used to guide the process of generating a new hypothesis, so that parametric models can be proposed and tested. When the objective of our study is prediction, we need to be aware of the edge effect of a nonparametric regression model. The edge effect refers to the disproportionate influence of data points near both ends of the time series on the fitted nonparametric model. Consequently, interpretation

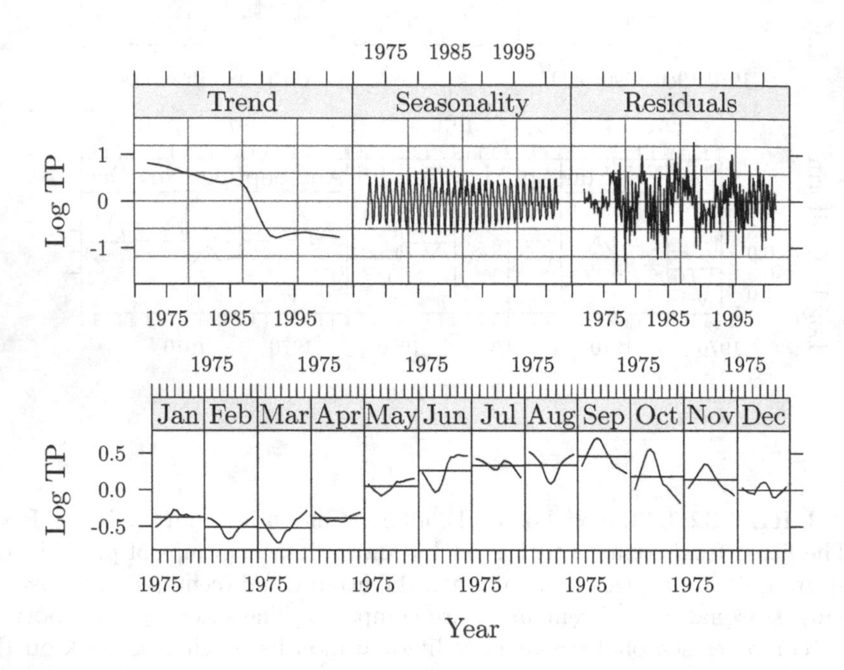

FIGURE 6.33: STL model of total phosphorus time series from the Neuse River – The fitted TP STL model shows a rapid drop in the long-term trend component responding to the phosphate detergent ban in 1987. The top row compares (from left to right) the fitted trend (centered at its overall mean), seasonality, and remainder. The bottom row compares seasonal patterns of individual months. Each tick mark on the x-axis represents a 5-year increment.

of a fitted model, especially patterns near both ends of a time series, must be done cautiously. To illustrate this point, the time series of Kjeldahl nitrogen (Figure 6.34, top panel) from the Clayton monitoring station is used to fit two STL models. The time series used in Qian et al. [2000a] ended in 1998. The phosphorus and nitrogen time series data from the monitoring station near Clayton retrieved from the EPA's STORET site in May 2008 ended in December 2001. When using the earlier data set ending in 1998, we concluded that nitrogen concentration had a generally stable trend, but likely decreasing in the last few years of the time series. This conclusion can be verified using Kjeldahl nitrogen concentration data from the Clayton site ending in 1998 (Figure 6.34, middle panel). When fitting the same model using data up to the end of 2001, our conclusion may not hold (Figure 6.34, bottom panel). Nutrient concentrations in rivers are correlated with stream flow. North Carolina is frequently affected by Atlantic hurricanes. With increased river flow accompanied by hurricanes, nutrient concentrations in rivers typically decrease if the main source of nutrient is from point sources such as wastewater treatment plant discharges. As the Clayton site is just downstream from the region's main metropolitan area, we expect higher flow is associated with low nutrient concentrations. The years 1996-1999 were unusually wet due to several strong hurricanes hitting the area. From 2000 to 2004 the area did not experience major hurricanes. As a result, the drop (in late 1990s) and subsequent rebound (2000–2001) of TKN are part of a lower frequency cyclical pattern that is not captured by the seasonal component of the STL model. The local peak between 1990-1995 was not reflected in the long-term trend. Should a longer time series become available, we would be able to see whether the dip in TKN concentrations in the late 1990s is a transient event.

6.4 Bibliographic Notes

This chapter again omitted statistical theories of nonlinear regression models, which is covered by Bates and Watts [2007]. Details of nonparametric regression methods can be found in Härdle [1991] (smoothing) and Hastie and Tibshirani [1990] (additive models). Muggeo [2003] presents the general use of piecewise linear models that is not limited to one threshold point. Cleveland [1993] presented a detailed analysis of the CO_2 data set using STL.

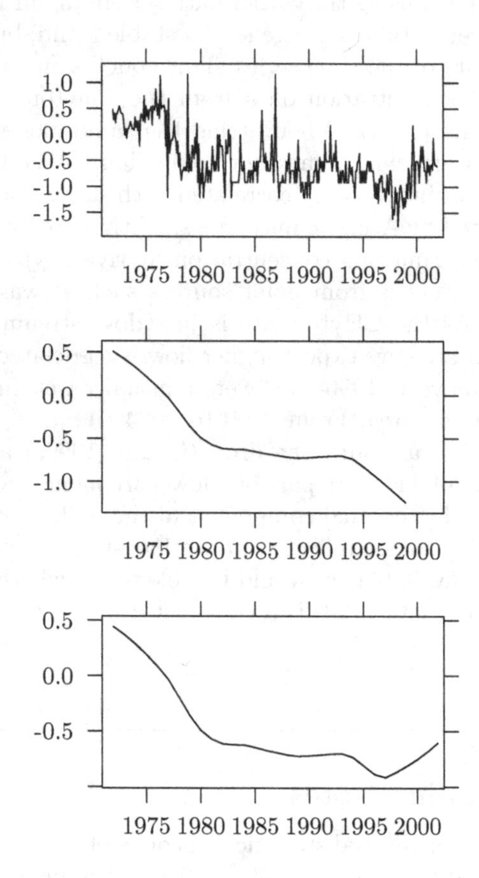

FIGURE 6.34: Long-term trend of TKN in the Neuse River – The TKN concentration time series (top panel) is compared to the fitted STL models using two different lengths. The middle panel is fitted using data up to 1998 and the bottom panel is fitted to the end of 2001.

6.5 Exercises

1. PCB in fish is a widespread problem, not only in the Great Lakes, but also in smaller lakes in the region. Bache et al. [1972] reported measured PCB concentrations of a number of lake trout from Cayuga Lake to the north of Ithaca, New York (data file `CayugaPCB.txt`). In their report, PCB concentrations were predicted by fish age using a log-linear regression model ($\log(PCB) = \beta_0 + \beta_1 age + \epsilon$). The data were later used as an example of nonlinear regression by Smyth [2002], but with a different model ($\log(PCB) = \theta_1 + \theta_2 age^{\theta_3} + \varepsilon$).

   (a) Fit both models and compare their fit to the data by analyzing the residuals from both models.

   (b) Fit a loess model and plot the resulting models on the scatter plot of log PCB against age.

   (c) In the Cayuga PCB data, we also have information on fish sex (juvenile, female, and male). Use the log-linear regression model to decide whether model coefficients vary by sex.

2. Write a self-starter function for the piecewise linear model.

3. Borsuk et al. [2001] used the following equation to describe the sediment oxygen demand in estuaries:

$$SOD = a\left(\frac{L_c}{[1+(n-1)kL_c^{n-1}h]^{1/(n-1)}}\right)^b$$

   where L_c is areal carbon loading, h is the depth of water column, and a, b, k, n are unknown parameters to be estimated. Use the data in file `SODdata.txt` to estimate the four unknown parameters, and discuss whether the model should be fit in the logarithmic scale.

4. Qian et al. [2003b] used a mixed order biological oxygen demand (BOD) decay model to illustrate the Bayesian Monte Carlo simulation. BOD is the amount of oxygen consumed by microorganisms, typically measured in a domestic sewage treatment plant. The model describes the oxygen consumed (BOD exerted) at time t or L_t:

$$L_t = \begin{cases} L_0 - [L_0^{1-N} - k_n t(1-N)]^{\frac{1}{1-N}} + \epsilon, & N \neq 1 \\ L_0(1 - e^{-k_n t}) + \epsilon, & N = 1 \end{cases}$$

   where L_0 is the ultimate BOD, k_n is the reaction rate constant, and N is the order parameter. Data used by Qian et al. [2003b] are in file `bodMCMC.s` (use function `source` to read it into R).

(a) Use `nls` to estimate the mixed order model parameters k_n, L_0, and N.

(b) Fit compare the resulting model to the first order model ($L_t = L_0(1 - e^{-k_n t}) + \epsilon$).

(c) Compare the two models based on their residuals.

5. The "one-gram" rule of P retention in wetland was based on an additive model reported in Reckhow and Qian [1994] (largely Figure 6.28). The response variable of the additive model is effluent TP concentration. When log transformed effluent TP concentration and input TP loading are plotted (Figure 6.24), TP response to loading rate can be more appropriately described by the 4-parameter logistic model. Fit a loess model using log effluent TP concentration as the response and log TP loading rate as the predictor to discuss whether a FPL is appropriate. If a FPL is appropriate, estimate the parameters using the appropriate self-starter function (`SSfpl` or `SSfpl2`).

6. Use the six ELISA tests data from the weekend of August 2, 2014 conducted by the Toledo Water Department (data on page 406) to determine whether the FPL should be defined on the logarithmic concentration scale or on the concentration scale.

7. Use the data from Exercise 4 in Chapter 2 to fit the following loess models:

   (a) Temporal changes in SRP (soluble reactive phosphorus): use SRP (or log SRP) as the response and time as the predictor;

   (b) Calculate the SRP load to Lake Erie (product of SRP concentration and flow) and model the temporal changes of SRP load.

8. Use the SRP concentration and loading data from the previous problem to document the long-term and seasonal trends using STL. Note that although the monitoring program aimed at collecting daily samples, there have been occasional missing days in the record. These missing values should be imputed using median polish with appropriate monthly or weekly averages.

Chapter 7

Classification and Regression Tree

7.1 The Willamette River Example 272
7.2 Statistical Methods 275
7.2.1 Growing and Pruning a Regression Tree 277
7.2.2 Growing and Pruning a Classification Tree 285
7.2.3 Plotting Options 289
7.3 Comments 293
7.3.1 CART as a Model Building Tool 293
7.3.2 Deviance and Probabilistic Assumptions 297
7.3.3 CART and Ecological Threshold 298
7.4 Bibliography Notes 300
7.5 Exercises 300

The statistical models we studied so far (linear regression, nonlinear regression, nonparametric smoothing, and additive models) have a common feature: to properly use these modeling techniques, we need to know which predictors to use in a model. Knowing which predictor variables to use is often the first step of a study. However, a variable selection result is always model-specific. For example, using a linear regression model, variables selected will likely be those having a linear relationship with the response variable. Even when the additive model is used, the additive assumption will affect the variable selection results. In environmental and ecological studies, the additive assumption is rarely realistic. As a result, both linear regression and the additive model are inefficient in selecting variables that may have strong interaction effects.

In this chapter an alternative to linear and additive models, the classification and regression tree (CART) model, is introduced. CART is a binary recursive partitioning method yielding a class of models called *tree-based models*. The method is attractive to many due to its capability of handling both continuous and discrete variables, its inherent ability to model interactions among predictors, and its hierarchical structure. Instead of using CART for its conventional functions (i.e., prediction and classification), I use CART primarily for identifying variables that contribute significantly to the variability of the outcome or response variable. This chapter starts with an example that motivated me to learn and use CART in Section 7.1, followed by a discussion of the statistical methods (Section 7.2). The chapter concludes with a discussion on the use of CART as a variable selection method for developing predictive models.

7.1 The Willamette River Example

In 1996, the U.S. Geological Survey's Portland, Oregon district conducted a survey of water quality in small streams in the Willamette River basin to study the distribution of dissolved pesticides and other water quality constituents and their relation to land use [Anderson et al., 1997]. Prior to this survey, the Oregon Department of Environmental Quality (ODEQ), through the Willamette River Basin Water Quality Study and data reported by the U.S. Geological Survey's National Water Quality Assessment (NAWQA) program, had concluded that synthetic organic compounds (pesticide residuals) from runoff are a potential concern for the area. The survey was primarily aimed at understanding the relationship between land use and dissolved pesticide concentrations using statistical modeling techniques. A total of 36 pesticides (29 herbicides and 7 insecticides) were detected basin-wide. The USGS report [Anderson et al., 1997] documented the frequency distributions of the measured pesticide concentrations and conducted simple tests to compare these concentrations from agricultural watersheds to those from urban watersheds. The relationship between pesticide concentrations and land use is indirectly inferred by using a cluster analysis. The resulting database consists of the following potential predictor variables: the percentage of sub-basin areas covered by agricultural (`Ag`), residential or urban (`Resid`), and forested (`Forest`) land uses, watershed size (`Size`), number of crops in the watershed (`NumCrop`), sampling site location (Latitude and Longitude), river water chemistry measurements (NH_4^+ or `NH4`; $NO_2^- + NO_3^-$ or `NOx`, `TKN`, 5-day biochemical oxygen demand or `BOD`; total phosphorus or `TP`; soluble reactive phosphorus or `SRP`), fecal bacteria, and month of the year (Month) when samples were taken (April, May, July, October, and November) to represent the seasonal effect. Water chemistry measurements can be used to represent the degree of human impact of the watershed. For instance, agricultural intensity may be partly represented by the concentration of nutrients, and BOD and fecal coliform bacteria are indicative of pollution from human and animal waste.

Developing a predictive model of pesticide concentrations using these predictors can be a tedious and difficult process because of the large number of potential predictors and the potentially nonlinear relationships. Furthermore, the response variables are measured with substantial amounts of data points with values below method detection limits (MDL). When a concentration value is below the MDL, the exact concentration value is only known to be less than a specific value. Data points with this characteristic are referred to as "nondetections" or "left censored." When the underlying probability distribution of the contaminant concentration is of interest, a censored value contributes less information than does an exactly measured value. Nevertheless, the censored value has information that should not be disregarded when estimating the probabilistic distribution parameters.

Many statistical methods are available for estimating distribution parameters when censored data are present (e.g., Gleit [1985]). The simple substitution method replaces the left-censored value by a specific number (e.g., 0, one-half of the MDL value, or the MDL value), which may lead to biased estimates. Alternatively, a regression on ordered statistics (ROS) method for estimating the mean and variance of a log-normal distribution has been proposed [Gilliom and Helsel, 1986, Helsel and Gilliom, 1986]. However, for the purpose of this example, these methods are not suitable because the objective of the survey is to develop a predictive model of pesticide concentrations, not estimating the probability distributions of pesticides.

Qian and Anderson [1999] used the graphical-based nonparametric tree-based models (CART). Unlike the nonparametric regression models based on smoothing techniques (e.g., the additive model in Chapter 6), a CART model derives simple prediction rules by dividing the predictor variable space into rectangular subspaces and each subspace is assigned a single response variable value. For example, Figure 7.6 on page 283 is a regression tree model for predicting diuron concentrations. The model is presented using a decision tree with multiple nodes. Each node represents a binary split of a predictor variable. The top node is the root node, where the variable `LU.Ag` (agriculture land use as a percentage of the watershed) is split at 74.5, that is, the data set is divided into two subgroups based on whether the agriculture land use is less than 74.5% (go to left) or not (go to right). When using the model in Figure 7.6 for prediction, we can view the model as a set of simple rules:

1. If agriculture land use exceeds 74.5% of the watershed,
   (a) the log diuron concentration is −0.87, if NH_4 concentration is below 0.0835
   (b) the log diuron concentration is 1.3 otherwise
2. If agriculture land use is less than 74.5% of the watershed,
   (a) log diuron concentration is −0.48, if NO_x concentration exceeds 1.735
   (b) log diuron concentration will be determined by forest coverage if NO_x concentration is less than 1.735
      i. it is −3.5 if forest cover is larger than 5.5%, and
      ii. −1.5 if the forest cover is less than 5.5%

These rules are easy to use and the hierarchy of the tree structure suggests the relative importance of the predictors used in the model.

When the response variable is categorical, for example, the species of the iris data (Figure 3.17, on page 71), the resulting tree is a classification model, and the classification tree in Figure 7.1 can also be expressed in terms of simple rules:

1. If petal length is less than 2.45, it is *Setosa*,

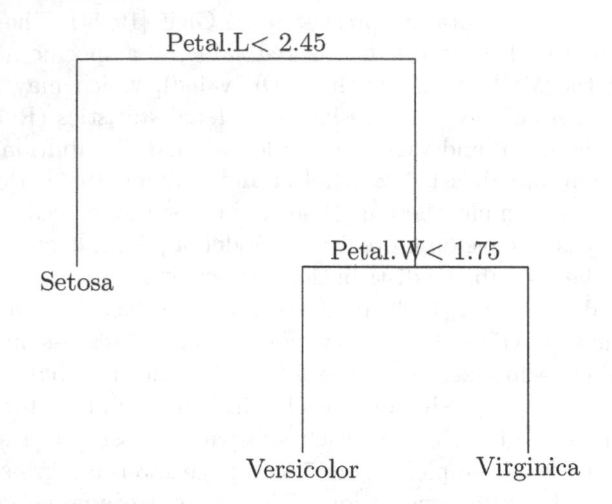

FIGURE 7.1: A classification tree of the iris data – The three species of iris are classified using a classification tree.

2. If petal length is larger than 2.45,
   (a) it is *Versicolor* if petal width is less than 1.75
   (b) otherwise, it is *Virginica*

Because only two predictor variables are used in the resulting tree model, the predictor space is two-dimensional. This set of rules can be presented as a set of rules that divides the predictor variable space (Figure 7.2).

In the Willamette River example, response variables (pesticide concentrations) are treated either as continuous or as categorical. A continuous concentration variable with substantial number of left-censored values is converted into a categorical variable by dividing the data into categories such as below MDL, low, mid-range, and upper-range, and a classification tree model can be used. Using CART, Qian and Anderson presented predictive models for five commonly used herbicides and three pesticides collected from small streams in the Willamette River Basin in Oregon to identify factors that affect the variation of their concentrations in the area.

CART can be used as an effective variable selection tool largely due to its hierarchical structure, which reveals the relative importance of each selected variable in the fitted tree structure. Using a tree-based model as an exploratory data analysis tool, we are able to explore the structure of data while imposing few prior assumptions. Furthermore, tree-based models are much easier to interpret and discuss than linear models when the set of predictors contains

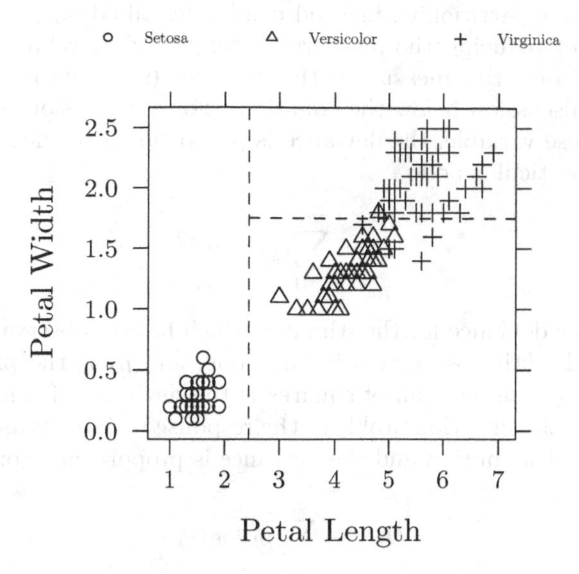

FIGURE 7.2: Classification rules for the iris data – The predictor space is divided into three rectangular subspaces for species classification.

a mix of numeric variables and factors. Because the predictors are split into subsets, tree-based models are invariant to a monotone transformation of a predictor so that the precise form in which these appear in a model is irrelevant. Tree-based models are more adept at capturing nonadditive behavior (the standard linear model does not allow interactions between variables unless they are prespecified and of a particular multiplicative form [Clark and Pregibon, 1992]).

7.2 Statistical Methods

Tree-based modeling, introduced by Breiman et al. [1984], is an exploratory technique for uncovering structure in data, increasingly used for (1) devising prediction rules that can be rapidly and repeatedly evaluated, (2) screening variables, (3) assessing the adequacy of linear models, and (4) summarizing large multivariate data sets. The models are fitted by binary recursive partitioning, whereby a data set is successively split into increasingly homogeneous subsets until it is infeasible to continue [Clark and Pregibon, 1992]. Tree-based models are so called because the primary method of displaying the fit is the form of a binary tree.

The recursive partitioning method can be described on a conceptual level as a process of reducing the measure of "impurity" [Breiman et al., 1984]. Typically, the impurity measure is the deviance (see equation (5.6) on page 156 and the discussion below the equation). For a regression problem with a normal response variable, the deviance is proportional to the residual sum of squares of a particular node i:

$$D_i = \sum_{k=1}^{m_i} (y_k - \mu_i)^2 \tag{7.1}$$

where D_i is the deviance for the ith node which has m_i observations (indexed by k), y_k is the kth observation in the node, and μ_i is the predicted mean for node i. The residual sum of squares is the deviance of a normal random variable. For a classification problem, the response variable is assumed to have a multinomial distribution and the deviance is proportional to:

$$D_i = -\sum_{k=1}^{g_i} p_k \log(p_k) \tag{7.2}$$

where g_i is the number of classes in the node, and p_k is the proportion of observations that are in class k. This measure is also known as the information index, because the entropy of Shannon information theory is $-\sum_{k=1}^{g_i} p_k \log_2(p_k)$. Often a "Gini impurity" is also used, defined as

$$D_i = \sum_{k=1}^{g_i} p_k(1 - p_k) \tag{7.3}$$

The Gini impurity is not related to the Gini index, popular measure of a country's income inequality.

The deviance is zero for a pure node, in which all the y_k's are the same (for a regression problem) or all the observations belong to one class (for a classification problem). At the beginning, all observations are assigned to the same "node." Each split puts observations into two child nodes (left and right), and the deviance after split is:

$$D_{i\ child} = D_{i,L} + D_{i,R} \tag{7.4}$$

The split that maximizes the reduction in deviance $\Delta D = D_i - D_{i\ child}$ is the split chosen at a given node. Specifically, the model visits each predictor in turn and splits the response variable into two groups. The splitting is based on the current predictor variable. If the predictor x_j is a continuous or ordered categorical variable, the split is to divide the response variable into two groups based on whether the predictor x_j is less than or greater than a specific value. If x_j has n_j unique values, there are $n_j - 1$ possible ways to split the response variable data. The model will try all of these possible splits and calculate the deviance reduction of each possible split. If the predictor x_j

is a categorical variable, the model will try all possible binary divisions and record the deviance reduction of each split. After exhausting all predictors, the predictor with the split that maximizes the deviance reduction is chosen as the best predictor. This approach is known as the greedy algorithm, in that the algorithm picks the best split for the current node only, without considering the performance of the overall tree. After each split the original data set is divided into two subsets. The same process will repeat for the two subsets. This process "grows" a tree. For a regression problem, this process is equivalent to choosing the split to maximize the between-groups sum-of-squares in a simple ANOVA problem.

7.2.1 Growing and Pruning a Regression Tree

In R, CART is implemented in two different libraries: The first is an R native library called `tree`, which implements the tree methodology described in Chambers and Hastie [1991]. The second is the `rpart` library [Therneau et al., 2015], which implements methodology closer to the CART of Breiman et al. [1984]. The library `rpart` is used in this book.

Functions included in the `rpart` library can be classified into two categories: modeling functions and plotting functions. The modeling functions are responsible for fitting a tree model, including `rpart` for fitting the actual model, `rpart.control` for tuning parameters that feed into `rpart`, `summary.rpart` for summarizing the fitted model, `snip.rpart` for interactive pruning features, and `prune.rpart` for pruning. Plotting functions responsible for producing the output as a nice graphical figure include `plot.rpart`, `text.rpart`, and `post.rpart` (for producing postscript versions of the fitted models).

In the Willamette River Basin example a regression tree model was used to develop a predictive model for predicting concentrations of eight pesticides. In this section, the herbicide diuron is used to illustrate the process of model fitting and assessment. Diuron is an urea herbicide used to control a wide variety of annual and perennial broadleaf and grassy weeds. It is used to control weeds and mosses on noncrop areas and among many agricultural crops such as fruit, cotton, sugar cane, and legumes. Diuron works by inhibiting photosynthesis. The data set consists of 94 measurements of diuron concentrations from 20 monitoring sites in 11 subwatersheds taken in spring (April and May), summer (July), and fall (October, November) of 1996. Figure 7.3 presents the logarithmic concentrations divided between agricultural and urban watersheds classified based on the predominant land-use patterns.

To construct a tree model, the function `rpart` is used. As in a linear regression model, a formula is specified:

```
#### R code ####
set.seed(12345)
diuron.rpart <- rpart(log(P49300) ~ NH4+NO2+TKN+NOx+
```

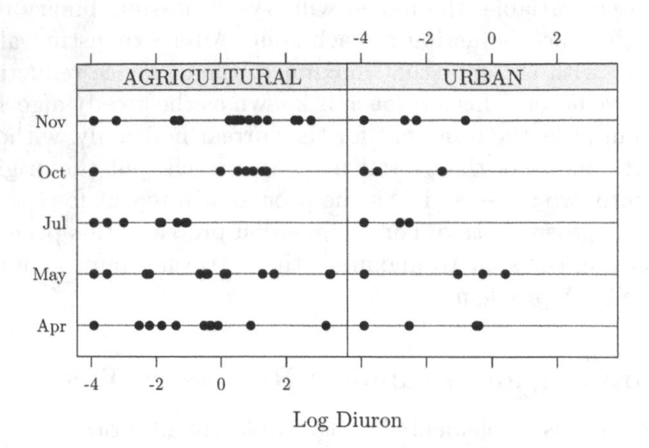

FIGURE 7.3: Diuron concentrations in the Willamette River Basin – The log diuron concentrations collected during the five months in 1996 are shown using a dot plot. Each dot in the plot represent a data point. Urban watersheds tend to have lower concentrations.

```
TOTP+SRP+BOD+ECOL+FECAL+Longitude+Latitude+Size+
LU.Ag+LU.For+LU.Resid+LU.Other+NumCrops+Month,
data=Willamette.data,
control=rpart.control(minsplit=4,cp=0.001)))
```

The line `set.seed(12345)` is used to ensure that the same result will be obtained by all. The fitted model can be presented by using `plot` and `text`:

```
#### R Code ####
 plot(diuron.rpart, margin=0.1)
 text(diuron.rpart, cex=0.5)
```

The result model (Figure 7.4) resembles an upside-down tree. The model is overly complicated and texts in the plot are illegible. The text on top of each node shows the variable and the criterion for splitting the (parent) node into two (child) nodes. The root node (at the top) represents the entire data set. The condition `LU.Ag < 74.5` indicates that the data are to be split into two subsets based on whether the variable `LU.Ag` is less than 74.5% (to the left) or not (to the right). The variable `LU.Ag` is the percentage of agriculture land use in the subwatershed represented by the monitoring station. The 63 data points on the left stem (`LU.Ag < 74.5`) are further split into two subsets based on whether `N2.3` ($NO_2^- + NO_3^-$) is less than 1.375 (to the left) or not (to the right). The fitted model can be seen as a set of rules for predicting log diuron concentrations. That is, for a given observation, we ask a series of questions posed by the splitting criteria, and the answers to these questions will lead us from the root node to one of the terminal nodes. The mean log diuron concentration of the terminal node is the estimated concentration. After the first split, the subset with `LU.Ag<74.5` has a sample size of 63 and the other subset (`LU.Ag>74.5`) has a sample size of 31. The deviances of the two subsets are 183.56 and 94.407, respectively. The sum of the two subset deviances (277.97) is about 62% of the total deviance before splitting (451.95). This represents a reduction of 38% of deviance. This information is summarized in the "Cp-Table":

```
#### R Output ####
> printcp(diuron.rpart)

Regression tree:
rpart(formula = log(P49300) ~ NH4 + NO2 + TKN + N2.3 + TOTP +
    SRP + BOD + ECOL + FECAL + Longitude + Latitude + Size +
    LU.Ag + LU.For + LU.Resid + LU.Other + NumCrops + Month,
    data = Willamette.data,
    control = rpart.control(minsplit = 4, cp = 0.005))

Variables actually used in tree construction:
 [1] BOD  FECAL  Longitude LU.Ag  LU.For  LU.Resid  Month
 [8] N2.3 NH4    Size      SRP    TKN     TOTP

Root node error: 451/94 = 4.8

n=94 (1 observation deleted due to missingness)

        CP nsplit rel error xerror   xstd
1  0.38360      0    1.0000  1.030 0.1030
2  0.13035      1    0.6164  0.819 0.1130
3  0.09846      2    0.4861  0.664 0.0996
4  0.07110      3    0.3876  0.611 0.1023
5  0.04023      4    0.3165  0.547 0.0959
6  0.02545      5    0.2763  0.656 0.1314
7  0.02276      6    0.2508  0.720 0.1392
8  0.02013      7    0.2281  0.733 0.1389
9  0.01798      8    0.2079  0.701 0.1394
10 0.01653     10    0.1720  0.701 0.1394
11 0.01604     12    0.1389  0.687 0.1378
12 0.01049     13    0.1229  0.683 0.1353
13 0.00987     14    0.1124  0.767 0.1437
14 0.00935     15    0.1025  0.778 0.1456
15 0.00835     16    0.0932  0.793 0.1458
16 0.00818     17    0.0848  0.791 0.1459
17 0.00653     18    0.0766  0.790 0.1447
18 0.00561     19    0.0701  0.798 0.1453
19 0.00500     21    0.0589  0.798 0.1451
```

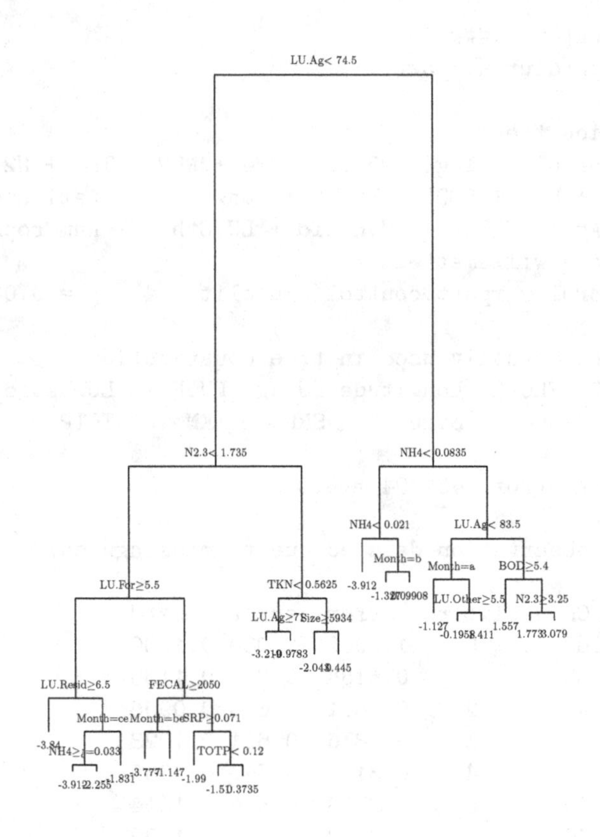

FIGURE 7.4: First diuron CART model – The model fitted by setting the complexity parameter to be 0.005. The fitted model is overly complicated. Most of the spltting variables are illegible, indicating a "tree" to be "pruned."

The function `printcp` shows the basic information about the fitted model – we used 18 predictor variables in the model formula and 13 were used in the fitted model. When fitting the model, we specified `cp=0.005`, which is the model *complexity parameter*. The smaller the `cp` value is, the more complex the model is. The specified `cp` value limits the complexity of the model, and the complexity of a model is directly related to the size of a tree model. The more branches (or splits) a tree has, the more complicated the model is. The summary table shows the complexity parameters for a series of trees. To evaluate a tree model's fit to the data, the root node error (mean deviance of the response variable data) is used as a reference. For this example, the mean deviance of logarithmic diuron concentration is 4.8. The mean deviance is also called "error." A model's relative error is defined as the ratio of the model's mean residual deviance and the root node error. For example, the model with only one split has a relative error of 0.6164 suggesting that the residual variance is only 62% of the root node error. In other words, the model with one split explains about 38% of the total deviance in the response variable data. A model's predictive accuracy is measured by the cross-validation error (`xerror`) and the cross-validation standard error (`xstd`). A cross-validation error is estimated by a simulation process where the original data set is randomly divided into 10 (default) subsets. One subset is set aside as a test data set. A tree model is fit using 9 sets and the set-aside test set is used to evaluate the model. This process is repeated for 10 times each time with a different test data set. The `xerror` is the sum of the 10 errors and the `xstd` is the standard error of the 10 errors.

From the Cp-table, we notice that the `xerror` reduces initially as the number of splits increases until the model with 4 splits. The next model (with 5 splits) has a larger `xerror`. In other words, a tree with more than 5 splits will have a predictive error larger than the model with 4 splits. The increased predictive error suggests that the increased complexity may be a result of "fitting noise." A commonly used method for selecting the "right size" of a tree is to choose the tree with the number of splits (or the cp-value) with the smallest `xerror`. Alternatively, Breiman et al. [1984] suggested to use the 1 standard error (SE) rule, which represents a tree, smaller in size but within 1 standard error to the tree yielding the minimum cross-validated error rate. We can use function `plotcp` to find the right size graphically:

```
#### R Code ####
 plotcp(diuron.rpart)
```

The resulting figure (Figure 7.5) shows that the model with size of 5 terminal nodes (or 4 splits) has the smallest `xerror` and the model with 4 terminal nodes (3 splits) has a `xerror` smaller than the smallest `xerror` plus 1 standard error.

To fit a model with the right size, we can either refit the model by specifying the appropriate Cp-value or use the function `prune`. To prune the model in Figure 7.4 into a model with 4 splits (with a Cp-value of 0.04023):

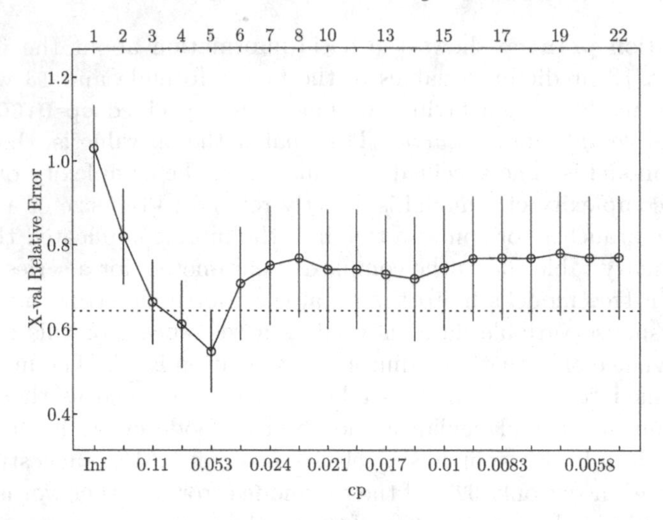

FIGURE 7.5: Cp-plot of the diuron CAR model – Cross-validation error is plotted against Cp-values. The size of a model (number of terminal nodes) is labeled on the top. The vertical line segments are the `xerror` plus/minus 1 standard error. The horizontal dashed line is the minimum `xerror` plus 1 SE.

```
#### R Code ####
diuron.rpart.prune <- prune(diuron.rpart, cp=0.05)
```

The Cp-value used for pruning can be any value between 0.04023 (the Cp-value for a model with 4 splits) and 0.0711 (with 3 splits). The resulting model divides the original data into five subsets, each with a relatively homogeneous distribution in log diuron concentrations. The following scripts produce a plot of the fitted model and boxplots of the corresponding log diuron concentration for each terminal node (Figure 7.6):

```
nf <- layout(matrix(c(1,2), nrow=2, ncol=1), 1, c(2,1))
par(mar=c(0,4,1,2))
plot(diuron.rpart.prune, compress=F, branch=0.4, margin=0.1)
text(diuron.rpart.prune, pretty=T, cex=0.55, use.n=T)
title(main="log diuron Concentration")
par(mar=c(0.5,4,0.5,2))
boxplot(split(predict(diuron.rpart.prune)+resid(diuron.rpart.prune),
    round(predict(diuron.rpart.prune), digits=4)),
        ylab="Diuron Concentrations",
        xlab=" ", axes=F, ylim=log(c(0.01, 50)))
axis(2, at=log(c(0.01, 0.1, 1, 10, 50)),
    labels=c("0.01","0.1","1","10","50"), las=1)
box()
```

If the plus 1 SE rule is used, the final model has 3 splits (Figure 7.7).

log diuron Concentration

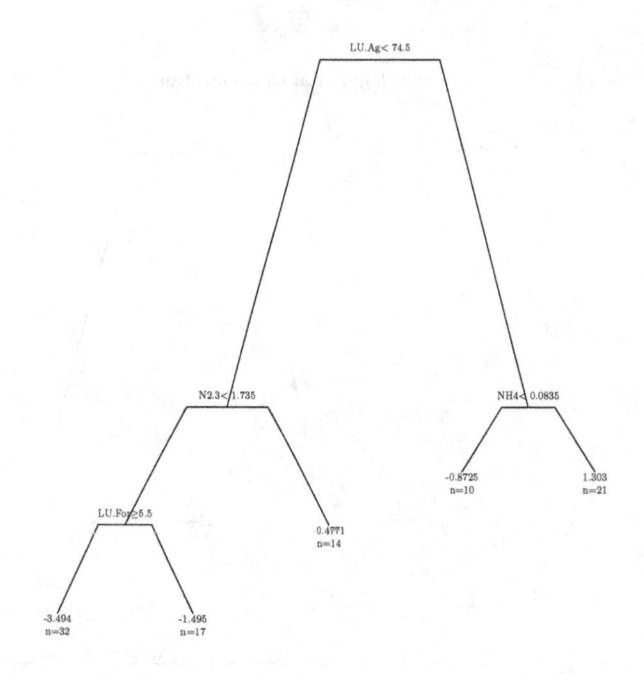

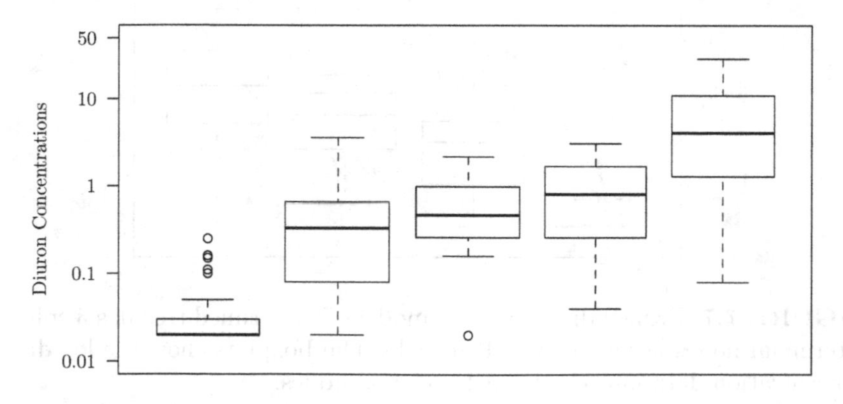

FIGURE 7.6: Pruned diuron CART model – The pruned tree has 4 splits or 5 terminal nodes. The boxplots shows the log diuron concentration data in each of the 5 terminal nodes.

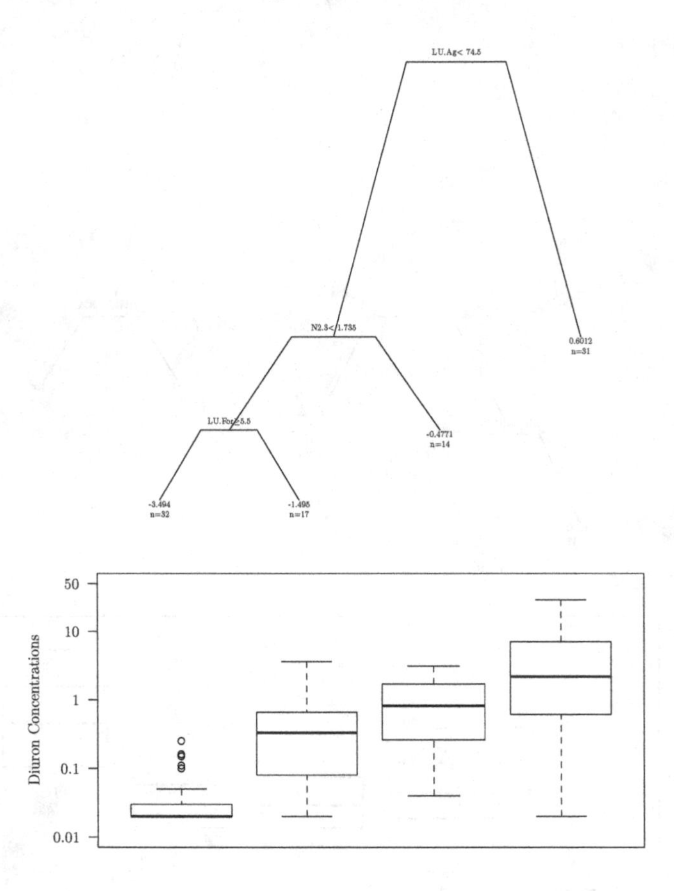

FIGURE 7.7: Pruned diuron CART model – The pruned tree has 3 splits or 4 terminal nodes using the plus 1 SE rule. The boxplots show the log diuron concentration data in each of the 4 terminal nodes.

Because there is no reason to use one rule over the other, whether the model with 3 splits (4 terminal nodes) or the model with 4 splits (5 nodes) should be used as the final model is entirely arbitrary. We also note that the fitted models (in Figures 7.6 and 7.7) are the result of using a specific random number seed, in other words, a specific cross-validation simulation. A different simulation (using a different random number seed and/or a different number of cross-validation subsets) may yield a different result. (The number of cross-validation subsets is set by using option `control = rpart.control(xval = xn)`, where `xn` is the number of cross-validations.) The nonparametric nature of a tree-based model makes it a good exploratory tool, but not a particularly good tool for model building. In Qian and Anderson [1999] CART was used as a tool for identifying variables associated with pesticide concentrations, rather than for predicting pesticide concentrations.

In the final (cross-validated) model, not all candidate predictor variables will be included. This particular feature may be undesirable for some problems. However, in exploratory studies, I view this feature as desirable, since tree-based models can serve as a means of variable selection. In other words, not all candidate variables are important in explaining the variation of the response variable. Using tree-based models, we will be able to identify important variables in predicting the response variable. We note that the binary recursive partition process works one variable at a time; as a result, the number of candidate predictor variables will not pose the problem of overfitting with too many variables as in a linear regression problem. Because a final tree model divides the predictor variable space into subregions, and within each subregion the response variable variance is relatively small, the recursive partitioning process can be viewed as the opposite of that of ANOVA. In other words, ANOVA tests whether the considered factors contribute to the response variance significantly, and the tree-model identifies those factors that contribute significantly to the response variable variation.

7.2.2 Growing and Pruning a Classification Tree

Another type of modeling problem is the classification problem, where the objective of a model is to predict class association. For example, in a water quality management problem, we are interested in determining whether a water body is in compliance of a specific designated use. Using a measured predictor variable, we predict either the water body is in compliance or in violation of a water quality standard. In some cases, continuous variables can be interpreted better if they are converted into categorical variables. For example, a common problem of many environmental studies is that a significant number of observations may be at values below the method detection limit (MDL), or “left censored.” There are many possible ways of dealing with left-censored data; for example, replacing those censored data with a fixed number is a common practice [Helsel, 1990]. However, when the portion of censored data is overwhelming (say, over 20%), replacing them with a constant may bias

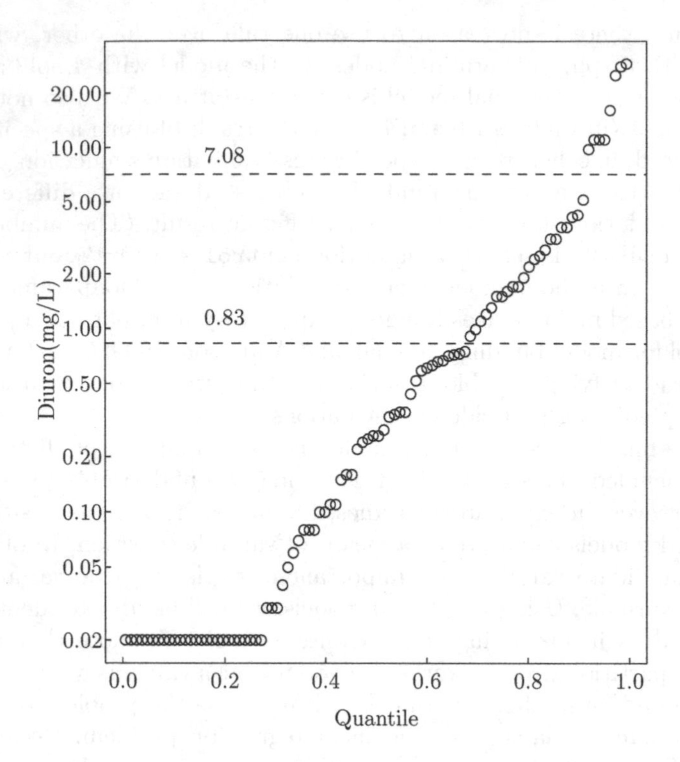

FIGURE 7.8: Quantile plot of the diuron data – Three natural breaks in data distribution (in a logarithmic scale) are visible. One separates values below detection limit from the rest; the other two (at concentration values of 0.83 and 7.08 μg/L) separate the uncensored data into three groups: low, medium, and high.

the resulting analysis. Instead, we may treat the concentration data as categorical, i.e., dividing the data into categories such as "Below MDL," "Low," "Medium," and "High," and a classification tree model can be used. In the Willamette River basin example, diuron concentration was also treated as categorical. The transformation of the continuous concentration values to categorical values is based on the quantile plot of the log diuron concentration (Figure 7.8). A classification tree model develops rules for predicting diuron concentration classes. The model fitting process is similar to the process of fitting a regression model. In R, the same function `rpart` can be used. In our example, we create a factor variable `Diuron`:

```
#### R Code ####
Willamette.data$Diuron <- "Below MDL"
Willamette.data$Diuron[Willamette.data$P49300>=7.08]
      <- "High"
```

```
Willamette.data$Diuron[Willamette.data$P49300<7.08 &
    Willamette.data$P49300>=0.83] <- "Medium"
Willamette.data$Diuron[Willamette.data$P49300<0.83 &
    Willamette.data$P49300>0.02] <- "Low"
Willamette.data$Diuron <- ordered(Willamette.data$Diuron,
    levels=c("Below MDL","Low","Medium","High"))
Willamette.data$Diuron[is.na(Willamette.data$P49300)] <- NA
```

As in the regression model, a model formula is to be specified with `Diuron` as the response variable:

```
set.seed(12345)
diuron.rpart2 <- rpart(Diuron ~ NH4+NO2+TKN+N2.3+TOTP+
    SRP+BOD+ECOL+FECAL+Longitude+Latitude+Size+LU.Ag+
    LU.For+LU.Resid+LU.Other+NumCrops+Month,
    data=Willamette.data, method="class",
    parms=list(prior=rep(1/4, 4), split="information"),
    control=rpart.control(minsplit=4,cp=0.005))
```

The option `method="class"` indicates a classification model is to be fit. In addition, we specifically specify two parameters, `prior` and the split method (`split`) (as part of the model parameter list in `parms`). The `prior` is a vector of prior probabilities of each class. These probabilities describe our knowledge of the population distribution of the categorical response variable. The distribution can be interpreted as the relative frequency of each level of the factor variable. In this example, the categorical variable was created based on the measured diuron concentration values. We have no specific information on the relative frequencies of the four levels in the Willamette River Basin (hence the equal probability). By default, if the prior is not specified, the relative frequencies of the levels observed in the data will be used. The split method is to tell R whether to use the information index (equation 7.2) or the Gini index (equation 7.3) to calculate the deviance.

As in the regression example, our initial model is overly complex (Figure 7.9), resulting in an illegible graphic model. The Cp-table and the Cp-plot (Figure 7.10) provide a basis for selecting the tree with a proper size (3 splits or 4 terminal nodes).

The final tree (Figure 7.11) has a Cp-value close to 0.06:

```
#### R Code ####
diuron.rpart2.prune <- prune(diuron.rpart2, cp=0.06)
```

The classification model has the same tree structure as the regression model in Figure 7.7. This coincidence was used as justification for treating a pesticide concentration variable with a large amount of concentration values below its detection limit as a factor variable.

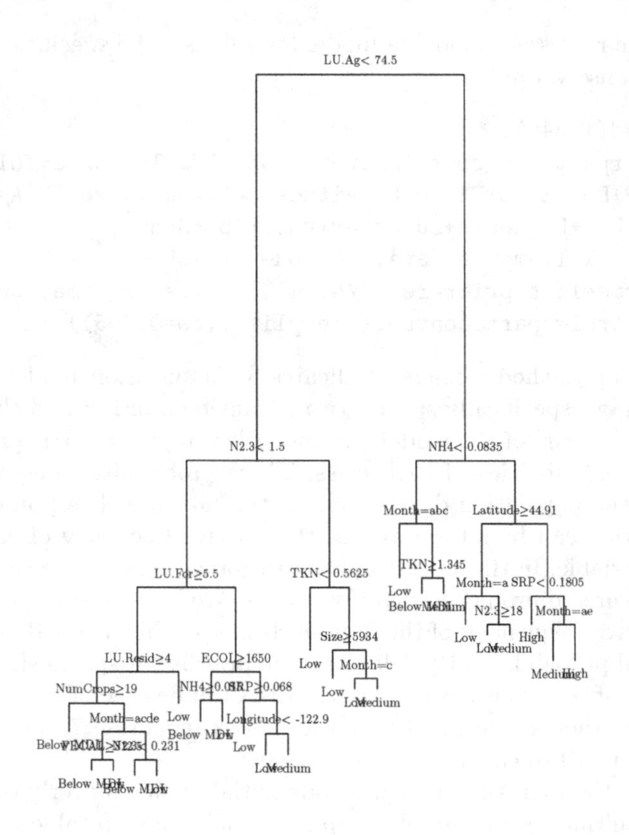

FIGURE 7.9: First diuron CART classification model – Setting the complexity parameter to be 0.005 resulted in an overly complicated model. The crowded and illegible tree suggested that pruning is necessary.

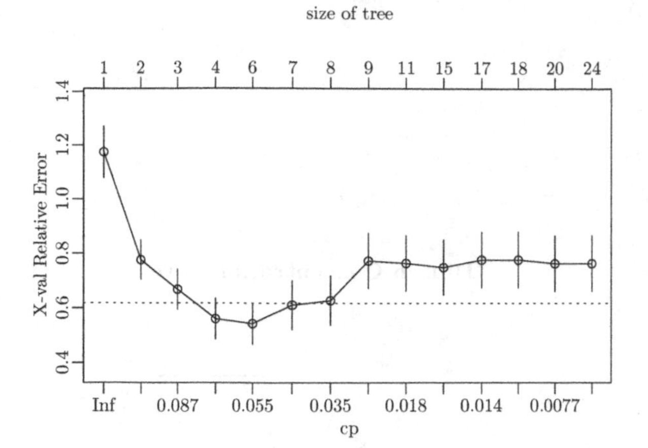

FIGURE 7.10: Cp-plot of the diuron classification model – Cross-validation error is plotted against Cp-values. The size of a model (number of terminal nodes) is labeled on the top. The vertical line segments are the `xerror` plus/minus 1 standard error. The horizontal dashed line is the minimum `xerror` plus 1 SE.

7.2.3 Plotting Options

The final CART model is always presented graphically. The R package `rpart` provides several options to present the fitted model. We use the model fit using the equal prior and Gini impurity to illustrate the various options.

By default, the use of `plot` and `text` will produce the basic output (Figure 7.12):

```
#### R Code ####
# Default
plot(diuron.rpart5.prune,margin=0.1,
   main="a. default")
text(diuron.rpart5.prune)
```

The argument `margin` provides an extra percentage of white space to leave around the borders of the tree. The default margin often cuts off labels. The main options of the `plot.rpart` function are `uniform`, and a logical argument specifying whether or not uniform vertical spacing of the nodes is used, and `branch`, a numeric number between 0 and 1 controlling the shape of the branches from parent to child nodes. A value of 1 gives square-shouldered branches, and a value of 0 gives inverted V-shaped branches, with other values being intermediate. Figure 7.13 is produced using the following scripts:

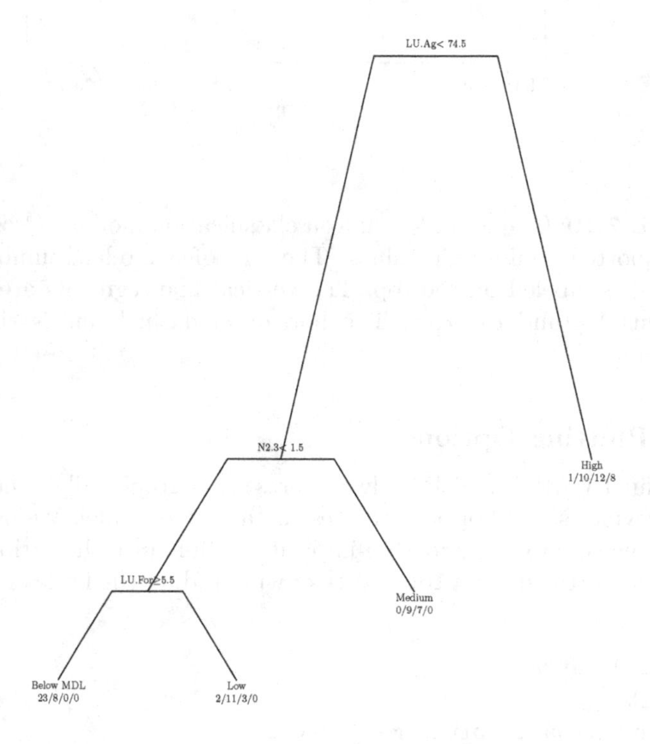

FIGURE 7.11: Pruned diuron classification model – The diuron classification model pruned to have 3 splits or 4 terminal nodes.

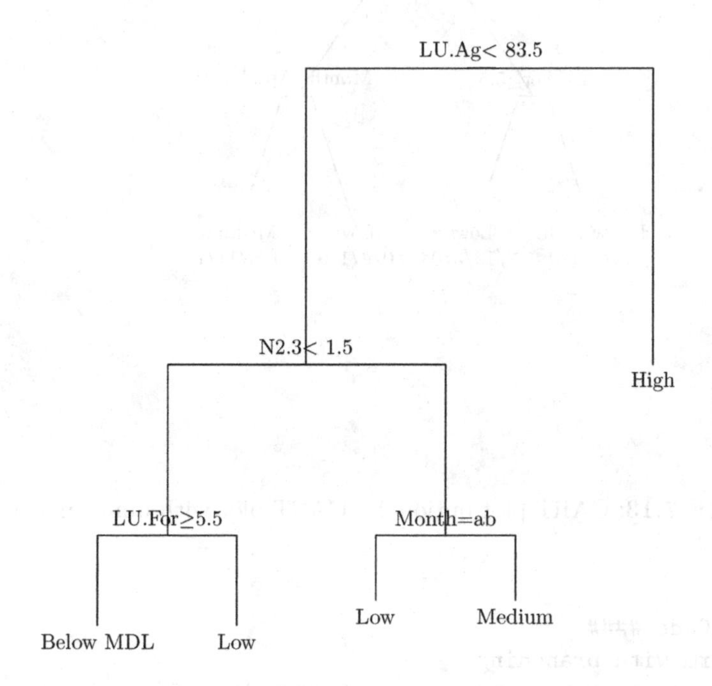

FIGURE 7.12: CART plot option 1 – Default CART plot.

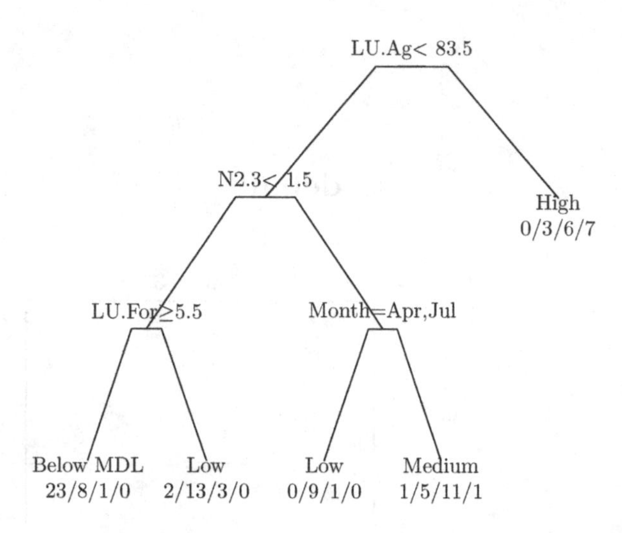

FIGURE 7.13: CART plot option 2 – CART plot with uniform spacing and branching.

```
#### R Code ####
# Uniform with branching
plot(diuron.rpart5.prune,uniform=T,branch=0.25,
margin=0.1, main="b. uniform with branching")
text(diuron.rpart5.prune,pretty=1,use.n=T)
```

The function `text.rpart` also provides options for presentation. In Figure 7.13, the argument `pretty` is an integer denoting the extent to which factor levels in split labels will be abbreviated. In Figure 7.12, the split on the factor variable `month` is labeled as `month=ab`. When using `pretty=1` the label is changed to `month=Apr,Jul`. The argument `use.n=T` adds sample sizes to the terminal node labeling.

The other two less-used arguments for `text.rpart` are `all=T` (labeling all nodes, instead of the default of labeling only the terminal nodes) and `fancy=T` (representing interior nodes by ellipses and terminal nodes by rectangles (Figure 7.14).

```
#### R Code ####
plot(diuron.rpart5.prune,uniform=T,branch=0,
    margin=0.1,main="c. fancy")
text(diuron.rpart5.prune,pretty=1,
    all=T,use.n=T,fancy=T)
```

7.3 Comments

7.3.1 CART as a Model Building Tool

The use of CART as a tool for developing a predictive model has its obvious advantages. For example, we don't have to decide which predictor variable and in what form (i.e., what transformation) to use. CART is also very efficient in displaying interaction effects. However, many applications of CART take the final (cross-validated) model literally. They often describe the final model as the definite description of the data structure. Because CART is a nonparametric exploratory tool, it should be used with caution. CART is often used to identify important predictors. Many applications simply list the variables used in the final model. This practice can be misleading for the following reasons:

- A tree model is fitted recursively and each split is selected to maximize the deviance reduction locally (the greedy algorithm). That is, the reduction is maximized for the current split only. It is possible that a locally less optimal split may lead to a better overall model. As a result, the selected variables in the final model are not always the most important predictors.

- The predictor variable selected in the final model may be a surrogate of a more important predictor.

- For a classification tree, different split methods and/or prior probabilities may lead to quite different trees.

When using CART for variable selections discussed in the previous section, it is important to consider not only the variables presented in the final model, but also their competing variables. Competing variables are presented in the summary statistics of an `rpart` object. For example, the summary for the first node in the `rpart` object for the model in Figure 7.11 shows that the variable `NH4` is almost as effective as the selected variable `LU.Ag`:

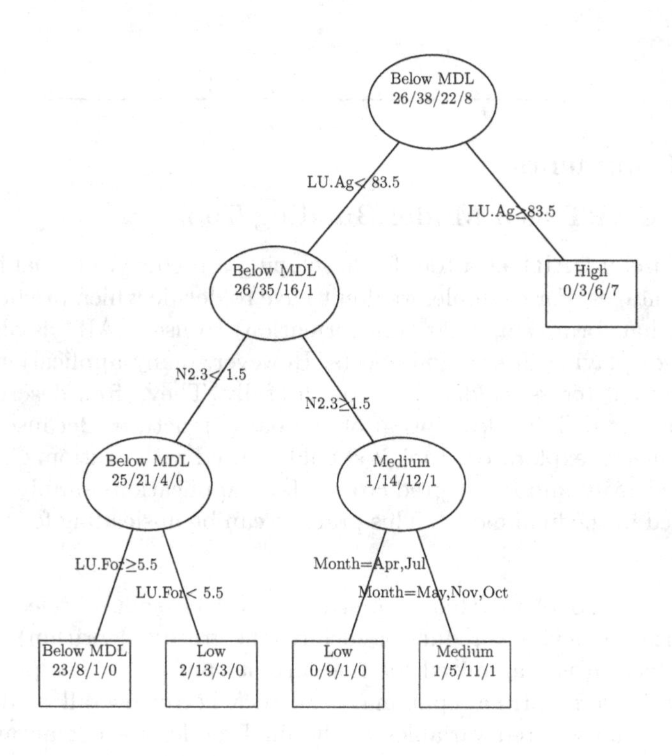

FIGURE 7.14: CART plot option 3 – Fancy CART plot with uniform spacing and branching.

```
#### R Code ####
> summary(diuron.rpart2.prune)
Call:
rpart(formula = Diuron ~ NH4 + NO2 + TKN + N2.3 + TOTP + SRP +
    BOD + ECOL + FECAL + Longitude + Latitude + Size + LU.Ag +
    LU.For + LU.Resid + LU.Other + NumCrops + Month,
    data = willamette.data, method = "class",
    parms = list(prior = rep(1/4, 4), split = "information"),
    control = rpart.control(minsplit = 4, cp = 0.005))
  n=94 (1 observation deleted due to missingness)

       CP nsplit rel error  xerror     xstd
1 0.32051      0   1.00000 1.17483 0.096700
2 0.10606      1   0.67949 0.77564 0.073316
3 0.07085      2   0.57343 0.66719 0.074220
4 0.06000      3   0.50258 0.57487 0.075417

Node number 1: 94 observations,    complexity param=0.32051
  predicted class=Low         expected loss=0.75
    class counts:    26    38    22     8
   probabilities: 0.250 0.250 0.250 0.250
  left son=2 (63 obs) right son=3 (31 obs)
  Primary splits:
      LU.Ag < 74.5     to the left,  improve=31.314, (0 missing)
      NH4   < 0.0835   to the left,  improve=31.256, (22 missing)
      TOTP  < 0.2015   to the left,  improve=29.062, (3 missing)
      N2.3  < 1.69     to the left,  improve=24.230, (3 missing)
      BOD   < 2.55     to the left,  improve=24.218, (30 missing)

......
```

The selected variable and split `LU.Ag` $<$ `74.5` will result in a reduction of 31.314% in deviance (the improvement). If `NH4` is used, the improvement would be 31.256%. The variable `LU.Ag` provides information on the extent of agricultural land use in the watershed, while the variable `NH4` may represent the intensity of agriculture in the watershed because ammonia nitrogen is likely from animal waste or fertilizer used for crops. Consequently, both variables may be important in deciding the variability of diuron in streams. If the objective of the study is to identify important predictors, it is imperative that both the final selected variables and their respective competing primary splits be considered at the same time. CART is considered, in this book, as an exploratory data analysis tool. Using CART as a model building tool can be problematic. For example, when using the classification model for the Willamette River data, specific prior probability and splitting method were used. There are two splitting methods (the information index and the Gini

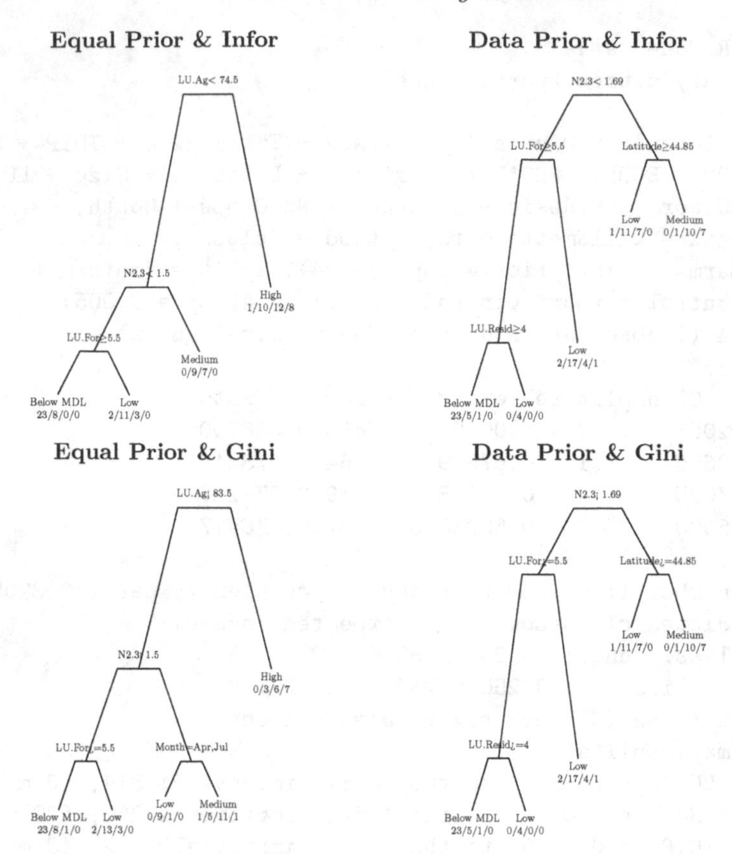

FIGURE 7.15: Alternative diuron classification models – Models fitted using different splitting methods and prior probability definitions are often very different.

impurity) and at least two different ways to specify the prior probabilities. The example in this section used equal prior probabilities, while the default is the relative frequency in the observed data. Just considering prior probability and splitting method, there are four alternative models. Figure 7.15 shows the four models selected using the plus 1 SE rule.

The difference between the two models using different prior probabilities can be avoided if we have an equal number of observations in all 4 levels of the response variable. When using observational data, this option is often infeasible. As with other nonparametric methods, CART is an exploratory tool. The four alternative models in this example suggest that agriculture is the main source of diuron in the Willamette River basin. This conclusion is obvious because diuron is mainly used for agricultural purposes. Ultimately, data analysis is aimed at guiding practitioners to develop the best management practices to reduce pesticide concentration in streams. If the model in Figure 7.11 is used and percent of agriculture land use is interpreted as the main cause of the increased diuron concentrations in streams, it is difficult to provide any useful guidance. The interpretation, that `N2.3` ($NO_2^- + NO_3^-$) and `NH4` (ammonia nitrogen) represent agricultural intensity in the watershed, is no more than a common sense understanding of the process because agriculture is the main reason for using pesticides. However, with this common-sense understanding verified, further studies can be designed to show whether agricultural best management practices such as a riparian buffer could be an effective management practice to reduce diuron concentration in these small streams.

I recommend that CART be used as a variable selection method for building parametric models, as an alternative to the stepwise variable selection procedure often used for selecting predictor variables for building a linear regression model. Building a CART model is an interactive process that requires user's interpretation of the results. This process allows a scientist to judge whether a variable should be included in a model based both on evidence shown from the data and on his/her scientific knowledge. In addition to the potential predictors, the CART model can also suggest whether interaction effects are present.

7.3.2 Deviance and Probabilistic Assumptions

Although CART is often described as a nonparametric method, the term nonparametric applies only to the "mean function," the right-hand side of the model formula. As in all statistical models, probabilistic assumptions about the response variable are necessary. When fitting a regression tree model, the `rpart` default method is to calculate the deviance using the sum of squares (equation 7.1), which implies that the response variable distributions within each of the terminal nodes are normal with a common variance. The generic definition of deviance is -2 times log-likelihood of a model. It is a function of parameters describing the response variable distribution. For a variable with a normal distribution, the likelihood is $\prod_{i=1}^{n} \frac{1}{\sqrt{2\pi}\sigma} e^{-\frac{(y_i-\mu)^2}{2\sigma^2}}$ or $\frac{1}{(2\pi)^{n/2}\sigma^n} e^{-\frac{\sum_{i=1}^{n}(y_i-\mu)^2}{2\sigma^2}}$. The log-likelihood is $\log(\frac{1}{(2\pi)^{n/2}\sigma^n}) - \frac{\sum_{i=1}^{n}(y_i-\mu)^2}{2\sigma^2}$. If the standard deviation σ is assumed to be a constant, the log-likelihood is proportional to $-\frac{\sum_{i=1}^{n}(y_i-\mu)^2}{2}$ and the deviance (-2 times log likelihood) is, hence, proportional to $\sum_{i=1}^{n}(y_i-\mu)^2$. If not specified, the `rpart` function has a set of rules for determining which deviance calculation method to use based on a set of characteristics of the response variable. The default for a numerical response is `method="anova"` or normal distribution deviance. The default implies that the three basic assumptions about the response variable discussed in Chapter 3 (normality, independence, and equal variance) hold.

In ecological and environmental studies, we often encounter response variables that are positive. By definition, these positive response variables cannot be approximated by the normal distribution. For example, in the Everglades study, one metric used for the study of algal level responses to elevated phosphorus concentrations is the relative abundance of diatom species in a sample of all algae. Not only is this variable nonnegative, it is also limited to be within 0 and 1, and its variance is related to its mean ($\sigma_p = \sqrt{p(1-p)/n}$). If the sample size is large, the normality assumption would be appropriate because of the central limit theorem. But the equal variance assumption will always be violated if the fraction varies along the phosphorus concentration gradient. Ideally, a classification tree should be used since the response variable is binary. However, the problem is often compromised because the raw binary data are not routinely reported. Only the relative composition of various species are reported because the counting process usually stops either when a predetermined total number is reached or all cells are counted. Likewise, when the response variable is a count variable, its variance is usually proportional to its mean. A count variable is often approximated by the Poisson distribution. The function `rpart` provides alternative splitting rules for Poisson response variables. For the commonly used microbial species composition data, specific splitting rules may be necessary and the `rpart` function can accommodate such needs.

Because CART is often intended as an exploratory tool, deriving new splitting rules for specific types of response variables may be an overkill. But we should know that the default splitting rule based on the sum-of-squares measure requires approximate normality and constant variance. Commonly used variable transformations should be applied to the response variable before fitting CART. For example, the logit transformation should be tried for percentages. (When there are 0 or 1 in the data, the `logit` function from package `car` can be used to re-scale percentages to between 0.025 and 0.975.)

7.3.3 CART and Ecological Threshold

Ecological threshold has been a topic of many environmental and ecological studies. However, the ecological threshold concept is somewhat new and different authors often have different interpretations of the concept. The commonly adopted definition used by the U.S. EPA relates a threshold to ecological resilience, which is the amount of disturbance that an ecosystem can withstand without changing its self-organizing processes and variables that control its structures, i.e., without shifting to an alternative stable state. An ecological threshold can be defined as a condition beyond which there is an abrupt change in a quality, property, or phenomenon of the ecosystem. Previous research has established that ecosystems often do not respond to a gradual change in forcing variables in a smooth way. Instead, they respond with abrupt, discontinuous shifts to an alternative state as the ecosystem exceeds a threshold in one or more of its key variables or processes. To translate this ecological problem into a statistical problem, a threshold problem must be defined in terms of a change in a probabilistic model. The statistical change point problem was first discussed by Smith [1975] in a Bayesian context, which fits to the ecological threshold concept nicely. In short, we can define the quantitative ecological threshold in two steps. First, an ecological threshold is defined with respect to a specific measure (or metric) of an ecosystem and this metric can be approximated by a probabilistic distribution parameterized by a set of parameters θ. Second, a threshold is a numeric value of a predictor variable at which the response variable distribution parameters change:

$$\begin{aligned} y_i &\sim \pi(y|\theta_j) \\ j &= \begin{cases} 1 & \text{if } x \leq \phi \\ 2 & \text{if } x > \phi \end{cases} \end{aligned} \tag{7.5}$$

where π represent a generic distribution function, and ϕ is the threshold of x. This definition includes both step change and piecewise linear model. For a step change threshold, that is, the ecological metric changes its value abruptly at the threshold, the model can be expressed as:

$$\begin{aligned} y_i &\sim N(\mu_j, \sigma_j^2) \\ j &= \begin{cases} 1 & \text{if } x \leq \phi \\ 2 & \text{if } x > \phi \end{cases} \end{aligned}$$

if the metric can be approximated by a normal distribution. For a piecewise linear regression problem, the model is:

$$\begin{aligned} y_i &\sim N(\beta_0 + \beta_{1j}(x - \phi), \sigma_j^2) \\ j &= \begin{cases} 1 & \text{if } x \leq \phi \\ 2 & \text{if } x > \phi \end{cases} \end{aligned}$$

Because the solution to this general threshold problem often requires advanced Bayesian computation techniques such as the Markov chain Monte Carlo simulation, Qian et al. [2003a] proposed to use the CART model as an alternative for estimating a step change threshold. In this approach, a single predictor x is used to fit a CART model y ~ x and the first split is reported as the likely threshold value. The method is unfortunately named as the nonparametric deviance reduction method, which gives an impression that the model is a distribution-free model. As we discussed in Section 7.3.2, deviance is distribution-specific. Although Qian et al. [2003a] discussed that different types of response variables require using different deviance calculation methods, almost all applications of this method in the literature have used the default sum-of-squares deviance measure in the R function distributed by the authors. Many of these applications have response variables of counts or fractions.

7.4 Bibliography Notes

Breiman et al. [1984] provides a complete description of CART. Much of the theories and practices are originated from this book. Guisan and Zimmermann [2000] introduced CART to the ecological modeling of habitat, and De'ath and Fabricius [2000] discussed the use of CART in some simple ecological data analysis problems. Both emphasize the predictive function of CART. The use of CART as a variable selection and explorative analysis tool is discussed by Qian and Anderson [1999]. CART is often used for pattern recognition and data mining [Ripley, 1996]. A main weakness of the tree-based models is their instability. A very different model may result from either a different model fitting method or a slightly different data set. Breiman [2001] introduced the bootstrapping aggregation (or bagging), where bootstrap samples were used to fit many tree models and the average prediction from these trees is used as the model prediction. This approach is often referred to as the random forest model.

7.5 Exercises

1. When analyzing data from the Finnish Lakes example (Section 5.3.7), Malve and Qian [2006] used CART to explore whether variables other than TP and TN should be used as predictors of chlorophyll a concentrations (`chla`). The data file includes the following potential predictor variables: `totp` (total phosphorus), `type` (lake type, 1–9), `year` (year sample was taken), `totn` (total nitrogen), `month` (month sample was taken), `depth` (average depth), `surfa` (lake surface area), and `color` (a numeric measure of color). Use these potential predictors and the TN:TP ratio to build a CART model for predicting `chla` and another to predict log `chla`. Briefly discuss why a log-transformation of `chla` is necessary but transforming TP and TN is not.

2. Siersma et al. [2014] reported the possible recovery of an environmentally sensitive burrowing mayfly (genus *Hexagenia*) in Saginaw Bay of Lake Huron. *Hexagenia* nymphs are an important food source for many Great Lakes fish. Their decline in density since the 1950s was attributed largely to land use practices in the watersheds that resulted in increased nutrient and sediment loadings. The recovery of these mayflies in recent years were linked to remediation efforts, as well as dreissenid mussel invasion. Siersma et al. [2014] reported a *Hexagenia* survey conducted in 2012 in Saginaw Bay. The study consisted of 48 sampling sites, at which sedi-

ment samples were taken to count the number of *Hexagenia* nymphs and measure other environmental variables. In the data file `SagBayHex.csv`, the number of nymphs are in the column `Hex`. Measured environmental variables include water depth (`depth`), dissolved oxygen concentration (`DO`), water temperature (`Temp`), conductivity, pH level, and percent of sand, clay, and silt in sediment.

(a) Nymphs were found only at 7 of the 48 sites. Use a classification tree model to find the environmental conditions (defined by the measured variables) that are likely associated with the presence of *Hexagenia*.

(b) Although nymphs were found only in 7 sampling sites, the number of nymphs found in each site may also convey information on environmental conditions favored by *Hexagenia*. Use a regression tree model for count data (i.e., `method='poisson'`) to infer the preferred environmental condition.

3. The state of Maine developed a biological condition-based method for evaluating a water body's compliance to its designated use. The method classifies a water into four categories: A (natural), B (unimpaired), C (maintaining structure and function), and NA (non-attainment). When classifying small streams, biological conditions are largely based on metrics calculated from benthic macroinvertebrate data. The Maine Department of Environmental Protection developed a multivariate clustering model. The model uses 25 metrics. The data file `MaineBCG.csv` includes the data used for developing the multivariate classification model. The number of metrics used in the model is large and these metrics are often redundant in that some of these metrics represent the same information. Use a classification tree model to determine whether all 25 metrics are needed for adequate classification. Because the model is for classifying streams not already classified, the model's predictive accuracy is of interest. If not all 25 metrics are necessary, how many (and which ones) should you use?

Chapter 8

Generalized Linear Model

8.1 Logistic Regression 305
8.1.1 Example: Evaluating the Effectiveness of UV as a Drinking Water Disinfectant 306
8.1.2 Statistical Issues 307
8.1.3 Fitting the Model in R 308
8.2 Model Interpretation 309
8.2.1 Logit Transformation 310
8.2.2 Intercept 310
8.2.3 Slope 311
8.2.4 Additional Predictors 312
8.2.5 Interaction 314
8.2.6 Comments on the Crypto Example 315
8.3 Diagnostics 316
8.3.1 Binned Residuals Plot 316
8.3.2 Overdispersion 316
8.3.3 Seed Predation by Rodents: A Second Example of Logistic Regression 319
8.4 Poisson Regression Model 332
8.4.1 Arsenic Data from Southwestern Taiwan 332
8.4.2 Poisson Regression 333
8.4.3 Exposure and Offset 340
8.4.4 Overdispersion 341
8.4.5 Interactions 344
8.4.6 Negative Binomial 351
8.5 Multinomial Regression 353
8.5.1 Fitting a Multinomial Regression Model in R 354
8.5.2 Model Evaluation 358
8.6 The Poisson-Multinomial Connection 361
8.7 Generalized Additive Models 367
8.7.1 Example: Whales in the Western Antarctic Peninsula .. 369
8.7.1.1 The Data 371
8.7.1.2 Variable Selection Using CART 371
8.7.1.3 Fitting GAM 374
8.7.1.4 Summary 378
8.8 Bibliography Notes 380
8.9 Exercises 381

In Chapter 5, we discussed that a linear regression model can be expressed in terms of probability distribution. In both linear and nonlinear regression problems, the normality (more specifically, conditional normality) assumption is used for the response variable:

$$y \sim N(f(x, \theta), \sigma^2) \tag{8.1}$$

where $f(x, \theta)$ represents the mean function. For linear regression models, $f(x, \theta) = X\beta$ is a linear function of predictors. This probabilistic assumption allows us to use the maximum likelihood estimator to estimate model coefficients θ. Model coefficients estimated using MLE are the same as the estimates from using the least squares method. As a result, when discussing linear and nonlinear regression models, we use the least squares method because it is conceptually easy to understand. In practice, when we have response variables that are known to have distributions other than the normal distribution, we often consider transformations to make the residual distribution approximately normal, so that we can use linear or nonlinear regression analysis.

When the response variable follows a different distribution, the least squares estimator may not be the same as the maximum likelihood estimator. Consequently, the maximum likelihood estimator is always used for models with a nonnormal response variable. The generalized linear model or GLM is a class of models for a set of specific response variable distributions from the exponential family of distributions. The exponential family includes many familiar distributions, including the normal distribution. The probability density function of the exponential family can be expressed in a general form (equation 8.2).

$$p(y|\theta) = h(y)e^{\left(\sum_{i=1}^{s} \eta(\theta)T_i(x) - A(\theta)\right)} \tag{8.2}$$

The normal, exponential, gamma, chi-squared, beta, Dirichlet, Bernoulli, binomial, multinomial, Poisson, negative binomial, geometric, and Weibull distributions are part of the exponential family of distributions. For example, the normal distribution $N(\mu, \sigma^2)$ density can be expressed in terms of equation 8.2 by setting:

$$\begin{aligned}
\theta &= \left(\frac{\mu}{\sigma^2}, \frac{1}{\sigma^2}\right)^T \\
h(x) &= \frac{1}{\sqrt{2\pi}} \\
T(x) &= \left(x, -\frac{x^2}{2}\right)^T \\
A(\theta) &= \frac{\mu^2}{2\sigma^2} - \log\left(\frac{1}{\sigma}\right), \text{ and} \\
\eta(\mu) &= \mu
\end{aligned}$$

Members of the exponential family of distributions can be used to describe most (if not all) response variable distributions in environmental and ecological studies. The *link function* η connects the mean parameter to a linear function of predictors:

$$\eta(\mu) = X\beta \tag{8.3}$$

A generalized linear regression problem can be described as follows. We are interested in the relationship between a response variable y and a set of predictors X. The probability distribution of the response is a member of the exponential family with mean μ and variance V. The mean μ is linked to the predictor variables through the link function η as described in equation 8.3. The generalized linear modeling approach provides a computational algorithm for solving for the maximum likelihood estimator of model coefficients (β). In this chapter, we discuss the two most commonly used GLMs for binary (logistic regression) and count (Poisson regression) response variables, as well as GLMs for multinomial distribution, a generalization of the binomial distribution. After discussing the connection between the Poisson regression and the multinomial/binomial regression, we conclude the chapter with the generalized additive model.

8.1 Logistic Regression

A logistic regression model is for a binary response variable y that takes only two values. In general, a binary response variable is represented as a variable taking either 0 or 1 modeled by the Bernoulli or binomial distribution. For example, we may visit a number of sites to survey a certain species of animal. If each site is visited only once, the resulting data for each site would be 0 (for absence) or 1 (for presence). If each site is visited multiple (n) times, the resulting data would be the number of times the animal is spotted. In statistical terms, the total number of visits n is the number of trials. For each trial, a success is when the animal is spotted and a failure is when the animal is not seen. The binomial distribution is the probabilistic model to describe the distribution of the number of successes.

$$p(y = k) = \binom{n}{k} p^k(1-p)^{n-k} \tag{8.4}$$

A Bernoulli distribution is a special case of the binomial distribution when $n = 1$. The distribution in equation 8.4 has one parameter p, the probability of success. The expected value of y is np and the variance of y is $np(1-p)$. The binomial distribution is a member of the exponential family (equation 8.2), with

$$\begin{aligned} \theta &= p \\ h(x) &= \tfrac{n!}{x!(n-x)!} \\ A(\theta) &= n\log(1-\theta) \\ \eta(\theta) &= \log\left(\tfrac{\theta}{1-\theta}\right) \end{aligned}$$

The likelihood function of a logistic regression is defined using the binomial probability distribution function (8.4) with a link function $\eta(p) =$

$\log(p/(1-p))$, the logit transformation of the probability of success, mean parameter of interest.

8.1.1 Example: Evaluating the Effectiveness of UV as a Drinking Water Disinfectant

We introduce the logistic regression model using an example from environmental engineering literature. Korich et al. [2000] reported a study of how laboratory mice respond to the exposure of microscopic parasites of the genus *Cryptosporidium*. The objective of the study was to develop a *dose-response* model, that is, a model to predict the probability of a mouse will be infected if it is exposed to a certain number of the parasites. The study is important because, according to the U.S. Center for Disease Control, *Cryptosporidium* is one of the most common causes of waterborne human diseases in the United States during the past two decades. The parasite causes a diarrheal disease known as *Cryptosporidiosis*. Once an animal or person is infected, the parasite lives in the intestine and passes in the stool. The parasite is protected by an outer shell that allows it to survive outside the body for long periods of time and makes it very resistant to chlorine-based disinfectants. Both the disease and the parasite are commonly known as "crypto." The parasite may be found in drinking water and recreational water throughout the world.

The U.S. EPA in a 2006 publication requires surface water systems in the United States to provide treatment using ultraviolet light (UV) disinfection as one treatment option to meet treatment requirements. Many drinking water utilities in the United States are looking at UV disinfection as the best available technology for meeting their crypto inactivation requirements and goals. The term "inactivation" refers to the fact that UV radiation typically does not kill the parasite, but alters the parasite's nucleic acids, thus preventing replication and infection. As a result, the effectiveness of UV radiation must be evaluated using mouse-infectivity studies.

In a typical mouse-infection study, a dose-response relationship is first determined. In this step, a number of mice are inoculated with a known number of crypto oocysts (d) to observe the number of infected mice. The resulting infectivity data are used to fit a dose-response model, typically a log-linear logistic model of the form:

$$logit(p_i) = \beta_0 + \beta_1 \log_{10}(d_i) \tag{8.5}$$

where p is the probability of infection, d is the number of oocyst ingested, $logit(p) = \log\left(\frac{p}{1-p}\right)$, and i indicates the ith dose level.

In engineering literature, the notation $\log(x)$ often represents the log base 10 of x or the common logarithm, and the natural log (base e) is denoted by $\ln(x)$. In statistics literature, the natural log is commonly denoted by $\log(x)$. In this book, the expression $\log(x)$ is the natural log, and any other based log is explicitly expressed as $\log_b(x)$.

8.1.2 Statistical Issues

There are two main issues. First, many studies equate p to the observed fraction of infected mice, and fit a simple linear regression (e.g., [Korich et al., 2000]). This practice ignores the binary structure of the data and uses a normal approximation. This normal approximation can often lead to misleading uncertainty estimations. On the one hand, the response variable is binary (a mouse is either infected or not infected); the observed fraction of infected mice is an estimate of the expected frequency of infection under a given oocyst dose. We expect the observed fraction of infected mice under a given oocyst dose will change from experiment to experiment because of inherited variations among test mice and experimental conditions. Because of the binary nature of the raw data, a simple regression approach as suggested in equation 8.5 cannot properly model the variability in the estimated probability $\hat{p}$. On the other hand, when the observed fraction of infected mice is 1 or 0, these data points are most likely removed because a logit transformation of 0 or 1 is impossible. Furthermore, using the observed fraction ignores the information in the number of mice tested, an important source of information on uncertainty. In other words, when a fraction, say 0.1, is estimated from 1 infected mouse out of the total of 10 tested, the uncertainty about the estimate is much larger than the same value that is estimated from a total of 100 mice. If we consider the probability of any given mouse being infected is an unknown constant under a given oocyst dose, the variance of the first estimate (0.1) is $0.1 \times (1 - 0.1)/10 = 0.009$, while the same for the second estimate is $0.1 \times (1 - 0.1)/100 = 0.0009$.

Second, the measured response variable is the number of mice infected, which is a count variable that cannot be approximated by the normal distribution effectively. The quantity of interest, the probability of infection, is not directly observed.

The generalized linear model (GLM) [McCullagh and Nelder, 1989] should be used. Because the response variable is binary (a mouse is either infected or not infected) and the probability of infection is only affected by the oocyst dose, a binomial model for the data is often appropriate. Under a binomial model, the variable p in equation 8.5 is the unobserved probability of infection. The binomial distribution model can be expressed as

$$y_i \sim bin(p_i, m_i) \tag{8.6}$$

for the ith dose level. The probability of infection (p_i) is modeled as a function of the oocyst dose (d_i) as in equation 8.5. The likelihood function for this model is proportional to:

$$\prod_{i=1}^{n} p_i^{y_i}(1 - p_i)^{m_i - y_i}$$

a function of unknown model coefficients β_0 and β_1. A fast computer algorithm for the maximum likelihood estimator of β_0 and β_1 is implemented in R.

8.1.3 Fitting the Model in R

When fitting a logistic regression model defined by equations 8.5 and 8.6, we use almost the same syntax as fitting a linear regression model. For the crypto data, the logistic model is fitted by calling the function `glm`:

```
#### R Code ####
  crypto.glm1 <- glm(cbind(Y,N-Y) ~ Dose,
      family=binomial(link="logit"))
```

The option `family=binomial(link="logit")` indicates that the response variable is from a binomial distribution and the *link function* is "logit." The formula `cbind(y,m-y) ~ dose` uses a two-column matrix as the response variable, where the first column is the number of success (infected mice) and the second column is the number of failure (mice not infected). Estimated model coefficients are summarized when calling the generic function `summary`:

```
#### R Output ####
> summary(crypto.glm1)

Call:
glm(formula = cbind(Y, N - Y)~ log10(Dose),
    family = binomial(link = "logit"), data = crypto.data)

Deviance Residuals:
    Min       1Q    Median       3Q       Max
-3.8111  -1.2590  -0.0883   1.7001   5.1206

Coefficients:
            Estimate Std. Error z value Pr(>|z|)
(Intercept)   -4.865     0.329   -14.8   <2e-16
log10(Dose)    2.616     0.162    16.2   <2e-16
---

(Dispersion parameter for binomial family taken to be 1)

    Null deviance: 692.99  on 97  degrees of freedom
Residual deviance: 368.05  on 96  degrees of freedom
AIC: 588
```

The summary table provides basic information about the fitted model. The deviance (-2 log-likelihood) is a measure of model error, similar to the residual sum of squares in a linear regression model. The estimated model coefficients (intercept and slope) are displayed along with their standard error. Whether the estimated model coefficient is different from zero is tested and the test result is presented by using the test statistic (`z value`) and the associated

p-value (`|Pr(>|z|)`). Alternatively, the function `display` from package `arm` can be used to show the most relevant information:

```
#### R Output ####
glm(formula = cbind(Y, N - Y) ~ log10(Dose),
    family = binomial(link = "logit"),
    data = crypto.data)
            coef.est coef.se
(Intercept) -4.865    0.329
log10(Dose)  2.616    0.162
  n = 98, k = 2
  residual deviance = 368.1,
  null deviance = 693.0 (difference = 324.9)
```

Although the output is presented in the familiar form as in a linear regression model, the linear relationship is between the logit of probability of infection and the log dose level:

$$logit(Prob(Infection)) = -4.865 + 2.616 \log_{10}(Dose)$$

We are interested in the probability and the relationship between the probability of infection and the log dose is not linear, however. For this model, the probability is a function of the dose:

$$Prob(Infection) = \frac{e^{-4.865+2.616 \log_{10}(Dose)}}{1 + e^{-4.865+2.616 \log_{10}(Dose)}}$$

a nonlinear relationship shown in Figure 8.1. The dose level of interest is the dose that will result in a probability of infection of 0.5 or LD_{50}. The commonly used method for estimating the LD_{50} in engineering and microbiology studies is to substitute $Prob(Infection) = 0.5$ into the fitted model and estimate the dose level. For this example, the estimated LD_{50} is $\widehat{LD}_{50} = 10^{-\hat{\beta}_0/\hat{\beta}_1} = 10^{4.865/2.616} = 77$. Uncertainty of $\widehat{LD}_{50}$ is often ignored. In this example, the standard error of $\widehat{LD}_{50}$ is, in fact, almost impossible to estimate. We will introduce a simple simulation approach for estimating the standard error in Chapter 9.

8.2 Model Interpretation

The fitted logistic model is presented in the familiar linear model form with intercept and slope. However, because of the logit transformation of the probability, the interpretation of the fitted model is more complicated than the interpretation of a linear regression model. To understand the fitted model, we need to understand the logit transformation.

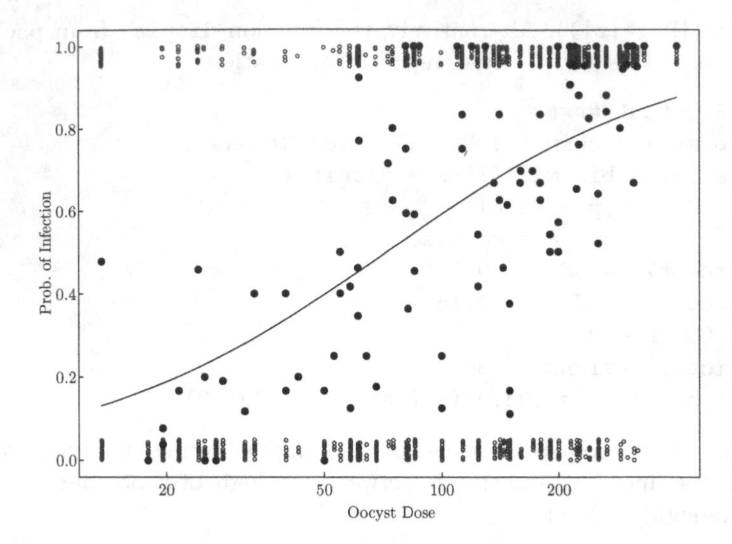

FIGURE 8.1: A dose-response curve – Logistic regression estimates the probability of infection as a function of the crypto oocyst dose (solid line). Black circles represents the outcome of each individual mouse (1=infected, 0=healthy), and gray dots are the fraction of mice infected under each oocyst dose.

8.2.1 Logit Transformation

The logit function, $logit(p) = \log\left(\frac{p}{1-p}\right)$, converts a probability (or fraction) variable p taking values between 0 and 1 to a continuous variable in the real line $(-\infty, +\infty)$. The inverse-logit function $logit^{-1}(x) = \frac{e^x}{1+e^x}$ transforms a continuous variable to the range $(0, 1)$ (Figure 8.2). Although a logistic regression model is presented in terms of a linear model of the predictor, the inverse-logit transformation is nonlinear, leading to a curved relationship between the probability and the predictors. This nonlinearity complicates the interpretation of the fitted logistic model.

8.2.2 Intercept

In a linear model, the estimated intercept is the estimated mean response variable value when the predictor is 0. In a logistic regression model, the intercept is the logit of the estimated probability when the predictor is 0. In the crypto example, the intercept -4.865 is the logit of probability of infection when $\log_{10}(Dose) = 0$ (or oocyst dose = 1). In other words, the probability of infection is 0.0077 if a mouse ingests 1 crypto oocyst. In many applications, centering the predictor variable on a particular dose level will lead to a more interpretable intercept.

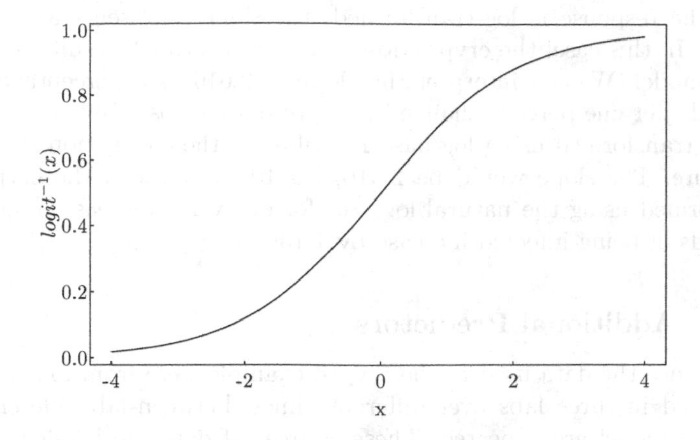

FIGURE 8.2: Logit transformation – The logit transformation converts a variable ranging between (0,1) (the y-axis) into a variable on the real line $(-\infty, +\infty)$. The transformation is nonlinear – the amount of stretch closer to both ends (0 and 1) of the original variable is larger.

8.2.3 Slope

The interpretation of a logistic regression model slope is less intuitive. Literally, the slope in our crypto example can be interpreted as the change in logit probability is 2.616 for every order of magnitude change in oocyst dose. This interpretation is mathematically accurate, but practically difficult to comprehend. The change in probability of infection is not a fixed rate for an order of magnitude change in oocyst dose. As shown in Figure 8.2, the rate of change in the probability of infection depends on the oocyst dose. In other words, the slope of the probability is a function of the predictor variable. The slope is small when $X\beta \to \pm\infty$, and increases as $X\beta$ increases, reaching the maximum at $X\beta = 0$.

Expressed in terms of probability, a simple logistic regression is

$$p = \frac{e^{\beta_0+\beta_1 x}}{1+e^{\beta_0+\beta_1 x}}$$

and the slope of p on x is $\frac{\partial p}{\partial x}$ and the maximum of the slope is $\beta_1/4$. That is, $\beta_1/4$ is the largest change in probability for a unit change in a predictor variable. For the crypto example, we say that the change of probability of infection is always less than 0.68 (2.737/4) for one order of magnitude change in an oocyst dose.

The ratio of the probability of success over the probability of failure is often known as the *odds*. The logit transformation is the log odds. Thus, we can interpret the model in terms of a log-linear model of the odds, if it is a familiar concept to the audience. As discussed in Section 5.4 on page 185, when the response is log-transformed, the slope represents a multiplicative change. In this case, the crypto dose is log-transformed, resulting in a log-log linear model. We can interpret the slope of 2.616 as a percentage change in the odds per one percent change in crypto oocyst dose. Note that the crypto dose is transformed using log base 10 following the convention in engineering literature. The slope would be 2.616/ log(10) or 1.136 if the crypto dose is transformed using the natural log. So, for every 1% increase in crypto dose, the odds of being infected increase by 1.136%.

8.2.4 Additional Predictors

Because the data used in the crypto example were from four experiments conducted in three labs over different times, between-lab differences in the resulting model are expected. These sources of data are labeled in our data as `Finch`, `SPDL-HE`, `SPDL-TH`, and `UA`. These are data reported in Finch et al. [1993] and Korich et al. [2000]. The Korich study reported experiments conducted at the University of Arizona (`UA`) and in the Scottish Parasite Diagnostic Laboratory in Glasgow, UK (`SPDL`). Two methods were used in SPDL for identifying infection. They are histological examination of H&E stained sections of terminal ileum (or HE) and tissue homogenization (or TH). To account for the between-lab/method differences, we can introduce a second predictor variable `Source` (Figure 8.3). Because there are two unknown coefficients in the model, two alternative models are often used. One assumes a common slope and varying intercept. Under this setting, we fit four parallel lines. The other assumes both intercept and slope are different among labs, or fitting four separate lines.

When assuming a common slope, we fit the model in R:

```
#### R Output ####
glm(formula = cbind(Y, N - Y) ~ log10(Dose) + factor(Source),
    family = binomial(link = "logit"), data = crypto.data)
                        coef.est coef.se
(Intercept)             -5.01     0.35
log10(Dose)              2.63     0.16
factor(Source)SPDL-HE   0.05     0.18
factor(Source)SPDL-TH   0.32     0.18
factor(Source)UA        0.07     0.16
  n = 98, k = 5
  residual deviance = 363.8,
  null deviance = 693.0 (difference = 329.1)
```

The R formula `cbind(Y,M-Y) ~ log10(Dose)+factor(Source)` indicates a model with a common slope for the continuous predictor `log10(Dose)` and distinct intercept for each of the four levels of the factor predictor `Source`. In the summary output, the common slope is clearly presented (2.63), but the intercepts are presented in terms of a baseline (`Finch`) intercept and the

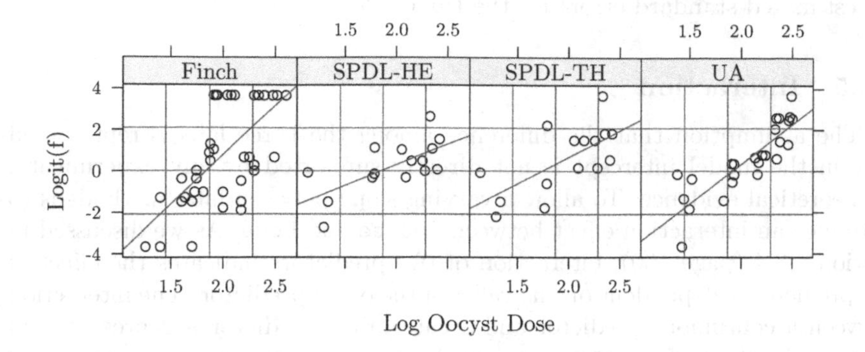

FIGURE 8.3: Mice infectivity data – Logit transformed fraction of infected mice (f) plotted against the corresponding oocyst dose.

differences in intercept between other sources and the baseline. In this case, the baseline, by default, is the first level (in alphabetical order) `Finch`. Using this output, intercept comparisons are straightforward. The differences in intercept between `SPLD-HE` and `Finch` and between `UA` and `Finch` are well within 1 standard error from 0; hence they are statistically not different from 0. Because the Finch study had a much larger sample size, Korich et al. compared their results to the model reported in Finch [Finch et al., 1993]. This output table addresses this comparison directly.

As in a multiple regression problem, we can force the model to have no intercept:

```
#### R Output ####
glm(formula = cbind(Y, N - Y) ~ log10(Dose) + factor(Source) - 1,
    family = binomial(link = "logit"), data = crypto.data)
                        coef.est coef.se
log10(Dose)              2.63     0.16
factor(Source)Finch     -5.01     0.35
factor(Source)SPDL-HE -4.96     0.34
factor(Source)SPDL-TH -4.69     0.34
factor(Source)UA        -4.94     0.34
  n = 98, k = 5
  residual deviance = 363.8,
  null deviance = 744.1 (difference = 380.3)
```

The "`-1`" in the model formula coerced the "intercept" to be 0. The summary table now shows the estimated intercepts for each level of `Source`. When forcing intercept to origin, the null deviance is no longer meaningful. Using this output table, we directly examine the estimated intercepts (along with the estimated standard error) for the three labs.

8.2.5 Interaction

The assumption that the difference among the three labs is represented only in the model intercept is not directly supported by any experimental or theoretical evidence. To allow a varying slope between labs/methods is to introduce an interaction effect between `Source` and `Dose`. As we discussed in Section 5.3.4 (page 160), interaction of two predictors indicates the effect of one predictor is dependent on the value of the other predictor. The interaction between a continuous predictor and a categorical predictor is represented in terms of the effect of the continuous predictor (slope) varying among the levels of the categorical predictor:

```
#### R Output ####
glm(formula = cbind(Y, N - Y) ~ log10(Dose) * factor(Source),
    family = binomial(link = "logit"), data = crypto.data)
                                  coef.est coef.se
(Intercept)                       -6.53    0.98
log10(Dose)                        3.39    0.49
factor(Source)SPDL-HE              3.35    1.13
factor(Source)SPDL-TH              2.37    1.15
factor(Source)UA                   0.06    1.17
log10(Dose):factor(Source)SPDL-HE -1.66    0.56
log10(Dose):factor(Source)SPDL-TH -1.03    0.57
log10(Dose):factor(Source)UA      -0.01    0.57
  n = 98, k = 8
  residual deviance = 344.5,
  null deviance = 693.0 (difference = 348.5)
```

Here we discover that the model from `SPDL-HE` is different from the `Finch` model, both in intercept and in slope. The difference in the intercept between `SPDL-HE` and `Finch` (3.39) is beyond 2 standard errors (2×1.13) away from 0. The difference in slope (-1.66) is also beyond 2 standard errors (2×0.56) from 0.

The estimated intercepts and slopes for the three models can be presented by using the same `-1` trick:

```
glm(formula = cbind(Y, N - Y) ~ log10(Dose) * factor(Source)
    - 1 - log10(Dose),
    family = binomial(link = "logit"), data = crypto.data)
                                  coef.est coef.se
factor(Source)Finch               -6.53    0.98
factor(Source)SPDL-HE             -3.18    0.57
```

```
factor(Source)SPDL-TH                 -4.16     0.60
factor(Source)UA                      -6.47     0.63
log10(Dose):factor(Source)Finch        3.39     0.49
log10(Dose):factor(Source)SPDL-HE      1.73     0.28
log10(Dose):factor(Source)SPDL-TH      2.36     0.31
log10(Dose):factor(Source)UA           3.38     0.30
  n = 98, k = 8
  residual deviance = 344.5,
  null deviance = 744.1 (difference = 399.6)
```

8.2.6 Comments on the Crypto Example

An obvious question from our analysis of the crypto data is why the conclusion we reached (four distinct models) is different from the conclusion reached by Korich et al. [2000]. The first difference in model fitting is that the response variable we used is the number of mice infected, while the response variable used in the Korich study is the logit transformed fraction of infected mice. Because Korich must remove observations with a fraction of 0 or 1, the data set we used is actually different from the ones used in their study. Secondly, the models reported in Korich et al. [2000] are fitted separately. Fitting models separately for each lab/method will not change the estimated model coefficients, but the estimated standard error will usually be larger than the same estimated jointly. In the Finch study, instead of using individual test results, the authors used average oocyst doses from several replicates. By doing so, the authors avoided the problem of fraction of infection being 0 or 1. However, they introduced the problem of error in variables, that is, uncertainty in predictor variable values.

Because the dose-response analysis presented in this example is often used as the first step of a study of the effectiveness of using UV as a disinfectant for inactivating crypto, small differences in estimated model coefficients can lead to large differences in the subsequent data analysis, where the standard errors of estimated coefficients are more important. See Qian et al. [2005a] for details. We note the large differences between the estimated model coefficients in our analysis and the same in the Korich study. When we fit the same linear regression model, the resulting model coefficients are somewhat different from the reported results:

```
#### R Output ####
lm(formula = I(logit(Y/N)) ~ log10(Dose) * factor(Source)
             - 1 - log10(Dose), data = crypto.data,
             subset = Y/N != 0 & Y/N !=1)

                                      coef.est coef.se
factor(Source)Finch                   -2.64     1.40
factor(Source)SPDL-HE                 -3.71     1.19
```

```
factor(Source)SPDL-TH                -3.97      1.21
factor(Source)UA                     -6.05      1.20
log10(Dose):factor(Source)Finch       1.12      0.72
log10(Dose):factor(Source)SPDL-HE     2.00      0.59
log10(Dose):factor(Source)SPDL-TH     2.23      0.61
log10(Dose):factor(Source)UA          3.23      0.57
  n = 76, k = 8
  residual sd = 0.95, R-Squared = 0.54
```

8.3 Diagnostics

8.3.1 Binned Residuals Plot

A logistic regression model predicts the probability of success, while the observation is binary. As a result, a plot of residuals (observed minus predicted) against the fitted probabilities is meaningless. When plotting the residual against the predicted probability of infection (Figure 8.4), we typically see two parallel lines. Gelman and Hill [2007] presented a binned residual plot for a binary regression model, which divide the x-axis of the residual plot (Figure 8.4) into bins. Within each bin, an average residual is calculated and plotted against the center of the bin. When a proper number of bins are used, we should see a typical shotgun pattern in the binned residual plot (Figure 8.5). In addition to the average binned residuals, we also estimate the standard error of the residuals within each bin. In the binned residual plot, we also plotted the approximate 95% confidence bounds ($0 \pm 2se$) to help assess the model performance.

8.3.2 Overdispersion

The logistic regression is based on the assumption that the response variable has a binomial distribution. When the probability of success p is known, the variance of the response count variable (number of success out of m trials) is known ($mp(1-p)$). If the binary trials are not independent (e.g., if mice from the same litter were used), or p's for the binary responses are not the same, or important predictor variables are not included in the model for p, then the response count variable typically will have a bigger variance than expected under a binomial model. This is the overdispersion problem.

To check for overdispersion, we calculate the residuals and the standardized residuals:

$$r_i = \hat{y}_i - y_i$$

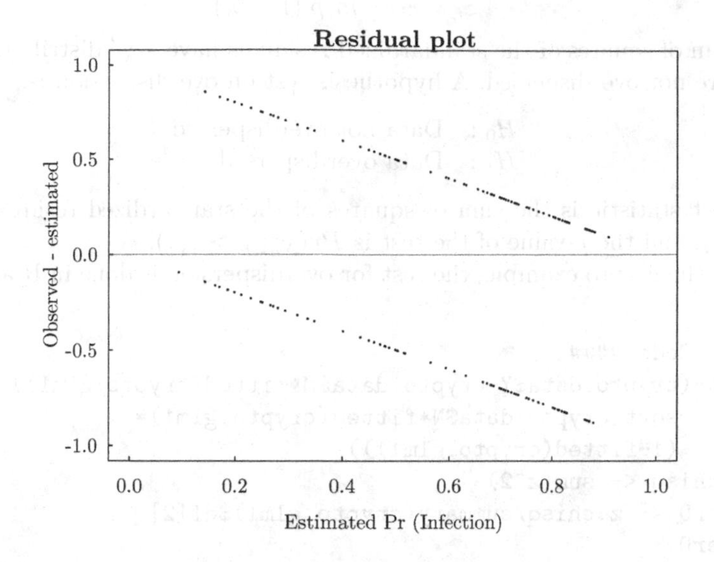

FIGURE 8.4: Logistic regression residuals – Residuals versus fitted plot from a logistic regression model is always hard to interpret.

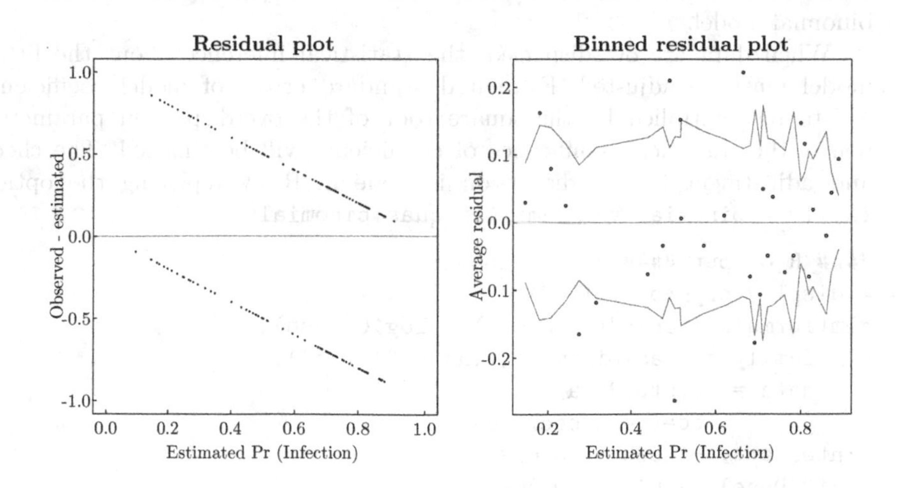

FIGURE 8.5: The binned residual plot – Comparison of the binned residual plot (right panel) and the residual plot (left panel).

where $\hat{y}_i = m_i\hat{p}_i$, and

$$z_i = r_i/\sqrt{m_i\hat{p}_i(1-\hat{p}_i)}$$

The sum of squares of the standardized residuals have a χ^2 distribution if the data are not overdispersed. A hypothesis test on overdispersion is:

$$\begin{array}{ll} H_0: & \text{Data not overdispersed} \\ H_1: & \text{Data overdispersed} \end{array}$$

The test statistic is the sum of squares of the standardized residuals $z_{\chi^2} = \sum_{i=1}^{n} z_i$, and the p-value of the test is $Pr(\chi^2_{n-k} > z_{\chi^2})$.

For the crypto example, the test for overdispersion is done in R as follows:

```
#### R Code ####
   z <-(crypto.data$Y-crypto.data$M*fitted(crypto.glm1)) /
        sqrt(crypto.data$N*fitted(crypto.glm1)*
        (1-fitted(crypto.glm1)))
   z.chisq <- sum(z^2)
   overD <- z.chisq/summary(crypto.glm1)$df[2]
   overD
    [1] 3.4784
   p.value <- 1-pchisq(z.chisq, df=summary(crypto.glm1)$df[2])
   p.value
   [1] 0
```

The estimated overdispersion parameter of 3.48 can be interpreted as the variation in the data is about 3.48 times of the expected variation under a binomial model.

When data are overdispersed, the statistical inference about the fitted model must be adjusted. Estimated standard errors of model coefficients are to be multiplied by the square root of the overdispersion parameter; hence, the statistical significance of coefficients will be changed. The check and adjustment for overdispersion is done in R by replacing the option `family = binomial` with `family = quasibinomial`:

```
#### R Output ####
> display(crypto.glm1, 3)
glm(formula = cbind(Y, M - Y) ~ log10(Dose),
    family = quasibinomial(link = "logit"),
    data = crypto.data)
            coef.est coef.se
(Intercept) -4.865    0.613
log10(Dose)  2.616    0.302
  n = 98, k = 2
  residual deviance = 368.1,
  null deviance = 693.0 (difference = 324.9)
  overdispersion parameter = 3.5
```

8.3.3 Seed Predation by Rodents: A Second Example of Logistic Regression

As an example of binary response variable model building and variable selection, we now present a second example based on a study of seed predation by small rodents conducted in a forest in southwest China (led by Dr. Zehao Shen of Peking University). The study was a randomized block design experiment. The researchers are interested in learning rodent foraging patterns, including (1) whether rodents are selective on seeds, (2) whether rodents prefer new seeds over leftover seeds from previous years, (3) do rodents only search for seeds on the surface, and (4) whether rodents are selective in where to forage (the role of topography). To answer these questions, researchers designed a four-factor factorial experiment. They selected 8 commonly seen tree species and collected seeds. Seeds were placed in small gauze bags and placed in 4 topographic locations, representing hilltop, sunny slope, shady slope, and valley bottom. Within each location, 6 seed bags were placed just below the leaf-litter layer and 6 were placed 4 cm below soil surface. These 6 bags were designed to be examined over time, once every other month for 11 months (from January to November, 2005). This pattern is replicated three times within each location, and the seed bags were buried in a matrix regime with an average distance of 1 m between each other. During each of the 6 field trips, researchers collected seed bags to check the status (whether seeds are intact or taken). In all cases, once a seed bag was discovered by a predator, the contents were all eaten. In addition to the field variables, researchers also measured average seed weight as a potential predictor variable.

The response variable (`Pred`) is the binary indicator of whether a bag of seeds is eaten (1) or not (0). As a result, we cannot easily use graphics to explore the likely predictors and the appropriate transformations. We can use CART to explore the relevant predictors as we did in Chapter 7. But in this case, the experiment was designed for answering a set of questions. For this reason, I will follow the ecological questions to explore the data – one question at a time.

First we examine whether predators favor certain seeds over the others. If rodents prefer one or more species of seeds, we may speculate and experiment later that these predators can sense nutritional values of their food. We can either model this preference by directly using `species` as a categorical predictor:

```
Pred ~ factor(species)
```

or using seed weight as a surrogate of nutritional/energy value:

```
Pred ~ log(seed.weight)
```

Based on the data plot (Figure 8.6), seed weight (in logarithm) was selected as the first predictor variable, which resulted in the following model:

```
#### R Output ####
```

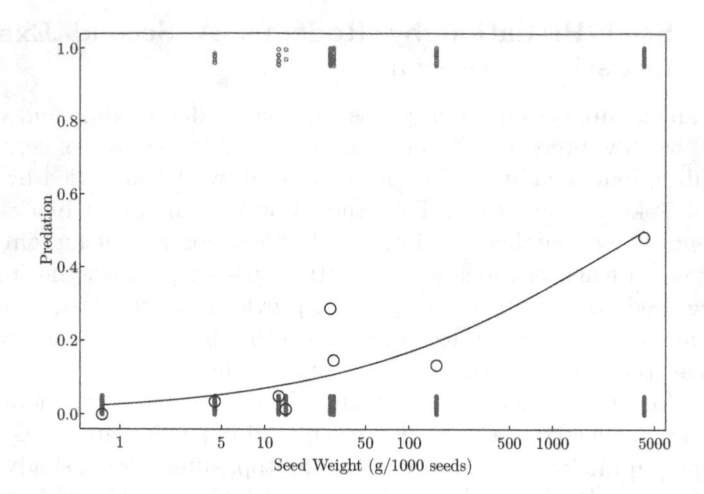

FIGURE 8.6: Seed predation versus seed weight – Proportions of seed bags eaten by predators are plotted against average seed weight of each species. The graph shows a positive association between seed weight and rate of predation. The solid line is the fitted logistic regression model using logarithmic seed weight as the only predictor.

```
glm(formula = Predation ~ log(seed.weight),
    family = binomial(link = "logit"),
    data = seedbank)
                  coef.est coef.se
(Intercept)       -3.55     0.20
log(seed.weight)  0.42      0.04
  n = 1142, k = 2
  residual deviance = 785.2,
  null deviance = 939.8 (difference = 154.6)
```

Using seed weight as the predictor not only reduces the number of coefficients, but also avoids a nonidentifiable problem when using the categorical predictor `species`. When `species` is used as a categorical predictor, the coefficient for `species` 1 is unidentifiable because all seeds of this species are intact at the end of the experiment.

```
#### R Output ####
glm(formula = Predation ~ species - 1,
    family = binomial(link = "logit"),
    data = seedbank)
         coef.est coef.se
species8  -0.07     0.17
species7  -1.88     0.25
species6  -1.76     0.24
```

```
species5  -0.91     0.18
species4  -4.26     0.71
species3  -2.97     0.39
species2  -3.32     0.46
species1 -18.57   549.31
  n = 1142, k = 8
  residual deviance = 721.1,
```

With 100% seeds intact, the MLE of the probability would be 0. However, the logit transformation of 0 (and 1) is not defined. Consequently, when using a logistic regression, we imply that the underlying probability of success does not equal to 0 or 1. As a result, the estimated probability of predation for species 1 has a large coefficient standard error.

Another consideration for selecting seed weight is model interpretation. Using seed weight, not only the fitted model can be easily presented graphically (Figure 8.6), but also subsequent models are much easier to interpret. For example, when examining the time effect, we include time as a categorical predictor and the subsequent model is straightforward to explain:

```
#### R output ####
glm(formula=Predation~factor(time)+log(seed.weight),
    family = binomial(link = "logit"), data = seedbank)
                  coef.est coef.se
(Intercept)        -6.33    0.64
factor(time)2       2.37    0.64
factor(time)3       2.70    0.64
factor(time)4       3.11    0.63
factor(time)5       2.98    0.63
factor(time)6       3.10    0.63
log(seed.weight)   0.46    0.04
  n = 1142, k = 7
  residual deviance = 726.3,
  null deviance = 939.8 (difference = 213.5)
```

The objective of this step is to examine whether the predation rate reached a stable value in the last three time periods. Ideally, we want to show that the fraction of seeds being eaten reach a stable number after the third time period when new seeds are available in the fall. In this model, the time effect is presented in terms of the difference between subsequent time periods and the initial one. This model can be interpreted as six parallel models for the six time periods, each with a different intercept and the same slope on `log(seed.weight)`. We re-fit the model by forcing the intercept to 0:

```
#### R Output ####
glm(formula = Predation ~ factor(time) + log(seed.weight) - 1,
    family = binomial(link = "logit"), data = seedbank)
                  coef.est coef.se
```

```
factor(time)1     -6.33      0.64
factor(time)2     -3.95      0.32
factor(time)3     -3.62      0.29
factor(time)4     -3.21      0.27
factor(time)5     -3.34      0.27
factor(time)6     -3.23      0.27
log(seed.weight)  0.46       0.04
  n = 1142, k = 7
  residual deviance = 726.3,
  null deviance = 1583.1 (difference = 856.9)
```

Now the six different intercepts are directly presented in the model summary table, which indicates an increasing probability of being eaten until time period 4 (July). Graphically, this result (Figure 8.7) is somewhat ambiguous because we see a clear pattern that indicates what we expected, but the high uncertainty in the estimated intercepts (represented by the 95% and 68% confidence intervals) suggests that the intercept of time period 1 is statistically different from the intercept of the rest time periods (which are themselves statistically not different). Admittedly, the regression coefficients are correlated and direct comparisons of the 95% confidence intervals may not be accurate. Furthermore, a comparison of the intercept does not guarantee the same result in probability. For a typical seed weight (28.5, the actual seed weight closest to the median), the estimated probabilities of a seed bag being discovered and eaten are

```
#### R Output ####
invlogit(coef(seedbank.glm4)[1:6] +
    coef(seedbank.glm4)[7] * log(28.5))

factor(time)1 factor(time)2 factor(time)3
       0.0082        0.0812        0.1094
factor(time)4 factor(time)5 factor(time)6
       0.1561        0.1399        0.1539
```

Also, because `time` is treated as a factor variable, the fitted model is actually six separate models, one for each sampling time. The difference in intercept (Figure 8.7) is presented in the logit scale. Given the nonlinear nature of the transformation, the difference in the probability scale may be different. To show these six models in the original probability scale, the following R code is used to produce Figure 8.8:

```
#### R Code ####
seedbank.glm4 <- glm(Predation~factor(time)+log(seed.weight),
    data=seedbank, family=binomial(link="logit"))
betas <- coef(seedbank.glm4)
par(mgp=c(1.5,.5,0), mar=c(3,3,1,0.25))
plot(jitter.binary(Predation) ~ log(seed.weight),
```

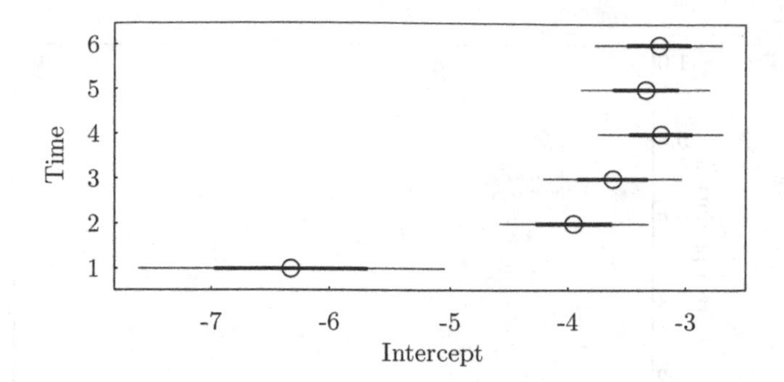

FIGURE 8.7: Seed predation over time – Estimated intercepts for the 6 sampling times are presented by the open circles. The thick lines are the estimated 50% confidence intervals and the thin lines are the estimated 95% confidence intervals.

```
    type="n", data=seedbank,
    xlab="log seed weight",
    ylab="prob. of predation")
points(jitter.binary(Predation)~log(seed.weight),
    col="gray")
curve(invlogit(betas[1]+betas[7]*x), add=T, col=gray(.1))
curve(invlogit(betas[1]+betas[2]+betas[7]*x), add=T,
    lty=2, col=gray(.2))
curve(invlogit(betas[1]+betas[3]+betas[7]*x), add=T,
    lty=3, col=gray(.3))
curve(invlogit(betas[1]+betas[4]+betas[7]*x), add=T,
    lty=4, col=gray(.4))
curve(invlogit(betas[1]+betas[5]+betas[7]*x), add=T,
    lty=5, col=gray(.5))
curve(invlogit(betas[1]+betas[6]+betas[7]*x), add=T,
    lty=6, col=gray(.6))
legend(x=0, y=0.9, legend=month.name[seq(1,11,2)],
    lty=1:6, col=gray((1:6)/10), cex=0.5, bty="n")
```

Directly estimating the standard errors of these predicted probabilities is difficult. But a simulation can provide this information easily. As we have discussed earlier, the function `sim` from package `arm` is designed for this purpose. Details of the simulation procedure will be explained in Chapter 9. The simulation results are shown in terms of the probability of a seed bag being eaten (Figure 8.9). These figures show that the basic relationship among the 6 time periods is the same as represented in Figure 8.7. Using simulation and expressing the model result in terms of probability, we have a more direct understanding of the effect of time and the difference in this effect between small and large seeds.

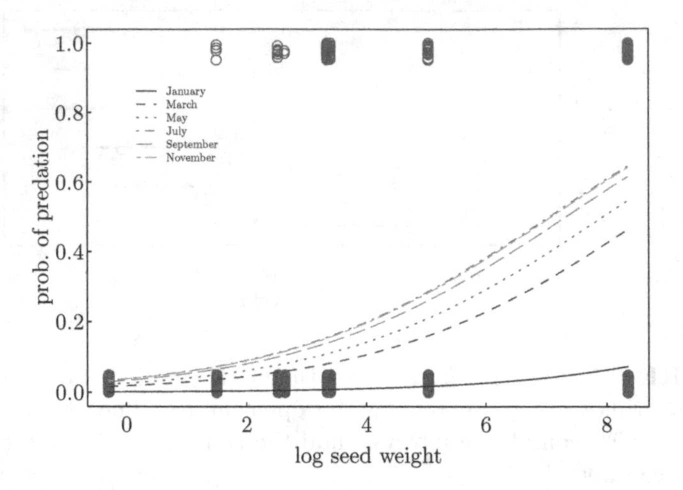

FIGURE 8.8: Time varying seed predation rate – The fitted models predict seed predation probability as a function of seed weight, one for each of the 6 sampling times. Because the seed bags were randomly assigned a sampling time, the difference between two times can be seen as an estimate of the increased predation between the two times.

Another statistical issue in this step is the different null deviances in the two models. When forcing the intercept to 0, we did not change the model because of the categorical predictor. However, the underlying statistical calculation implemented in the generic function `glm` uses the null model with a fixed intercept of 0, as in the linear regression case (section 5.3.6).

After the two factors that we know will affect the probability of predation are considered, we add the factor of interest, the topographic effect, into the model:

```
#### R Output ####
glm(formula=Predation~factor(time)+factor(topo)
    +log(seed.weight),
    family = binomial(link = "logit"),
    data = seedbank)
                  coef.est coef.se
(Intercept)       -5.72     0.67
factor(time)2      2.68     0.67
factor(time)3      3.06     0.67
factor(time)4      3.53     0.66
factor(time)5      3.38     0.67
factor(time)6      3.59     0.67
factor(topo)2     -2.21     0.31
factor(topo)3     -1.79     0.28
```

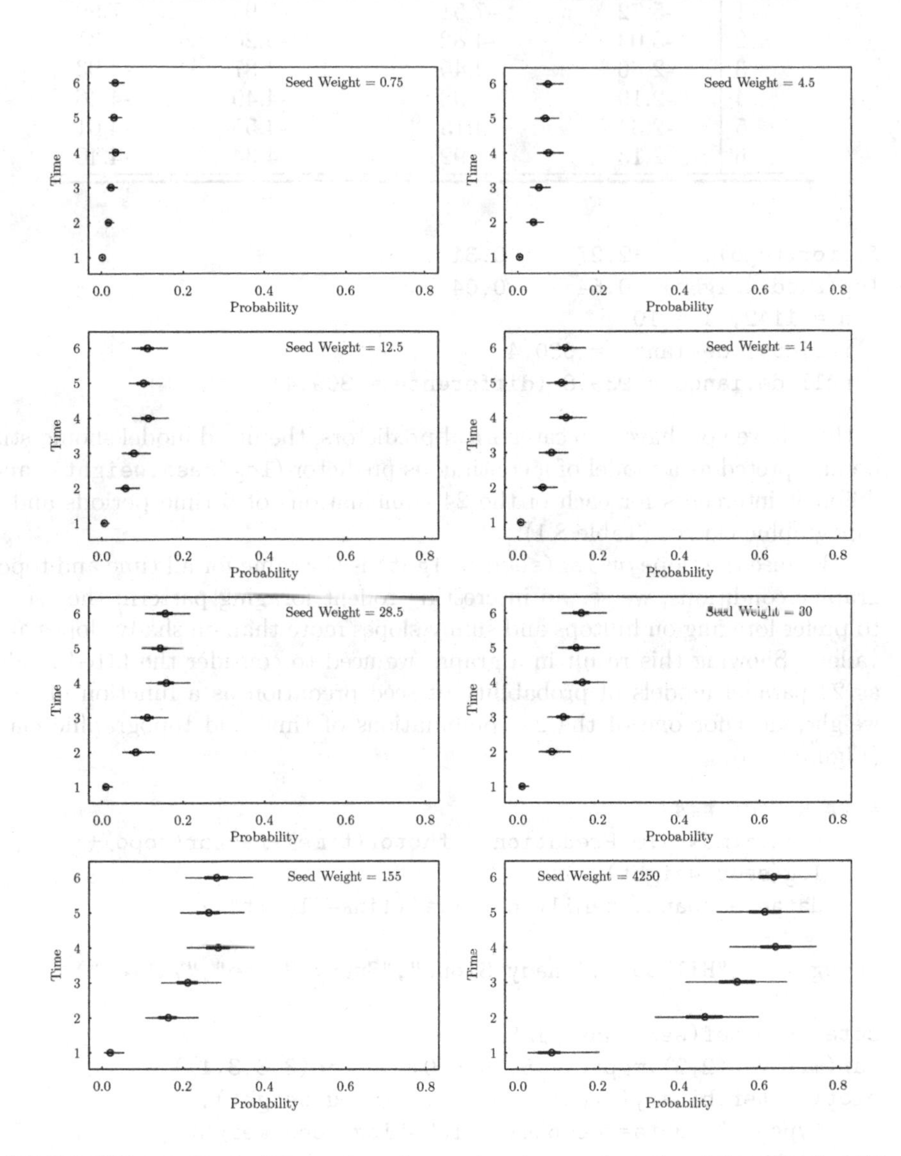

FIGURE 8.9: Probability of predation by time and seed weight – Probability of seed predation (black dot) and its 50% and 95% confidence intervals (thick and thin lines, respectively) are estimated by using simulation.

TABLE 8.1: Intercepts for 24 time-topography combinations

Time	Hilltop (1)	Sunny Slope(3)	Shady Slope (2)	Valley
1	-5.72	-7.51	-7.93	-7.99
2	-3.03	-4.82	-5.25	-5.31
3	-2.66	-4.45	-4.87	-4.93
4	-2.19	-3.98	-4.40	-4.46
5	-2.34	-4.13	-4.55	-4.61
6	-2.13	-3.92	-4.34	-4.40

```
factor(topo)4     -2.27      0.31
log(seed.weight)  0.54      0.04
  n = 1142, k = 10
  residual deviance = 630.4,
  null deviance = 939.8 (difference = 309.4)
```

Although we now have two categorical predictors, the fitted model should still be interpreted as a model of a continuous predictor (`log(seed.weight)`) and different intercepts for each of the 24 combinations of 6 time periods and 4 topographic classes (Table 8.1).

Because the slope on `log(seed.weight)` is the same for all time and topographic conditions, we see an interesting rodent foraging pattern: they seem to prefer foraging on hilltops and sunny slopes more than on shady slopes and valleys. Showing this result in a graph, we need to consider the fitted model as 24 parallel models of probability of seed predation as a function of seed weight, each for one of the 24 combinations of time and topographic class (Figure 8.10).

```
#### R Code ####
seedbank.glm5<-glm(Predation ~ factor(time)+factor(topo)+
    log(seed.weight),
    data=seedbank, family=binomial(link="logit"))

topog <- c("Hilltop","Shady Slope","Sunny Slope","Valley")

betas <- coef(seedbank.glm5)
par(mfrow=c(2,2),mgp=c(1.5, 0.5,0), mar=c(3,3,3,1))
plot(jitter.binary(Predation) ~ log(seed.weight),
    type="n", data=seedbank, xlab="log seed weight",
    ylab="prob. of predation")
points(jitter.binary(Predation) ~ log(seed.weight),
    col="gray", subset=topo==1)
curve(invlogit(betas[1]+betas[10]*x), add=T, col=gray(.1))
curve(invlogit(betas[1]+betas[2]+betas[10]*x), add=T,
    lty=2, col=gray(.2))
curve(invlogit(betas[1]+betas[3]+betas[10]*x), add=T,
```

```
    lty=3, col=gray(.3))
curve(invlogit(betas[1]+betas[4]+betas[10]*x), add=T,
    lty=4, col=gray(.4))
curve(invlogit(betas[1]+betas[5]+betas[10]*x), add=T,
    lty=5, col=gray(.5))
curve(invlogit(betas[1]+betas[6]+betas[10]*x), add=T,
    lty=6, col=gray(.6))
legend(x=0, y=0.9, legend=month.name[seq(1,11,2)],
    lty=1:6, col=gray((1:6)/10), cex=0.5, bty="n")
title(main=topog[1], cex=0.75)
for (i in c(3, 2, 4)){
    plot(jitter.binary(Predation) ~ log(seed.weight),
        type="n", data=seedbank,
        xlab="centered log seed weight",
        ylab="prob. of predation")
    points(jitter.binary(Predation) ~ log(seed.weight),
        col="gray", subset=topo==i)
    curve(invlogit(betas[1]+betas[10]*x), add=T,
        col=gray(.1))
    curve(invlogit(betas[1]+betas[2]+betas[i+5]+betas[10]*x),
        add=T, lty=2, col=gray(.2))
    curve(invlogit(betas[1]+betas[3]+betas[i+5]+betas[10]*x),
        add=T, lty=3, col=gray(.3))
    curve(invlogit(betas[1]+betas[4]+betas[i+5]+betas[10]*x),
        add=T, lty=4, col=gray(.4))
    curve(invlogit(betas[1]+betas[5]+betas[i+5]+betas[10]*x),
        add=T, lty=5, col=gray(.5))
    curve(invlogit(betas[1]+betas[6]+betas[i+5]+betas[10]*x),
        add=T, lty=6, col=gray(.6))
    title(main=topog[i], cex=0.75)
}
```

Since both `topo` and `log(seed.weight)` are statistically different from 0, we further consider their interaction:

```
#### R Output ####
glm(formula = Predation ~ factor(time) +
    factor(topo) * log(seed.weight),
    family = binomial(link = "logit"), data = seedbank)
                               coef.est coef.se
(Intercept)                    -7.07     0.95
factor(time)2                   3.21     0.78
factor(time)3                   3.60     0.78
factor(time)4                   4.08     0.77
factor(time)5                   3.93     0.77
factor(time)6                   4.14     0.78
```

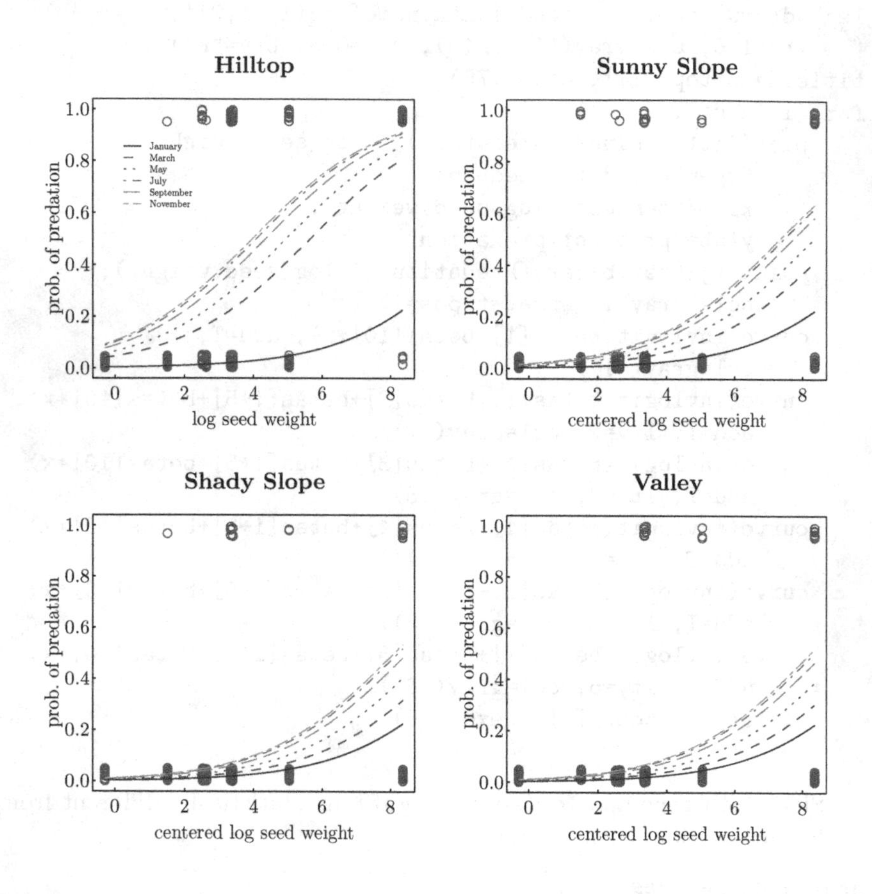

FIGURE 8.10: Probability of seed predation as a function of seed weight – The relationship varies by time and by topographic class.

```
factor(topo)2                   -0.97      0.67
factor(topo)3                   -0.62      0.62
factor(topo)4                   -1.05      0.68
log(seed.weight)                 0.77      0.11
factor(topo)2:log(seed.weight) -0.31      0.14
factor(topo)3:log(seed.weight) -0.30      0.13
factor(topo)4:log(seed.weight) -0.30      0.14
  n = 1142, k = 13
  residual deviance = 622.4,
  null deviance = 939.8 (difference = 317.3)
```

When foraging on hilltops, the seed weight effect is much stronger (larger slope) than the seed weight effect when foraging elsewhere. Is there an ecological explanation for the stronger seed weight effect on hilltops? The fitted model is presented in Figure 8.11.

Finally, we add **ground** to the model:

```
#### R Output ####
glm(formula=Predation~factor(time)+factor(topo)*log(weight) +
    factor(ground)-log(weight),
    family = binomial(link = "logit"), data = seedbank)
                          coef.est coef.se
(Intercept)               -6.8470   0.9855
factor(time)2              3.3878   0.8119
factor(time)3              3.8065   0.8097
factor(time)4              4.3153   0.8088
factor(time)5              4.1574   0.8089
factor(time)6              4.4228   0.8119
factor(topo)2             -1.0222   0.6847
factor(topo)3             -0.6749   0.6324
factor(topo)4             -1.1058   0.6967
factor(ground)2           -1.3135   0.2294
factor(topo)1:log(weight)  0.8174   0.1101
factor(topo)2:log(weight)  0.4822   0.0934
factor(topo)3:log(weight)  0.4969   0.0866
factor(topo)4:log(weight)  0.4867   0.0950
  n = 1142, k = 14
  residual deviance = 586.1,
  null deviance = 939.8 (difference = 353.7)
```

This result confirms what the researchers expected: when a seed bag is buried in the soil it is less likely to be discovered and eaten.

The final model provides the following conclusions:

- Rodents in this part of China favor large seeds
- They tend to forage more often in hilltops and sunny slopes than other locations

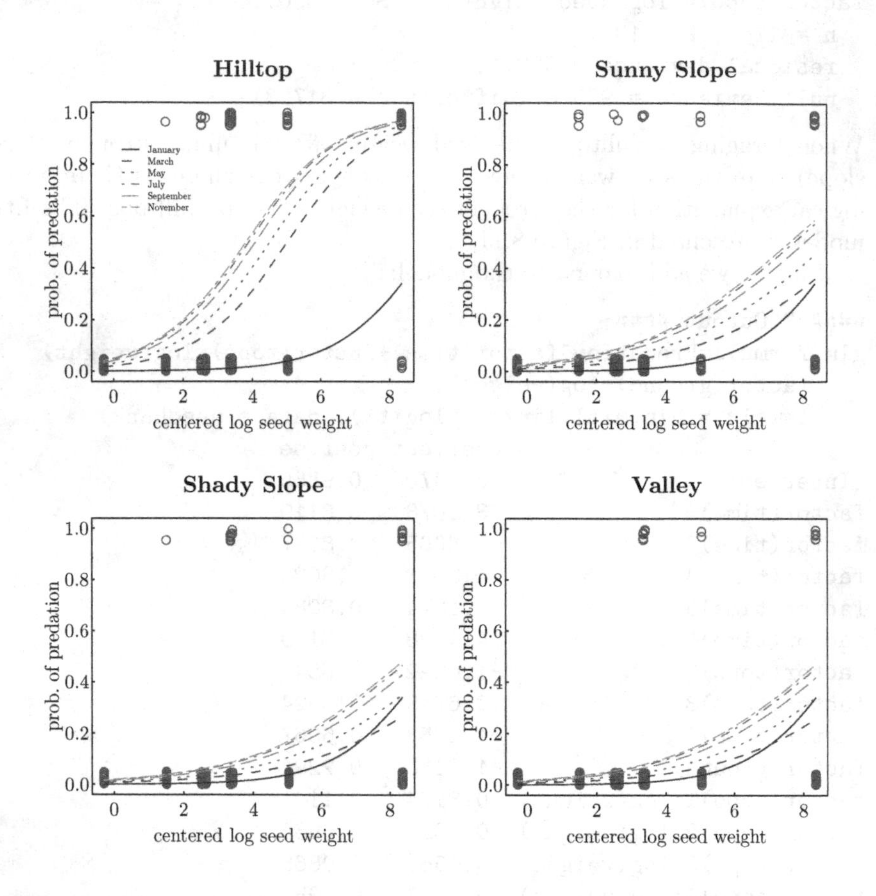

FIGURE 8.11: Seed weight and topographic class interaction – Probability of seed predation are predicted as a function of seed weight. The relationship varies by time and by topographic class

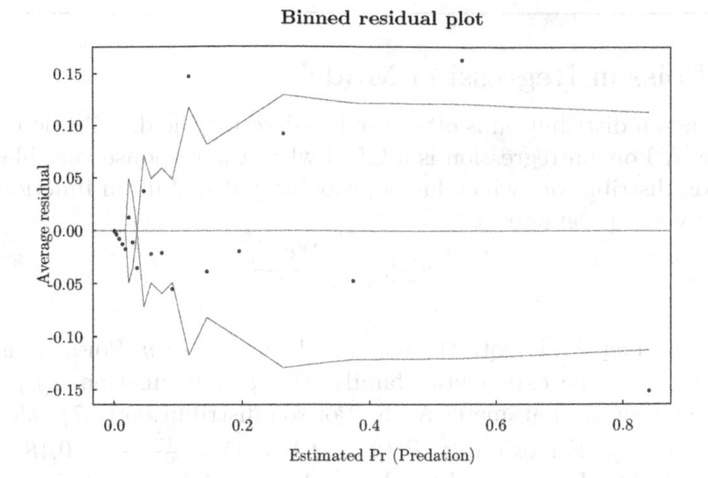

FIGURE 8.12: Binned residual plot of the seed predation model – the final model.

- They favor new seeds
- When foraging on hilltops, their preference for large seeds is more obvious, and
- They don't like to dig around if not necessary.

The binned residual plot is used to evaluate the model fit (Figure 8.12).

Another statistic for evaluating a model's fit to the data is the mean error rate, which is the average error rate when using the fitted probability values as a classifier. In a typical logistic regression case, we often predict a success (1) when the probability of success is over 0.5, and predict a failure (0) when the probability is below 0.5. The mean error rate for our final model is calculated as follows:

```
error.rate <- mean((fitted > 0.5 & observed==0) |
                   (fitted < 0.5 & observed==1))

> error.rate
[1] 0.106
```

Using 0.5 as a separating point, the model missclassified 10.6% of the observations.

8.4 Poisson Regression Model

A Poisson distribution is often used to describe the distribution of a count variable. A Poisson regression is a GLM when the response variable y follows a Poisson distribution, which has a probability distribution function parameterized by one parameter:

$$p(y|\lambda) = \frac{\lambda^y e^{-\lambda}}{y!} \tag{8.7}$$

The parameter λ is both the mean and variance of y. Poisson distribution is a member of the exponential family with a link function $\eta(\mu) = \log(\mu)$. Given the intensity parameter λ, the Poisson distribution (8.7) calculates the probability of y. For example, $\Pr(y = 2|\lambda = 1) = \frac{1^2 e^{-1}}{2!} = 0.18$. As in the previous section, I will introduce the model using an example.

8.4.1 Arsenic Data from Southwestern Taiwan

Arsenic is a naturally occurring substance known to be toxic if ingested in a large quantity. Arsenic in drinking water became front page news in 2001 when the newly inaugurated U.S. president George W. Bush chose to withdraw a new standard for arsenic in drinking water, reversing the U.S. drinking water standard for arsenic from 10 ppb back to 50 ppb. The move, the first environmental related action of the new administration, was widely used as evidence of the Bush administration choosing the interests of the mining industry and some small water suppliers over the protection of public health.

Arsenic in drinking water is widely known to cause skin cancer. In 1999, a U.S. National Academy of Sciences study concluded that arsenic in drinking water causes bladder, lung and skin cancer, and may cause kidney and liver cancer [National Research Council, 1999]. The study also found that arsenic harms the central and peripheral nervous systems, as well as heart and blood vessels, and causes serious skin problems. It also may cause birth defects and reproductive problems [National Research Council, 1999]. The U.S. drinking water standard for arsenic was 50 ppb, established in 1942. This concentration level was found by many studies to be associated with a substantial increased risk of cancer and is not sufficiently protective of public health [Morales et al., 2000]. The most compelling data arose from a rural population in southwestern Taiwan that had been exposed to high levels of arsenic in drinking water after primitive "tube wells" had been sunk, in a postwar effort to increase supplies of fresh drinking water in the region. The data reported by Chen et al. [1985] (`arsenic.csv`) were analyzed by Ryan [2003] as an example of an epidemiologically based environmental risk assessment. The response variable is the number of cancer deaths in 44 villages. The main predictors are village-specific median arsenic levels and person-years at risk in each age group. The data sets were grouped by gender and cancer types. Table 8.2 shows the female bladder cancer data for two villages.

8.4.2 Poisson Regression

A count variable such as the cancer death incident can rarely be transformed to normality. As a result, the generalized linear model is used. A count variable is often assumed to have a Poisson distribution:

$$Y_i \sim Pois(\lambda_i) \tag{8.8}$$

The Poisson distribution variable λ_i is modeled by

$$\log(\lambda_i) = X_i\beta \tag{8.9}$$

That is, the logarithmic expected number of incidents is modeled by a linear function of potential predictors. As in the logistic regression, equation 8.8 defines the probabilistic assumption on the response variable and the form of likelihood function. The likelihood function is a function of the regression coefficients defined in equation 8.9. The estimated model coefficients are the ones that maximize the likelihood function. The maximum likelihood estimator is implemented in R function `glm`. For example, suppose that we use only the village-specific median arsenic concentration as the only predictor. The model is fitted by calling the function and specifying `family="poisson"`:

```
#### R code ####
ar.m1 <- glm(events ~ conc, data=arsenic,
    family="poisson")

#### R Output ####
display(ar.m1, 4)
glm(formula = events ~ conc,
    family = "poisson", data = arsenic)
            coef.est coef.se
(Intercept) 3.0569   0.0189
conc        -0.0284  0.0005
---
  n = 2236, k = 2
  residual deviance = 34167.2,
  null deviance = 48869.5 (difference = 14702.3)
```

The interpretation of model coefficients is similar, in this case, to the interpretation of a log-linear regression model.

The intercept (3.0569) is the logarithmic expected number of cancer deaths (or ~21 deaths) at an arsenic concentration of 0 ppb.

TABLE 8.2: The arsenic in drinking water example data – A subset of the data from Chen et al. [1985] showing entries of village, arsenic concentration, age group, person-years at risk, and the number of bladder cancer deaths

Village	Arsenic conc. (ppb)	Age group midpoint (years)	Person-years at risk	No. bladder cancer deaths
1	0	22.5	2595529	0
1	0	27.5	1846189	2
1	0	32.5	1402764	0
1	0	37.5	1215899	2
1	0	42.5	1191615	8
1	0	47.5	1111810	14
1	0	52.5	957985	36
1	0	57.5	774836	52
1	0	62.5	634758	77
1	0	67.5	492203	68
1	0	72.5	342767	70
1	0	77.5	199630	43
1	0	82.5	96293	21
2	10	22.5	934	0
2	10	27.5	489	0
2	10	32.5	276	0
2	10	37.5	317	0
2	10	42.5	374	0
2	10	47.5	435	1
2	10	52.5	342	0
2	10	57.5	277	1
2	10	62.5	203	2
2	10	67.5	175	0
2	10	72.5	105	1
2	10	77.5	78	1
2	10	82.5	38	0

The slope (−0.0284) suggests that for every 1 ppb increase in arsenic concentration we expect a reduction in cancer death rate of approximately 2.8%. This is clearly counterintuitive. There may be many reasons for this unexpected result. First, the data we used combined both male and female cohorts (specified by the variable `gender`) and both bladder and lung cancer deaths (specified by variable `type`). We can introduce `gender` and `type` as two additional predictors. Because both `gender` and `type` are binary variables, they are converted to numeric: `gender=0` for female and `gender=1` for male, and `type=0` for bladder cancer and `type=1` for lung cancer.

```
#### R Code ####
ar.m3 <- glm(events ~ conc + gender + type,
    data=arsenic, family="poisson")

display(ar.m3, 3)
glm(formula = events ~ conc + gender + type,
    family = "poisson", data = arsenic)
            coef.est coef.se
(Intercept)  1.752    0.036
conc        -0.028    0.000
gender       0.672    0.027
type         1.383    0.032
---
  n = 2236, k = 4
  residual deviance = 31168.3,
  null deviance = 48869.5 (difference = 17701.2)
```

The results indicate that the effects of both `gender` and `type` are statistically different from 0. But the slope of arsenic concentration is still negative! We can further explore by adding interaction terms. But it is time to take a look at the data.

Many more cancer deaths are recorded for villages with median arsenic concentrations of 0 (Figures 8.13 and 8.14). The log-log scale plot (Figure 8.14) illustrates this finding more clearly.

This is because there are more people in the data set that are not exposed to positive arsenic concentrations. Figure 8.15 plots the at-risk population (total number of person-years exposed to a specific concentration) against the median concentration. The plot shows a much larger population was used in the data set as the "control" population that is not exposed to elevated arsenic concentrations in their drinking water. When comparing the number of cancer deaths as a fraction of the at-risk population (Figure 8.16), we are interested in the increase in cancer death *rate*.

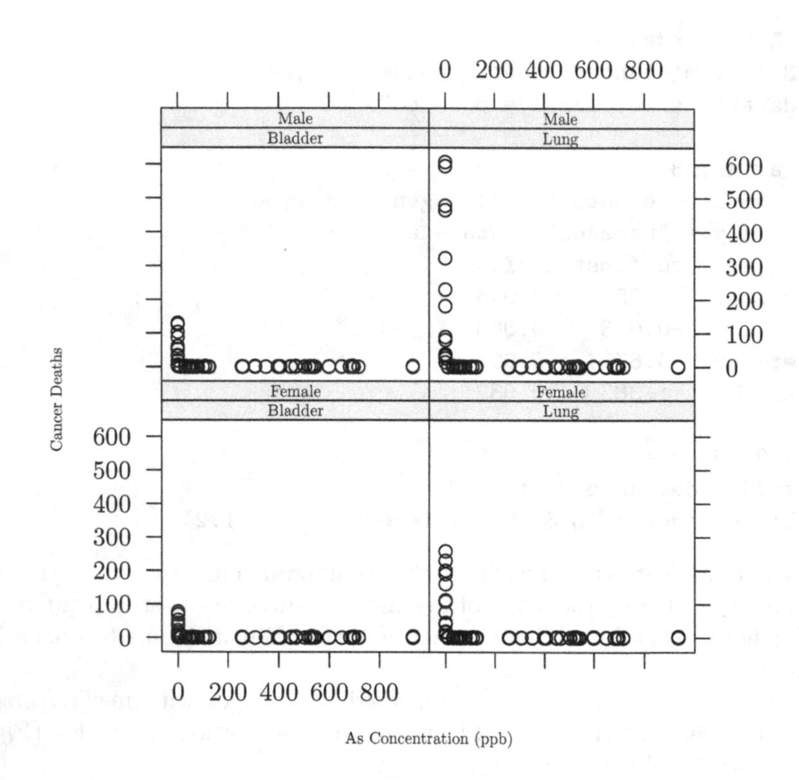

FIGURE 8.13: Arsenic in drinking water data 1– Cancer deaths are plotted against village-specific median arsenic concentrations.

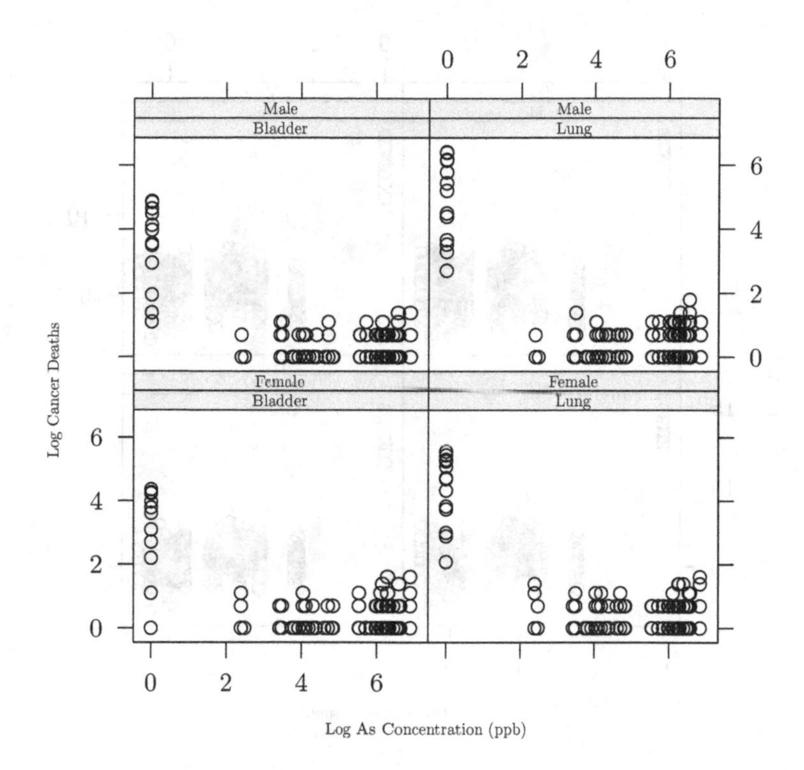

FIGURE 8.14: Arsenic in drinking water data 2 – Log cancer deaths are plotted against log village-specific median arsenic concentrations. The $\log(x+1)$ transformation is used to include 0 concentration values.

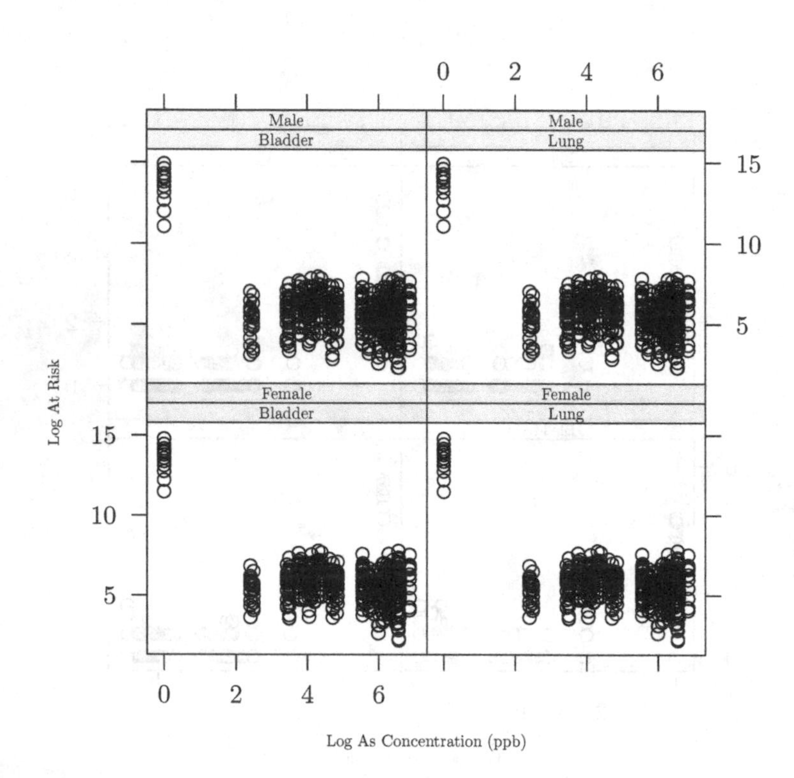

FIGURE 8.15: Arsenic in drinking water data 3 – The at-risk person-years are plotted against the log arsenic concentrations. The population exposed to 0 arsenic concentration (the control) is much larger than the population exposed to any positive arsenic concentrations.

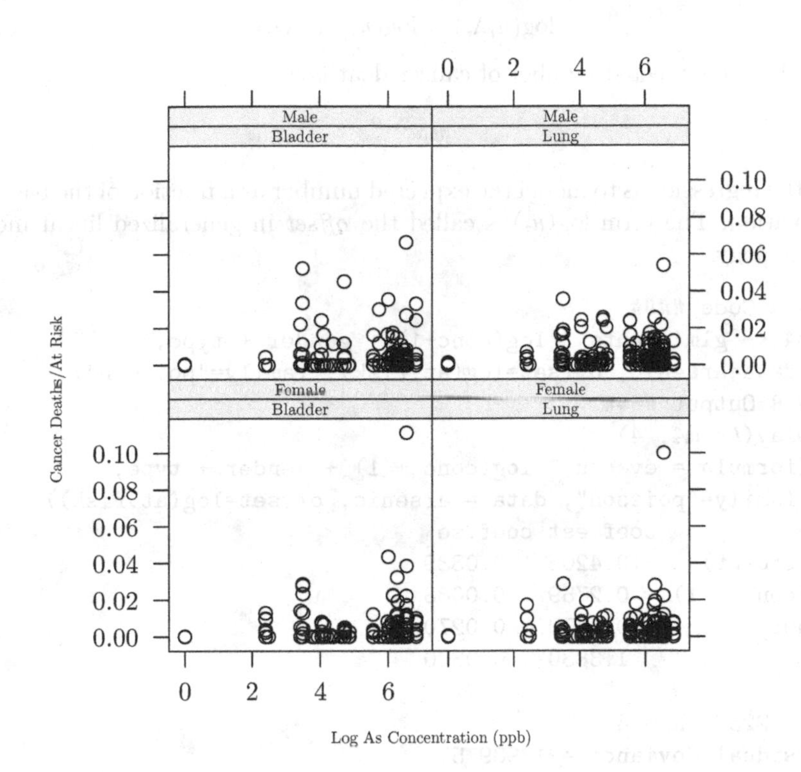

FIGURE 8.16: Arsenic in drinking water data 4 – Cancer deaths as a percentage of total population are plotted against village-specific median arsenic concentrations.

8.4.3 Exposure and Offset

Based on the observation of Figure 8.16, it is reasonable to model the number of cancer deaths as a fraction of the total population and model the fractional constant as a function of arsenic concentration. This total population or the baseline is often called the *exposure* and the Poisson model is written as:

$$Y_i \sim Poisson(u_i \lambda_i)$$

and

$$\log(u_i \lambda_i) = \log(u_i) + X_i \beta \tag{8.10}$$

That is, the expected number of cancer deaths is

$$u_i e^{X_i \beta}$$

and the regression is to model the expected number as a fraction of the baseline population. The term $\log(u_i)$ is called the *offset* in generalized linear model terminology.

```
#### R Code ####
As.m4 <- glm(events ~ log(conc+1) + gender + type,
    data=arsenic, offset=log(at.risk), family="poisson")
#### R Output ####
display(As.m4, 4)
glm(formula = events ~ log(conc + 1) + gender + type,
    family="poisson", data = arsenic, offset=log(at.risk))
                coef.est coef.se
(Intercept)    -10.4205   0.0339
log(conc + 1)    0.2759   0.0088
gender           0.5420   0.0270
type             1.3830   0.0320
---
  n = 2236, k = 4
  residual deviance = 13989.5,
  null deviance = 17398.7 (difference = 3409.3)
```

The intercept of this model (-10.42) is the logarithm of the bladder cancer (`type=0`) death rate for females (`gender=0`) when arsenic median concentration is 0. This translates into a baseline bladder cancer rate among females of $e^{-10.42}$ or 0.00002983, or slightly less than 3 deaths per 100,000 people per year. For each 1% increase in arsenic concentration in drinking water, the cancer death rate increases by a 0.2759%. At a given arsenic concentration, males have a higher cancer death rate than females (a factor of $e^{0.542}$ or an increase of 72%), and lung cancer death rate is $e^{1.383}$ or almost 4 times as high as the bladder cancer rate. To evaluate the effect of the Bush administration's withdrawal of the new arsenic standard for drinking water, we can compare the cancer death rates at a concentration of 10 ppb and at a concentration of

TABLE 8.3: The arsenic standard effect in cancer death rates – The estimated cancer death rates (death per 100,000 people per year) when population is exposed to an arsenic concentration of 10 ppb are compared to the death rates exposing to a concentration of 50 ppb (in parentheses)

	Female	Male
Bladder cancer	5.8(8.8)	9.9(15.2)
Lung cancer	23.0(35.2)	39.6(60.5)

50 ppb (Table 8.3). Using the fitted model directly, the ratio of the expected cancer death rates is $e^{\beta_0+\beta_1 \log(50+1)}/e^{\beta_0+\beta_1 \log(50+1)}$ of $(51/11)^0.2759 = 1.53$. In other words, an increase of 53% in cancer death rates is expected when a population's exposure to arsenic increased from 10 to 50 ppb.

8.4.4 Overdispersion

In linear regression, we evaluate the model by examining residuals. A normal response model would have a residual distribution with mean 0 and a constant standard deviation. The Poisson distribution variance is the same as its mean. Because the fitted value is the estimated mean of the Poisson distribution, residuals from a Poisson regression model should have a variance equal to the predicted mean. As a result, when plotting residuals against the fitted values, we expect to see a wedge-shaped pattern, that is, the residual variance increases at the same rate as the predicted mean increases. A commonly seen problem with a Poisson regression on count variables is that the variance increases at a faster rate than the predicted mean increases. This phenomenon is known as *overdispersion*. When data are overdispersed, the Poisson model will underestimate the uncertainty in the regression coefficients, leading to potentially misleading inferences. For example, the conclusion that arsenic in drinking water may lead to an increased risk of cancer is based on a positive slope of the arsenic concentration that is statistically different from 0, a conclusion contingent on the assumption of a Poisson model. If overdispersion is present, the estimated coefficient standard deviation is too small. To avoid misleading results, checking for overdispersion is necessary.

Because under the Poisson distribution the variance equals the mean (or the standard deviation equals the square root of the mean), the fitted value of a Poisson regression model of equation 8.10, $\hat{y}_i = u_i\hat{\lambda}_i$, is the estimated variance of $\hat{y}_i$. The standardized residual is then calculated to be:

$$z_i = \frac{y_i - \hat{y}_i}{sd(\hat{y}_i)} = \frac{y_i - \hat{y}_i}{\sqrt{u_i\hat{\lambda}_i}}$$

If there is no overdispersion, z_i should be approximately independent with mean 0 and standard deviation 1. Furthermore, $\sum_{i=1}^{n} z_i^2$ should follow a χ^2 distribution with degrees of freedom of $n-k$, where n and k are the number of data points and the number of regression coefficients, respectively. The χ^2 distribution with degrees of freedom of $n-k$ has a mean of $n-k$. We would expect $\sum_{i=1}^{n} z_i^2$ to be close to $n-k$ if there is no overdispersion. If there is overdispersion, $|z_i|$ (hence $\sum_{i=1}^{n} z_i^2$) would be larger than expected. We use $\omega = \frac{1}{n-k}\sum_{i=1}^{n} z_i^2$ to represent the estimated amount of overdispersion (the overdispersion parameter). To test whether there is overdispersion, we can use $\sum_{i=1}^{n} z_i^2$ as the test statistic and compare the calculated value to the χ^2 distribution with degrees of freedom of $n-k$ to calculate the p-value. For example, in the model As.m4 in Section 8.4.3, the estimated overdispersion and the p-value are calculated as follows:

```
#### R Code ####
As.yhat <- predict(As.m4, type="response")
As.z <- (arsenic$events - As.yhat)/sqrt(As.yhat)
overD <- sum(As.z^2)/summary(As.m4)$df[2]
p.value <- 1-pchisq(sum(As.z^2), summary(As.m4)$df[2])
```

The estimated overdispersion is $\omega = 14.18$ and the p-value is close to 0. This result suggests that the calculated $\sum_{i=1}^{n} z_i^2$ of 31650.4 is 14 times of the expected value under a $\chi^2_{df=2232}$ distribution. The probability that a randomly drawn number from this distribution is as large or larger than 31650.4 is essentially 0. For any reasonable sample size, an overdispersion of 2 can be considered large.

Graphically, we can diagnose overdispersion by plotting the standardized residuals z_i against the predicted values. If there is no overdispersion, the plot should resemble a typical residual versus fitted plot we used in regression analysis (e.g., Figure 5.11 on page 173), on which points should be randomly scattered around 0 on the y-axis direction and a standard deviation of 1. If we plot two horizontal lines at $\pm$ 2, we expect about 95% of the data points to be bounded by these two lines. If there is overdispersion, we expect more than 5% of the data points to be outside the bounds formed by the two lines.

As expected, the raw residuals from the arsenic model (As.m4) show increased variance as predicted values increase (Figure 8.17, left panel). The standardized residuals should have mean 0 and a standard deviation of 1 if there is no overdispersion. The two dashed lines at ± 2 in Figure 8.17 (right panel) show more data points lay outside the bounds.

One way to model overdispersion is to use the "quasi-Poisson" distribution in R. The quasi-Poisson is a count variable distribution with the same mean function as the Poisson distribution but a variance equal to ω times the mean. The resulting overdispersed Poisson regression model has the same estimated regression coefficients, but the estimated coefficient standard errors are larger. To correct for overdispersion, we can simply multiply the estimated regression coefficient standard errors by the square root of the estimated overdis-

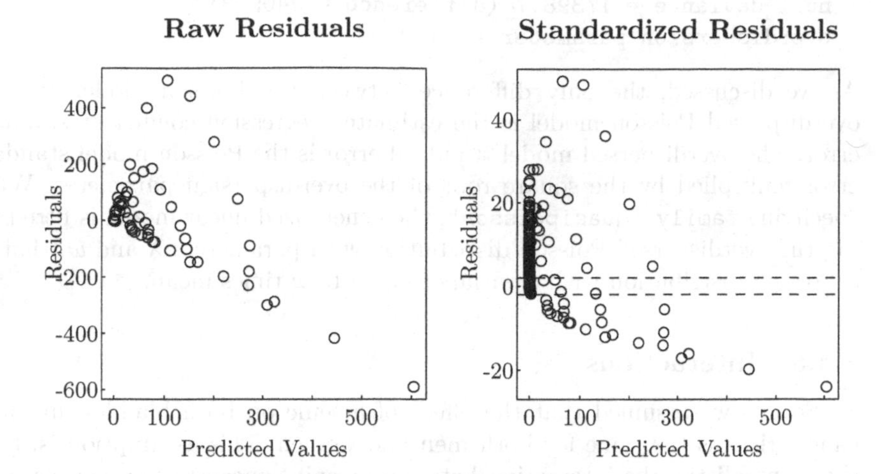

FIGURE 8.17: Raw versus standardized residuals of an additive Poisson model – Residual plots are used for testing overdispersion. The left panel shows the raw residuals versus the predicted values and the right panel shows the standardized residuals versus predicted values. The Poisson regression model uses arsenic concentration as the only continuous predictor.

persion. For the model considered in Section 8.4.3, the correction factor is $\sqrt{14.18} = 3.766$. In model `As.m4`, all the estimated regression coefficients are still statistically different from 0, after the overdispersion effect is accounted for. Our conclusion is not affected by the overdispersion.

```
#### R Code ####
> As.m5 <- glm(events ~ log(conc+1) + gender + type,
    data=arsenic, offset=log(at.risk),
    family="quasipoisson")
#### R Output ####
> display(As.m5, 4)
glm(formula = events ~ log(conc + 1) + gender + type,
    family = "quasipoisson",
    data = arsenic, offset = log(at.risk))
               coef.est coef.se
(Intercept)   -10.4205   0.1277
log(conc + 1)   0.2759   0.0330
gender          0.5420   0.1019
type            1.3830   0.1203
---
  n = 2236, k = 4
  residual deviance = 13989.5,
```

```
  null deviance = 17398.7 (difference = 3409.3)
  overdispersion parameter = 14.2
```

As we discussed, the only difference between the Poisson model and the overdispersed Poisson model is the estimated regression coefficient standard error. The overdispersed model standard error is the Poisson model standard error multiplied by the square root of the overdispersion parameter. When specifying `family="quasipoisson"`, the generalized linear model is fitted using the overdispersed Poisson distribution with parameters λ and ω, that is, a Poisson distribution with a variance equal to ω times mean.

8.4.5 Interactions

So far, we assumed that the effect of arsenic on both bladder and lung cancer risks is the same for both men and women. This assumption is, however, unrealistic; the interaction between arsenic concentration, cancer type, and gender should be considered. Because arsenic concentration is the only continuous predictor and there are four different cancer type–gender combinations, considering interactions among the three predictors is the same as fitting four models with arsenic concentration as the only predictor, one for each different cancer type–gender combinations. Statistical inference will be focused on testing whether intercepts and slopes are different among the four models. For this example, we can start from the most complicated model and simplify it backwards:

```
#### R Code ####
> As.m6 <- glm(events ~ log(conc+1)*gender*type,
    data=arsenic, offset=log(at.risk),
    family="poisson")

#### R Output ####
> display(As.m6, 4)
glm(formula = events ~ log(conc + 1) * gender * type,
    family = "poisson",
    data = arsenic, offset = log(at.risk))
                             coef.est coef.se
(Intercept)                  -10.3889   0.0501
log(conc + 1)                  0.4563   0.0203
gender                         0.3732   0.0635
type                           1.3070   0.0565
log(conc + 1):gender          -0.0858   0.0285
log(conc + 1):type            -0.1761   0.0264
gender:type                    0.2628   0.0709
log(conc + 1):gender:type     -0.0098   0.0364
---
  n = 2236, k = 8
```

```
  residual deviance = 13844.2,
  null deviance = 17398.7 (difference = 3554.5)
```

The three-way interaction is not necessary. This model is fitted using the standard Poisson model (i.e., `family="poisson"`) without considering overdispersion. When a slope is statistically not different from 0 under the Poisson model, it will also be not different from 0 under the overdispersed Poisson model.

```
#### R Code ####
> As.m7 <- update(As.m6, .~. -log(conc+1):gender:type)
> display(As.m7)
glm(formula = events ~ log(conc + 1) + gender +
              type + log(conc + 1):gender +
              log(conc + 1):type + gender:type,
    family = "poisson",
    data = arsenic, offset = log(at.risk))
                      coef.est coef.se
(Intercept)           -10.39    0.05
log(conc + 1)           0.46    0.02
gender                  0.38    0.06
type                    1.31    0.05
log(conc + 1):gender  -0.09    0.02
log(conc + 1):type    -0.18    0.02
gender:type             0.26    0.07
---
  n = 2236, k = 7
  residual deviance = 13844.3,
  null deviance = 17398.7 (difference = 3554.4)
```

Now all coefficients are statistically different from 0. The output can be easily interpreted by creating a table of intercepts and slopes (of `log(conc+1)`) for men and women and for lung and bladder cancers (Table 8.4).

TABLE 8.4: Interactions between gender and cancer type – The estimated model coefficients (intercept and slope of the log concentration plus 1) for men, women, lung cancer, and bladder cancer

	Female		Male	
	Intercept	Slope	Intercept	Slope
Bladder	-10.39	0.46	-10.39+0.38	0.46 - 0.09
Lung	-10.39+1.31	0.46-0.18	-10.39+0.38+1.31+0.26	0.46-0.09-0.18

We should now check for overdispersion and seek further simplification if possible. The simplest way of checking for overdispersion is to use the overdispersion Poisson regression:

```
#### R Code ####
```

```
> As.m8 <- update(As.m7, .~., family="quasipoisson")

#### R Output ####
> display(As.m8, 4)
glm(formula = events ~ log(conc + 1) + gender +
              type + log(conc + 1):gender +
              log(conc + 1):type + gender:type,
              family = "quasipoisson",
    data = arsenic, offset = log(at.risk))
                      coef.est coef.se
(Intercept)           -10.3920   0.1686
log(conc + 1)           0.4593   0.0580
gender                  0.3782   0.2095
type                    1.3110   0.1885
log(conc + 1):gender   -0.0918   0.0614
log(conc + 1):type     -0.1812   0.0629
gender:type             0.2565   0.2311
---
  n = 2236, k = 7
  residual deviance = 13844.3,
  null deviance = 17398.7 (difference = 3554.4)
  overdispersion parameter = 11.9
```

The estimated overdispersion parameter of 11.9 suggests that the actual variance of the response variable is about 12 times as large as the variance predicted by a Poisson model. The relatively high standard error of the `gender:type` interaction slope suggests that the gender difference is not affected by cancer type and vice versa. The `gender:type` interaction term can be removed:

```
#### R Output ####
> As.m9 <- update(As.m8, .~.-gender:type)
> display(As.m9, 4)
glm(formula = events ~ log(conc + 1) + gender +
                       type + log(conc + 1):gender +
                       log(conc + 1):type,
    family = "quasipoisson",
    data = arsenic, offset = log(at.risk))
                      coef.est coef.se
(Intercept)           -10.5265   0.1249
log(conc + 1)           0.4691   0.0587
gender                  0.5855   0.0977
type                    1.4789   0.1178
log(conc + 1):gender   -0.1012   0.0607
log(conc + 1):type     -0.1873   0.0626
---
```

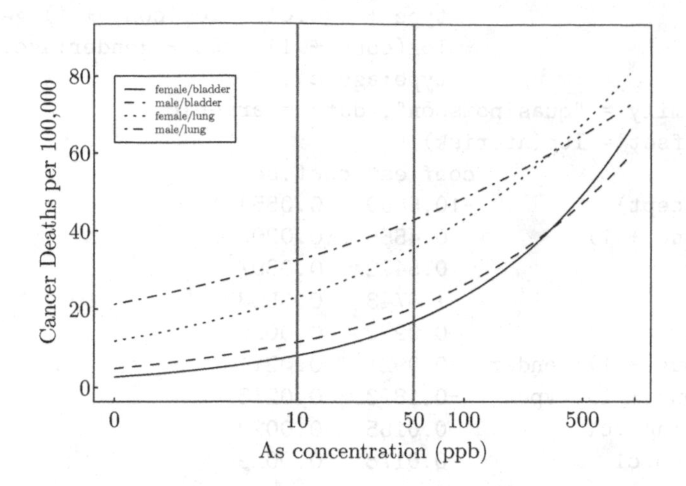

FIGURE 8.18: Fitted overdispersion Poisson model – The overdispersed Poisson model predicts bladder and lung cancer death rates for men and women in southwestern Taiwan.

```
  n = 2236, k = 6
  residual deviance = 13858.8,
  null deviance = 17398.7 (difference = 3539.9)
  overdispersion parameter = 12.0
```

The simplified model suggests that the difference between the slopes of men and women (-0.1012) is weak. However, the difference of -0.1 is about 20% of the baseline (female) slope and the sign of the slope seems to make sense. That is, men have a higher baseline (concentration equals 0) cancer rate ($e^{0.5855}$ or about 80% higher than the cancer rate of women), and they would be less sensitive to the exposure to arsenic in drinking water. So we may keep this term in the model. The fitted model is presented in Figure 8.18.

The model presented in Figure 8.18, however, did not account for the effect of age, an important cancer risk factor that should be considered. Including age as a linear predictor, we assume that a fixed factor increase in cancer rate increases for every year increase in age. Because `age` is introduced into the model as a continuous variable, we center the age variable by subtracting the mean age in the data (52.5) from each age value. When comparing the intercepts of the resulting model, we are comparing the cancer rates for people at the age of 52.5 years old.

```
> arsenic$age.c1 <- arsenic$age - mean(arsenic$age)
> As.m10<-update(As.m9, .~.+age.c1*gender+age.c1:type)
> display(As.m10, 4)
glm(formula = events ~ log(conc + 1) + gender +
```

```
                          type + age.c1 + log(conc + 1):gender +
                          log(conc + 1):type + gender:age.c1 +
                          type:age.c1,
    family = "quasipoisson", data = arsenic,
    offset = log(at.risk))
                        coef.est coef.se
(Intercept)             -10.5859   0.0554
log(conc + 1)             0.4556   0.0205
gender                    0.5422   0.0397
type                      1.6743   0.0533
age.c1                    0.0922   0.0029
log(conc + 1):gender     -0.0921   0.0211
log(conc + 1):type       -0.1873   0.0218
gender:age.c1             0.0155   0.0022
type:age.c1              -0.0178   0.0029
---
  n = 2236, k = 9
  residual deviance = 2193.1,
  null deviance = 17398.7 (difference = 15205.6)
  overdispersion parameter = 1.4
```

Slopes for all predictors are statistically different from 0 and the overdispersion parameter is now reduced from the previous model's 12 to the current estimate of 1.4. Compared to the previous model, we now have two continuous predictor variables. We can still build a table such as Table 8.4 to present the 4 different models. But a graph is better. Figure 8.19 is produced in two steps. First, model predictions are produced over the range of arsenic concentrations seen in the area for three ages (37.5, 52.5, 67.5, representing the 25th, 50th, and 75th percentile of the age variable in the southwestern Taiwan data set):

```
#### R Code ####
pred.data <- data.frame(
    conc=rep(seq(0, 1000, 10), 4),
    type=rep(rep(c(0,1), each=101), 2),
    gender=rep(c(0,1), each=202))
pred.age1 <- predict(As.m10, newdata=
    data.frame(pred.data, age.c1= rep(-15, 404),
               at.risk=rep(100000, 404)),
    type="response")
pred.age2 <- predict(As.m10, newdata=
    data.frame(pred.data, age.c1= rep(0   , 404),
               at.risk=rep(100000, 404)),
    type="response")
pred.age3 <- predict(As.m10, newdata=
    data.frame(pred.data, age.c1= rep( 15, 404),
               at.risk=rep(100000, 404)),
```

```
    type="response")
```

Then, using the lattice function xyplot to produce a multipanel plot:

```
#### R Code ####
plot.data1 <- data.frame(
    events=c(pred.age1, pred.age2, pred.age3),
    rbind(pred.data, pred.data, pred.data),
    age=rep(c(-15+52.5, 52.5, 52.5+15), each=404))
plot.data1$Type <- "Lung Cancer"
plot.data1$Type[plot.data1$type==0]
    <- "Bladder Cancer"
plot.data1$Gender <- "Male"
plot.data1$Gender[plot.data1$gender==0]<-"Female"

trellis.par.set(theme =
    canonical.theme("postscript", col=FALSE))
trellis.par.set(list(fontsize=list(text=8),
                par.xlab.text=list(cex=1.25),
                    add.text=list(cex=1.25),
              superpose.symbol=list(cex=1)))
key <- simpleKey(unique(as.character(plot.data1$age)),
    lines=T, points=F, space = "top", columns=3)
key$text$cex <- 1.25
xyplot(events~log(conc+1)|Type*Gender,
    data=plot.data1, type="l",
    group=plot.data1$age,
    key=key,
    xlab="As concentration (ppb)",
    ylab="Cancer deaths per 100,000",
    panel=function(x,y,...){
        panel.xyplot(x,y,lwd=1.5,...)
        panel.grid()
    },
    scales=list(x=list(
        at=log(c(0, 10, 50, 100, 500, 1000)+1),
        labels=as.character(c(0, 10, 50, 100, 500, 1000))))
    )
```

The figure suggests that the arsenic effects are most pronounced in older people, presumably reflecting a lifetime exposure to elevated arsenic concentrations in their drinking water. Compared to the residual plot of the initial model without considering interaction and the age effect (Figure 8.17), this model has a much smaller overdispersion (Figure 8.20). As shown in Figure 8.19, the expected number of cancer deaths increases as arsenic concentration increases. However, the difference in cancer death risks between older and

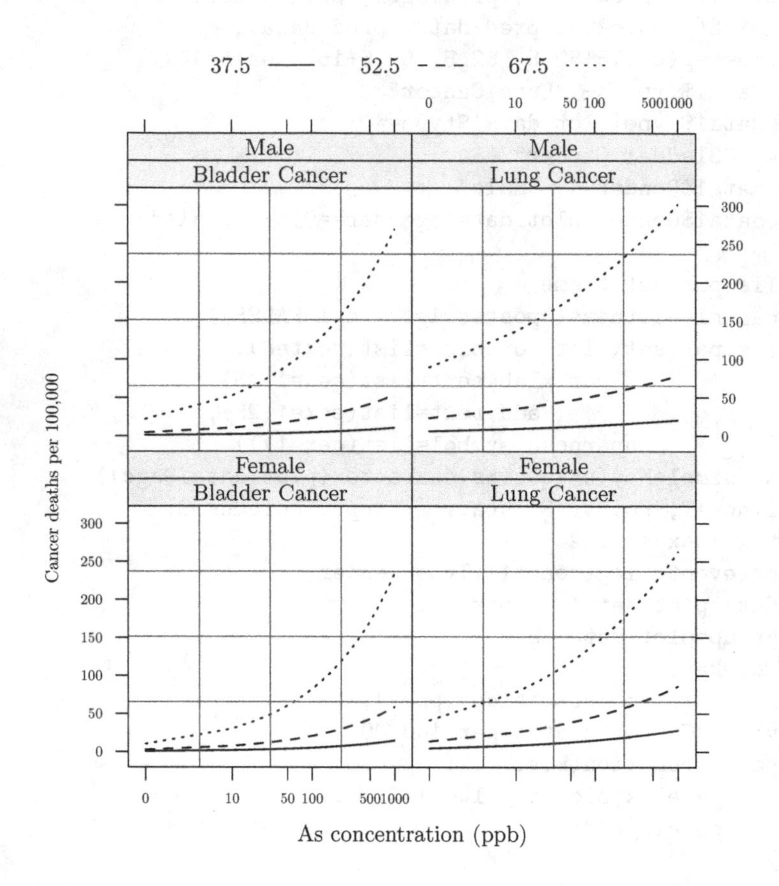

FIGURE 8.19: Fitted overdispersion Poisson model with age as a covariate – The overdispersed Poisson model predicts bladder and lung cancer death rates for men and women in southwestern Taiwan using drinking water arsenic concentration and age as predictors.

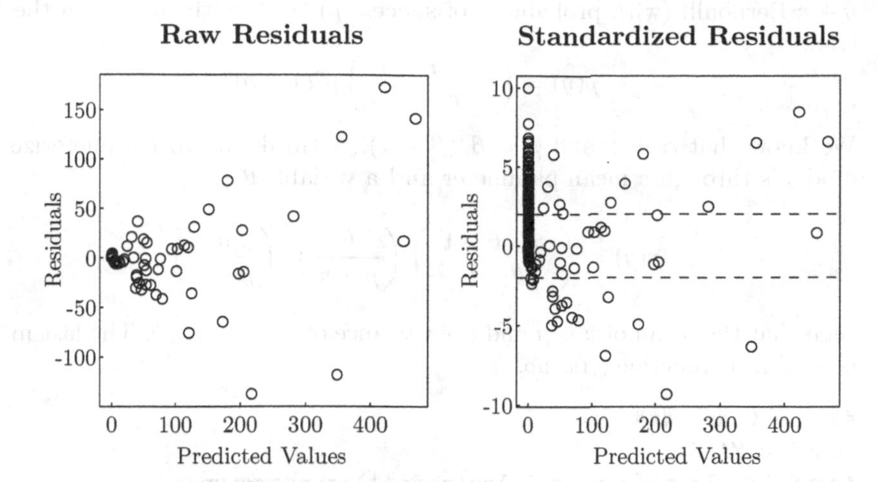

FIGURE 8.20: Residuals of a Poisson model – Residual plots are used for testing overdispersion. The left panel shows the raw residuals versus the predicted values and the right panel shows the standardized residuals versus predicted values. The Poisson regression model uses arsenic concentration and age as continuous predictors and allows intercept and slopes to vary between men and women and lung and bladder cancers.

younger populations increases as well. As a result, not only higher cancer risk due to higher arsenic concentration lead to higher variability in the number of cancer deaths, but also the increased differences in number of cancer deaths among different age groups.

8.4.6 Negative Binomial

When using the option `family="quasipoisson"`, we assume the variance of the data is ω times the mean. Statistically, the use of ω changes the likelihood function and drops the exponential family structure between mean and variance. Another frequently used model for overdispersed count data is the negative binomial model, another member of the exponential family. There are several different ways for parameterizing the negative binomial distribution. The first uses parameters α, β, instead of the mean:

$$p(y) = \begin{pmatrix} y+\alpha-1 \\ \alpha-1 \end{pmatrix} \left(\frac{\beta}{\beta+1}\right)^{\alpha} \left(\frac{1}{\beta+1}\right)^{y}$$

where the expected value of y is α/β and the variance is $\alpha(\beta+1)/\beta^2$. That is, the variance is mean times $1+1/\beta$, the overdispersion. The second model is the distribution describing the probability of y failures and r successes in

$y + r$ Bernoulli (with probability of success p) trials with success on the last trial:

$$p(y) = \binom{y+r-1}{r-1} p^r (1-p)^y$$

We know that $\alpha = r$ and $p = \beta/(\beta+1)$. A third way to parameterize the model is through a mean parameter and a variable θ:

$$p(y) = \binom{y+\theta-1}{\theta-1} \left(\frac{\theta}{\mu+\theta}\right)^\theta \left(\frac{\mu}{\mu+\theta}\right)^y$$

such that the mean of y is μ and the variance of y is $\mu + \mu^2/\theta$. The last model is used in R function `glm.nb`.

```
#### R Code ####
require(MASS)
As.m5nb<-glm.nb(events ~ log(conc+1)+gender+type+
                offset(log(at.risk)), data=arsenic)
summary(As.m5nb)
```

Note that the offset is specified as a term in the model formula, rather than an argument as in function `glm`. The option `family` is no longer necessary. The model output includes the estimated $\hat{\theta}$:

```
> summary(As.m5nb)

glm.nb(formula = events ~ log(conc + 1) + gender + type +
    offset(log(at.risk)), data = arsenic,
    init.theta = 0.228840989818068, link = log)

Deviance Residuals:
   Min       1Q  Median       3Q      Max
-1.838  -0.678  -0.532  -0.303    3.192

Coefficients:
              Estimate Std. Error z value Pr(>|z|)
(Intercept)    -8.8636     0.2306  -38.44  < 2e-16
log(conc + 1)   0.2432     0.0395    6.16  7.4e-10
gender          0.1810     0.1255    1.44  0.14904
type            0.4545     0.1259    3.61  0.00031
---

    Null deviance: 1239.9  on 2235  degrees of freedom
Residual deviance: 1197.6  on 2232  degrees of freedom
AIC: 3328

Number of Fisher Scoring iterations: 1
```

```
              Theta:  0.2288
          Std. Err.:  0.0228
 2 x log-likelihood:  -3318.3190
```

The negative binomial model resulted in a very different interpretation of the data. The gender effect is not significant. Obviously, we must decide which model is appropriate for this data.

8.5 Multinomial Regression

Both Poisson and logistic regression models are models for count response variables. In a Poisson regression model, the response variable value is, in theory, not limited by a total count. When fitting a logistic regression, we use two counts – the number of successes and number of failures. Because the outcome in a logistic regression problem is binary, only one parameter is needed to describe the relative likelihood of occurrence of the two outcomes. In practice, outcome of a problem often has more than two values. For example, ecologists often use biological measures to represent the ecological condition of streams. Some of the most frequently used measures are based on counts of various benthic macroinvertebrates, fish, or other organisms. In the U.S., many macroinvertebrates have been assigned a pollution tolerance score and researchers often group them into four groups: intolerant, moderate-tolerant, tolerant, and unknown. When an individual specimen is identified, it can be assigned into one of these four groups. After all collected individuals are examined, the data set consists of counts of individuals belonging to each of the four groups. The probability distribution describing the variation of these counts is the multinomial distribution. For the four-group example, we observe counts y_1, y_2, y_3, y_4 (number of individuals belonging to the four groups). The multinomial distribution of these counts has four parameters, representing the probabilities of observing an individual belonging to one of the four groups, $\pi_1, \pi_2, \pi_3, \pi_4$. As a macroinvertebrate specimen belongs to one of the four groups, these four probabilities sum to 1 ($\pi_1+\pi_2+\pi_3+\pi_4 = 1$). The probability of observing y_1, y_2, y_3, y_4 is

$$\Pr(y_1, y_2, y_3, y_4) = \frac{n!}{y_1!y_2!y_3!y_4!}\pi_1^{y_1}\pi_2^{y_2}\pi_3^{y_3}\pi_4^{y_4} \tag{8.11}$$

where $n = y_1 + y_2 + y_3 + y_4$ is the total number of individuals in the sample. Because the four probabilities sum to 1, only three need to be estimated.

In general, for a multinomial problem with r possible outcomes, $r - 1$ free parameters are needed to specify the distribution. These probabilities are the mean frequencies of each outcome. The binomial distribution is a special case of the multinomial distribution when $r = 2$, and only one free probability is needed. In ecological studies, these frequencies are known as relative abundances. Relative abundances are measures of community structure and are often used as measures of the stream ecosystem condition.

In a statistical modeling problem, when the response is a collection of count variables from a multinomial distribution, the modeling objective is to estimate the multinomial distribution model parameters. In a binomial regression, the link function is the logit transformation $logit(\pi) = \log(\pi/(1-\pi))$. That is, the logit transformation is the log ratio of the two probabilities (π and $1-\pi$) in the binomial distribution. In a multinomial model, there are $r-1$ free parameters, e.g., $\pi_2, \cdots, \pi_r$, and the remaining probability can be calculated: $\pi_1 = 1 - \pi_2 - \cdots - \pi_r$. We can call group 1 the reference group. The logit transformation for a multinomial model is the log ratio of the $r-1$ free probabilities over the reference probability:

$$logit(\pi_j) = \log\left(\frac{\pi_j}{\pi_1}\right) \tag{8.12}$$

for $j = 2, \cdots, r$. The logit transformation in equation (8.12) is the link function of the multinomial distribution. When modeling multinomial response variables, the logit transformed probabilities are modeled as a linear function of predictors.

$$logit(\pi_j) = X\beta_j \tag{8.13}$$

for $j = 2, \cdots, r$, where $X\beta_j$ represents a linear function of predictors (i.e., $X\beta_j = \beta_{0j} + \beta_{1j}x_1 + \cdots + \beta_{pj}x_p$, where p is the number of predictors). Because $\sum_{j=1}^{r} \pi_j = 1$, we have

$$\pi_1 = \frac{1}{1 + \sum_{i=2}^{p} e^{X\beta_i}},$$

and

$$\pi_j = \frac{e^{X\beta_j}}{1 + \sum_{i=2}^{r} e^{X\beta_i}},$$

for $j = 2, \cdots, r$. Letting $\eta_j = X\beta_j$ and setting $\beta_1 = 0$ (i.e., $\beta_{01} = \beta_{11} = \cdots = \beta_{p1} = 0$), the estimated multinomial probabilities are:

$$\pi_j = \frac{e^{\eta_j}}{\sum_{i=1}^{r} e^{\eta_i}} \tag{8.14}$$

8.5.1 Fitting a Multinomial Regression Model in R

The MLE of the model in equation (8.13) is implemented in R function `multinom` from package `nnet`. The first argument of the function is the model formula. Depending on the data format, the formula can be specified in two different ways, just like in a logistic regression model. In the crypto example, the response variable data consist of two numbers, the number of infected mice (successes) and the number of uninfected mice (failures). The R formula for the response variable is a matrix of two columns. In some cases, the number of trials for each observation is one (e.g., recording whether a certain species is present or not at a number of locations) and the resulting data are recorded in a single column of 0s (failure) and 1s (success). When the response variable is represented by a vector, `glm` will take the total number of trials to be 1 and the number of success is either 1 or 0. For the multinomial regression, the response variable can be a matrix of r columns, each representing the number of occurrences of the respective group. In the macroinvertebrate data, we have four groups and the response variable can be a matrix of four columns. The response variable can also be represented by a factor variable (a vector) when, for example, each observation identifies only one subject for its group association.

In the EUSE example, benthic macroinvertebrate samples were taken from 30 watersheds in each of the nine metropolitan regions. As these 30 watersheds in each region were selected to represent an urban gradient, one objective of the study was to understand how macroinvertebrate community structure changes along the urban gradient. Using the multinomial model, we can study the changes of relative abundances of various species or species groups along the gradient. The four tolerance groups were used for an illustration purpose by Qian et al. [2012]. I will use the data collected from the Boston region as an example. The data file (in data file `usgsmultinomial.csv`) includes individual taxa counts as well as counts of the tolerance groups.

```
#### R Code ####
euseSPdata <- read.csv(paste(dataDIR, "usgsmultinomial.csv",
                             sep="/"), header=TRUE)
```

The counts of the four tolerance groups are in columns 30 to 33 (intolerant, moderate-tolerant, tolerant, and unknown, respectively). In the EUSE study, the urban gradient often is represented by the percentage of a watershed's land cover that is classified as "developed," reported in the National Land Cover Database (NLCD) (the column `NLCD2`). The multinomial formula can be specified as `as.matrix(euseSPdata[,30:33] ~ NLCD2`:

```
#### R Code ####
multinom.BOS1 <- multinom(as.matrix(euseSPdata[,30:33])~NLCD2,
                          data=euseSPdata, subset=City=="BOS")
```

By default, the reference group is the first column. In this case it is the intolerant group. In this case, we may want to use the `Unknown` group as the reference. We can fit the model as follows.

```
PID <- c(33, 30:32)
multinom.BOS1 <- multinom(as.matrix(euseSPdata[,PID])~NLCD2,
                          data=euseSPdata, subset=City=="BOS")
#### R Output ####
summary(multinom.BOS1)
```

```
Call:
multinom(formula = as.matrix(euseSPdata[, PID]) ~ NLCD2,
    data = euseSPdata, subset = City == "BOS")

Coefficients:
             (Intercept)        NLCD2
Intol.rich      2.500278 -0.044351300
Modtol.rich     2.708246 -0.005198547
Tol.rich        1.199572  0.004172518

Std. Errors:
             (Intercept)       NLCD2
Intol.rich     0.2265808 0.008565843
Modtol.rich    0.2205117 0.007271081
Tol.rich       0.2418887 0.007827359

Residual Deviance: 2368.06
AIC: 2380.06
```

The model summary includes the estimated model coefficients and their estimation standard errors. For the intolerant group, the model is:

$$\log\left(\frac{\pi_I}{\pi_U}\right) = 0.22658 - 0.04435x$$

where π_I (π_U) is the probability (or relative abundance) of the intolerant (Unknown) group. Because the linear model is in the log probability ratio scale, the resulting model is difficult to explain in terms of changes in relative abundance. Given the model uses only one predictor, I opt to present the result by plotting the estimated relative abundances (π's) directly. The estimated relative abundances can be calculated using the function `predict`:

```
pp <- predict(multinon.BOS1, type="probs",
              newdata=data.frame(NLCD2=0:100))
```

These estimated relative abundances are plotted against the observed relative abundance (number of individuals belonging to a group divided by the total number of individuals in the sample). As in the nonlinear regression case, the option `se.fit` is ignored. As a result, the estimation uncertainty in model coefficients cannot be readily converted into the uncertainties in the estimated relative abundances (Figure 8.21).

I wrote a function to implement a Monte Carlo simulation approximation of the uncertainty in the estimated probabilities. Details of the simulation approximation are discussed in Chapter 9. Using the simulation function, I draw random samples of model coefficients and calculate the relative abundances for each set of coefficients as the likely outcome from the model. I used the package `rv` for better processing simulation results.

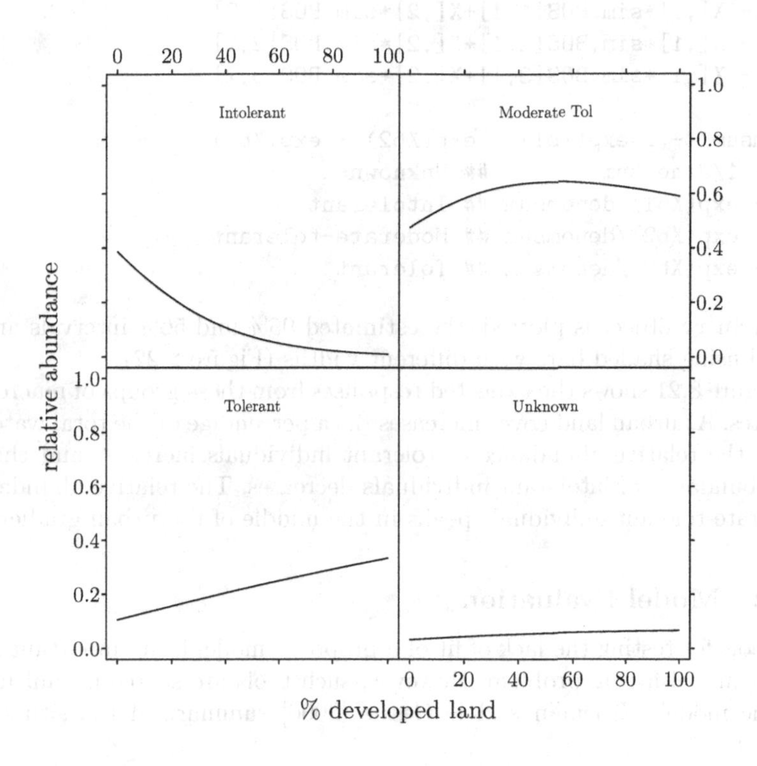

FIGURE 8.21: Multinomial model – Estimated probability of occurrences or relative abundances are plotted as functions of urban intensity measured by the % developed land cover in the watershed.

```
#### R Code ####
sim.BOS <- rvsims(sim.multinom(multinom.BOS1, 2500))
## generating random samples of model coefficients and
## store them as an rv object

sim.BOS <- rvmatrix(sim.BOS, nrow=3,ncol=2, byrow=T)

## calculating probabilities
X <- cbind(1,seq(0,100,1))
Xb1 <- X[,1]*sim.BOS[1,1]+X[,2]*sim.BOS[1,2]
Xb2 <- X[,1]*sim.BOS[2,1]+X[,2]*sim.BOS[2,2]
Xb3 <- X[,1]*sim.BOS[3,1]+X[,2]*sim.BOS[3,2]

denomsum <- 1+exp(Xb1) + exp(Xb2) + exp(Xb3)
p3 <- 1/denomsum          ## Unknown
p0 <- exp(Xb1)/denomsum ## Intolerant
p1 <- exp(Xb2)/denomsum ## Moderate-tolerant
p2 <- exp(Xb3)/denomsum ## Tolerant
```

When an `rv` object is plotted, the estimated 95% and 50% intervals are presented using shaded bars with different widths (Figure 8.22).

Figure 8.21 shows the expected responses from these groups of macroinvertebrates. As urban land cover increases (as a percentage of the total watershed area), the relative abundance of tolerant individuals increases and the relative abundance of intolerant individuals decreases. The relative abundance of moderate-tolerant individuals peaks in the middle of the urban gradient.

8.5.2 Model Evaluation

Tools for testing the lack of fit of a proposed model is an important aspect of any model fitting problem. However, such tools are scarce in multinomial logistic models. Goeman and le Cessie [2006] summarized the situation as follows:

> Hosmer and Lemeshow [2000] suggested looking at the multinomial model as if it were a set of independent ordinary logistic models of each outcome against the reference outcome, and testing the fit of each of these separately. Lesaffre and Albert [1989] give diagnostics for detecting influential, leverage, and outlying samples in multinomial logistic regression, but provide no explicit goodness-of-fit test. The only actual test for the fit of the multinomial logistic regression model is given by Pigeon and Heyse [1999]. It is an extension of the test of Hosmer and Lemeshow [2000] for binary regression, which is well known to have low power for detecting quadratic effects [Le Cessie and Van Houwelingen, 1991].

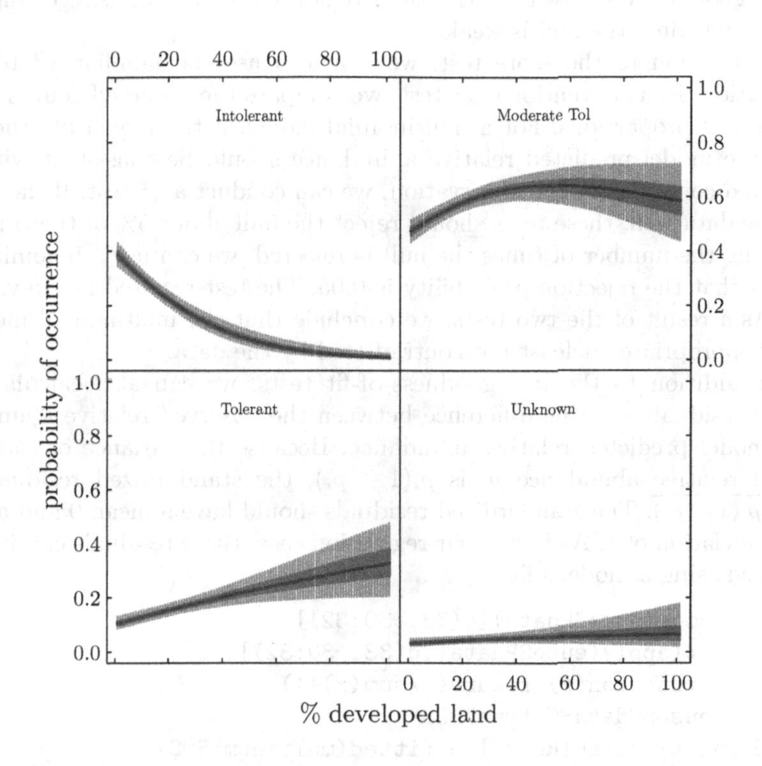

FIGURE 8.22: Uncertainty of the estimated probability of occurrences or relative abundances are estimated using Monte Carlo simulation.

Goeman and le Cessie [2006] proposed a score test to check the fit of a multinomial regression model. The null hypothesis is that the model fits well and the alternative is that residuals of samples close to each other in covariate space tend to deviate from the model in the same direction. The test statistic is a sum of smoothed residuals. Because the test is based on approximate asymptotic distribution of the test statistic, the result may not be entirely reliable if the sample size is small.

Details of the Geoman and le Cessie test can be found in the reference. Using the R functions for the test distributed by the lead author, the p-value of the goodness-of-fit test for the BOS region model is 0.32, suggesting that evidence against the null is weak.

In addition to the score test, we can also use the familiar χ^2 test for proportion. In a conventional χ^2 test, we compare the observed counts to the theoretical proportions. For a multinomial model, if the model fits the data well, the model predicted relative abundance should be consistent with the observed counts. For each observation, we can conduct a χ^2 test. If the model fits the data well, these tests should reject the null about 5% of the time. By counting the number of times the null is rejected, we can use a binomial test to test that the rejection probability is 0.05. The test resulted in a p-value of 0.5. As a result of the two tests, we conclude that the multinomial model is likely appropriate, at least not contradicted by the data.

In addition to the two goodness-of-fit tests, we can also calculate the model residuals r_i – the difference between the observed relative abundance and model predicted relative abundance. Because the variance of each predicted relative abundance p_i is $p_i(1-p_i)$, the standardized residuals are $r_i/\sqrt{p_i(1-p_i)}$. The standardized residuals should have a mean 0 and a standard deviation of 1. As in a linear regression case, these residuals can be used for diagnosing a model's fit.

```
txs <- names(euseSPdata)[c(33, 30:32)]
dataP <- t(apply(euseSPdata[,c(33, 30:32)], 1,
         function(x) return(x/sum(x))))
Rows <- euseSPdata$City=="BOS"
resid.mat <- dataP[Rows,] - fitted(multinom.BOS1)
Resid <- unlist(resid.mat)
Fitted <- unlist(fitted(multinom.BOS1))
Group <- rep(txs, each=dim(resid.mat)[1])

stdResid <- as.vector(Resid/sqrt(Fitted*(1-Fitted)))
FittedV <- as.vector(Fitted)
GroupV <- rep(c("Intol","Modtol","Tol","Unknown"),
              each=dim(Fitted)[1])
key <- simpleKey(c("Intol","Modtol","Tol","Unknown"),
                 lines=F, points=T,
                 space = "top", columns=4)
key$text$cex <- 1.2
```

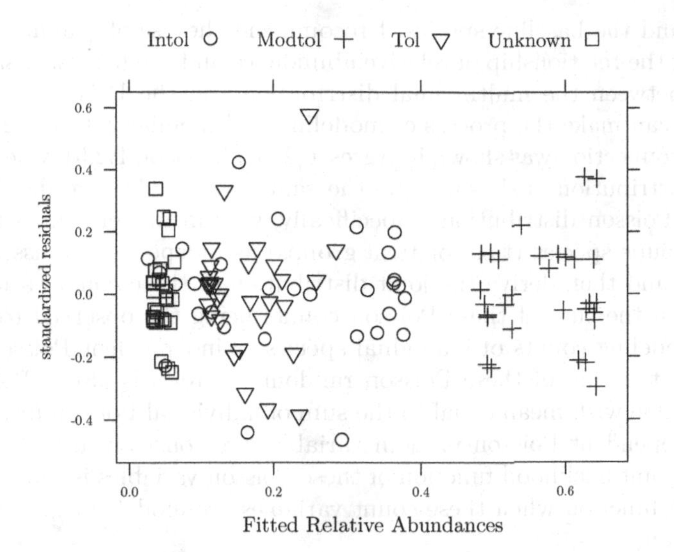

FIGURE 8.23: Standardized "residuals" from the multinomial regression model are plotted against the fitted relative abundances

```
xyplot(stdResid~FittedV, group=GroupV,
       ylab="standardized residuals",
       xlab="Fitted Relative Abundances",
       key=key)
```

Plots of standardized model residuals are often more informative. We plot standardized residuals against the fitted values (relative abundance) as in a typical linear regression situation (Figures 8.23). The plot show no apparent pattern along the relative abundance gradient.

8.6 The Poisson-Multinomial Connection

A multinomial regression is an example of multivariate regression, where the response variable is a multivariate variable of compositions of multiple groups. Traditionally, count data from individual taxa or taxa groups are analyzed independent of each other. When using the multinomial regression, we are able to properly estimate the relative abundances of species in an assemblage. A difficulty in using the multinomial regression is the interpretation of model coefficients. In fact, the multinomial regression model coefficients are mostly meaningless because the multinomial logistic regression coefficients are defined with respect to the ratio of the relative abundance of the species of interest and the baseline species. I recommend the use of graphical tools for exploring the relationship of relative abundance and predictors. A simple connection between the multinomial distribution and the Poisson distribution, however, can make the process of modeling multinomial data more intuitive.

This connection was shown in Agresti [2002] (Section 1.2.5), where a multinomial distribution is shown to be the same as a conditional distribution of multiple Poisson distributions. Specifically, we can model counts from individual groups species (taxa or taxa groups) as independent Poisson random variables and then derive the joint distribution of these count variables conditional on the sum of these Poisson counts being the observed total count. When modeling counts of individual species as independent Poisson random variables, the sum of these Poisson random variables is also a Poisson random variable with mean equal to the sum of individual Poisson means. When these independent Poisson random variables are constrained to have a fixed sum, the joint likelihood function of these Poisson variables is the same as the likelihood function when these count variables are modeled as a multinomial distribution.

When using the Poisson regression, the mean of the count variable is linked to a linear model through a logarithmic link function:

$$\begin{array}{rcl} y_{ij} & \sim & Pois(\lambda_{ij}) \\ \log(\lambda_{ij}) & = & X_i\alpha_j \end{array} \tag{8.15}$$

For a given predictor variable value of X_i, we observed a total of $Y_i = \sum_j y_{ij}$ organisms from taxa $j = 1, \cdots, J$. The link between the Poisson and multinomial distribution suggests that we can derive the relative abundance as follows:

$$p_{ij} = \frac{e^{\lambda_{ij}}}{\sum_j e^{\lambda_{ij}}} = \frac{e^{X_i\alpha_j}}{\sum_j e^{X_i\alpha_j}} \tag{8.16}$$

When analyzing count data from multiple taxa, we need only fit independent Poisson models for each taxon to model the change of each taxon's abundance along the gradient of the environmental predictor. The resulting model can be used directly to derive information on relative abundance. The advantage of this approach is that we can fit simple univariate Poisson models which we know well both in terms of model interpretation and diagnostics. In addition, when analyzing individual taxon abundance, we should not be limited to a log-linear model (equation (8.15)). If the environmental predictor is nutrient concentration or pH, changes of the total abundance of an individual taxon along the predictor gradient may be better approximated by a unimodal model. That is, a taxon has an optimum along the gradient and a range which the taxon can tolerate. Different taxa may use different total abundance models. As a result, the relative abundance model is not limited to the logit-linear model of equation (8.13). Unimodal models have been discussed in the literature. Early studies used the Gaussian model to describe how taxon abundances change along an environmental gradient [ter Braak,

1996]. Oksanen and Minchin [2002] compared the Gaussian model to 3 alternative unimodal model forms. Qian and Pan [2006] discussed a versatile gamma model, where the Poisson mean is modeled by a function similar to the (log) gamma distribution density:

$$\begin{array}{rcl} y_i & \sim & Pois(\lambda_i) \\ \log(\lambda_i) & = & \gamma_0 + \gamma_1 x_i + \gamma_2 \log(x_i) \end{array} \quad (8.17)$$

where y_i is the ith observed count and x_i is the respective covariate value. The count variable is modeled by a Poisson distribution with parameter λ. Because the gamma model can capture many typical taxon response patterns, we can use the gamma model as the default model form for abundance data from individual taxa. The gamma model requires only 2 non-zero counts to quantify model coefficients, whereas the alternative models by Oksanen and Minchin [2002] need 5 non-zero counts.

Furthermore, count variables are often overdispersed with respect to the Poisson distribution, binomial distribution, and the multinomial distribution. Count data can also be zero-inflated [Lambert, 1992, Martin et al., 2005]. Identifying overdispersion or zero-inflation in a univariate count variable is relatively simple, but not so for multinomial count data. As a result, exploring univariate count data can serve as a first step in analyzing multinomial count data such as species compositional data common in ecological studies. For this reason, I recommend that the analysis of multinomial count data should start with fitting appropriate univariate models. When models for individual taxa or taxa groups are properly developed, models of relative abundances can be derived accordingly.

Figure 8.24 compares the estimated relative abundances of the four tolerance groups as functions of the urban gradient (% developed land) using the multinomial regression to the same derived from four independent Poisson models. The two sets of estimates are identical. The result is expected because we know that the Poisson linear model fits the abundance data well (Figure 8.25).

The tolerance group example is an exception in that Poisson regression models for individual groups lead to the logit linear multinomial model. When a linear Poisson model fits the data poorly, we expect to see the estimated relative abundances from the Poisson model differ from the same multinomial regression model. As an example, I use the data of 10 mayfly species (taxa) from the same data set.

Mayflies are mostly sensitive to pollution. As a result, their abundance decreases in general when urban land use in the watershed increases. But some species are less sensitive than others and these less sensitive ones can actually thrive in moderately developed watersheds because of the vacated habitat by those more sensitive ones. As a result, the abundance of some mayfly species may increase as urban land cover increases initially, before the eventual decline. Figure 8.26 shows the abundances of the 10 mayfly taxa as a function of % developed land. Two Poisson models are fit for each taxon. The solid lines

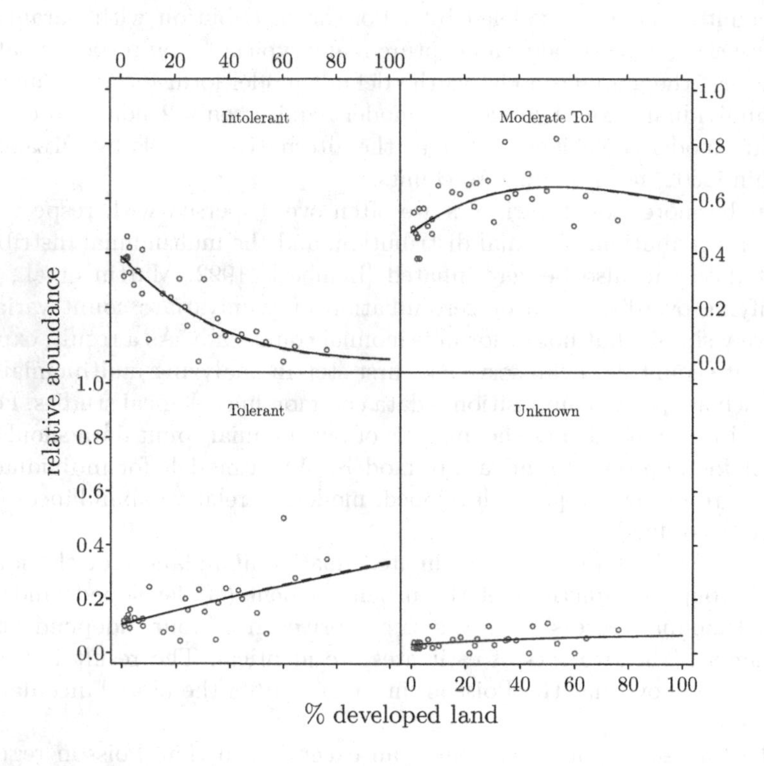

FIGURE 8.24: Relative abundances of the four tolerance groups predicted by the multinomial regression model (solid lines) are compared to the relative abundances derived from the independently fit Poisson regression models (dashed lines). These two sets of predictions are identical, as expected.

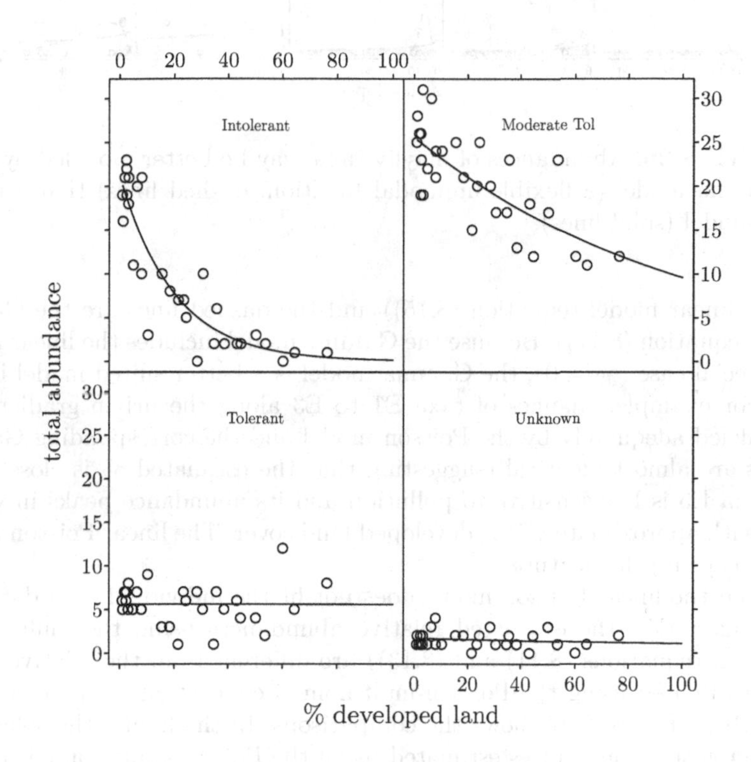

FIGURE 8.25: Independently fit univariate Poisson regression models can be used to predict abundances of each tolerance group that can be compared against respective data.

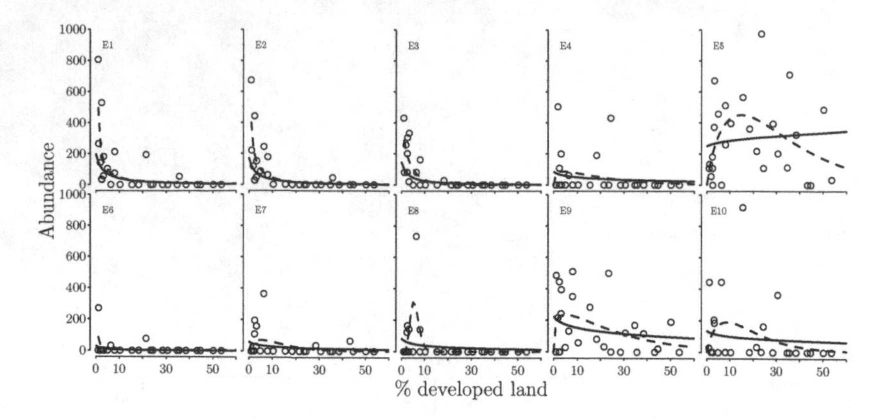

FIGURE 8.26: Abundances of mayfly taxa may be better modeled by using the gamma model (a flexible unimodal function, dashed lines) than the log-linear model (solid lines).

are the linear model (equation (8.15)) and the dashed lines are the Gamma model (equation (8.17)). Because the Gamma model includes the linear model as a special case ($\gamma_2 = 0$), the Gamma model is a better suited model in this case. For example, changes of taxa E1 to E3 along the urban gradient can be modeled adequately by the Poisson model and the corresponding Gamma models are almost identical (suggesting that the estimated γ_2 is close to 0). The taxa E5 is less sensitive to pollution and its abundance peaks in watersheds with approximately 10% developed land cover. The linear Poisson model cannot capture this feature.

When the linear Poisson model does not fit the individual taxa data well (e.g., taxon E5), the estimated relative abundances using the multinomial regression (equations (8.11) and (8.13)) are different from the relative abundance estimated using the Poisson-multinomial connection (equations (8.15) and (8.16)). Figure 8.27 shows the comparisons. In the figure, the solid lines are the relative abundances estimated using the Poisson-multinomial connection based on the Gamma model, the dashed lines are the same based on the linear model, and the dotted lines are the relative abundances estimated using the logit-linear multinomial regression model. Estimated relative abundances for taxa E1 to E3 are the same from all three methods, while the relative abundance estimates are different among the three methods for taxon E5.

The Poisson-multinomial connection is likely the most helpful tool for exploring the appropriate model form. On the one hand, evaluating a model's fit to a univariate count data is relatively simple. By fitting Poisson models to data from individual taxa or taxa groups, we can select the most appropriate model forms for individual taxa. On the other hand, the logit-linear multinomial model cannot be easily checked against data, but it is applied to all taxa (or taxa groups). It is difficult to justify the use of a single model for different

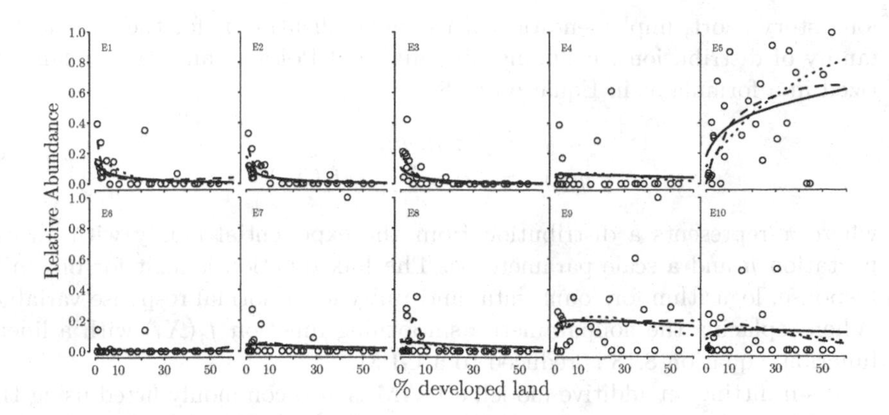

FIGURE 8.27: The multinomial regression estimated relative abundances (dotted line) are not always the same as the Poisson regression derived relative abundances (dashed lines). The relative abundances derived from the gamma model (solid line) may fit the data better for some taxa (e.g., E5).

taxa. As a result, using the Poisson-multinomial connection is the preferred practice in analyzing species composition data.

8.7 Generalized Additive Models

The additive models described in Section 6.3.1 are for normal response variables. When the response variable Y is not normal, we use the generalized additive model (GAM), just as the generalization from linear model to generalized linear model. This process of generalization is mathematically more challenging than the same transition from linear regression to generalized linear model. In linear regression, the least square estimator coincides with the maximum likelihood estimator for a normal response variable. As a result, the transition from LM to GLM is largely conceptual – we need to consciously make a probabilistic distribution assumption on the response variable. Instead of thinking of a model fitting process as a process of finding parameter values that minimize the error measured by the sum of squares of residuals (which does not need to have a probabilistic assumption), the model fitting process is now specifically associated with a probabilistic assumption and the objective of the process is to estimate model parameters to maximize the likelihood function. Numerically, both least squares estimator and MLE are a problem of optimization. The generalization from the additive model to GAM has the same conceptual shift but the computation is more complicated. To make a long story short, implementation of the generalization is for the exponential family of distributions, including binomial and Poisson, and can be summarized in a formula as in Equation 8.18:

$$\begin{aligned} y &\sim \pi(\mu, \phi) \\ \eta(\mu) &= \alpha + \textstyle\sum_{j=1}^{k} f_j(X_j) \end{aligned} \tag{8.18}$$

where π represents a distribution from the exponential family with an expectation μ and a scale parameter ϕ. The link function is logit for binomial response, logarithm for count data, and unity for a normal response variable. When replacing the nonparametric smoothing function $f_j(X_j)$ with a linear function, equation 8.18 is reduced to a GLM.

As in fitting an additive model, a GAM is also commonly fitted using the backfitting algorithm. In the additive model, the backfitting algorithm is a repeated fitting of a smoothing. In GAM, a model is fitted using the maximum likelihood estimator and numerically using the Fisher scoring procedure – a Newton–Raphson algorithm for solving multiple equations:

$$\sum_{i=1}^{n} x_{ij} \left(\frac{\partial \mu_i}{\partial \eta_i}\right) V_i^{-1}(y_i - \mu_i) = 0, j = 0, 1, \cdots, k$$

When fitting a GLM, the Fisher scoring procedure is implemented by using an iteratively reweighted least-squares–fitting weighted least-squares using an adjusted response variable, which is defined as:

$$z_i = \eta_i^0 + (y_i - \mu_i^0) \left(\frac{\partial \mu_i}{\partial \eta_i}\right)_0$$

where μ_i^0 is the initial mean based on a set of initial value of model parameters. The weights are:

$$w_i^{-1} = \left(\frac{\partial \mu_i}{\partial \eta_i}\right)_0^2 V_i^0$$

with V_i^0 being the variance of Y at μ_i^0. The weighted least squares regression is of the form:

$$z_i = X\beta$$

At each iteration, the estimated parameter values from the previous iteration are used to evaluate μ_i^0 and V_i^0. Convergence of the process is determined by the change in deviance of each iteration. For a GAM, the iteratively reweighted least-squares is modified to a local scoring procedure. In the iteratively reweighted least-squares for GLM, $\eta = X\beta$. For a GAM, $\eta = \alpha + \sum_{j=1}^{k} f_j(x_{ij})$. The process starts with a set of initial values of $f_j^0 : f_1^0, \cdots, f_k^0$, which lead to an initial estimate of $\mu_i^0 = \eta^{-1}\left(\alpha + \sum_{j=1}^{k} f_j^0(x_{ij})\right)$. The adjusted response variable is then:

$$z_i = \eta_i^0 + (y_i - \mu_i^0) \left(\frac{\partial \mu_i}{\partial \eta_i}\right)_0,$$

with weights:

$$w_i^{-1} = \left(\frac{\partial \mu_i}{\partial \eta_i}\right)_0^2 V_i^0$$

which lead to a weighted additive model:

$$z_i = \alpha + \sum_{j=1}^{k} f_j(X_j)$$

The last step resulted in a set of new estimates: $f_j^1 : f_1^1, f_2^1, \cdots, f_k^1$, which will replace the initial values for the next iteration. Convergence of the process is evaluated by comparing f_j^1 and f_j^0.

Because GAM is implemented in R, fitting a GAM is as simple as fitting an additive model with a normal response variable. However, model interpretation is more complicated because the resulting model is presented in the link function scale. For example, the GAM output presents the estimated functional form in the logarithmic scale when the response is a Poisson variable. For a binary response variable, the GAM plots are shown in the logit scale. Model assessment becomes a more complicated process.

8.7.1 Example: Whales in the Western Antarctic Peninsula

To understand the effect of global climate change on the abundance, diversity, and productivity of marine populations, the southern ocean GLOBEC (Global Ocean Ecosystem Dynamics) research program (part of the International Geosphere-Biosphere Program, or IGBP) conducted two surveys of marine mammals (cetaceans) from early April to late May in 2001 and 2002, in and around the continental shelf waters of Marguerite Bay on the western side of the Antarctic Peninsula (WAP). A typical marine mammal survey is conducted by trained observers aboard survey vessels cruising along predetermined sampling stations or transects. Numbers of animal sightings and environmental conditions are recorded for each station/transect. Figure 8.28 shows the location of the study area and the survey stations for the three cruises in the Marguerite Bay area. Two vessels were involved in the WAP survey. One vessel transited between stations spaced 40 km apart and oriented perpendicular to the coast, and the other vessel transited between process sites for fine-scale sampling and experimental activities. On board both vessels, visual surveys for cetaceans were conducted by trained observers. Each observer searched in a 90° sweep from the bow of the ship to the perpendicular beam with the naked eye or binoculars. When a cetacean was sighted, its species was identified. Observers also recorded environmental and sighting conditions (weather, visibility, glare, swell height, Beaufort sea state, sea ice concentrations) and tracked changes as these occurred. As noted by Thiele et al. [2004], most whales in the Antarctic are medium to large species, and thus can be detected at relatively high Beaufort sea states (a measure of sea roughness).

FIGURE 8.28: Antarctic whale survey locations – The Marguerite Bay area in the Western Antarctic Peninsula is the study area of the southern ocean GLOBEC program. The 2001–2002 cetacean survey tracks and stations are shown as open circles (where cetaceans were sighted) and gray crosses (where no cetaceans were sighted).

The WAP is a biologically rich area supporting large standing stocks of krill and top predators (including whales, seals, and seabirds). Friedlaender et al. [2006] described a study of predicting whale distribution patterns using GAM with predictors selected based on a CART model using all available predictors, including various environmental variables and acoustic backscattering as an indicator of prey abundance. The unique feature of this study is the availability of a variable, acoustic backscatter, indicating prey abundance. Acoustic backscattering measures the strength of acoustic signals reflected back from, presumably, krill and other zooplankton at a given depth.

8.7.1.1 The Data

The response variable of interest is the number of whales (humpback *Megaptera novaeangliae* and minke *Balaenoptera acutorstrata*) at a given location. The objective of the study is to link the number of whales and other environmental conditions. Our first step is to plot the observed number of whale against potential predictor variables suggested by scientific knowledge of marine mammals. The marine system is, however, a dynamic system. The commonly used variables for describing environmental conditions are often variables of convenience (e.g., sea surface temperature, distance to shore, and bathymetric depth), not necessarily variables reflecting or describing attributes of habitat. The resulting models are often descriptive in nature. Exploratory plots (Figure 8.29) show that almost all whales were sighted in areas that are relatively shallow (less than 1000 meters deep), or are relatively flat (bathymetric slope less than 6%), or have a high acoustic backscatter (> -85 dB). Chlorophyll *a* concentration (`chla`) is used to represent the primary productivity. It is often hypothesized that marine animals are attracted to regions with a high primary productivity that are likely to support a healthy amount of zooplankton and other predators. The scatter plot of number of whales sighted against chlorophyll a shows an inverse relationship. The two whale species included in this data set prey on krill, a type of shrimp-like swarming marine invertebrate animal, and Antarctic krill feed directly on phytoplankton (or algae). It is tempting to speculate that areas with a high density of krill are often with a smaller chlorophyll a because of the high grazing pressure. The scatter plot of acoustic backscatter against chlorophyll a hints of such a relationship.

8.7.1.2 Variable Selection Using CART

As we discussed in Section 7.3.2, CART is ideally suited for variable selection in a model building process. For this data set, we are not entirely clear on what are the key factors affecting whale distribution in the region. Using all available predictor variables, we fit a CART model and the pruned CART model (Figures 8.30 and 8.31) suggesting the following predictors: acoustic backscattering between 25 and 100 meters of depth, distance to ice edge, distance to shore, chlorophyll a, and bathymetric depth.

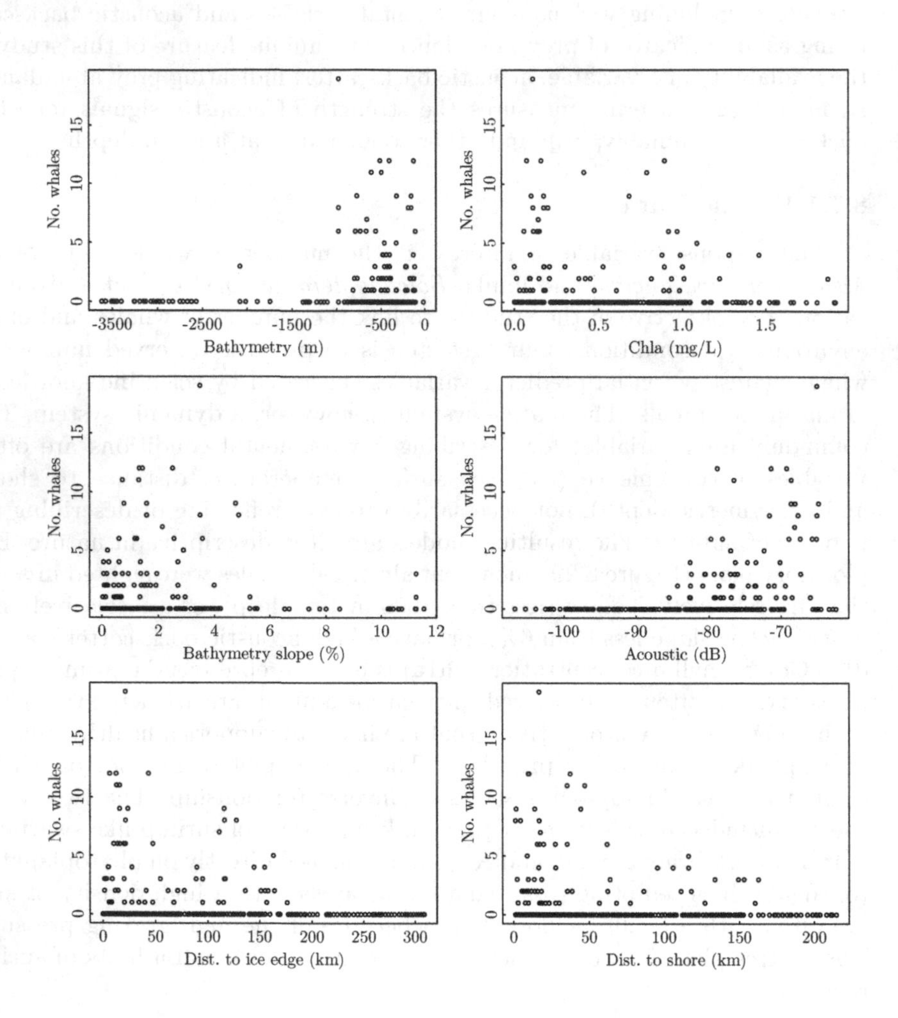

FIGURE 8.29: Antarctic whale survey data scatter plots – Selected scatter plots show the noisy relationships between the observed number of whales and other potential predictors.

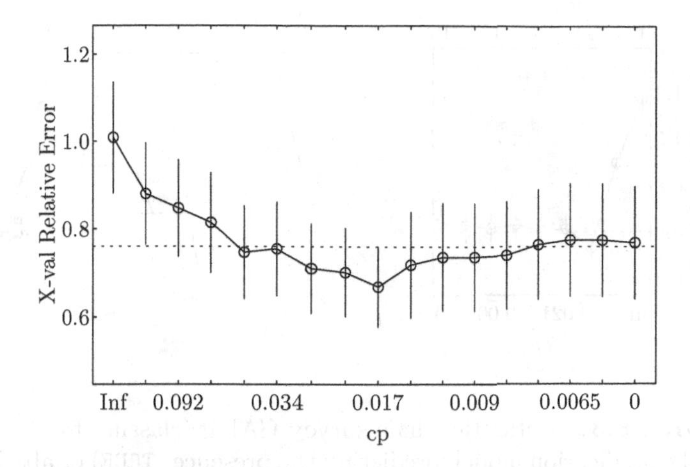

FIGURE 8.30: Antarctic whale survey CART model Cp plot – The Cp plot of the regression model shows a tree with 5 splits is optimal based on the cross-validation error plus one standard deviation rule.

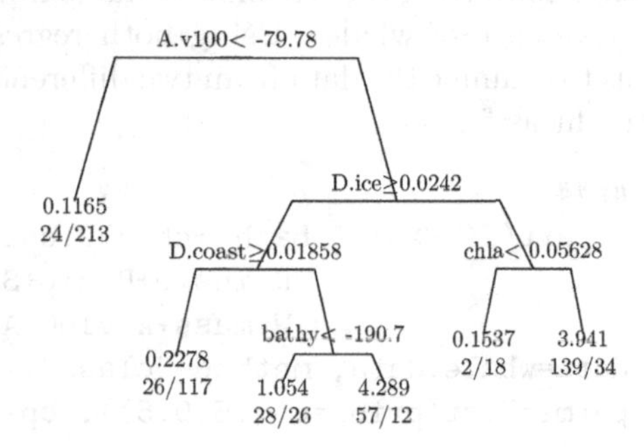

FIGURE 8.31: Antarctic whale survey CART (regression) model – The pruned CART model suggests that acoustic backscattering from 25 to 100 meters in depth (`A.v100`) is the most significant predictor, followed by distance to ice edge (`D.ice`), distance to shore (`D.coast`), chlorophyll a, and bathymetric depth (`Bathy`).

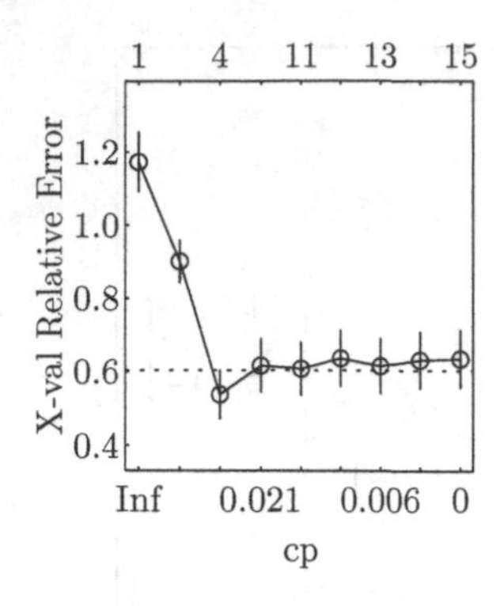

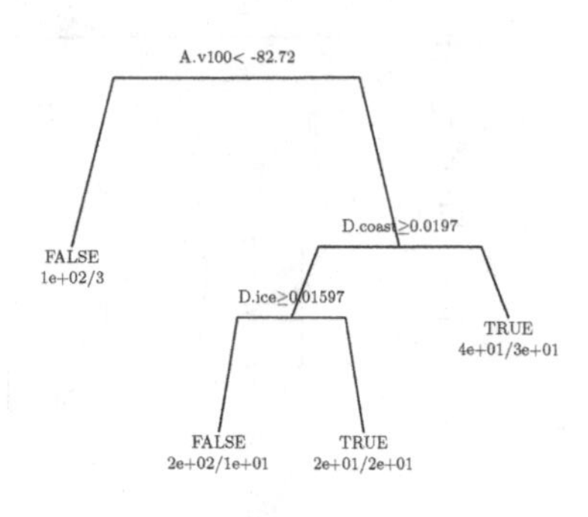

FIGURE 8.32: Antarctic whale survey CART (classification) model – The pruned classification model, predicting the presence (`TRUE`) or absence (`FALSE`) of whales, suggests that acoustic backscattering from 25 to 100 meters in depth (`A.v100`), distance to ice edge (`D.ice`), and distance to shore (`D.coast`) are the three primary predictors. Pruning is based on the cross-validation error plus one standard deviation rule (left panel).

A Poisson regression and a logistic regression can be linked through the probability of absence. As an exploratory step, we should also consider fitting a CART model after converting the response variable from the number of whales to the absence/presence of whales. Using both regression and classification models allows us to examine the data from two different angles. A classification model can easily fit as follows:

```
#### R Code ####
TW.rpart2 <- rpart(I(TW>0) ~ bathy+chla+D.coast+D.ice+
                              D.inswb+D.slp+S.bathy+
                              W.mass+A.v100+A.v300.2,
            data=whale.data, method="class",
            parms=list(prior=c(0.5,0.5)), cp=0.00)
```

The results (Figure 8.32) are slightly different in that the pruned model has only three predictors (`A.v100`, `D.ice`, and `D.coast`). Bathymetric depth and chlorophyll a are not included in the pruned model.

8.7.1.3 Fitting GAM

Poisson Response Models

CART suggests five predictors should be considered. A generalized additive model was used by Friedlaender et al. [2006] to explore the functional relationship between whale abundance and selected environmental variables. We use the five predictors suggested by our CART model as a starting point.

When using the package mgcv, the resulting model can be plotted as follows:

```
#### R Code ####
whale.gam1 <- gam(TW~s(A.v100,bs="ts")+
                    s(chla,bs="ts")+
                    s(bathy,bs="ts")+
                    s(D.ice,bs="ts")+
                    s(D.coast,bs="ts"),
           data=whale.data, family="poisson")

par(mfrow=c(3,2), mar=c(4,4,0.5,0.5))
plot(whale.gam1, scale=0, pages=0, select=1,
    xlab="Backscatter 25-100m", ylab="f(x)",
    residuals=T, shade=T, lwd=2, pch=1, cex=0.5)
plot(whale.gam1, scale=0, pages=0, select=2,
    xlab="Chlorophyll a", ylab="f(x)",
    residuals=T, shade=T, lwd=2, pch=1, cex=0.5)
plot(whale.gam1, scale=0, pages=0, select=3,
    xlab="Bathymetry", ylab="f(x)",
    residuals=T, shade=T, lwd=2, pch=1, cex=0.5)
plot(whale.gam1, scale=0, pages=0, select=4,
    xlab="Dist. to ice edge", ylab="f(x)",
    residuals=T, shade=T, lwd=2, pch=1, cex=0.5)
plot(whale.gam1, scale=0, pages=0, select=5,
    xlab="Dist. to shore", ylab="f(x)",
    residuals=T, shade=T, lwd=2, pch=1, cex=0.5)
```

The fitted model (Figure 8.33) suggests a piecewise linear relationship between (log) whale abundance and all predictors except chlorophyll a. The relationship between whale abundance and chlorophyll a is difficult to interpret. Because krill graze on phytoplankton and whale prey on krill, the presence of high chlorophyll a does not necessarily imply a high concentration of krill. A large swarm of krill may consume all phytoplankton and result in low chlorophyll a. A high chlorophyll a may indicate that krill have not yet arrived. Consequently, the fitted GAM function of chlorophyll a should not be taken too seriously. Furthermore, the interactive relationship between the concentration of krill and chlorophyll a suggests that the additive assumption should not be applied on these two predictors. Because whales are attracted to krills, not chlorophyll a, chlorophyll a is no longer relevant when a measure of krill density is available.

The apparent piecewise linear relationships between log whale abundance and acoustic backscattering, distance to ice edge, and distance to shore are tempting, but somewhat suspicious. The data (Figure 8.29) seem to suggest step functions for these three predictors. There were no whales observed when A.v100 is less than -87 dB, or distance to ice edge is larger than 180 km, or

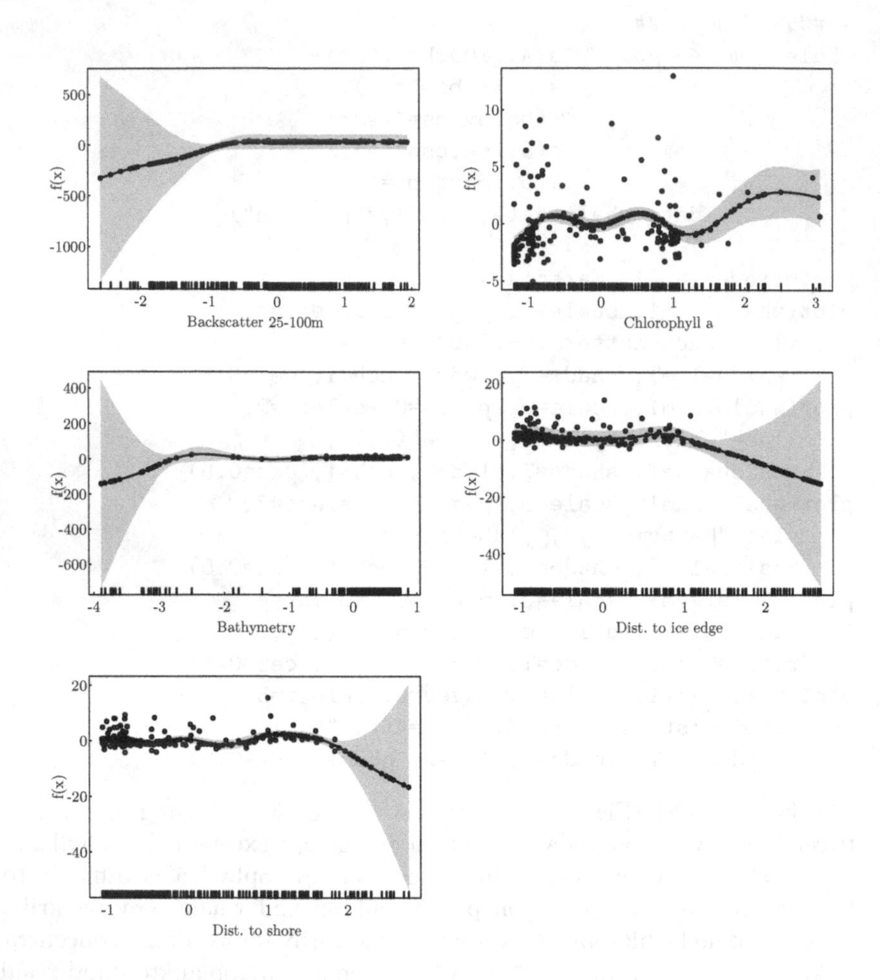

FIGURE 8.33: Antarctic whale survey Poisson GAM – Fitted GAM functions show the effects of acoustic backscatter (top left, likely a piecewise linear model), chlorophyll a (top right, somewhat erratic), bathymetric depth (middle left, likely linear), distance to ice edge (middle right, likely a piecewise linear model), and distance to shore (bottom, another piecewise linear model). The model is fitted assuming the response variable follows a Poisson distribution.

distance to shore is greater than 170 km. The scientific question here is whether a continuous function is reasonable to describe the relationship between whale density and environmental variables used in this work. If a continuous and smooth relationship is expected, the use of GAM is reasonable. Otherwise, a Poisson additive model is inappropriate.

As in a Poisson regression problem, overdispersion is also a potential problem for an additive model. The option of fitting an overdispersed Poisson additive model is available by using `family="quasipoisson"`. The underlying statistical problem is the same as explained in Section 8.4.4. For this example, we can diagnose the problem by calculating the overdispersion parameter and by using residual plots (Figure 8.34):

```
#### R Code ####
yhat <- predict(whale.gam1, type="response")
z <- (whale.data$TW - yhat)/sqrt(yhat)
overD <- sum(z^2)/summary(whale.gam1)$residual.df
p.value <- 1-pchisq(sum(z^2),
   summary(whale.gam1)$residual.df)

plot(yhat, whale.data$TW - yhat,
   xlab="Predicted Values", ylab="Residuals",
   main="Raw Residuals")
abline(h=c(-2,2), lty=2)
plot(yhat, z, xlab="Predicted Values",
   ylab="Residuals", main="Standardized Residuals")
abline(h=c(-2,2), lty=2)
```

The estimated overdispersion parameter (`overD`) is 1.8 ($p < 0.00001$).

Because overdispersion affects only the estimated variances, the fitted model is not affected. When using GAM as an exploratory tool, we will see no visible difference when fitting a GAM using either `family="poisson"` or `family="quasipoisson"`.

Binary Response Model

In this example, the number of whales can be highly uncertain. The uncertainty on the actual count can be effectively removed if we convert the count response variable into a binary (presence/absence) variable. Because whales travel in pods, what conditions exist when a pod of whales is present is likely the relevant information we are seeking. In other words, by using a binary response variable we examine the same problem from a different angle.

```
#### R Code ####
whale.gam3 <- gam(I(TW>0)~s(A.v100,bs="ts")+
                          s(chla,bs="ts")+
                          s(bathy,bs="ts")+
                          s(D.ice,bs="ts")+
                          s(D.coast,bs="ts"),
   data=whale.data, family="binomial")
```

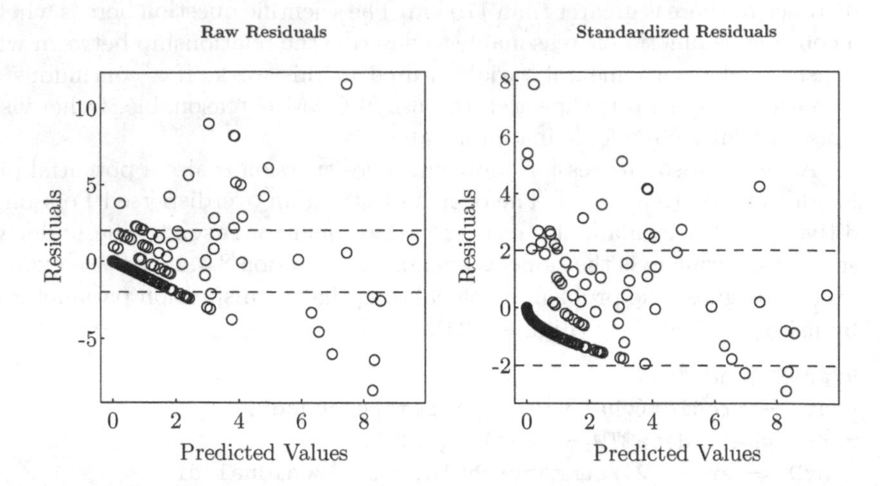

FIGURE 8.34: Residuals from GAM show overdispersion – Residual plots show typical overdispersion symptoms. The left panel shows the residuals against the fitted, with a wedge-shaped data cloud indicating increasing variance as predicted mean increases. The standardized residuals still have a variance larger than 1.

The fitted model (Figure 8.35) shows similar effects of acoustic backscatter and distance to ice edge. The chlorophyll a effect is still erratic. The effect of distance to shore does not exist any more. The most interesting one is the reversal of the effect of bathymetric depth. When fitting a Poisson model, the effect is positive, and the effect is negative when a binary model is fitted.

8.7.1.4 Summary

Friedlaender et al. [2006] used 6 predictors, including 2 acoustic backscattering variables (`A.v100` for the top 100 meters and `A.v300` for 100–300 meters), chlorophyll a, 2 distance variables (`D.ice` and distance to inner shelf water), and bathymetric slope. The choice was based on both competing splits of the CART model and authors' knowledge on the subject. As in many analyses of marine mammal survey data, GAM is most useful as an exploratory data analysis tool. Using GAM as a tool for developing a predictive model is not recommended. Friedlaender et al. [2006] interpreted the model results as an indication that whales were consistently and predictably associated with the distribution of zooplankton, and humpback and minke whales may be able to locate physical features and oceanographic processes that enhance prey aggregation. The analysis presented in this section further suggests that the likely physical feature that enhances prey aggregation is the distance to ice edge. In fact, studies have shown that the Antarctic ice edge is an important habitat

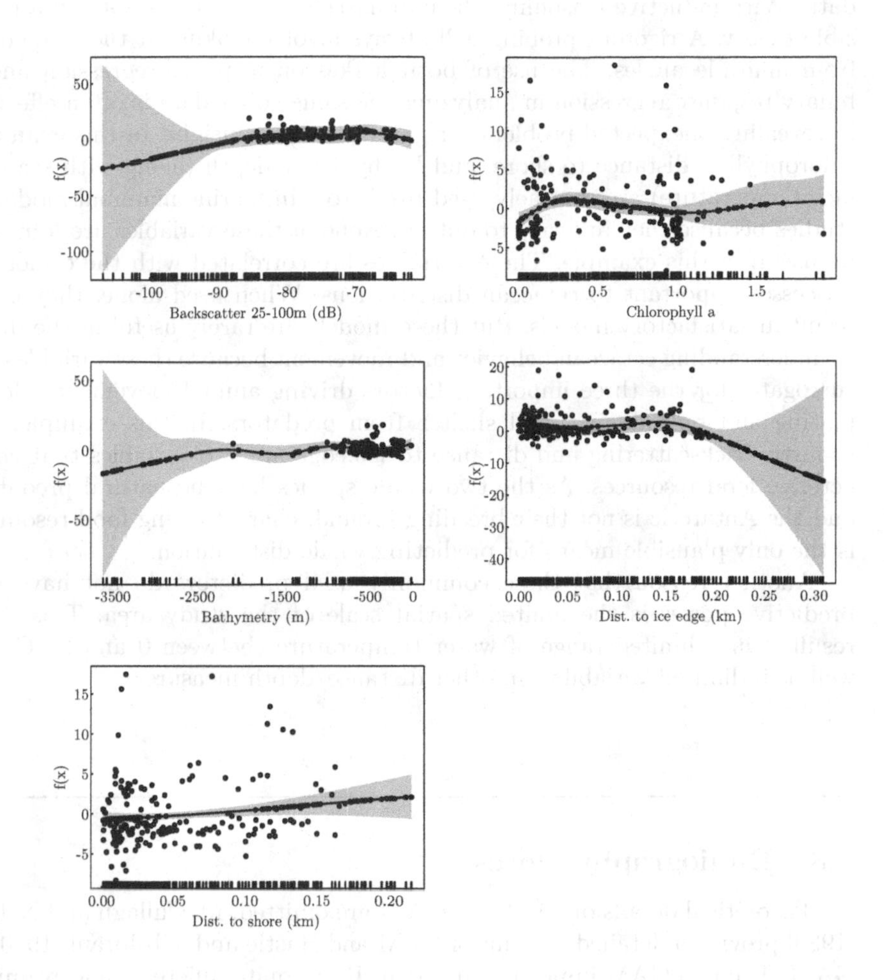

FIGURE 8.35: Antarctic whale survey logistic GAM – Fitted GAM functions show the effects of acoustic backscatter (top left), chlorophyll a (top right), bathymetric depth (middle left), distance to ice edge (middle right), and distance to shore (bottom).

for ice algae, and seasonal blooms of sea algae are important food resources for krill.

Because statistical analysis is a tool for inductive reasoning, fitting a model is a process of finding the likely model that can best explain the observed data. With inductive reasoning, the emphasis is on casting doubts on a plausible theory. A rigorous probing will always involve looking at the same data from multiple angles. The use of both a Poisson response regression and a binary response regression in analyzing the same count data is often effective in revealing unexpected problems or gaining further insight. In this example, chlorophyll a, distance to shore, and bathymetric depth (along with sea surface temperature) are routinely used predictors in marine mammal modeling studies because they are easy to obtain. None of these variables are found to be useful in this example. These variables are correlated with the ecological processes important to cetacean distributions. When used alone, they often result in satisfactory models. But these models are rarely useful in the quest of understanding cetacean behavior and movement because these variables are surrogates for the three important factors driving animal movement – food, mating and reproducing, and shelter from predators. In this example, the acoustic backscattering and distance to ice edge are two variables that characterize food resources. As the two whale species have no natural predators and the Antarctic is not their breeding ground, characterizing food resources is the only plausible means for predicting whale distribution.

Another reason that these commonly used predictors do not have any predictive power is the limited spatial scale of the study area. This limit resulted in a limited range of water temperature (between 0 and 2 °C), as well as in limited variability in other distance/depth measures.

8.8 Bibliography Notes

Theoretical details of GLM and GAM are omitted. McCullagh and Nelder [1989] provide a detailed account of GLM and Hastie and Tibshirani [1990] of GAM. The two GAM implementations in R are quite different. For example, the `gam()` function from `mgcv`, by default, estimates the degree of smoothness of model terms, while the `gam()` function from the package `gam` requires a user-supplied smoothness parameter. The two packages also differ in variance estimation method, which can lead to somewhat different confidence intervals. Readers should consult Hastie and Tibshirani [1990] and Wood [2006] for details. Applications of GLM and GAM in ecological modeling are discussed in Guisan and Zimmermann [2000].

8.9 Exercises

1. Dodson [1992] used a multiple regression analysis to build a predictive model for zooplankton species richness in North American lakes. Data used in the paper were from 66 North American lakes. The chosen lakes have a range from 4 m^2 to 80×10^9 m^2 surface area, range from ultra-oligotrophic to hypereutrophic, and have zooplankton species lists based on several years of observation. The abstract of the paper states that "the number of crustacean zooplankton species in a lake is significantly correlated with lake size, average rate of photosynthesis (parabolic function), and the number of lakes within 20 km." Furthermore, "prediction of species richness was not enhanced by knowledge of lake depth, salinity, elevation, latitude, longitude, or distance to the nearest lake."

   As many have found regression analysis conducted in the literature before the time of R is often inadequate. Using the general principle we discussed in Chapter 5, build a model to predict the species richness (number of species) and compare your model to the model discussed in the paper. If your model is different from the model Dodson published, explain why yours is better.

   The data set is available from package `alr3`, named `lakes`. This data frame contains the following variables:

   - Species – Number of zooplankton species
   - MaxDepth – Maximum lake depth, m
   - MeanDepth – Mean lake depth, m
   - Cond – Specific conductance, microsiemans
   - Elev – Elevation, m
   - Lat – N latitude, degrees
   - Long – W longitude, degrees
   - Dist – distance to nearest lake, km
   - NLakes – number of lakes within 20 km
   - Photo – Rate of photosynthesis, mostly by the 14C method
   - Area – Lake area, in hectares
   - Lake – Name of Lake

2. The Galapagos Islands provided evidence of evolution to Charles Darwin, and continuously serves as a laboratory for studying factors influencing the development of species. Johnson and Raven [1973] provided data on plant species richness for 29 islands in the Galapagos Islands to establish the relationship between species richness and island area.

The data set is available from R package `alr3` (`galapagos`). It is a data frame with row names being island names and eight columns: `NS` (number of species), `ES` (number of endemic species), `Area` (area of the island in km^2, `Anear` (area of the closest island, in hectares), `Dist` (distance to the nearest island, in km), `DistSC` (distance to Santa Cruz Island, in km), `Elevation` (elevation in m), and `EM` (whether elevation is observed, 1, or missing, 0).

- Fit a Poisson regression model with species richness as the response variable. Because the data were used to establish the relationship between species richness and island size, the variable `Area` should be included as a predictor. Explore other potential predictors and appropriate transformations. Explain the final model in plain language.
- In the previous step, we found that log-transforming `Area` results in a better model, which suggests that species richness is proportional to size of the island (explain why). Another way of looking at this problem is to use the size of island as an offset to model unit area species richness. Explore all potential predictors to develop a model for unit area species richness.
- `Area` and `Elevation` are highly correlated. Can the data be used to reason which factor is more responsible to the variation of species richness?

3. The data set in file `rodents.csv` contains data from a 2002 New York City Housing and Vacancy Survey (`http://www.census.gov/housing/nychvs/data/2002/nychvs02.html`). Among the variables surveyed is `rodent2`, indicating whether rodents were in the building.

   (a) Build a model to predict the presence of rodents given indicators for ethnic groups (`race`). Combine categories as appropriate and discuss the estimated coefficients in the model.

   (b) Add to the model some other potentially relevant predictors describing the apartment, building, and community district. Build the model using the general principles explained in Section 5.4, and discuss the coefficients for ethnicity indicators in your model.

4. Schoener [1968] collected information on the distribution of two *Anolis* lizard species (*A. opalinus* and *A. grahamii*) to see if their ecological niches were different in terms of where and when they perched to prey on insects. Data are in file `lizards.txt`. Perches were classified by twig diameter, their height in the bush, whether the perch was in sun or shade when the lizard was counted, and the time of day at which they were foraging. The response variable is a count of the number of times a lizard of each species was seen under each of the contingencies. GLM was not yet available when the study was published. Obviously, a Poisson

regression can be appropriate for analyzing the data. Develop a model using the general principles of Section 5.4 to predict the expected number of times a lizard would be seen. Interpret the model result in terms of habitat niches of each of the two species. Note that all predictors are categorical. Consider developing one model using species as a factor predictor.

5. Although Schoener [1968] reported four species of *Anolis* lizard in the paper, only two species were included in the data set. A Poisson regression does not constrain the total number of lizards in a given habitat condition, which may be misleading because the number of lizard cannot be unlimited. A different way of analyzing the same data is using logistic regression. Assuming that there are only two competing species, we can model one species habitat preference as the probability of seeing one species over the other. The data set we used has two parts. The first 24 rows are the number of times of species *A. opalinus* and the second 24 rows are the number of times of species *A. grahamii*. Use logistic regression to predict the probability of seeing one species (e.g., *A. grahamii*). That is, use response of *A. grahamii* as success and the response of *A. opalinus* as failure, and develop a model to predict the probability of success. If the probability of success is high for one condition, the habitat defined by this condition is preferred by *A. grahamii*; if the probability is low, the habitat is preferred by *A. opalinus*; if the probability is close to 0.5, the habitat is shared by both. Discuss whether your interpretation of species habitat preference changes from the results using a Poisson regression model.

6. As the binomial distribution is a special case of a multinomial distribution, the previous two questions are connected (Section 8.6). Use the Poisson models from Problem 4 to derive a binomial model of relative abundances of the two species of lizards and compare the model to the results from Problem 5.

7. A storm on July 4, 1999 with wind speeds exceeding 90 miles per hour hit the Boundary Waters Canoe Area Wilderness (BWCAW) in northeastern Minnesota, causing serious damage to the forest. A study of the effects of the storm surveyed the area and counted over 3600 trees to determine whether each of them was dead or alive (data `blowdown` from package `alr3`). One of the objectives of the study is to learn the dependence of survival on species, size of the tree, and on local severity. The data set includes results from 3666 trees, including whether a tree was dead or alive (`y=1` or `y=0`), its diameter (`D` in cm), local severity (`S` proportion of trees killed), and species (`SPP: BF= balsam fir, BS= black spruce, C= cedar, JP= jackpine, PB= paper birch, RP= red pine, RM= red maple, BA = black ash, A= aspen`).

Fit a logistic regression model and discuss the dependence of survival on the three potential predictors (size, local severity, and species).

8. As an example of field observation in evidence of theories of sexual selections, Arnold and Wade [1984] presented the data on size (mm) and number of mates in 38 male bullfrogs (*Rana catesbeiana*) (in file `bullfrogs.csv`). Is there evidence that the distribution of number of mates in this population is related to body size? If so, supply a quantitative description of that relationship, along with an appropriate measure of uncertainty. Write a brief summary of statistical findings.

9. Atlantic sturgeon (*Acipenser oxyrychus*) is a long-lived, estuarine-dependent, anadromous fish. They were once a valuable and abundant resource along North America's east coast. Habitat degradation, direct harvesting, and by-catch resulted in substantial declines in Atlantic sturgeon stock. In 2012 a segment of Atlantic sturgeon (New York Bight distinct population segment) was listed as a U.S. endangered species. Monitoring changes in sturgeon population is often done by sampling juvenile populations because juvenile sturgeons stay in their natal reproductive habitat for two to six years before migrating as mixed stocks along Atlantic coastal areas. The New York State Department of Environmental Conservation monitors juvenile sturgeon abundance in the tidal portion of the Hudson River. The data (`sturgeon.csv`) used in this problem are from 2006 to 2015, including counts of juvenile sturgeons caught (`CATCH`), effort (`Effort`), water chemistry (dissolved oxygen (`DO`), conductivity (`COND`), salinity (`SALINITY`)), tidal stage (`S_Tide`), distance to salt front (`DTSF`), and sampling month (`MON`) and year (`YEAR`).

   (a) Use Poisson regression to model the changes in sturgeon abundance over time (year, i.e., `Catch ~ YEAR`) using effort as the offset.

   (b) Use GAM to explore the nature of the temporal trend, after the effects of other factors (e.g., temperature, salinity, distance to salt front) are accounted for.

   (c) Revise the Poisson regression model based on GAM output and check if overdispersion is a problem.

Part III

Advanced Statistical Modeling

Chapter 9

Simulation for Model Checking and Statistical Inference

9.1	Simulation	388
9.2	Summarizing Regression Models Using Simulation	390
9.2.1	An Introductory Example	390
9.2.2	Summarizing a Linear Regression Model	392
9.2.2.1	Re-transformation Bias	396
9.2.3	Simulation for Model Evaluation	397
9.2.4	Predictive Uncertainty	405
9.3	Simulation Based on Re-sampling	408
9.3.1	Bootstrap Aggregation	410
9.3.2	Example: Confidence Interval of the CART-Based Threshold	411
9.4	Bibliography Notes	414
9.5	Exercises	414

This chapter introduces the use of simulation for model checking and inference. Simulation, often known as the Monte Carlo simulation, is a technique widely used in statistics and in environmental modeling. In environmental and ecological modeling, Monte Carlo simulation is primarily used for assessing model uncertainty in response to uncertain model parameters and other inputs. In statistics, simulation represents a class of computational algorithms that rely on repeated random sampling to compute results. We use these methods when computing an exact result with a deterministic algorithm that is infeasible or impossible. In this chapter, I emphasize the concept of using simulation for model checking. The chapter starts with an introduction of the basic concepts of simulation, followed by introductions on model-based simulation for estimation problems and for regression model checking. The use of simulation generated predictive distributions and their tail-areas as a tool for model checking is largely borrowed from the Bayesian p-value concept. The chapter concludes with a resampling method-based simulation method.

9.1 Simulation

Statistical inference relies largely on integration and differential computations. For example, the calculation of probability is frequently an integration problem and maximization of a likelihood function involves solving a set of differential equations. In a one-sample one-sided t-test problem, the main computation is the calculation of the p-value, which is the probability of observing a random variable from the null distribution (a t-distribution) that is larger than the calculated test statistic. In mathematical terms, this problem has the following steps:

- The test statistic $T = \frac{\bar{x}-\mu_0}{s}$ follows the null distribution, a t-distribution with degrees of freedom $\nu = n - 1$, with a probability density function of the form $\pi(x|\nu) = \frac{\Gamma(\frac{\nu+1}{2})}{\Gamma(\frac{\nu}{2})} \frac{1}{\nu\pi} \frac{1}{\left(1+\frac{x^2}{\nu}\right)^{(\nu+1)/2}}$
- With the observed test statistic of T^*, calculate the p-value:

$$p = \int_{T^*}^{\infty} \pi(x|\nu)dx \tag{9.1}$$

The integral has no closed form solution. Statistical inference is possible because of the availability of tabulated results (or fast computer algorithms) of commonly used probability distributions.

This integral can be approximated by using simulation, a process of repeatedly drawing random numbers from its distribution and computing the result. In this case, using the definition of a p-value as a long-run frequency, we can approximate the value by drawing random samples from the null distribution and calculating the fraction of these random samples that are larger than the calculated test statistic. Suppose the null distribution has $\nu = 23$, and $T^* = 2.34$. The corresponding p-value is 0.014 (`1-pt(2.34, df=23)`). Using simulation, we draw, for example, 10,000 random samples from the null, and calculate the fraction of these numbers larger than T^*:

```
#### R Code ####
set.seed(1)
t.sample <- rt(10000, df=23)
p.value <- mean(t.sample > 2.34)
```

Setting the random number seed to 1 is to ensure that readers will obtain the same estimate of the p-value (0.014).

Using simulation is unnecessary for this example because the test statistic distribution (t-distribution) is known and a fast computer algorithm for computing the integral in equation (9.1) is readily available. But this example brings home the central idea of simulation, that is, quantities of interest can either be calculated by mathematical means or by simulation. When the distribution of a random variable x is represented in terms of a probability distribution function $\pi(x)$, we can derive its mean, standard deviation, or, in this case, the tail area. These, and almost all, quantities of interest are derived using integration/differential calculation (e.g., $E(x) = \int x\pi(x)dx$ and $\Pr(x > \tilde{x}) = \int_{\tilde{x}}^{\infty} \pi(x)dx$). When the required calculation is mathematically intractable for quantities that are not in the few commonly used distribution families, simulation is often the convenient alternative.

In general, simulation is a method which solves a problem by directly using random numbers drawn from relevant distributions. The method is useful for obtaining numerical solutions to problems which are too complicated to solve analytically. The general method is known as the Monte Carlo method, a name coined by S. Ulam, who in 1946 became the first mathematician to dignify this approach with a name, in honor of a relative who would often borrow money because he "just had to go to Monte Carlo," the famed Monaco casino. The idea of using statistical sampling to solve difficult mathematical problems first came to Ulam when he "was convalescing from an illness and playing solitaire. The question was what are the chances that a Canfield solitaire laid out with 52 cards will come out successfully?" [Eckhardt, 1987]. Because the probability of winning is mathematically intractable, Ulam started to think a more practical method, which eventually led to the Monte Carlo method.

In statistical inference, the Monte Carlo method is used in two general areas. One is the calculation of difficult integrals. The mean of random variable x with a known distribution function $f(x)$ is $E(x) = \int xf(x)dx$. If random numbers from $f(x)$ can be easily obtained, $E(x)$ can be approximated by the average of these random numbers. The integral of an arbitrary function $g(x)$ can be approximated by simulation if this function can be factored into a product of two functions and one of them is a known probability distribution function. That is, if $g(x) = h(x)f(x)$ and $f(x)$ is a probability distribution function, the integral $I = \int g(x)dx = \int h(x)f(x)dx$ can be approximated in two steps: (1) drawing random numbers from $f(x) : x_i \sim f(x)$, and (2) calculating the average of $h(x_i)$s: $I \approx \frac{1}{n}\sum_{i=1}^{n} h(x_i)$. Numerical integration using the Monte Carlo method is a frequently used alternative to other numerical methods.

The other area of application of the Monte Carlo method is the calculation of one or more properties of a probability distribution. This is achieved by drawing random samples of the target distribution and using these samples to repeatedly calculate parameters defining these characteristics. With these samples any property of a distribution that is analytically intractable can be directly calculated. We have repeatedly used simulation in this book. In Chapter 4 we used simulation to estimate the 75th percentile of TP concentration distribution, to calculate a test's type I error probability, and evaluate the behavior of sample mean distributions based on samples from different population distributions. The emphasis of this chapter is the use of simulation for generating regression model predictive distributions for model assessment.

9.2 Summarizing Regression Models Using Simulation

9.2.1 An Introductory Example

Assessing water quality standard compliance in Section 4.8.3 with a known water quality variable (e.g., TP concentration) distribution is an example. Suppose the TP concentration distribution is log-normal. The task of assessing water quality compliance is to estimate the log mean and log standard deviation from data. Once the log mean and standard deviation are known, the probability that the water quality standard of, say, 10 μg/L is exceeded is a problem of a simple integration. However, the true mean and standard deviation are rarely known. Using sample mean and sample standard deviation of the logarithm of phosphorus concentrations will run into the problem encountered by William Gosset [Student, 1908]: using sample mean and sample standard deviation will introduce error in inference especially when the sample size is small. Let Y represent the random variable (i.e., logarithmic TP concentration, hence $Y \sim N(\mu, \sigma^2)$), and $y = \{y_1, \cdots, y_n\}$ denote the observed data. The quantity of interest is the probability of Y exceeding the water quality standard, or $\Pr(Y \geq \log(10))$. Because μ, σ^2 are unknown and must be estimated from the data, the probability must be evaluated through the *predictive distribution* of Y given the observed data. We use the common notation $\tilde{y}$ for a future value. The predictive distribution is denoted as $f(\tilde{y}|y)$. The unbiased estimates of μ and σ are the sample mean ($\bar{y}$) and sample standard deviation ($\hat{\sigma}$), respectively. Using $\bar{y}$ and $\hat{\sigma}$ as substitutes for μ and σ can give us an approximate estimate of the probability. This estimate is, however, associated with uncertainty. The sample mean $\bar{y}$ is a random variable, and its sampling distribution is normal according to the central limit theorem. The sample standard deviation is also a random variable, and its distribution is a scaled inverse χ^2 distribution, described by the following relationship:

$$(n-1)\frac{\hat{\sigma}^2}{\sigma^2} \sim \chi^2(n-1) \tag{9.2}$$

By chance, $\bar{y}$ may be smaller or larger than μ and $\hat{\sigma}^2$ may also be smaller or larger than σ^2. As a result, the probability of exceeding the standard can be underestimated or overestimated. How can this uncertainty be properly evaluated? We can certainly get more data if possible, because a large sample size will reduce the sample mean standard deviation (the standard error). But most likely additional data is impossible before we can justify the need in terms of the value of information. One way to quantify the uncertainty is to use $\bar{y}$ and $\hat{\sigma}$ as a reference for generating possible values of μ and σ. For example, if a random sample θ^* is drawn from the χ^2 distribution, we can generate a likely value of σ using the relationship in equation 9.2: $\sigma^* = \hat{\sigma}\sqrt{\frac{n-1}{\theta^*}}$. Likewise, we can draw samples of μ by using the relationship defined by the central limit theorem: $\bar{y} \sim N(\mu, \sigma^2/n)$ or $\frac{\bar{y}-\mu}{\sigma/\sqrt{n}} \sim N(0,1)$. Letting z^* be a random number drawn from $N(0,1)$, a likely value of the mean is $\mu^* = z^*\sigma^*/\sqrt{n}+\bar{y}$. The pair of likely mean (μ^*) and standard deviation (σ^*) can then be used to draw likely values of y. By repeating this process of drawing likely values of μ, σ, and then y many times, we obtain many values of y. These values are from the predictive distribution of y. Therefore, the Monte Carlo simulation for generating samples from the predictive distribution has the following 3 steps:

1. Calculate sample mean $\bar{y}$ and sample standard deviation $\hat{\sigma}$
2. For $i = 1, \cdots, k$,
   (a) draw a sample θ^i from $\chi^2(n-1)$
   (b) calculate $\sigma^i = \hat{\sigma}\sqrt{\frac{n-1}{\theta^i}}$
   (c) draw a sample μ^i from $N(\bar{y}, \sigma^i/\sqrt{n})$
3. Draw a sample $\tilde{y}^i$ from $N(\mu^i, \sigma^i)$

Using $\{\tilde{y}^i\}$ we can calculate the probability of exceeding the standard as a fraction of the generated numbers that are larger than the standard: $\Pr(Y > \log(10)) \approx \frac{1}{k}\sum_{i=1}^{k} I\left(y^i > \log(10)\right)$, where

$$I(x) = \begin{cases} 1 & \text{if } x > 0 \\ 0 & \text{if } x \leq 0 \end{cases}$$

In fact, we can calculate any statistics of the predictive distribution using these random samples.

In R, random number generation is part of each distribution function. For example, the procedure for generating the predictive distribution of y:

```
#### R Code ####
 y <- log(rlnorm(25, 1.9, 1))
 n.sim <- 5000
 n <- length(y)
 y.bar <- mean(y)
 s.hat <- sd(y)
 theta.i <- s.hat * sqrt((n-1)/rchisq(n.sim, n-1))
 mu.i <- rnorm(n.sim, y.bar, theta.i/sqrt(n))
 y.tilde <- rnorm(n.sim, mu.i, theta.i)
```

In this example, a sample of 25 numbers is drawn from a log-normal distribution with a log-mean 1.9 and a log standard deviation 1. To calculate the probability of standard violation:

```
#### R Code ####
 Pr <- mean(y.tilde > log(10))
```

A frequently discussed problem in water resources literature is the re-transformation bias. This is because many water quality standards are defined with respect to population means of pollutant concentrations. However, because most concentration variables can be approximated by the log-normal distribution, environmental variables such as flow and concentration are often log-transformed before any analysis. When analyzing data using log-transformed concentration data, we estimate log-mean and log-variance. If a concentration variable X has a log-normal distribution, its logarithm $Y = \log(X)$ has a normal distribution. A log-normal distribution is defined by the log-mean μ and log standard deviation σ. If $X \sim LN(\mu, \sigma^2)$, then $Y \sim N(\mu, \sigma^2)$. We know that the mean of Y is μ, but the mean of X is not e^{μ}, rather $E(X) = e^{\mu+\sigma^2/2}$ or $e^{\mu}e^{\sigma^2/2}$. Because $\sigma^2 > 0$, the multiplicative factor $e^{\sigma^2/2}$ is larger than 1. That is, when log-mean is estimated, the exponent of log-mean is always smaller than the mean in the original scale. For example, if the concentration or flow data are from a log-normal distribution with a log-mean of 1.9 and a log standard deviation of 1, the mean (expected value) of the variable is $e^{1.9+1/2} = 11.023$, but the exponent of the log mean is $e^{1.9} = 6.686$. We may use the estimated sample log-mean and log standard deviation to calculate the mean in the original scale. However, uncertainty (standard error) is difficult to re-transform back to the original scale. For a simple problem of estimating the mean, an analytic solution exists. But Monte Carlo simulation is often the most direct way to get to the answer. For calculating the standard error of the mean $\tilde{x} = e^{\tilde{y}}$, we can first convert `y.tilde` back to the concentration scale

```
#### R Code ####
 x.tilde <- exp(y.tilde)
```

as samples from the predictive distribution of the concentration variable. The population mean and standard deviation of x is now:

```
#### R Code ####
 mu.x <- mean (x.tilde)
 sigma.x <- sd (x.tilde)
```

and the confidence interval is then

```
#### R Code ####
 CI.x <- mu.x + qt(c(0.025,0.975), df=25-1) * sigma.x/sqrt(25-1)
```

The resulting 95% confidence interval is approximately `(4.26, 16.36)`, which is wider than the exponential of the 95% confidence interval calculated based on the log-transformed data (`(4.41, 9.35)`).

9.2.2 Summarizing a Linear Regression Model

As in the problem of estimating sample mean and sample standard deviation, the problem of using a linear regression model for prediction is a problem of finding the predictive distribution of the response variable at a predictor variable value ($\tilde{x}$) that is not included in the data. For a simple linear regression problem $y = \beta_0 + \beta_1 x + \varepsilon$ with the estimated model coefficients $\hat{\beta}_0$ and $\hat{\beta}_1$, the predicted mean is $\hat{\beta}_0 + \hat{\beta}_1\tilde{x}$ and equation (5.13) (on page 190) gives the prediction standard error. When the response variable is transformed, it is often not a simple task to transform the estimated predictive standard error into the original scale of the response variable. A simple and straightforward method for summarizing model uncertainty is Monte Carlo simulation. The predictive distribution of $\tilde{y}$ is a conditional normal distribution with mean $\hat{\beta}_0 + \hat{\beta}_1\tilde{x}$ and the standard deviation $\hat{\sigma}$. As in the sample mean example, the distribution of σ can be obtained by the relationship

$$(n-p)\frac{\hat{\sigma}^2}{\sigma^2} \sim \chi^2(n-p)$$

where p is the number of model coefficients. The joint distribution of β_0 and β_1 is a multivariate normal distribution with mean $(\hat{\beta}_0, \hat{\beta}_1)$ and a variance-covariance matrix estimated to be the product of $\hat{\sigma}^2$ and the unscaled estimation covariance matrix,[1] which is stored in the fitted linear model object (`summary(lm.obj)[["cov.unscaled"]]`). One way to draw samples from the predictive distribution of $\tilde{y}$ is as follows:

- Fit the linear model and save the result in an R object, e.g., `lm.obj`. Useful items from the linear model object are the estimated model coefficients, estimated coefficient standard error, residual standard deviation, the unscaled covariance matrix, sample size, and the number of coefficients:

```
summ <- summary(lm.obj)
coef <- summ$coef[,1:2]
sigma.hat <- summ$sigma
beta.hat <- coef[,1]
V.beta <- summ$cov.unscaled
n <- summ$df[1] + summ$df[2]
p <- summ$df[1]
```

- With the linear model object, we draw random samples of $\tilde{y}$ by repeating the following steps:

  1. draw a sample from a χ^2 distribution with degrees of freedom $n-p$:

  ```
  chi2 <- rchisq(1, n-p)
  ```

  2. draw a sample of σ:

[1] The unscaled covariance matrix is often denoted as $V_\beta = (X^TX)^{-1}$; see Weisberg [2005].

```
sigma <- sigma.hat*sqrt((n-p)/chi2)
```

3. draw a sample β_0^i, β_1^i from $MVN(\hat{\beta}, \sigma^i V\beta)$

```
beta <- mvrnorm(1, beta.hat, V.beta*sigma^2)
```

4. draw a sample of $\tilde{y}$:

```
y.tilde <- rnorm(1, beta[1]+beta[2]*x.tilde, sigma)
```

- Store the resulting random samples of $\beta_0, \beta_1, \sigma, \tilde{y}$.

This simulation procedure is included in the R function `sim` (in package `arm`). It takes the number of simulations and a linear model (or generalized linear model) object as inputs and returns random samples of model coefficients and residual variance.

In Section 5.5 (page 189), we used the predictive 95% confidence interval of predicted log PCB concentration in fish in 2007, which is $(-2.121, 1.363)$ estimated using equation 5.13 on page 190. The corresponding interval in the original scale is easily obtained using simulation:

```
#### R Code ####
n.sims<-1000
sim.results <- sim(lake.lm1, n.sims)
```

The resulting object `sim.results` is a list of two objects: `coef` and `sigma`. The object `beta` is a matrix of `n.sims` rows and p (number of model coefficients) columns. Each row represents a possible combination of β_0 and β_1. The first 10 rows of `beta` are shown below:

```
#### R Output ####
 sim.results@coef[1:10,]
      (Intercept) I(year - 1974)
 [1,]      1.7053      -0.063053
 [2,]      1.5584      -0.058869
 [3,]      1.5538      -0.051409
 [4,]      1.6369      -0.057951
 [5,]      1.7868      -0.073736
 [6,]      1.6514      -0.059717
 [7,]      1.5680      -0.055303
 [8,]      1.5634      -0.056510
 [9,]      1.6884      -0.062708
[10,]      1.5310      -0.054117
```

The object `sigma` is a vector of `n.sims` random samples of σ.

With random samples of β's and σ, the predictive distribution of log PCB can be drawn:

```
#### R Code ####
log.PCB <- rnorm(n.sims, sim.results@coef[,1] +
                         sim.results@coef[,2]*(2007-1974),
                         sim.results@sigma)
```

These samples can be used to calculate the 95% confidence interval:

```
#### R Code ####
d.f <- summary(lake.lm1)$df[2]
sigma.hat <- summary(lake.lm1)$sigma
mean(log.PCB) + qt(c(0.025,0.975), d.f)*sigma.hat

[1] -2.0954  1.3544
```

We can also use the middle 95% range to represent the uncertainty:

```
#### R Output ####
quantile(log.PCB, prob=c(0.025,0.25,0.5,0.75,0.975))
    2.5%       25%       50%       75%     97.5%
-2.02503 -0.92517 -0.36001  0.25094  1.39737
```

A programming note

The R package rv [Kerman and Gelman, 2007] provides a set of functions that can be used for linear regression model simulation. The advantage of using the rv package is the computation efficiency. The function posterior from the package rv carries out the same simulation as the function sim from package arm. When using the rv package, the simulated model coefficients are stored in "random variable" objects. For example, the simulation of the model lake.lm1 can be done using the posterior function:

```
#### R Code ####
packages(rv)
setnsims(5000)
lake1.sim <- posterior(lake.lm1)

#### R Output ####
> print(lake1.sim)
$beta
              name  mean     sd   2.5%    25%   50%    75%  97.5%
[1]     (Intercept)  1.60 0.0728  1.457  1.551  1.60  1.648  1.741
[2] I(year - 1974) -0.06 0.0055 -0.071 -0.064 -0.06 -0.056 -0.049

$sigma
    mean    sd 2.5%  25%  50% 75% 97.5%
[1] 0.88 0.025 0.83 0.86 0.88 0.9  0.93
```

The results are stored in a list of two objects. Simulated model coefficients are in the "vector" beta, and the simulated residual standard deviation is in sigma. When using rv outputs, instead of a vector of random numbers for each parameter of interest, we use an rv variable as if it is a scaler. For example, the code used for deriving the predictive distribution of log PCB in 2007 can be written as follows:

```
log.PCB <- rvnorm(1, lake1.sim$beta[1]+lake1.sim$beta[2]*
                    (2007 - 1974),
                  lake1.sim$sigma)
```

The advantage of using rv is more obvious when we want to predict log PCB for more than one year:

```
log.PCB2 <- rvnorm(1, lake1.sim$beta[1]+lake1.sim$beta[2]*
                    ((2000:2007) - 1974),
                   lake1.sim$sigma)
```

```
> log.PCB2
      mean   sd   1% 2.5%   25%    50%  75% 97.5% 99% sims
[1]  0.038 0.87 -2.0 -1.6 -0.56  0.057 0.63   1.7 2.1 5000
[2] -0.026 0.88 -2.1 -1.7 -0.64 -0.031 0.57   1.7 2.0 5000
[3] -0.068 0.88 -2.1 -1.8 -0.67 -0.045 0.52   1.6 1.9 5000
[4] -0.129 0.88 -2.2 -1.8 -0.74 -0.136 0.46   1.6 2.0 5000
[5] -0.186 0.89 -2.4 -2.0 -0.77 -0.177 0.42   1.5 1.8 5000
[6] -0.256 0.88 -2.3 -1.9 -0.86 -0.261 0.33   1.5 1.8 5000
[7] -0.308 0.88 -2.3 -2.0 -0.92 -0.305 0.28   1.4 1.7 5000
[8] -0.410 0.89 -2.5 -2.2 -1.02 -0.415 0.20   1.3 1.6 5000
```

In the rest of this chapter, I will use functions from the rv package when possible.

9.2.2.1 Re-transformation Bias

The re-transformation bias has been a topic of interest in the literature for many years. Stow et al. [2006] summarized the problem and solutions in the literature, and recommended an alternative approach using a Bayesian regression analysis based on Markov chain Monte Carlo simulation. The simulation approach described in this section is the same as the Bayesian regression model. As a result, we can correct the re-transformation bias by drawing random samples from the predictive distribution and calculating the mean concentration after re-transforming these samples into the concentration scale:

```
predict.PCB <- exp(log.PCB)
```

The simulated distribution of log PCB concentrations has a mean of -0.326 and a standard deviation of 0.882. The predictive distribution of PCB concentrations (predict.PCB) has a mean of 1.071 (not $e^{-0.326}$ or 0.722) and a standard deviation of 1.192. Note that the re-transformation bias occurs only when the mean is needed. The median and other quantiles can be directly re-transformed back from the logarithmic scale to the concentration scale:

```
quantile(predict.PCB, prob=c(0.025,0.25,0.5,0.75,0.975))
   2.5%     25%     50%     75%   97.5%
0.13199 0.39646 0.69767 1.28524 4.04462
```

With random samples of model coefficients and residual standard deviation, we can calculate quantities that are based on the fitted model. For example, the question "Will Lake Michigan lake trout meet the Great Lakes Strategy 2002 PCB reduction goal?" raised in Stow et al. [2004] can be addressed in a few simple steps:

- Run `sim` to obtain `n.sims` pairs of model coefficients as shown above;
- Predict the mean PCB concentration distributions for year 2000 and year 2007, assuming the PCB concentration follows a log normal distribution;
- For each predicted pair of mean concentrations for 2000 and 2007, calculate the percent reduction;
- Summarize the distribution of these percentages

```
#### R Code ####
n.sims<-5000
sim.results <- sim(lake.lm1, n.sims=1000)
predict.PCB07 <- exp(sim.results@coef[,1] +
                     sim.results@coef[,2]*(2007-1974) +
                     0.5* sim.results@sigma)

predict.PCB00 <- exp(sim.results@coef[,1] +
                     sim.results@coef[,2]*(2000-1974) +
                     0.5 * sim.results@sigma)

percentages <- 1-predict.PCB07/predict.PCB00
hist(percentages)
```

The resulting predicted distribution of percent reduction of PCB concentration from 2002 to 2007 (Figure 9.1) suggested that the goal of 25% reduction in PCB concentration from 2000 to 2007 can be achieved.

The model we used here is the model 1 in Stow et al. [2004], a model that was recognized as unrealistic. One potential problem with this model is the imbalance in fish sizes before and after 1987 (Figure 9.2). Before 1987, fish specimens were collected through field samples conducted by the Wisconsin Department of Natural Resources, and fish specimens after 1987 were mostly donated by anglers.

9.2.3 Simulation for Model Evaluation

Simulation is a useful tool for model evaluation and assessment. In our discussions of model diagnostics, we focused on the analysis of model residuals to check whether assumptions about the data are met. A model with a

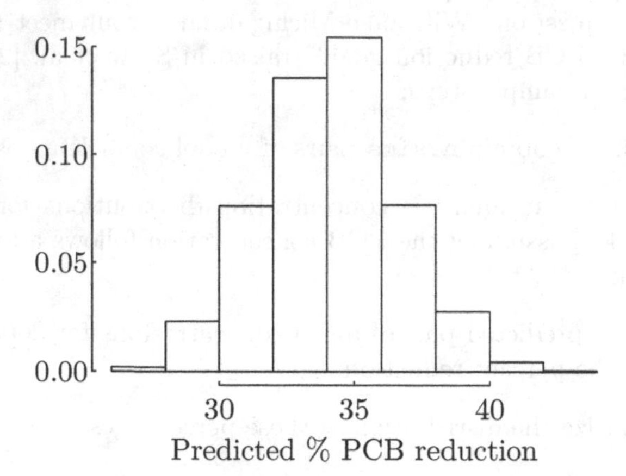

FIGURE 9.1: Fish tissue PCB reduction from 2000 to 2007 – Predicted PCB reduction (in percent) from 2002 to 2007 using the simple log-linear regression model.

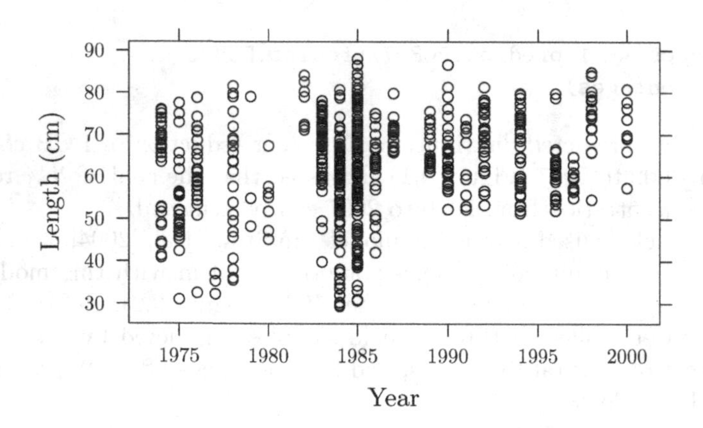

FIGURE 9.2: Fish size versus year – A potential problem in the data is that the fish size distribution over the study period is not random. After 1987, all fish collected were larger than 125 cm.

proper residual distribution is not necessarily a good model. A model's predictive characteristics are not represented in residuals. Using simulation, we can perform model evaluation to check whether the fitted model can replicate the data and features we know to be representative of the data. Two examples will be used in this section, PCB in fish and Cape Sable seaside sparrow, to illustrate the process of model evaluation. Because a modeling problem is uniquely associated with a specific problem and the mechanism that resulted in the data, a general procedure for model evaluation is illusive.

The basic idea of model evaluation is to see whether the model can replicate the observations or some features of the data with a reasonable accuracy. The difficulty in assessing a model's predictive performance lies in the mismatch between the observed data and model prediction. A model's prediction is in the form of a predictive distribution, which is an estimate of the underlying distribution that resulted in the observed data. A good model is a model that describes the distribution of the data accurately. Whether the estimated distribution is close to the true underlying distribution is, however, impossible to verify. Our goal of model evaluation is to assess how well the model captures the underlying process that generated the data. A commonly used method for evaluating a model's predictive accuracy is the jackknife method, where a model is fit repeatedly each time leaving one observation out. The fitted model is then used to predict the response variable value of the left-out observation. A more general approach is the cross-validation simulation where a fraction of data points are randomly chosen and set aside for assessing a model's predictive accuracy. These methods rely on the predictive residuals and statistics of these residuals for quantifying predictive accuracy. A residual, the difference between the predicted mean and the observation, tells us how close is the predicted mean to the observation, but does not provide information on how well the predictive distribution fits the observed data point. Figure 9.3 illustrates this problem by comparing two hypothetical predictive distributions to the same observed data point. The means of the two predictive distributions are the same, resulting in the same residual value for both models. The observed data point sits at the 98th percentile of one distribution and 69th percentile of the other. The observed data are more likely to be from the latter distribution (the solid line).

Because the predictive distribution represents the estimated underlying probability distribution of the response variable, we expect that the observed response variable values are covered by the predictive distribution reasonably well if the model is adequate. To describe how well the model prediction is, we compare the observed response variable value to the predictive distribution and calculate the area under the curve to the right of the observed value. This area is the probability of observing a value equal to or larger than the observed value under the predictive distribution. A good model would yield a probability reasonably close to 0.5. If the probability is smaller than 0.05 or larger than 0.95, we have reasons to believe that there is a discrepancy between the predictive distribution (or the model) and the observed number. One way

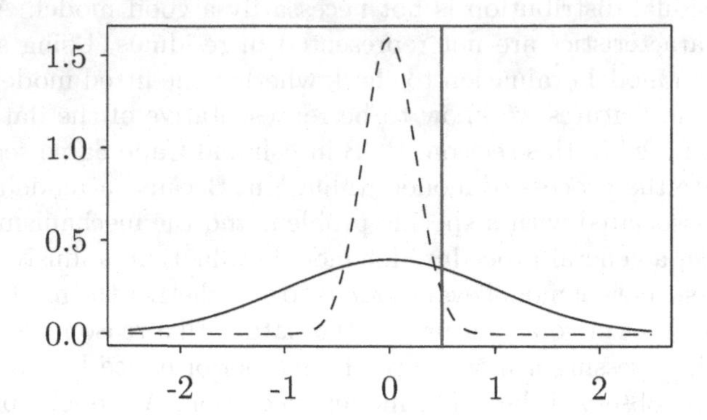

FIGURE 9.3: Residuals as a measure of goodness of fit – Residuals represent one aspect of a model's goodness of fit. The two curves represent two hypothetical predictive distributions that share the same mean. Comparing the two distributions as a likely parent distribution of the data, the observed data point (the gray vertical line) is very unlikely from the predictive distribution represented by the dashed line.

to summarize the model evaluation result is to produce predictive distributions for all data points and calculate tail areas for all observations (Figure 9.4). The collection of tail areas can be presented using a histogram. However, we are not really interested in how well a model can replicate individual data points. Rather we want the model to be useful in that it can be used to summarize important features in the data so that meaningful inferences can be made. As a result, assessing a regression model should also consider the context and objectives of the problem. For example, the crypto in drinking water example in Section 10.6.2 requires an accurate assessment of the fraction of drinking water systems in the U.S. with high concentrations of crypto in their source water. A model that can represent the high end of the crypto concentration distribution is a useful model (Figure 10.28).

The simple linear regression model of the PCB in the fish example (Section 5.3.2, page 156) is known to be problematic. Fish specimens collected after 1986 were mostly large fish (Figure 9.2). As a result, the fitted model underpredicts PCB concentrations for the last few years. From earlier studies of this data set, we know the simple linear regression model is inadequate because of the imbalance of fish sizes over time. Using simulation we expose this problem from a different angle.

Because statistical modeling is a process of searching for the model that has the most support from the data, assessing a model using multiple methods will give us confidence on the final model. We use simulation to predict specific features of the data. A good model should produce features that are consistent

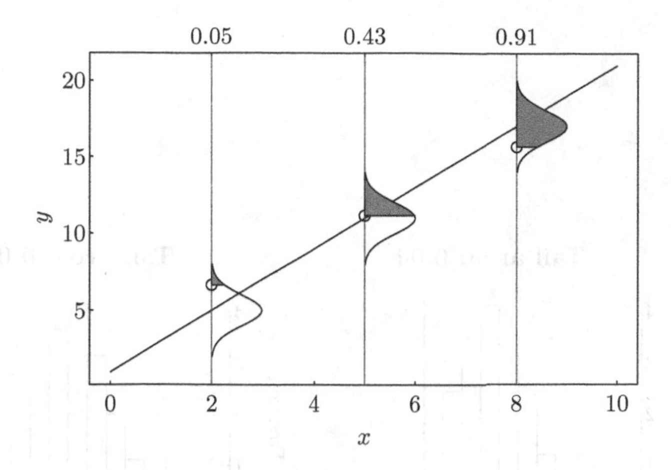

FIGURE 9.4: Simulation for model evaluation – Model evaluation using simulation is achieved by calculating the tail area of the observed data under the predictive distribution.

with the data or known values. Different applications often have different emphases. If we are interested in predicting the mean, we can simulate mean by repeatedly replicating the data and calculating the mean each time. Figure 9.5 compares some frequently used statistics from the PCB concentration data to the same statistics replicated by the model. Again, the tail areas of the 5th and 95th percentiles (0.01 and 0.04, respectively) suggest that the model is underpredicting the smallest and largest concentration values.

To close this section, we use the population survey data of the Cape Sable seaside sparrow, an endangered species found only in the Everglades National Park in south Florida as an example of simulation for the generalized linear model.

The National Parks Service conducted an initial survey in 1981 and estimated its population to be 6656. The annual surveys since 1992 indicate a decline to an estimated 2624 birds by 2001. A subset of the data consists of survey sites with vegetation covers consistent to known habitats of the bird. The survey used a helicopter to drop observers at sites along a 1-km grid that covers all sparrow habitats. Observers recorded the number of sparrows seen or heard within a 7-min interval for up to 3 hours each morning. Because each site was visited the same amount of time and observers were highly trained experts, annual average numbers per site were used as an indicator of the total population (Figure 9.6). To model the year-to-year variation, we initially used the Poisson regression model using `year` as the only (categorical) predictor. The objective is to test whether the population changes over time are significant.

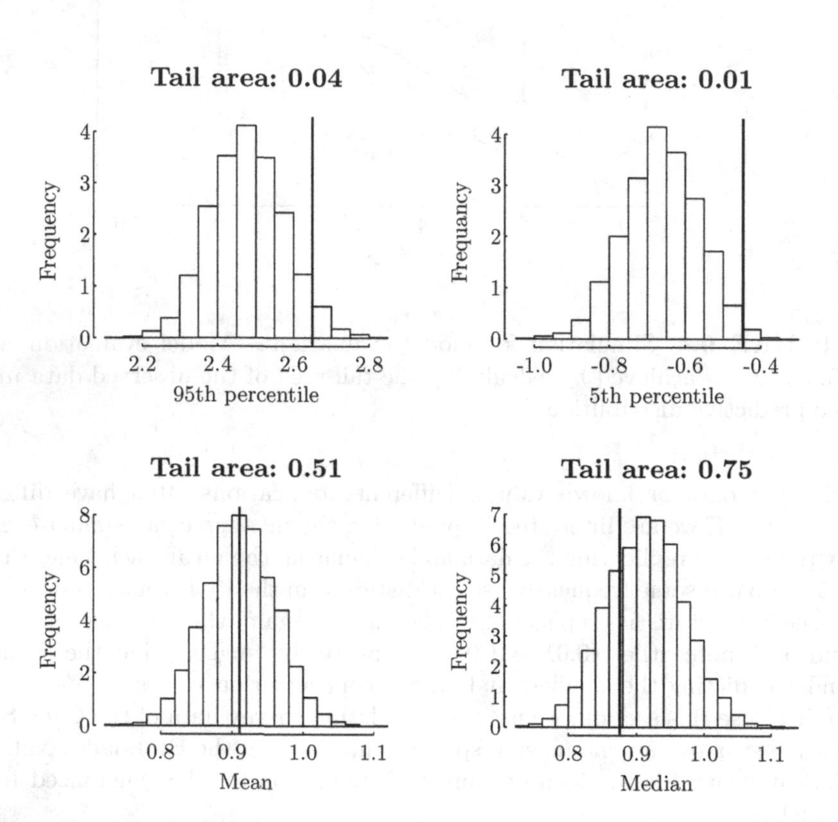

FIGURE 9.5: Tail areas of selected PCB statistics – Histograms show the replicated log PCB concentration statistics (clockwise from top left, 95th percentile, 5th percentile, median, and mean) and the corresponding statistics calculated from the observed log PCB concentrations (the vertical lines). The tail areas of the 5th and 95th percentiles of 0.01 and 0.04, respectively, suggest that the model is unable to replicate extremely large or small data values well.

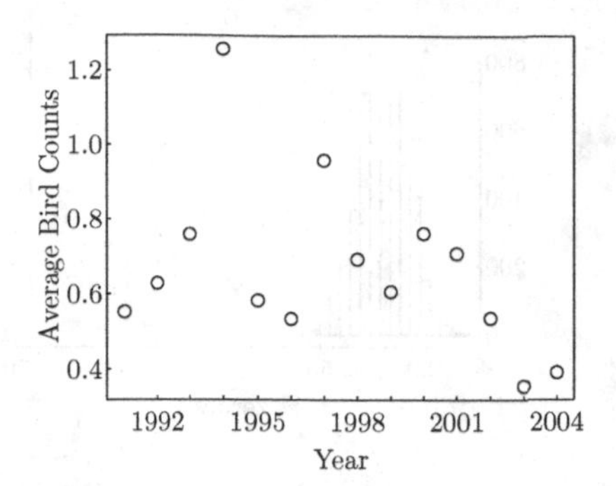

FIGURE 9.6: Cape Sable seaside sparrow population temporal trend – Annual averages of Cape Sable seaside sparrow counts.

```
#### R Code & Output ####
spar.glm1 <- glm(Bird.Count ~ factor(year),
    data=sparrow, family=poisson)
display(spar.glm1)

glm(formula = Bird.Count ~ factor(year),
    family = poisson, data = sparrow)
                 coef.est coef.se
(Intercept)      -0.59     0.11
factor(year)1992  0.13     0.14
factor(year)1993  0.32     0.14
factor(year)1994  0.82     0.19
factor(year)1995  0.05     0.17
factor(year)1996 -0.03     0.14
factor(year)1997  0.55     0.13
factor(year)1998  0.23     0.15
factor(year)1999  0.09     0.14
factor(year)2000  0.32     0.17
factor(year)2001  0.25     0.19
factor(year)2002 -0.03     0.28
factor(year)2003 -0.44     0.32
factor(year)2004 -0.34     0.35
---
  n = 1723, k = 14
  residual deviance = 2947.8,
  null deviance = 3008.8 (difference = 61.0)
```

For this example, I will not interpret the fitted model or comment on the conclusion of a declining trend in sparrow population. I want to evaluate whether the use of a Poisson regression is appropriate. One feature of the data

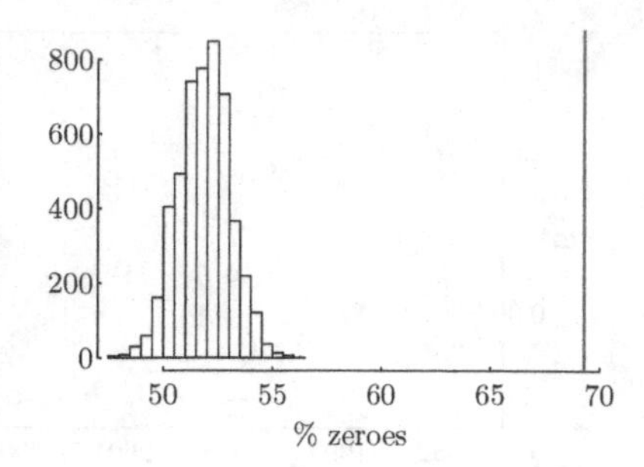

FIGURE 9.7: Cape Sable seaside sparrow model simulation – Histogram of 5000 simulated percent of zeros is compared to the observed percentage of zeros (gray line).

is that about 69% of all observed counts are 0s. Can the fitted model predict as many 0s? We can use the fitted model to replicate the observed counts and count the fraction of 0s:

```
#### R Code ####
 n <- dim(sparrow)[1]
 set.seed(123)
 y.rep <- rpois(n, predict(spar.glm1,type="response"))
 zeros <- mean(y.rep==0)
```

The fraction of zeros is 0.52. To answer the question "How close is 0.52 to the observed fraction of 0 in the data (0.69)?" we can repeat the process many times to capture the variability of the fraction of 0. Figure 9.7 shows the histogram of 5000 simulated fractions of 0s. The observed fraction is far greater than the simulated fraction, suggesting that the Poisson model cannot replicate the number of zeros in the data. One explanation is that an observed 0 can either be that there was no bird or there were birds but the observer missed them. When using a Poisson regression, we assume that the expected number of birds at each site is larger than 0. Because of the possibility of a false negative, the probability of observing a 0 is larger than the Poisson model can predict. A different kind of model (zero-inflated Poisson model) should be used.

Zero inflation is common in ecological count data. When fitting a Poisson regression model, using simulation to check whether the fitted model can replicate the fraction of zeros in the data can serve as a simple diagnostic method for zero inflation.

9.2.4 Predictive Uncertainty

In Section 5.5 we discussed the process of measuring microcystin (MC) concentration in drinking water using ELISA. The measurement process is a statistical modeling and prediction process. A number of water samples with known MC concentrations are placed in testing wells to record the changes in color measured as optical density (OD). The resulting data (OD values versus MC concentrations) are used for developing a regression model. Two regression models were recommended by the manufacturer of the ELISA kit used in the Toledo Water Department and Ohio EPA. One is the log-linear regression model discussed in Section 5.5. The other is a nonlinear regression model based on a four-parameter logistic function (equation (9.3)). As discussed in Section 5.5, predictive uncertainty on the concentration scale is difficult to derive for a log-linear model. I used the ELISA example in that section to illustrate the use of simulation for deriving predictive uncertainty. In Chapter 6, I mentioned that the predictive uncertainty for a nonlinear regression model is unavailable from the function `predict`. In this section, we discuss the use of simulation for approximating nonlinear regression prediction uncertainty.

The ELISA method uses data from standard solutions to fit a regression model and predict the MC concentration from a water sample based on the sample's optical density measurement. As a result, the uncertainty of the reported MC concentration is the regression model predictive uncertainty.

For the log-linear regression model, we use the simulation discussed in Section 9.2.2.

```
#### R Code ####
mc <- c(0.167, 0.444, 01.110, 2.220, 5.550)
rOD <- c(0.784, 0.588, 0.373, 0.270,0.202)
stdcrv <- lm(log(mc) ~ rOD)
stdcrv.sims <- posterior(stdcrv, n.sims=5000)
```

The simulation results in 5000 sets of model coefficients and residual standard deviations. Each set can be seen as a potential model (standard curve). For each set of model coefficients, we have an estimated log mean concentration ($\beta_0 + \beta_1 r\tilde{O}D$). Coupled with the residual standard error, we have a predictive distribution of log MC concentration at the given relative OD value $r\tilde{O}D$. We draw a random sample from this predictive distribution as a sample of a likely log MC concentration value.

```
#### R Code ####
mc.sims <- exp(rvnorm(1, stdcrv.sims$beta[1] +
                      stdcrv.sims$beta[2]*0.261,
                      stdcrv.sims$sigma))

#### R Output ####
> mc.sims
    mean sd  1% 2.5% 25% 50% 75% 97.5% 99% sims
```

```
[1]  3.8 22 0.6 0.91 2.2 2.8 3.6   8.1  13 5000
```

The estimated predictive 95% interval is (0.91, 8.1), very close to the analytical result in Section 5.5.1 (Figure 5.21).

The other recommended regression model for ELISA standard curve is a four-parameter nonlinear regression model, recommended by the kit manufacturer (Abraxis), with OD as the response variable:

$$OD = \frac{\alpha_1 - \alpha_4}{1 + \left(\frac{x}{\alpha_3}\right)^{\alpha_2}} + \alpha_4 + \varepsilon \tag{9.3}$$

where ε is the model error term (and $\varepsilon \sim N(0, \sigma^2)$), $\alpha_1 - \alpha_4$ are unknown regression coefficients to be estimated, OD is the observed OD value, and x is standard solution concentration. Note that the response variable of this model is the measured OD while the predictor variable is the MC concentration. As the regression model is fit to minimize the error in OD, estimating MC concentrations using the inverse model of equation (9.3) will lead to larger than expected estimation uncertainty (based on regression model summary statistics such as residual variance) because model coefficients are estimated to minimize the prediction error in OD, not in x. Conceptually, equation (9.3) is the right model, in that concentration determines the optical density. Furthermore, data used for developing the standard curve are based on the measured OD from standard solutions with known concentration values. However, when measuring the MC concentration of a water sample, we use the measured OD for estimating the concentration.

Although to fully account for the predictive uncertainty in a nonlinear regression model may require more than the simple Monte Carlo simulation, we can often use the simulation approach described in this chapter to approximate nonlinear predictive uncertainty. Using the ELISA test data from Toledo, I illustrate this approximation. During the weekend of August 1, 2014, a total of six ELISA tests were conducted, all with the same six standard solutions, each with two replicates:

```
stdConc8.1<- rep(c(0,0.167,0.444,1.11,2.22,5.55), each=2)
```

The measured optical density (OD) was published in a report issued on August 4, 2014 (available at `http://goo.gl/m4m0ky`)

```
#### R Code ####
Abs8.1.0<-c(1.082,1.052,0.834,0.840,0.625,0.630,
            0.379,0.416,0.28,0.296,0.214,0.218)
Abs8.1.1<-c(1.265,1.153,0.94,0.856,0.591,0.643,
            0.454,0.442,0.454,0.447,0.291,0.29)
Abs8.1.2<-c(1.051,1.143,0.679,0.936,0.657,0.662,
            0.464,0.429,0.32,0.35,0.241,0.263)
Abs8.2.0<-c(1.139,1.05,0.877,0.914,0.627,0.705,
            0.498,0.495,0.289,0.321,0.214,0.231)
```

```
Abs8.2.1<-c(1.153,1.149,0.947,0.896,0.627,0.656,
            0.465,0.435,0.33,0.328,0.218,0.226)
Abs8.2.2<-c(1.124,1.109,0.879,0.838,0.61,0.611,
            0.421,0.428,0.297,0.308,0.19,0.203)
```

As in all ELISA tests, a regression model is fit for each test. We follow the practice to fit six standard curves.

```
#### R Code ####
toledo <- data.frame(stdConc=rep(stdConc8.1, 6),
                     Abs=c(Abs8.1.0,Abs8.1.1,Abs8.1.2,
                         Abs8.2.0,Abs8.2.1,Abs8.2.2),
                     Test=rep(1:6, each=12))

TM1 <- nls(Abs ~ (A-D)/(1+(stdConc/C)^B)+D,
           control=list(maxiter=200),
           data=toledo[toledo$Test==1,],
           start=list(A=0.2,B=-1,C=0.5,D=1))
TM2 <- nls(Abs ~ (A-D)/(1+(stdConc/C)^B)+D,
           control=list(maxiter=200),
           data=toledo[toledo$Test==2,],
           start=list(A=0.2,B=-1,C=0.5,D=1))
TM3 <- nls(Abs ~ (A-D)/(1+(stdConc/C)^B)+D,
           control=list(maxiter=200),
           data=toledo[toledo$Test==3,],
           start=list(A=0.2,B=-1,C=0.5,D=1))
TM4 <- nls(Abs ~ (A-D)/(1+(stdConc/C)^B)+D,
           control=list(maxiter=200),
           data=toledo[toledo$Test==4,],
           start=list(A=0.2,B=-1,C=0.5,D=1))
TM5 <- nls(Abs ~ (A-D)/(1+(stdConc/C)^B)+D,
           control=list(maxiter=200),
           data=toledo[toledo$Test==5,],
           start=list(A=0.2,B=-1,C=0.5,D=1))
TM6 <- nls(Abs ~ (A-D)/(1+(stdConc/C)^B)+D,
           control=list(maxiter=200),
           data=toledo[toledo$Test==6,],
           start=list(A=0.2,B=-1,C=0.5,D=1))
```

Once a model is fit, I will use the same algorithm described for the linear regression model to draw random variates of model coefficients. The code is written in a function named `sim.nls`:

```
#### R Code ####
sim.nls <- function (object, n.sims=100){
    object.class <- class(object)[[1]]
```

```
    if (object.class!="nls") stop("not an nls object")

    summ <- summary (object)
    coef <- summ$coef[,1:2,drop=FALSE]
    dimnames(coef)[[2]] <- c("coef.est","coef.sd")
    sigma.hat <- summ$sigma
    beta.hat <- coef[,1]
    V.beta <- summ$cov.unscaled
    n <- summ$df[1] + summ$df[2]
    k <- summ$df[1]
    sigma <- rep (NA, n.sims)
    beta <- array (NA, c(n.sims,k))
    dimnames(beta) <- list (NULL, names(beta.hat))
    for (s in 1:n.sims){
        sigma[s] <- sigma.hat*sqrt((n-k)/rchisq(1,n-k))
        beta[s,] <- mvrnorm (1, beta.hat, V.beta*sigma[s]^2)
    }
    return (list (beta=beta, sigma=sigma))
}
```

To use the rv package, I also convert the resulting vector (sigma) and matrix (beta) to random variable object:

```
#### R Code ####
test1.sim <- sim.nls(TM1, 4000)
test1.beta <- rvsims(test1.sim$beta)
test1.sigma <- rvsims(test1.sim$sigma)
```

To summarize the uncertainty in the estimated MC concentration, we can derive the inverse function of equation (9.3) and calculate a concentration value for each set of model coefficients.

The simulations of these six ELISA tests show that the predictive uncertainty can be quite high because of the small sample size. In addition, we see the problem of using the inverse function of this nonlinear regression model for estimating MC concentrations: the calculated MC concentration is associated with a much higher uncertainty than the regression model predictive uncertainty with respect to the response variable (Figure 9.8).

9.3 Simulation Based on Re-sampling

In Section 4.2.1, the bootstrapping methods are introduced for estimating standard deviations of statistics of interest. The bootstrapping methods are a general re-sampling technique that can be applied to many modeling problems.

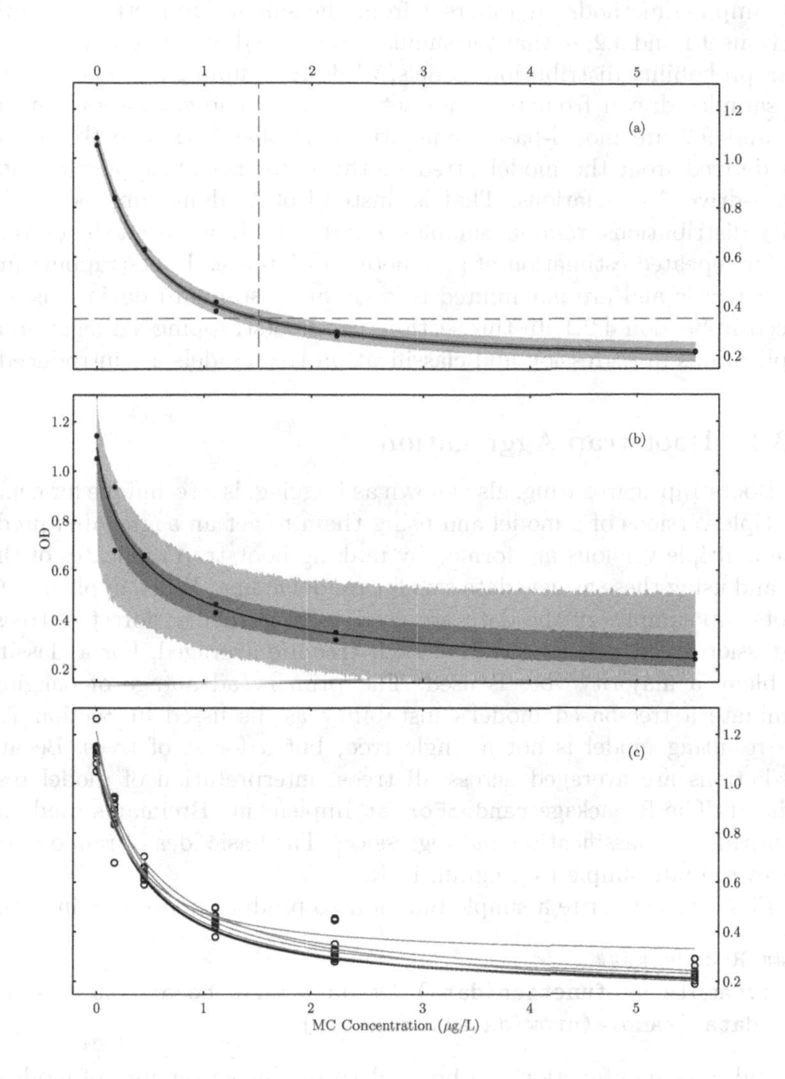

FIGURE 9.8: Uncertainty associated with the standard curve that detected the high MC concentration on August 1, 2014 in Toledo's drinking water is still considerable even with a nearly perfect fit (a). The within test variation of a subsequent test is, however, much larger, even though the model fits the data well (b). The 6 curves developed on August 1 and 2 show a large among test variation reflecting the measurement uncertainty in OD at various MC concentrations (c). The model's predictive error for MC concentration (x-axis) at a given OD value is higher (the dashed horizontal line in (a)) as the model minimizes the prediction error in OD (y-axis direction, the dashed vertical line in (a)). (Used with permission from Qian et al. [2015a]

Re-sampling methods are different from the simulation methods described in Sections 9.1 and 9.2, in that the simulation is based on random numbers drawn from probability distribution models, while re-sampling approaches are based on samples drawn from the data set at hand. The simulations in Sections 9.1 and 9.2 are model-based simulations. Probabilistic distribution models are derived from the model fitted to the data. Bootstrapping methods are "data-driven" simulations. That is, instead of random samples from probability distributions, random samples of data are drawn from the current data set for repeated estimation of parameters of interest. Bootstrapping methods are versatile and are not limited to estimating standard deviations as introduced in Section 4.2.1. In this section, the bootstrapping aggregation and its applications in regression and classification tree models are introduced.

9.3.1 Bootstrap Aggregation

Bootstrap aggregating, also known as bagging, is a technique for generating multiple versions of a model and using them to get an aggregated prediction. The multiple versions are formed by making bootstrap replicates of the data set and using these as new data sets for model fitting. When applied to CART, bootstrap samples of the data are used to construct a "forest of trees." For regression trees, predictions from each tree are averaged. For a classification problem, a majority vote is used. The primary advantage of bagging is to eliminate a tree-based model's instability as discussed in Section 7.3. But the resulting model is not a single tree, but a forest of trees. Because the predictions are averaged across all trees, interpretation of model results is difficult. The R package `randomForest` implements Breiman's random forest algorithm for classification and regression. The basic idea of random forest is, however, quite simple to program in R.

First, we can write a simple function to produce bootstrapping samples:

```
#### R code ####
boot.sample <- function(data) ## data must be a data frame
    data [sample(nrow(data), rep=T),]
```

Second, a simple function can be used to produce a number of models using the generated bootstrapping samples:

```
#### R Code ####
my.bagging <- function(obj,
      data=eval(obj$call$data), n.bags=500, ...){
   bags.list <- list()
   for (i in 1:n.bags)
      bags.list[[i]] <- update(obj,
         data=boot.sample(data))
   oldClass(bags.list) <- "bagrpart"
   return(bags.list)}
```

This function takes a fitted CART model object (using `rpart`) and uses the function `update` to repeatedly fit the same CART model each time with a different data generated by the function `boot.sample`.

Predictions from these trees can be summarized using the functions `apply` and `sapply`:

```
#### R Code ####
predict.bagrpart <- function(obj, newdata, ...)
apply(sapply(obj, predict, newdata=newdata), 1, mean)
```

9.3.2 Example: Confidence Interval of the CART-Based Threshold

Qian et al. [2003a] proposed a method for estimating an ecological threshold based on the deviance reduction approach of a regression tree model. Specifically, the model uses a gradient variable of interest as the only predictor to fit a regression tree model. The first split is taken as the ecological threshold. The confidence interval of the estimated threshold is estimated using the bootstrap method (the percentile method). A $(1-\alpha)\times 100\%$ confidence interval is used to describe the certainty we have on an estimate of an unknown parameter. The confidence interval is defined in terms of long-run frequency. That is, if the experiment is repeated many (infinite) times, each time the same $(1-\alpha)\times 100\%$ confidence interval is calculated, $(1-\alpha)\times 100\%$ of the resulting intervals will contain the true population parameter. The probability a confidence interval will contain the target statistics is known as the coverage probability. Breiman [1996] showed that bootstrapping is unable to produce a confidence interval for a change point problem with the correct coverage probability. This claim can be shown by using a simulation. We will revisit this example in Chapter 11.2 for another statistical issue with this method.

We will simulate the definition of the confidence interval by repeatedly sampling from a model with a known change point:

$$y_i \sim \begin{cases} N(-1,1) & \text{if } x < 25 \\ N(0.5,1) & \text{if } x \geq 25 \end{cases} \tag{9.4}$$

The model defines that the response variable y has a normal distribution $N(-1,1)$ when the predictor x is less than 25, and $N(0.5,1)$ when $x \geq 25$. If the threshold is estimated based on $n = 20$ observations, we repeatedly sample 20 x values from a uniform distribution between 5 and 45, and for each sample of x generate a y based on the model (9.4). Figure 9.9 shows typical data sets with 6 different sample sizes.

For each set of generated samples, the change point method is used for estimating the threshold and bootstrapping is applied to calculate the 90% confidence interval of the estimated threshold. This process was repeated 5000

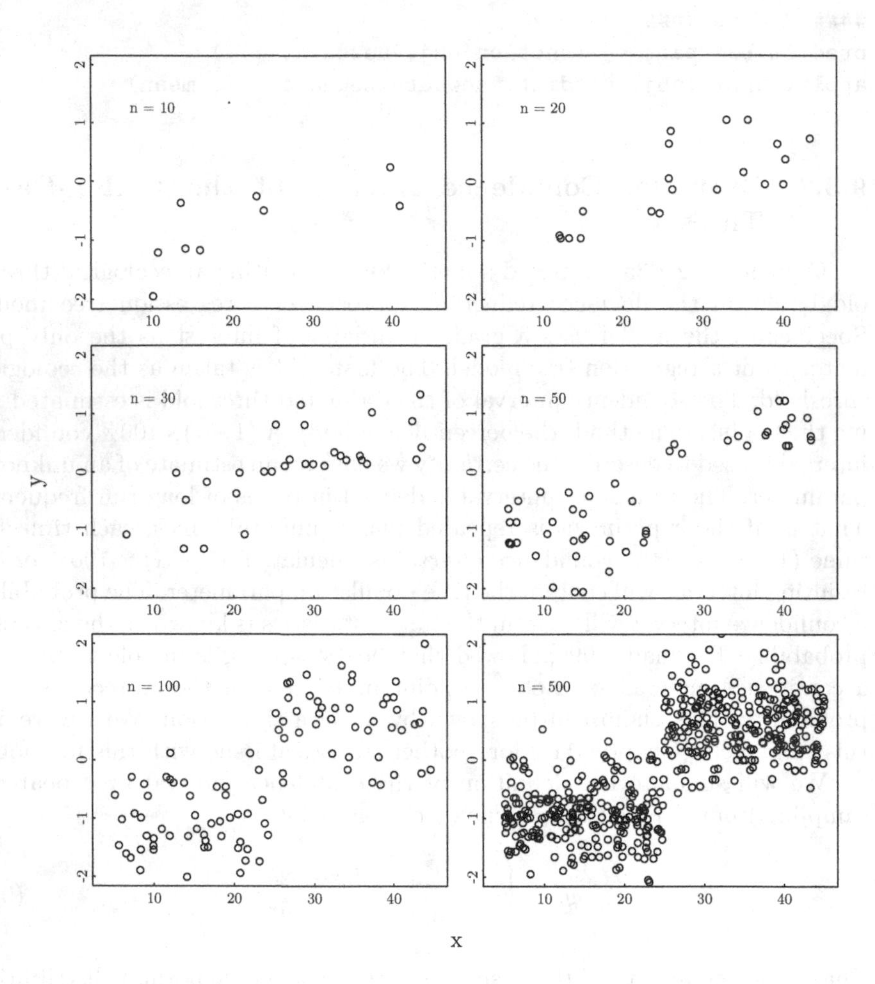

FIGURE 9.9: Bootstrapping for threshold confidence intervals – A set of data generated based on equation 9.4 with 6 different sample sizes (n).

times, resulting in 5000 sets of data and 5000 confidence intervals of the threshold. The number of those 5000 intervals containing the true threshold of 25 is then counted and the percent coverage (percent of the intervals containing the true threshold or the coverage probability) is calculated. If the bootstrapping procedure is appropriate for estimating the confidence interval of the threshold, approximately 90% of these 5000 intervals should contain the true threshold value (25).

Several functions are needed for this simulation. First, a simple function to replicate the CART modeling process with one predictor:

```
#### R Code ####
chngp <- function(infile)
{ ## infile is a data frame with two columns
  ##  Y and X
    temp <- na.omit(infile)
    yy <- temp$Y
    xx <- temp$X
    mx <- sort(unique(xx))
    m <- length(mx)
    vi <- numeric()
    vi [m] <- sum((yy - mean(yy))^2)
    for(i in 1:(m-1))
            vi[i] <- sum((yy[xx <= mx[i]] - mean(yy[xx <=
                mx[i]]))^2) + sum((yy[xx > mx[i]] - mean(
                yy[xx > mx[i]]))^2)
    thr <- mean(mx[vi == min(vi)])
    return(thr)
}
```

Second, a boostrap confidence interval of the threshold can be calculated using the following function:

```
#### R Code ####
my.bootCIs<-
function (x, nboot, theta, ...,
          alpha = c(0.05, 0.95))
{
    n <- length(x)
    thetahat <- theta(x, ...)
    bootsam <- matrix(sample(x, size = n * nboot,
        replace = TRUE), nrow = nboot)
    thetastar <- apply(bootsam, 1, theta, ...)
    confpoints.percent <- quantile(thetastar, alpha)
    return(confpoints.percent)
}
```

With a given sample size, we can generate a data set and estimate the bootstrapping confidence interval as follows:

```
#### R Code ####
size <- 25
x.unif <- runif(size, 5, 45)
data.file <- data.frame(
      X=x.unif,
      Y=ifelse(x.unif<25,
               rnorm(sum(x.unif<25), -1),
               rnorm(sum(x.unif>=25),0.5)))
CIs <- my.bootCIs(1:size, nboot=5000,
                  theta=function(x, infile){
                     chngp(infile[x, ])},
                  infile=data.file)
```

To calculate the coverage probability of this confidence interval, we do the following to convert the confidence interval into a binary variable:

```
#### R Code ####
cover <- CIs[1] < 25 & CIs[2] > 25
```

Repeating these steps many times (e.g., 5000), we can calculate the coverage probability of the 90% confidence interval.

This simulation is repeated for 6 sample sizes (n = 10, 20, 30, 50, 100, 500). The percent coverage for the 6 sample sizes are 64.08%, 72.28%, 73.60%, 73.84%, 71.24%, and 68.54%, respectively. This result indicates that the bootstrapping method failed to generate the correct confidence interval, even for a sample size of 500.

9.4 Bibliography Notes

Monte Carlo simulation is a vast field with many specialized techniques. This chapter includes only a small fraction. Robert and Casella [2004] provide a summary of the techniques. The simulation in Section 9.3.2 is also discussed in Banerjee and McKeague [2007]. The concept of Bayesian p-value is discussed in Gelman et al. [2014].

9.5 Exercises

1. Consider the model you developed in question 8 of Chapter 8 and perform a simulation to see if the model you developed adequately describes the response variable data distribution. A potential problem with this

data set is the limited variability in the response variable. This could be caused by the difficulty in accurately recording the number of mates a frog had; either the duration of observation is too short, or there might be mates that were not observed. The consequence of this problem is the underreporting of the number of mates, and the resulting model is likely to underestimate the number of mates (and producing too many 0s). Arnold and Wade [1984] discussed other problems with such data.

2. Use simulation to evaluate the revised Poisson model in problem 9 in Chapter 8. A potential problem of such data is the excess number of zeroes, a phenomenon known as zero-inflation [Lambert, 1992].

3. Qian et al. [2003a] proposed a "nonparametric deviance reduction" method for detecting ecological threshold. The method is based on the CART model, but uses only one predictor representing the environmental gradient. The first split point is used as the threshold. In the paper, the authors suggested that a χ^2 test can be used to test whether the resulting split point is "statistically significant." Because the split point is the point that results in the largest difference in deviance, it is highly likely that such a test will have a highly inflated type I error probability. Design a simulation to estimate the type I error probability of such a test. In the simulation, we can assume that the response variable is a normal random variable, such that the χ^2 test is reduced to a two-sample t-test. As the method is used to detect a threshold, the null hypothesis should be that a threshold does not exist, or the response variable distribution does not change along the gradient.

Chapter 10

Multilevel Regression

10.1 From Stein's Paradox to Multilevel Models 417
10.2 Multilevel Structure and Exchangeability 421
10.3 Multilevel ANOVA 425
10.3.1 Intertidal Seaweed Grazers 426
10.3.2 Background N_2O Emission from Agriculture Fields 431
10.3.3 When to Use the Multilevel Model? 434
10.4 Multilevel Linear Regression 436
10.4.1 Nonnested Groups 447
10.4.2 Multiple Regression Problems 453
10.4.3 The ELISA Example—An Unintended Multilevel Modeling Problem 464
10.5 Nonlinear Multilevel Models 465
10.6 Generalized Multilevel Models 469
10.6.1 Exploited Plant Monitoring—Galax 470
10.6.1.1 A Multilevel Poisson Model 471
10.6.1.2 A Multilevel Logistic Regression Model 474
10.6.2 Cryptosporidium in U.S. Drinking Water—A Poisson Regression Example 478
10.6.3 Model Checking Using Simulation 482
10.7 Concluding Remarks 486
10.8 Bibliography Notes 489
10.9 Exercises 489

10.1 From Stein's Paradox to Multilevel Models

An important task of statistics is to estimate unobservable model parameters from observed data [Fisher, 1922]. Because the parameter of interest cannot be observed directly, an estimator (a formula used to calculate an estimate) must be selected based on its performance and mathematical assumptions about the distribution of the data. Statistical models discussed in this book are examples of such estimators. Throughout the book, we used the maximum likelihood estimator because it is unbiased and is often the least variable among all unbiased estimators.

Theoretical basis of the maximum likelihood estimator can be traced back to Gauss, who derived the probability law (later known as the normal or Gaussian distribution) to justify the use of the least squares method for estimating a mean. Pierre-Simon Laplace's central limit theorem [Stigler, 1975], which states that the distribution of sample averages of independent random variables can be approximated by the normal distribution regardless of the original distribution from which these random variables were drawn, cements the normal distribution as the most important distribution in statistics.

Many environmental variables (concentration variables in particular) can be approximated by the log-normal distribution [Ott, 1995]. Thus, a rule of thumb in environmental statistics is that we should log-transform concentration variables before statistical analysis [van Belle, 2002], so that properties of normal distributions can be used advantageously. An important result of normal distribution theory is that the "best" estimator of the normal distribution mean is the sample average. It is the best because it is unbiased, and least variable among all unbiased estimators, and it is also a maximum likelihood estimator. Consequently, sample average and standard deviation are commonly reported in scientific studies. The normal distribution results also have a wide practical implication. For example, in environmental standard assessment, these normal distribution properties helped justify the use of a hypothesis testing approach instead of the raw score assessment approach [Smith et al., 2001]. Even when the variable of interest is not a normal random variable (e.g., in generalized linear model), estimated model coefficients are normal random variables.

The central role of the maximum likelihood estimator was challenged by Stein's paradox, which refers to the surprising features of a family of estimators originally introduced in the 1950s [Stein, 1955] and revised in 1961 [James and Stein, 1961]. These estimators are paradoxical because they suggest that the best method for estimating the mean of one variable (calculating sample average, the MLE) is not the best approach when the means of several variables are to be estimated simultaneously. Specifically, James and Stein [1961] showed that the overall accuracy (defined as the sum of squared differences between the estimated and true means) can be improved if we "shrink" the individually estimated averages towards the overall average – increasing those below and decreasing those above the overall average.

For example, in the context of assessing nutrient criterion compliance, Stein's paradox implies that the best estimator of a lake's mean nutrient concentration for assessing nutrient criterion compliance of a single lake, sample average, is no longer the best if we are to assess multiple lakes at the same time.

In the 1970s, Efron and Morris published a series of papers discussing the James–Stein estimator (and its modifications) and its role in various estimation problems [Efron, 1975, Efron and Morris, 1973a,b, 1975]. In their work, Efron and Morris used Bayes risk as a measure of estimation accuracy. Bayes risk is the average of the sum of squared differences:

$$R(\theta, \delta) = E_\theta \sum_{j=1}^{J} (\delta_j - \theta_j)^2 \tag{10.1}$$

where θ_j are unknown means (e.g., the true annual mean concentration of TP in a lake), δ_j is an estimator of θ_j (e.g., annual average of monthly monitoring data), and E_θ represents averaging over the distribution of θ_j. Bayes risk is often seen as the Bayesian version of the mean squared errors. A small Bayes risk is a good feature of an estimator. Efron and Morris showed that the Bayes risk of the James–Stein estimator is always lower than the Bayes risk of the corresponding maximum likelihood estimator.

In explaining this paradox, Efron [1975] used a mathematical theorem about sample averages of multiple variables $y_j, j = 1, \cdots, J$. The theorem compares sample averages $\bar{y}_j$ to the true underlying means θ_j. When y_j's are from a multivariate normal distribution with an overall mean μ, the following relationship holds:

$$\Pr\left[\sum_j (\bar{y}_j - \mu)^2 > \sum_j (\theta_j - \mu)^2\right] > 0.5 \tag{10.2}$$

That is, on average, sample averages $\bar{y}_j$ are more likely farther away from the overall mean μ than the true means θ_j are. Equation (10.2) implies that if we know the overall mean μ, we can improve the estimates by moving sample averages towards the overall mean. Note that the theorem states that the probability that the sum of squares of sample averages with respect to the overall mean is larger than the sum of squares of the true means is larger than 0.5. In other words, the sample averages are more likely, not necessarily always, farther away from the overall mean than the true means are. As a result, by moving sample averages towards the overall means (increasing the ones below the overall mean and decreasing the ones above the overall average, or shrinking towords the overall mean) will improve upon sample averages on average, but not necessarily every time. In other words, the statement that a shrinkage estimator will improve upon its MLE counterpart is analogous to the statement that a sample mean is an unbiased estimator of population mean – the average of many sample averages are equal to the underlying population mean, not necessarily a specific sample average.

Intuitively, the improvement of a shrinkage estimator is achieved by making use of additional information. Specifically, when estimating the mean of one variable (and we don't have data from other similar variables), we have no knowledge of the overall mean μ. As a result, we don't know towards which direction to shrink the resulting sample average. When we have observations from several similar variables, the average of sample averages $\hat{\mu}$ is a good estimate of the overall mean. As a result, we know how to shrink individual sample averages.

Given the theoretical justification of equation (10.2), the question now is how much shrinkage for each sample average. The James–Stein estimator is one such estimator where a level of shrinkage is derived for each sample average.

$$\hat{\theta}_j^{js} = \mu + m_j^{js}(\bar{y}_j - \mu) \tag{10.3}$$

where μ, the mean of θ_j, is often estimated by the average of $\bar{y}_j$, i.e., $\hat{\mu} = \frac{1}{J}\sum_j \bar{y}_j$, σ_1 is the standard deviation of individual variables, and

$$m_j^{js} = 1 - \frac{\sigma_1^2/n_j}{\sum_j(\hat{\theta}_j - \hat{\mu})^2/(J-2)}.$$

The coefficient m_j^{js} represents the level of shrinkage. When m_j^{js} approaches 1, $\hat{\theta}_j^{js}$ approaches $\bar{y}_j$, and $\hat{\theta}_j^{js}$ approaches $\hat{\mu}$ when m_j^{js} approaches 0. In most cases, m_j^{js} is between 0 and 1. As a result, the James-Stein estimator is between the overall mean $\hat{\mu}$ and the sample average $\bar{y}_j$. James and Stein [1961] showed that the Bayes risk of the James-Stein estimator is always smaller than the Bayes risk of the likelihood estimator (sample averages).

The James–Stein estimator can be seen as a generalization of an ANOVA model, where our interest is quantifying the means of several variables. In a one-way ANOVA model, data from individual variables are assumed to have normal distributions with different means but with a common variance:

$$y_{ij} \sim N(\theta_j, \sigma_1^2) \tag{10.4}$$

The common within-group variance is compared to the between-group variance in order to determine whether the means θ_j are the same. In ANOVA, the within-group and between-group variances are estimated as sums of squares, $\sum_j(\sum_i(y_{ij} - \bar{y}_j)^2)$ and $\sum_j n_j(\hat{y}_j - \hat{\mu})^2$, respectively. Just as the within-group variance can be summarized as a parameter of a probabilistic model, the between-group variance can also be represented as the variance of θ_j:

$$\theta_j \sim N(\mu, \sigma_2^2) \tag{10.5}$$

We can jointly solve equations (10.4) and (10.5) for θ_j by assuming that μ, σ_1, and σ_2 are known. The result

$$\hat{\theta}_j \approx \frac{\frac{n_j}{\sigma_1^2}\hat{y}_j + \frac{1}{\sigma_2^2}\mu}{\frac{n_j}{\sigma_1^2} + \frac{1}{\sigma_2^2}} \tag{10.6}$$

is a weighted average of the group sample average $\hat{y}_j$ and the overall average μ. Under this model, $\hat{\theta}_j$ approaches to $\hat{y}_j$ when $\sigma_2^2 \to \infty$, and $\lim_{\sigma_2^2 \to 0} \hat{\theta}_j = \mu$. In a sense, the ANOVA model is a special case of the probabilistic model represented in equations (10.4) and (10.5). We note that the ANOVA computation assumes a near infinity between-group variance σ_2^2, while the null hypothesis of ANOVA assumes that $\sigma_2^2 = 0$. A more reasonable value of σ_2^2 is likely neither 0 nor infinity, but something between 0 and infinity.

The weighted average in equation (10.6) is a shrinkage estimator and it can be rearranged to compare to the James–Stein estimator:

$$\hat{\theta}_j = \hat{\mu} + m_b(\hat{y}_j - \hat{\mu})$$

where $m_j^b = 1 - \frac{\sigma_1^2/n_j}{\sigma_2^2+\sigma_1^2/n_j}$.

Judge and Bock [1978] showed that m_j^{js} is an unbiased estimator of m_j^b. In other words, the James–Stein estimator can be seen as deriving the overall mean, between and within-group variances from the data. When we know μ, σ_2^2, and σ_1^2, the estimator of equation (10.6) has the smallest Bayes risk among all estimators [Lehmann and Casella, 1998]. Because we normally don't know μ and σ_2^2, we can view the James-Stein estimator as the "next best thing."

We note that the level of shrinkage (m_j^b) is largely determined by (1) the ratio of σ_1/σ_2 and (2) sample size n_j. A large standard deviation ratio (the standard deviation of individual variables, or within-group standard deviation, is large in comparison to the standard deviation among variable means) suggests a low confidence on the hypothesis that θ_js are different. It leads to a small m_j^b, thereby a large level of shrinkage towards the overall mean. A small n_j (indicating a low confidence on $\hat{y}_j$) leads to a small m_j^b, thereby a high level of shrinkage. In other words, using a shrinkage estimator is an effective way of addressing high uncertainty associated with a sample average estimated with a small sample size.

Although the modern multilevel models were developed independent of the Stein's paradox, mathematically, a simple multilevel model is the same as represented by equations (10.4) and (10.5). The multilevel model uses maximum likelihood estimator of μ, σ_1^2, and σ_2^2 and typically based on approximate programs such as the ones implemented in the R package `lme4` [Bates, 2010].

10.2 Multilevel Structure and Exchangeability

Multilevel or hierarchical structure is a common feature of many environmental and ecological problems. It can be a result of spatial, temporal, or organizational factors. An example of spatial multilevel data is reported by Qian et al. [2015b], where water quality monitoring data from small streams and rivers around the Great Lakes are used to discuss the advantage of using a shrinkage estimator when assessing environmental standard compliance of multiple waters. In that data set, data points are from individual streams and streams are grouped based on either ecoregions or states in the Great Lakes watersheds. Stow and Scavia [2009] used the temporal multilevel structure in data to study changes in the extent of hypoxia in the Gulf of Mexico over time. Qian and Cuffney [2014] present an example of organizational multilevel structure, where biological responses to changes in watershed urbanization intensity are grouped by taxa or taxon groups. The multilevel structure in the Finnish Lakes example represents a combination of spatial and organizational factors. Lakes in Finland are grouped into nine types based on their size and morphological features. In all cases, the multilevel structure is constructed based on our understanding of the underlying processes that resulted in the variation of the data. As a result, the multilevel structure is a conceptual construct.

Grouping data based on certain characteristics of the subject or environmental and biological conditions is often essential to understand the key relationship of interest. When we know or want to test potential underlying processes, we conduct studies by collecting data based on the multilevel structure. In the mangrove example (Section 4.8.4), observed data are grouped by treatment because we want to understand the effect of the live sponge on mangrove root growth. The multilevel structure is based on the hypothesized relationship between mangrove and sponge. In observational study, we also group data based on one or more multilevel structures. Grouping Finnish lakes into nine types recognizes the potential differences among lake types in the chlorophyll a–nutrient relationship. In other cases, we explore the data to find likely groupings. Various tree-based models are often the most convenient tools for such exploration. In the Willamette River example (Section 7.1), we used a simple tree-based model for exploring factors affecting the variation in pesticide concentration in the Willamette River. In that example, we described CART as the opposite of ANOVA – finding relatively homogeneous groups. Data points within each group have relatively smaller variation, compared to between group variation. Yuan and Pollard [2015] used a variant of the classical CART model to group lakes in U.S. EPA's National Lake Assessment into three groups for the purpose of developing nutrient criteria.

Whether the multilevel structure is based on existing knowledge or exploratory analysis of the data, the multilevel structure in the data is a conceptual construct. We use the multilevel structure to better organize the data and facilitate the development of meaningful models. Whether the structure can be "obvious" (e.g., grouping streams by state or ecoregion in Qian et al. [2015b]) or derived through more complicated exploratory analysis [Yuan and Pollard, 2015], the goal is almost always to create groups with "similar" units. In Qian et al. [2015b], units are streams and their similarity is defined by nutrient concentrations – streams in a group should have similar mean concentrations. In Yuan and Pollard [2015], units are lakes in the National Lake Assessment study and similarity is measured by the relationship between chlorophyll a concentration and nutrient (TP and TN) concentrations. Although the need to group similar units is quite intuitive, the statistical reason for grouping is often opaque. Through the learning of Stein's paradox, I realized that one implication of the paradox is how to group data properly.

Stein's paradox shows that pooling data is advantageous. The James–Stein estimator, however, does not require that only closely related variables (e.g., variables with similar means) should be grouped because the overall Bayes risk is reduced no matter what are the underlying true means. Efron and Morris [1977] discussed the problem of how to group data in length. On the one hand, the benefit of using a shrinkage estimator (reducing overall Bayes risk) diminishes as the difference among units increases. On the other hand, the shrinkage estimator cannot be demonstrated to be superior over the maximum likelihood estimator for any particular variable. Requiring that data from variables sharing similar features be grouped is, therefore, reasonable. In this regard, the multilevel structure of a typical environmental data is such a construct for grouping similar variables.

In a study of water quality of drinking water source waters in China, Wu et al. [2011] grouped source waters in three different ways in the absence of necessary information on how source waters should be grouped because the study was the first water quality assessment study of drinking water source water in China. The objective of the study was to derive a distribution of source water mean pollution concentrations across all source waters in China. From a management perspective, the study is to estimate the extent of water quality problems in the country (e.g., estimating the fraction of source waters with mean concentration of a particular pollutant exceeding a water quality standard). If the study is to summarize the water quality status for the country as a whole, we can treat each source water as an entity and estimate their mean concentrations separately. Such an approach was indeed routinely used in water quality assessment in the U.S. For example, Section 305(b) of the 2002 U.S. Clean Water Act amendment requires states to report the water quality status of all waters of the state (including rivers/streams, lakes, estuaries/oceans, and wetlands) every two years. A typical report from a state includes a summary of individual waters, typically the estimated mean concentrations of regulated pollutants. By pooling data and using a shrinkage estimator, we know that we can improve the overall accuracy of the assessment.

The concept of *exchangeability* is the mathematical formalization of grouping similar units. Formally, exchangeability is defined in terms of "symmetry" – the parameters $\alpha_1, \alpha_2, \cdots, \alpha_n$ are exchangeable in their joint distribution if $p(\alpha_1, \alpha_2, \cdots, \alpha_n)$ is invariant to permutations in the index $1, 2, \cdots, n$. Intuitively, units are exchangeable when we know that they are likely different, but we don't know the nature of the differences among units. In Wu et al. [2011], source water systems are grouped according to management needs. Their first model groups source water systems based on administrative borders so that various levels of governments can have a clear picture of the water-quality status within their jurisdictions. Source water systems were also grouped based on major drainage basins to better separate natural and anthropogenic contributions of pollutants. Finally, source water systems were grouped based on their source water type (groundwater or surface water) and size (characterized by population served) to facilitate a meaningful human exposure risk assessment. In this case, we assume that source water systems within a group are exchangeable, that is, system means within a group can be modeled as random variables with the same probability distribution even though we know that these means are probably different. By imposing the exchangeable assumption within each group, we summarize the between group difference in terms of the group-specific common probability distribution of system means. Once systems are grouped, we most likely view the between-group differences as more prominent compared to within-group differences, thereby ignoring the patterns within a group. Such intentional or unintentional ignorance is often harmless and imposing the exchangeability assumption is a natural way of modeling this ignorance, rather than ignoring this source of information.

In grouping source water systems into groups, we assume that source water systems will have different mean concentrations within a group but the difference is undefined. The exchangeability assumption of a specific group (k) is to impose a common probability distribution on system means:

$$\begin{array}{rcl} \log(y_{ijk}) & \sim & N(\theta_{jk}, \sigma_1^2) \\ \theta_{jk} & \sim & N(\mu_k, \sigma_2^2) \end{array}$$

where y_{ijk} is the ith observed pollutant concentration in source water system j from group k, θ_{jk} is the system mean, and σ_1^2 and σ_2^2 are the within-system and between-system variances. The common distribution suggests that θ_{jk}'s are different, but we do not know *á priori* how they differ from each other. Therefore, the exchangeability assumption is imposed on θ_{jk}.

In grouping Finnish lakes into nine types, Malve and Qian [2006] imposed the exchangeability assumption on lakes within each type, with respect to the chlorophyll a–nutrient concentration relationship. In their work, the chlorophyll a response model coefficients from lakes within a type are assumed to be exchangeable using the following model,

$$\begin{array}{rcl} \log(chla_{ijk}) & \sim & N(X\beta_{jk}, \sigma_1^2) \\ \beta_{jk} & \sim & N(\beta_k, \sigma_2^2) \end{array}$$

where $chla_{ijk}$ is the ith observed chlorophyll a concentration in lake j from type k, $X\beta_{jk}$ is the linear regression model with TP and TN (and their interaction) as predictors, and β_{jk} is the coefficient vector for lake j. Model coefficients for lakes in type k are assumed exchangeable with a common distribution $N(\beta_k, \sigma_2^2)$.

In practice, assuming exchangeability on units of parameter means that we can impose the same a priori distributional assumption on these parameters. The concept is closely related to the idea of identically and independently distributed (iid) random variables. If we treat each data point as a special case of a parameter, exchangeability is the generalization of iid. For the Finnish lakes example, this assumption is probably correct because the purpose of dividing lakes into types is to create relatively homogeneous groups of lakes so that a common management strategy can be developed for lakes in each type.

When data points are iid, we can use the maximum likelihood estimator for parameter estimation. The iid concept is essential for statistical inference. Likewise, when parameters from multiple units can be assumed exchangeable, we can use multilevel modeling to pool data together to better estimate these parameters.

Implicit multilevel model structure is surprisingly common in environmental and ecological studies, yet it is rarely exploited in statistical modeling in these fields. A multilevel model is likely preferred over a standard multiple regression model in analyzing cross-sectional data when there are logical hierarchical groupings (e.g., samples within lakes, lakes within ecoregions, etc.) or categorical variables (natural versus manmade lakes, run-of-river lakes, etc.).

10.3 Multilevel ANOVA

Analysis of variance (ANOVA) is widely used in scientific research for testing multiple, often complicated, hypotheses, by comparing the means of a response variable from several groups (or populations). As we discussed in Chapter 4, Fisher's hypothesis testing is a tool for inductive reasoning by providing evidence against the null hypothesis. As originally presented in Fisher's seminal work [Fisher, 1925], ANOVA can be viewed as the collection of the calculus of sums of squares and the associated models and significance tests. These tests and models have had a profound impact on science in general. In ecological studies, ANOVA provides the computational framework for the design and analysis of ecological experiments [Gotelli and Ellison, 2004]. The use of ANOVA is, however, quite limited in that it is designed specifically for randomized experimental data. When ANOVA is used in other situations, interpretation of ANOVA results can be problematic. These situations include cases where the normality and independence assumptions of the response data are not met, where the experimental design is nested or unbalanced, or where there are missing values. More importantly, ANOVA results are difficult to explain when applied to observational data where the significance test is often not very informative [Anderson et al., 2000]. Furthermore, when an experiment is proposed, we almost always have reasons to believe that a treatment effect exists. Therefore, we often want to know the strength of the effect a treatment has on the outcome rather than whether the treatment has an effect or not on the outcome. By using a significance test where the inference is based on the assumption of no treatment effect, we emphasize the type I error rate (erroneously rejecting the null hypothesis of no treatment effect) often at the expense of statistical power, especially when multiple comparisons are used.

We illustrate the multilevel ANOVA approach using a one-way ANOVA example. For a one-way ANOVA problem, we have a treatment with several levels, and the statistical model is:

$$y_{ij} = \beta_0 + \beta_i + \epsilon_{ij} \tag{10.7}$$

where β_0 is the overall mean, β_i is the treatment effect for level i and $\sum \beta_i = 0$, and j represents individual observations in treatment i. The residuals are assumed to have a same normal distribution with mean 0 and an unknown variance $[\epsilon_{ij} \sim N(0, \sigma^2)]$. This model is equivalent to

$$y_{ij} \sim N(\beta_0 + \beta_i, \sigma^2) \tag{10.8}$$

The MLE of β_0 is the overall mean $(\frac{1}{N}\sum_i \sum_j y_{ij})$ and the MLE of β_i is the treatment mean $(\frac{1}{n_i}\sum_j y_{ij})$. The total variance of y_{ij} is partitioned to the within- and between-group variances. Statistical inference is based on the comparison of the two variance components. When the emphasis is on the estimation and comparison of the treatment effects, the ANOVA emphasis of testing the null hypothesis is less effective and the estimated group means are often unstable when sample sizes are small. If the null hypothesis is true, treatment effects β_i are expected to be 0 and are otherwise exchangeable. As a result, a common prior distribution is used in a multilevel model of the same problem:

$$\beta_i \sim N(0, \sigma_\beta^2) \tag{10.9}$$

With this assumption explicitly used, the treatment effects must be estimated differently. This setup (equations 10.8 and 10.9) explicitly parameterizes the between-group standard deviation (σ_β) and within-group standard deviation (σ). For a simple one-way ANOVA problem, model parameters $(\beta_0, \beta_i, \sigma, \sigma_\beta)$ can be estimated using the maximum likelihood estimator. The likelihood of observing y_{ij} is defined by the normal distribution in equation 10.8 – a conditional normal distribution. The full likelihood function is then the product of the normal density of equation 10.8 and the normal density of equation 10.9. The computation is implemented in R function `lmer()` from package `lme4`.

We introduce two more examples to illustrate the use of `lmer()` for fitting a multilevel ANOVA model and the use of the fitted model for multiple comparisons.

10.3.1 Intertidal Seaweed Grazers

This example was used in the text by Ramsey and Schafer [2002] (Case Study 13.1, p. 375), describing a randomized experiment designed to study the influence of three ocean grazers, small fish (`f`), large fish (`F`), and limpets (`L`), on the regeneration rate of seaweed in the intertidal zone along the coast of Oregon, USA. The experiments were carried out in eight locations to cover a wide range of tidal conditions and six treatments were used to determine the effect of different grazers (`C`: control, no grazer allowed; `L`: only limpets allowed; `f`: only small fish allowed; `Lf`: large fish excluded; `fF`: limpets excluded; and `LfF`: all allowed). The response variable is the seaweed recovery of the experimental plot, measured as percent of the plot covered by regenerated seaweed. The standard approach illustrated in Ramsey and Schafer [2002] is a two-way ANOVA (plus the interaction effect) on the logit transformed percent regeneration rates. The logit of percent regeneration rate (y) is the logarithm of the regeneration ratio (% regenerated over % not regenerated).

Using the multilevel notation, this two-way ANOVA model can be expressed as:

$$Y_{ijk} = \beta_0 + \beta_{1i} + \beta_{2j} + \beta_{3ij} + \epsilon_{ijk} \tag{10.10}$$

where Y is the logit of regeneration rate, β_{1i} is the treatment effect ($i = 1, \cdots, 6$ and $\sum \beta_{1i} = 0$), β_{2j} is the block effect ($j = 1, \cdots, 8$ and $\sum \beta_{2j} = 0$), and β_{3ij} is the interaction effect ($\sum \beta_{3ij} = 0$). The residual term ϵ_{ijk} is assumed to have a normal distribution with mean 0 and a constant variance, where $k = 1, 2$ is the index of individual observations within each Block–Treatment cell. The total variance in Y is partitioned into four components: treatment, block, interaction effects, and residual.

In general, fitting a multilevel model in R is similar to fitting a linear regression model with varying intercept and/or slope. The one-way ANOVA problem is a linear regression with no continuous predictor and the intercept varies by treatment levels. First, we consider the simple one-way ANOVA situation where only the treatment effects are modeled:

$$Y_{ik} = \beta_0 + \beta_{1i} + \epsilon_{ik}$$

This is specified in R formula as:

```
y ~ Treatment
```

The variable `Treatment` identifies group association of each data point. This formula is the same as:

```
y ~ 1 + Treatment
```

where the common intercept is explicitly specified. In specifying a multilevel model in R, the R formula is mostly the same except for the specification of group indicator:

```
y ~ 1 + (1|Treatment)
```

The right-hand side of the formula is a sum of two parts: the main model structure (`1`) and the group (`(1|Treatment)`). The second part (in parentheses) specifies which part of the main model (`1`, the intercept) will change by group (`Treatment`).

```
#### R code ####
seaweed.lmer <- lmer(y ~ 1+(1|Treatment), data=seaweed)
```

Model results (treatment effects and variance decomposition) are stored in the R object of class mer. The summary function will bring out some basic information:

```
#### R output ####
> summary(seaweed.lmer)

 Linear mixed model fit by REML ['lmerMod']
 Formula: y ~ 1 + (1 | Treatment)
    Data: seaweed

 REML criterion at convergence: 303.7

 Scaled residuals:
      Min       1Q   Median       3Q      Max
 -1.80695 -0.72417 -0.03866  0.56969  2.62582

 Random effects:
  Groups    Name        Variance Std.Dev.
  Treatment (Intercept) 1.139    1.067
  Residual              1.178    1.085
 Number of obs: 96, groups:  Treatment, 6

 Fixed effects:
             Estimate Std. Error t value
 (Intercept)  -1.2326     0.4495  -2.742
```

The "random effects" part shows the estimated variance components. The estimated between-group variance σ_β^2 is 1.139 and the estimated within-group variance σ^2 is 1.178. These two variances sum to the total variance (variance of the response variable). The "fixed effect" is the common intercept (or overall mean of the response). The terms *fixed* or *random* effects are somewhat confusing. Gelman and Hill [2007] (sections 1.1 and 11.4) discussed the reasons for not using these terms. The use of these terms in lmer output can be interpreted as follows. A multilevel model has coefficient(s) common to all groups and group-specific coefficients. Fixed effects are the estimates of the common coefficients and random effects are the group-specific coefficients. The estimated "fixed effects" are shown in the summary output. The estimated group-specific coefficients (or random effects) can be extracted by using the function ranef:

```
#### R output ####
> ranef(seaweed.lmer)

 $Treatment
```

```
            (Intercept)
CONTROL            1.33
f                  0.86
fF                 0.39
L                 -0.45
Lf                -0.72
LfF               -1.40
```

The listed numbers are the estimates of β_{1i}. The estimation uncertainty (standard error of $\hat{\beta}_{1i}$) is extracted by using `se.ranef` (from package `arm`):

```
#### R output ####
> se.ranef(seaweed.lmer)

$Treatment
     [,1]
[1,] 0.26
[2,] 0.26
[3,] 0.26
[4,] 0.26
[5,] 0.26
[6,] 0.26
```

Understanding the difference between the model fitted using `lmer` and the model fitted using `lm` is the key to appreciate the advantages of multilevel modeling. The first difference is the estimated treatment effects (Figure 10.1). The multilevel estimates are always closer to the overall average than the linear model estimates (group means). This is often referred to as the "shrinkage" effect. Mathematically, the shrinkage effect is a direct result of using the common á priori distribution for β_{1i} (equation 10.9). The analytical solution of the treatment effects (when the between and within-group variances are known) is a weighted average between the overall mean and group mean:

$$\hat{\beta}_{1i} = \frac{\frac{n_i}{\sigma^2}\bar{y}_{i\cdot} + \frac{1}{\sigma_\beta^2}\bar{y}_{\cdot\cdot}}{\frac{n_i}{\sigma^2} + \frac{1}{\sigma_\beta^2}} \tag{10.11}$$

The standard error of $\hat{\beta}_{1i}$ is $1/\sqrt{\frac{n_i}{\sigma^2} + \frac{1}{\sigma_\beta^2}}$. From this analytical result, we know that the multilevel estimate $\hat{\beta}_{ij}$ is closer to the group mean $\bar{y}_{i\cdot}$ when the group sample size n_i is large, or the within-group variance σ^2 is small, or when the between-group variance β_β is large. Under all three conditions, we would trust the group means as a reliable estimate of treatment effects because the uncertainty in group means is small. The group means and the overall mean represent two pieces of information we have about the response variable. When using group means as the estimate of treatment effects, we ignore the information represented in the overall mean. This information tells

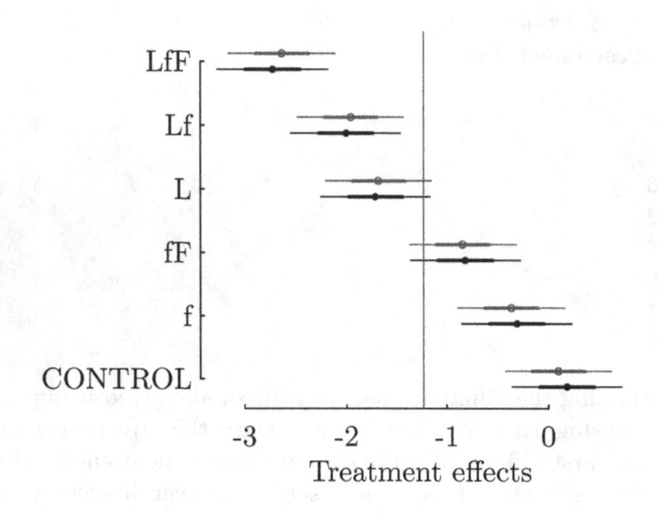

FIGURE 10.1: Seaweed grazer example comparing `lm` and `lmer` – Estimated treatment effects from a linear model (black lines) and a multilevel model (gray lines) are compared. The estimated means are shown by the solid dots and the horizontal line segments represent the means plus and minus one (thin lines) and two (thick lines) standard error. The vertical gray line is the overall response average.

us where the response variable value should be centered. If an extremely large or small group mean is based on a small sample size, we may have reason to believe that the true mean is somewhat less extreme. The multilevel estimate of treatment effect is a compromise between the two pieces of information. It automatically calculates the relative strengths of each and yields a weighted average as the estimate. One advantage of the shrinkage effect is that the estimated treatment effects can be directly used for multiple comparison without various adjustments discussed in Section 4.7.3. In multiple comparisons, there are several independent hypothesis tests. If each is based on a significance level, that is, a probability of making a type I error, of 0.05, the probability that at least one test resulted in erroneous rejection of the null is much higher than the stated $\alpha = 0.05$. The Bonferroni and other corrections are to adjust the significance level such that the family-wise type I error probability is limited to be 0.05. These adjustments are often at the expense of statistical power. The use of multilevel models for multiple comparison is discussed by Gelman et al. [2012]. On the one hand, the shrinkage effect has moved the estimated treatment effects toward the center. They are conservative estimates of the treatment effect. On the other hand, the amount of shrinkage each treatment effect receives is based on the comparison of the specific treatment and the effects of other treatment levels. Consequently, comparing the multilevel estimated confidence intervals is often a more effective means for multiple comparisons.

The multilevel estimated treatment effects and their standard errors in Figure 10.1 are not very different from the estimated treatment effects using the conventional ANOVA model. This is because with the sample sizes for all six treatment levels the same (16), the between-group variation is relatively large compared to the within-group variance. In other words, the multilevel model is unnecessary under some situations. The conventional linear model approach will yield comparable results. But for most observational studies, the multilevel model is often a better choice.

10.3.2 Background N_2O Emission from Agriculture Fields

Carey [2007] conducted a meta analysis of the effect of fertilizer input on N_2O emission from agriculture soils by pooling data from field scale studies reported in 164 peer-reviewed publications. N_2O is a greenhouse gas often associated with nitrogen fertilizer use. As nitrogen fertilizer usage is expected to increase rapidly, understanding the magnitude (and the variation) of the fertilizer effect on emissions is important in developing effective mitigation strategies for combating global climate change. In the 164 studies, various forms of linear modeling analysis were used. Results from these studies are difficult to compare because of the difference in local climate and soil conditions, as well as different experimental designs. We use only the emission data from fields without fertilizer application, as an illustration of the typical situation where multilevel modeling is helpful.

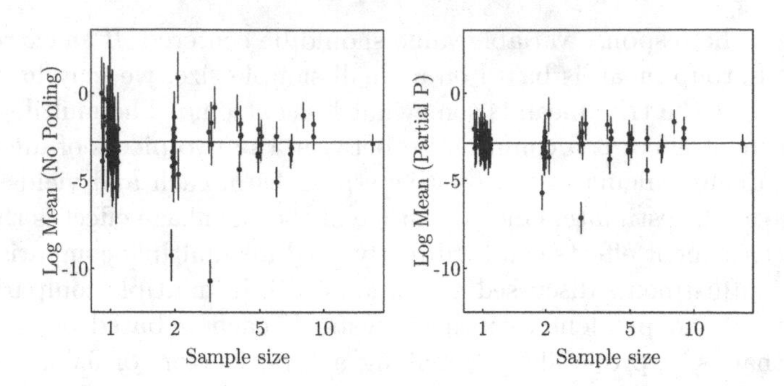

FIGURE 10.2: Comparisons of three data pooling methods in the N_2O emission example – Background N_2O emissions are estimated using no pooling (left panel), complete pooling (horizontal line, both panels), and partial pooling (right panel). The dots are the estimated means, and the vertical line segments are the mean plus/minus one standard error.

When analyzing the control group data, two methods are commonly used to study the same problem from different angles.

1. Assuming homogeneity among studies, observed N_2O emissions from different studies are treated as replicates and pooled together to obtain a single estimate. This method is called *complete pooling.*

2. Assuming heterogeneity among studies, observations from different studies are treated as incomparable and analyzed separately to obtain study-specific estimates (no pooling).

The homogeneity assumption is difficult to justify because N_2O emission is related to many factors. Pooling data together will lead to an overestimation of the uncertainty and oversimplification of the problem. Analyzing data separately often results in reduced sample sizes leading to a high variability in the estimated mean emission among studies. In this case, there are many studies that reported only one observation for the control, making the estimation of standard deviation impossible unless a linear regression model is used by assuming a common within study variance (equation 10.7). The emission data used in this section is the monthly average emission.

The multilevel model is the compromise of the two approaches, which not only reports the overall pattern, but also retains group-specific features. The multilevel modeling approach is also known as partial pooling. Figure 10.2 compares the estimated average N_2O emission using no pooling, complete pooling, and partial pooling.

The introduction of a common prior distribution in the multilevel model resulted in a "partial pooling" effect: the estimated study mean N_2O emission is a weighted average between the estimates from complete pooling and no pooling (Figure 10.2, right panel). As a result, the partial pooling results are always closer to (or pulled towards) the overall mean than the no pooling results (the shrinkage effect). Shrinkage represents a form of information discounting. Results from complete pooling and no pooling represents two pieces of information obtained from the data. Partial pooling is a mathematical means for reconciling the differences between the two. When the sample size of a specific group (study) is small or the group-specific variance is large, the amount of information represented in the group-specific no pooling estimate is small. The corresponding partial pooling result will be closer to the overall mean than the no-pooling mean. When the sample size is large or the estimated no-pooling standard deviation is small, the amount of pulling will be small (Figure 10.2, right panel). The no pooling estimated study mean N_2O emissions are highly variable because of the small sample sizes used in many studies. These estimates are pulled toward the overall mean using partial pooling. The amount of shrinkage is larger when the study mean is far away from the overall mean and/or is estimated based on a small sample size. Compared to the no pooling estimates, the partial pooling estimated group means are less variable. This is because the no pooling is a special case of the partial pooling when we set the between-group variance to be infinity ($\sigma^2_\beta = \infty$). The complete pooling result corresponds to the partial pooling result when the between-group variance is set to 0 ($\sigma^2_\beta = 0$). The multilevel model includes the no pooling and complete pooling as special cases. With partial pooling, we estimate the between-group variance from data. In most cases, between-group variance is neither 0 nor infinity. Partial pooling will almost always result in more reasonable estimates than no pooling or complete pooling can. This conclusion can be extended to linear regression and generalized linear model cases [Gelman and Hill, 2007].

Furthermore, with a multilevel model, we can include group level predictors to explore the reasons for the between-group variation. In this case, we suspect that soil organic carbon may be a factor affecting the emission of N_2O because N_2O is a product of microbial activities in soil and organic carbon is a main source of energy for microbes. The relationship between N_2O emission and soil organic carbon is often impossible to quantify using study-specific data because soil carbon measured in a given study does not vary by much from field to field. Soil carbon represents a large spatial scale variable that cannot easily be manipulated. By pooling data from multiple studies, we can model the study-mean as a function of soil carbon:

$$
\begin{array}{rcl}
y_{ij} & = & \theta_i + \epsilon_{ij} \\
\theta_i & = & \alpha_0 + \alpha_1 x_i + \eta_i
\end{array}
\qquad (10.12)
$$

where x is the logit transformed average percent soil carbon for each study. The logit transformation is used to make the distribution of the data less skewed (Figure 10.3). The estimated slope α_1 is positive (Figure 10.4), suggesting a positive relationship between N_2O emission and soil carbon concentration.

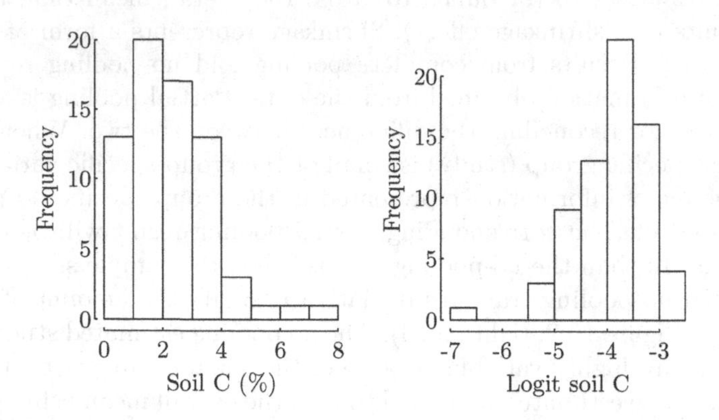

FIGURE 10.3: Logit transformation of soil carbon – The group (study) level average soil carbon (%) distribution is skewed and a logit transformation resulted in approximate symmetry.

The association is weak, as expected, because other factors (e.g., soil moisture, temperature) are not considered in this model.

The model in equation 10.12 is fitted in R by introducing a column in the data set representing group (study) average soil carbon:

```
carbon.group <- tapply(N2O.control$carbon/100,
                       N2O.control$group,
                       mean, na.rm=T)
carbon.full <- carbon.group[N2O.control$group]

bckg.lmer2 <- lmer(log(y) ~ 1 + logit(carbon.full) + (1|group),
                   data=N2O.control)
```

To properly write the R formula, it is helpful to rewrite equation 10.12 as

$$y_{ij} = \alpha_0 + \alpha_1 x_i + \eta_i + \epsilon_{ij}$$

The mean function $(\alpha_0 + \alpha_1 x_i)$ has an intercept and one predictor, which is included in the R model formula as `1 + logit(carbon.full)`. There are two error terms. ϵ_{ij} is the usual model residual term and η_i is indexed by i only, representing group level uncertainty not explained by the group level predictor, represented in R model formula as `(1|group)`.

10.3.3 When to Use the Multilevel Model?

The seaweed grazers example shows a situation where a multilevel model shows no advantage over the conventional linear modeling approach, whereas

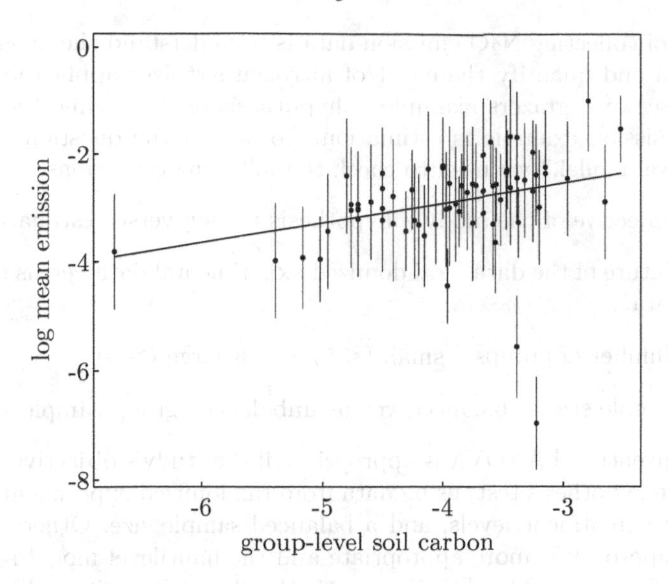

FIGURE 10.4: N_2O emission as a function of soil carbon – Estimated model intercept (study mean N_2O emission) is weakly correlated with study level soil carbon content. The dots are the estimated means and the vertical line segments are the mean plus/minus one standard error.

the N_2O emission example shows the advantages of multilevel regression. Whether or not a multilevel regression is beneficial can be decided by two factors: sample size n_i and between and within group variances (σ^2 and σ^2_β, respectively). If the sample size $n_i \to \infty$, the partial pooling estimate is the same as no pooling estimate. In fact, when sample size is large enough, the difference between partial pooling and no pooling is negligible. The term "large enough" is relative, just as we discussed in Chapter 4 (Figure 4.1 on page 83). The weight of the no pooling estimate ($\bar{y}_{i\cdot}$) is the ratio of n_i and within-group variance σ^2. Its contribution to the partial pooling estimate is determined by the size of the weight of the complete pooling estimate $(1/\sigma^2_\beta)$. If the between-group variance σ^2_β is large, the weight of the complete pooling estimate is small and vice versa. The seaweed grazers example shows that a multilevel model may not give us more than a simple ANOVA model would give in a well-designed experiment with balanced sample size and a limited number of treatment levels. The experimental design of the study used treatment levels that are expected to have a strong effect. Consequently, the between-group variance is large and the complete pooling estimate is designed to have a low importance. The N_2O emission example shows a case where the multilevel model is a better choice than the conventional linear model. The data from multiple studies are of different and often of small sample sizes. The between study variation is not controlled to contrast the difference. Above all, the objective of collecting N_2O emission data is to understand the magnitude of the problem and quantify the effect of nitrogen fertilizer application. The focus of the seaweed grazers example is hypothesis testing, while the focus of the N_2O emission example is estimation. To answer the question "when to use multilevel model," we need to check the following conditions:

- Objective of the study – hypothesis testing versus estimation
- Nature of the data – randomized experimental data versus observational data
- Number of groups – small (≤ 5) versus large (> 5)
- Sample sizes – balanced versus unbalanced group sample sizes

The conventional ANOVA is appropriate if the study's objective is to perform a simple hypothesis test, using data from randomized experiments, with fewer than five treatment levels, and a balanced sample size. Otherwise, the multilevel approach is more appropriate and the multilevel model results will be more informative. The decision on whether to use a multilevel model instead of ANOVA is always an easy one. Whenever possible, the estimation oriented multilevel model should be used over the hypothesis testing oriented ANOVA. ANOVA results are difficult to explain in ecological terms because the significance test is not always informative [Anderson et al., 2000]. On the one hand, when an experiment is proposed, we almost always have reasons to believe that a treatment effect exists. Therefore, we often want to know the strength of the effect a treatment has on the outcome rather than whether the treatment has an effect or not. By using a significance test where the inference is based on the assumption of no treatment effect, we emphasize the type I error rate (erroneously rejecting the null hypothesis of no treatment effect) often at the expense of statistical power, especially when multiple comparisons are used. On the other hand, a nonexistent treatment effect can be shown to be statistically significant if one tries often enough [Ioannidis, 2005].

Although we can fit multilevel models to all cases where ANOVA is intended, the maximum likelihood algorithm is not efficient in estimating the variances (σ^2 and σ_β^2) when the number of groups is too small. In cases where the maximum likelihood estimator is less efficient, a more computing intensive simulation method should be used. Qian and Shen [2007] discussed the use of the Bayesian approach for multilevel ANOVA.

10.4 Multilevel Linear Regression

We introduce the multilevel linear regression model by first introducing the example data from a study conducted by the United States Geological Survey

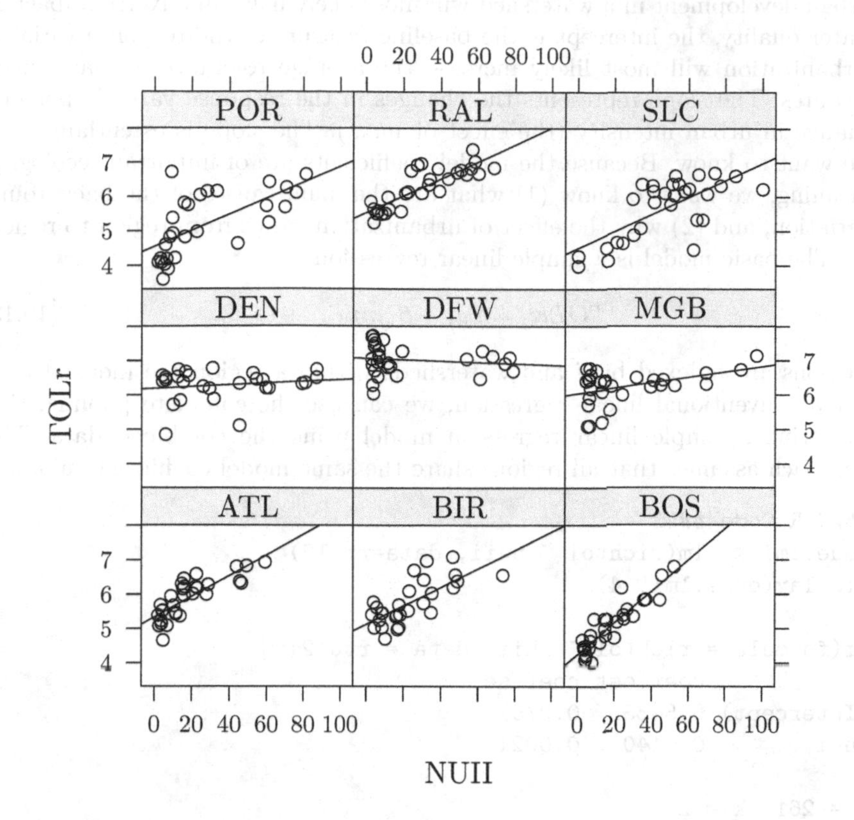

FIGURE 10.5: The EUSE example data – Mean taxa tolerance (TOLr) is plotted against urbanization intensity measured by the National Urbanization Intensity Index (NUII).

(USGS) on the effects of urbanization on stream ecosystems (EUSE) (see Section 1.3). One of the response variables used in the study is the average tolerance of taxa (`TOLr`) of macroinvertebrates [Cuffney et al., 2005]. The tolerance measure is an indicator of whether a taxon can survive in a polluted environment. The higher the tolerance, the "tougher" the bugs represented by the taxon. In general, higher tolerance taxa are found in waters with poorer water quality.

When plotted against the national urban intensity index (`nuii`), the relationship between `TOLr` and `nuii` appears to be linear (Figure 10.5). But the relationship varies from region to region, each with a different intercept and slope. The intercept represents the average tolerance of taxa at 0 urban intensity (the background `TOLr`). This value is the estimated tolerance of a watershed with no urban development. If the underlying assumption is that urban development in a watershed will most likely have an adverse impact on water quality, the intercept is the baseline measure of the response variable. Urbanization will most likely increase the average tolerance of macroinvertebrates. The slope represents the changes in the response variable per unit change in urban intensity (the effect of `nuii`). The slope is essentially what we want to know. Because the model coefficients are of important ecological meaning, we want to know (1) what are the main causes of the background variation, and (2) why the effect of urbanization varies from region to region.

The basic model is a simple linear regression:

$$TOLr_{ij} = \beta_{0j} + \beta_{1j} nuii_{ij} + \epsilon_{ij} \tag{10.13}$$

Regions are indexed by j and watersheds within a region are indexed by i. For a conventional linear regression, we can use the complete pooling, that is, fitting a simple linear regression model using the combined data. This approach assumes that all regions share the same model coefficient values:

```
#### R Code ####
esue.lm1 <- lm(richtol ~ nuii, data=rtol2)
display(euse.lm1, 4)

lm(formula = richtol ~ nuii, data = rtol2)
            coef.est coef.se
(Intercept) 5.5433   0.0752
nuii        0.0140   0.0021
---
n = 261, k = 2
residual sd = 0.7963, R-Squared = 0.15
```

The resulting model is unsatisfactory, not because only 15% of the total variation in the response variable is explained by including `nuii` as a predictor, but rather the model cannot explain the clear variation in intercept and slope (Figure 10.5). Alternatively, we can use the no pooling model, that is, fitting separate models for the nine regions:

```
#### R Output ####
lm(formula = richtol ~ nuii * factor(city) - 1 - nuii,
             data = rtol2)
                         coef.est coef.se
factor(city)ATL           5.3318   0.1355
factor(city)BIR           5.1228   0.1544
factor(city)BOS           4.2486   0.1392
factor(city)DEN           6.1978   0.1499
factor(city)DFW           7.0704   0.1167
factor(city)MGB           6.0501   0.1227
factor(city)POR           4.5529   0.1305
factor(city)RAL           5.5340   0.1543
```

```
factor(city)SLC        4.5080    0.1936
nuii:factor(city)ATL   0.0301    0.0056
nuii:factor(city)BIR   0.0269    0.0053
nuii:factor(city)BOS   0.0455    0.0062
nuii:factor(city)DEN   0.0025    0.0034
nuii:factor(city)DFW  -0.0019    0.0033
nuii:factor(city)MGB   0.0078    0.0033
nuii:factor(city)POR   0.0233    0.0035
nuii:factor(city)RAL   0.0250    0.0044
nuii:factor(city)SLC   0.0248    0.0037
---
n = 261, k = 18
residual sd = 0.4744, R-Squared = 0.99
```

The no-pooling estimated model coefficients have larger estimation variances compared to the complete pooling ones (Figure 10.6). It is easy to understand the variation in the intercept because macroinvertebrates are also influenced by features such as temperature and pH, which vary from region to region over a larger spatial scale (beyond a region). The variation in slope is somewhat puzzling because we believe that urbanization will inevitably bring a disturbance to the watershed and cause changes in water quality, thereby inducing changes in the macroinvertebrate community. The complete pooling results are expected as the estimated intercept and slope are at the center of the respective no pooling estimates. If the complete pooling model is used, the model is of little predictive power. The high level of variability in the no pooling estimated model coefficients suggests that other factors not included in the model are more important in predicting the response variable value.

The multilevel model is a compromise between the no pooling and the complete pooling methods. The no pooling model is represented by equation 10.13, and the complete pooling model is of the form:

$$TOLr_i = \beta_0 + \beta_1 nuii_i + \epsilon_i$$

The partial pooling model can be expressed as:

$$\begin{array}{rcl} TOLr_{ij} & \sim & N(\mu_{ij}, \sigma^2) \\ \mu_{ij} & = & \beta_{0j} + \beta_{1j} nuii_{ij} + \epsilon_{ij} \\ \begin{pmatrix} \beta_{0j} \\ \beta_{1j} \end{pmatrix} & \sim & MVN\left[\begin{pmatrix} \beta_0 \\ \beta_1 \end{pmatrix}, \begin{pmatrix} \sigma^2_{\beta_0} & \rho\sigma_{\beta_0}\sigma_{\beta_1} \\ \rho\sigma_{\beta_0}\sigma_{\beta_1} & \sigma^2_{\beta_1} \end{pmatrix} \right] \end{array} \quad (10.14)$$

That is, the partial pooling model recognizes the difference among the regions by modeling the intercept and slope as region-specific. However, at this moment we have no knowledge as of what may influence the model coefficients. As a result, assuming all intercepts and slopes are from a common prior distribution (i.e., assuming exchangeability on model coefficients) is reasonable. This formal specification can be expressed informally as

$$y_{ij} = (\beta_0 + \delta_{0j}) + (\beta_1 + \delta_{1j})x_{ij} + \epsilon_{ij}$$

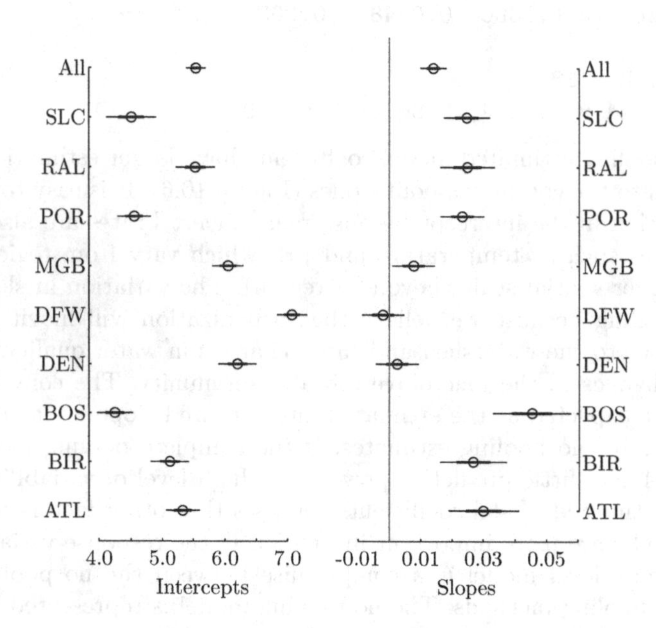

FIGURE 10.6: EUSE example linear model coefficients – Estimated linear regression model coefficients using complete pooling (labeled as "`All`") and no pooling (labeled by region names) are compared. The open circles are the estimated values and the thin and thick line segments are the mean plus/minus 2 and 1 times estimated standard error, respectively.

which translates to R formula as

```
y ~ x + (1+x|group)
```

For the EUSE data:

```
#### R Code ####
euse.lmer1 <- lmer(richtol ~ nuii+(1+nuii|city),
                   data=rtol2)
```

The fitted partial pooling model is stored in R object `euse.lmer1`. The `summary` function reports some basic information:

```
#### R Output ####
summary(euse.lmer1)

Linear mixed model fit by REML
Formula: richtol ~ nuii + (1 + nuii | city)
   Data: rtol2
 AIC BIC logLik deviance REMLdev
 424 445   -206      401     412
Random effects:
 Groups   Name        Variance Std.Dev. Corr
 city     (Intercept) 0.817228 0.9040
          nuii        0.000188 0.0137   -0.893
 Residual             0.225311 0.4747
Number of obs: 261, groups: city, 9

Fixed effects:
            Estimate Std. Error t value
(Intercept)  5.41839    0.30516   17.76
nuii         0.01943    0.00479    4.06

Correlation of Fixed Effects:
     (Intr)
nuii -0.877
```

The output summary includes all the necessary parameter values to specify the multivariate normal distribution. The estimated means $(\hat{\beta}_0, \hat{\beta}_1)$ are the "fixed" effects (5.418 and 0.0194, respectively). The variance-covariance matrix is defined by the variances of β_0, β_1 ($\sigma^2_{\beta_0} = 0.817$ and $\sigma^2_{\beta_1} = 0.000188$, respectively) and their correlation coefficient ρ (-0.893). The residual variance of 0.2253 is the estimated σ^2. In theory, these six coefficients are sufficient to form the model for predictive purposes. The model output also includes fitted regression coefficients – the estimated intercept and slope for each group ($\beta_0+\delta_{0j}$ and $\beta_1+\delta_{1j}$). The estimated group-specific intercept and slope are in two parts, the coefficients that are common to all regions (or "fixed effects,"

$\hat{\beta}_0, \hat{\beta}_1$) and the region-specific coefficients (or "random effects," $\hat{\delta}_{0j}, \hat{\delta}_{1j}$). The fixed effects are extracted by the function `fixef()`:

```
#### R Output ####
fixef(euse.lmer1)

(Intercept)         nuii
   5.418387     0.019431
```

The standard errors of the estimated fixed effects are:

```
#### R Output ####
se.fixef(euse.lmer1)

(Intercept)          nuii
  0.3051636     0.0047898
```

Information about the random effects are extracted by the functions `ranef()` and `se.ranef()`:

```
#### R Output ####
> ranef(euse.lmer1)
$city
    (Intercept)         nuii
ATL    -0.030748  0.0067451
BIR    -0.266135  0.0060253
BOS    -1.079855  0.0206173
DEN     0.744879 -0.0156565
DFW     1.626480 -0.0210608
MGB     0.614451 -0.0109547
POR    -0.864952  0.0049036
RAL     0.147261  0.0038998
SLC    -0.891381  0.0054811

> se.ranef(euse.lmer1)
$city
         [,1]      [,2]
  [1,] 0.12689 0.0046299
  [2,] 0.14515 0.0045649
  [3,] 0.12779 0.0049203
  [4,] 0.14732 0.0032192
  [5,] 0.11545 0.0031238
  [6,] 0.12096 0.0031427
  [7,] 0.12838 0.0032621
  [8,] 0.14854 0.0039813
  [9,] 0.18847 0.0035250
```

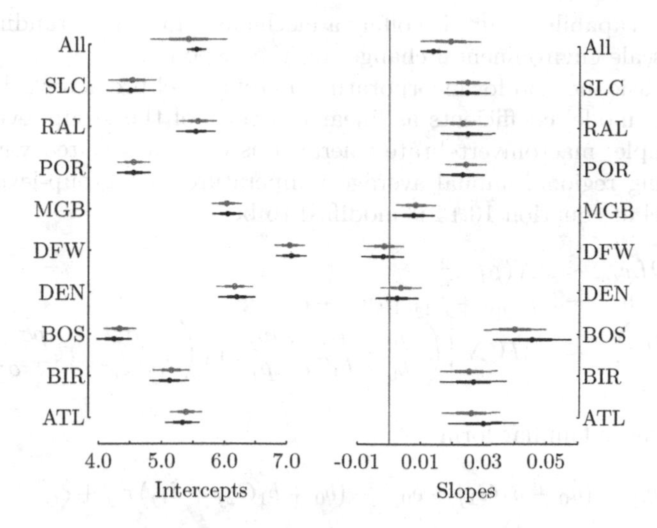

FIGURE 10.7: Comparison of linear and multilevel regression – Intercepts and slopes estimated using no pooling (black dots [mean] and line segments [plus/minus one standard error]) are compared with the same estimated using multilevel (partial pooling) model (gray dots [mean] and line segments [plus/minus one standard error]).

What is the gain of fitting a multilevel model in this example? The one not-so-obvious advantage is the estimated correlation between intercept and slope. When used for prediction, we can use this information to generate random samples of pairs of intercept and slope, which will reduce the predictive uncertainty, compared to the complete pooling model. Compared to the no pooling model in terms of the estimated region-specific intercepts and slopes (Figure 10.7), partial pooling estimated model coefficients are not very different.

This example is typically considered as not worthwhile for multilevel modeling because of the large differences in model coefficients and roughly even sample sizes among regions. As a result, the amount of pulling toward the overall mean is small for all groups. The advantage of a multilevel regression in this case can be realized when group (region) level predictors are available. Group level predictors can be physical characteristics of a region, representing processes in a larger spatial or temporal scale. For example, soil carbon content in the N_2O emission example is a group level predictor with limited within-group variation. Such group level predictors are often difficult to include in a modeling study because of the limited variability within a given group. Under a multilevel modeling setting, a group level predictor can be used to describe the changes of model coefficients (intercept or slope, or both) as a function of the group level predictor. The resulting model not only improves the model's predictive capability, but also offers a mechanism for understanding the effect of large scale environmental changes on the response.

The basic method for incorporating a group level predictor is to model the regression model coefficients as linear functions of the group level predictor. For example, macroinvertebrate tolerance is often associated with temperature. Using regional annual average temperature as a group-level predictor, the model of equation 10.14 is modified to be:

$$\begin{array}{rcl} TOLr_i & \sim & N(\mu_i, \sigma_i^2) \\ \mu_i & = & \beta_{0j[i]} + \beta_{1j[i]} nuii + \epsilon_i \\ \left(\begin{array}{c} \beta_{0j} \\ \beta_{1j} \end{array}\right) & \sim & MVN\left[\left(\begin{array}{c} a_0 + a_1 Temp_j \\ b_0 + b_1 Temp_j \end{array}\right), \left(\begin{array}{cc} \sigma^2_{\beta_0} & \rho\sigma_{\beta_0}\sigma_{\beta_1} \\ \rho\sigma_{\beta_0}\sigma_{\beta_1} & \sigma^2_{\beta_1} \end{array}\right)\right] \end{array} \tag{10.15}$$

Or, in a more familiar form

$$y_{ij} = (a_0 + a_1 G_{1j} + \delta_{0j}) + (b_0 + b_1 G_{2j} + \delta_{1j})x_{ij} + \epsilon_{ij} \tag{10.16}$$

Rearranging the terms:

$$y_{ij} = (a_0 + \delta_{0j}) + (b_0 + \delta_{1j})x_{ij} + a_1 G_{1j} + b_1 G_{2j} x_{ij} + \epsilon_{ij}$$

which is the model in equation 10.14 plus two additional terms associated with the group level predictor(s). It is often convenient to use the same group level predictor in both the intercept and slope term. However, the ecological meaning of the intercept and slope terms is often different and allowing different group level predictors for these coefficients is often scientifically necessary or more reasonable.

The joint distribution of model coefficients in equation 10.15 can also be expressed as:

$$\left(\begin{array}{c} \beta_{0j} \\ \beta_{1j} \end{array}\right) = \left(\begin{array}{c} a_0 + a_1 Temp_j \\ b_0 + b_1 Temp_j \end{array}\right) + \left(\begin{array}{c} \delta_{0j} \\ \delta_{1j} \end{array}\right) \tag{10.17}$$

where

$$\left(\begin{array}{c} \delta_{0j} \\ \delta_{1j} \end{array}\right) \sim MVN\left[\left(\begin{array}{c} 0 \\ 0 \end{array}\right), \left(\begin{array}{cc} \sigma^2_{\beta_0} & \rho\sigma_{\beta_0}\sigma_{\beta_1} \\ \rho\sigma_{\beta_0}\sigma_{\beta_1} & \sigma^2_{\beta_1} \end{array}\right)\right]$$

To implement this model in R, a new group level predictor variable of the same length as the response variable is needed. In the EUSE example, we have a vector of annual average temperature (in °C) for the nine regions:

```
#### R Output ####
> AveTemp
  ATL   BIR   BOS   DEN   DFW   MGB   POR   RAL   SLC
16.27 16.00  8.71  9.19 18.30  7.63 10.81 14.93  9.73
```

Since the vector `AveTemp` is sorted alphabetically, we can use the following code to create a group level predictor object:

```
> site <- as.numeric(ordered(rtol2$city))
> temp.full <- AveTemp[site]
```

The R model formula for the multilevel model with group level predictor is then:

```
y ~ x + G1 + G2:x + (1+x|group)
```

In the EUSE example, we use annual average temperature as the sole group level predictor first, and the model is fit using the following script:

```
euse.lmer2 <- lmer(richtol ~ nuii+temp.full+nuii:temp.full+
                   (1+nuii|city), data=rtol2)
```

When a group level predictor is used, the regression model coefficients (slopes and intercepts) are no longer exchangeable because we now assume that the joint distribution of β_{0j} and β_{1j} are different for each region. However, if we now look at the model as expressed in equation 10.16, the means of model intercepts and slopes are region-specific, but the error terms δ_{0j}, δ_{1j} are exchangeable – they are from a bivariate normal distribution with mean of 0 and the variance-covariance matrix expressed in equation 10.15.

The use of a group level predictor may or may not improve the model's fit. To explore the model fit, we look at both the summary statistics and plots.

```
#### R Output ####
> summary(euse.lmer2)
Linear mixed model fit by REML
Formula: richtol ~ nuii + temp.full + nuii:temp.full +
                   (1 + nuii | city)
   Data: rtol2
 AIC BIC logLik deviance REMLdev
 440 469   -212      396     424
Random effects:
 Groups   Name        Variance Std.Dev. Corr
 city     (Intercept) 0.789371 0.8885
          nuii        0.000207 0.0144   -0.932
 Residual             0.225589 0.4750
Number of obs: 261, groups: city, 9

Fixed effects:
                Estimate Std. Error t value
(Intercept)     4.319222   1.039736    4.15
nuii            0.023160   0.017280    1.34
temp.full       0.088587   0.080268    1.10
nuii:temp.full -0.000298   0.001338   -0.22

Correlation of Fixed Effects:
```

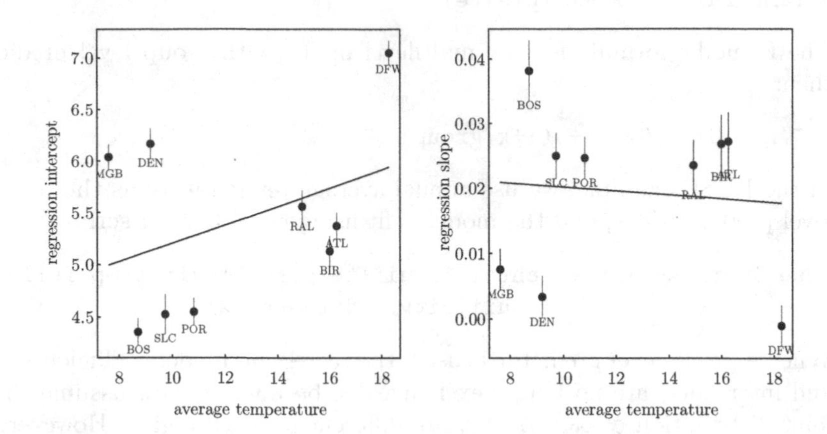

FIGURE 10.8: Multilevel model with a group level predictor – The regional annual average temperature (°C) is used as a group level predictor to describe the variation in the fitted region-specific intercept (left panel) and slope (right panel).

```
              (Intr) nuii   tmp.fl
nuii          -0.918
temp.full     -0.957  0.879
nui:tmp.fll    0.877 -0.957 -0.916
```

The residual variance of this model is 0.225589. Compared to the model without using the group level predictor which has a residual variance of 0.225311, we would suggest that the group level predictor did not improve the model's fit. Furthermore, the estimated slopes of `temp.full` and the interaction term `nuii:temp.full` are not statistically different from 0. From this angle, we would suggest that regional annual average temperature is not a good group level predictor. This impression is further strengthened by the graphical display of the fitted group level model (Figure 10.8).

However, further examination of the plots suggests that the nine regions should be separated into two groups – MGB, DFW, and DEN versus the rest. These three regions are known to have high levels of agriculture activities in their watersheds. To reflect this difference, we extracted background agriculture land-use (in percentage of the total watershed area) using subbasins with minimum urban development. This variable represents the preurbanization agriculture land-use as a fraction of the total watershed, or antecedent agriculture land cover. If we use the antecedent agriculture land cover as a group level predictor, the model fit to the data is not much better measured by the model's residual standard variance:

```
#### R Code and Output ####
> euse.lmer3 <- lmer(richtol~nuii+ag.full+nuii:ag.full+
                            (1+nuii|site), data=rtol2)
> summary(euse.lmer3)
Linear mixed model fit by REML
Formula: richtol ~ nuii + ag.full + nuii:ag.full +
         (1 + nuii | site)
   Data: rtol2
 AIC BIC logLik deviance REMLdev
 421 449   -202      384     405
Random effects:
 Groups   Name        Variance Std.Dev. Corr
 site     (Intercept) 1.84e-01 0.42944
          nuii        2.74e-05 0.00524  -0.343
 Residual             2.25e-01 0.47462
Number of obs: 261, groups: site, 9

Fixed effects:
             Estimate Std. Error t value
(Intercept)   4.45845    0.24164   18.45
nuii          0.03412    0.00366    9.31
ag.full       2.52938    0.49390    5.12
nuii:ag.full -0.03884    0.00707   -5.50

Correlation of Fixed Effects:
            (Intr) nuii   ag.fll
nuii        -0.419
ag.full     -0.781  0.319
nuii:ag.fll  0.337 -0.792 -0.406
```

However, the estimate regression coefficients for the term `ag.full` and `nuii:ag.full` are statistically different from 0. The graphical display of the group-level model (Figure 10.9) shows that the antecedent agriculture land-use is a group level predictor with two clusters – MGB, DFW, and DEN have antecedent agriculture land-use more than 70% of their respective watershed area, while the rest of the six regions have less than 30%.

10.4.1 Nonnested Groups

Comparing Figures 10.8 and 10.9, it seems that the antecedent agriculture land-use should be used as a factor predictor to divide the nine regions into two groups. Together with the region, the low and high antecedent agriculture land-uses form two nonnested groups. With both region and antecedent agriculture land-use groups, we can check whether regional annual temperature is still a viable group level predictor. A nonnested model can consist of simple additive group effects, in which the binary antecedent agriculture land-use

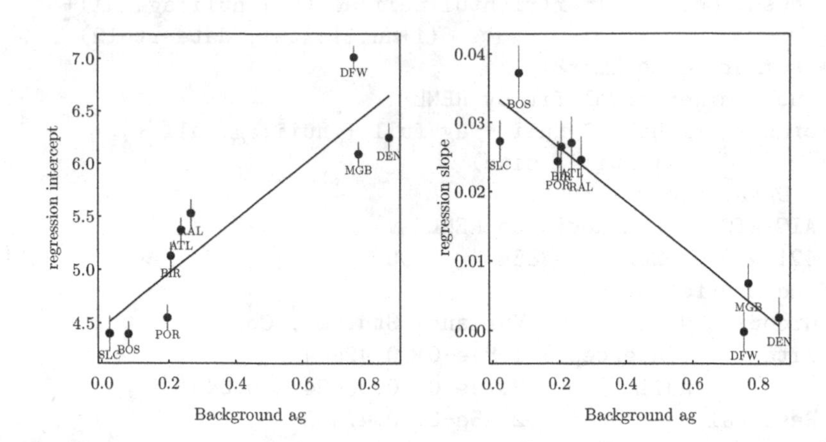

FIGURE 10.9: Antecedent agriculture land-use as a group level predictor – The regional antecedent agriculture land-use as a percentage of the total watershed area is used as a group level predictor to describe the variation in the fitted region-specific intercept (left panel) and slope (right panel).

Ag_j can be simply added as a group level predictor:

$$\begin{array}{rcl} TOLr_{ijk} & \sim & N(\mu_{ijk}, \sigma^2) \\ \mu_{ijk} & = & \beta_{0jk} + \beta_{1jk} nuii_{ijk} \\ \begin{pmatrix} \beta_{0jk} \\ \beta_{1jk} \end{pmatrix} & = & \begin{pmatrix} a_0 + a_1 Temp_j \\ b_0 + b_1 Temp_j \end{pmatrix} + \begin{pmatrix} \delta_a^{Ag_k} \\ \delta_b^{Ag_k} \end{pmatrix} + \begin{pmatrix} \delta_a^{Region_j} \\ \delta_b^{Region_j} \end{pmatrix} \end{array} \tag{10.18}$$

where

$$\begin{pmatrix} \delta_a^{Ag_k} \\ \delta_b^{Ag_k} \end{pmatrix} \sim MVN\left[\begin{pmatrix} 0 \\ 0 \end{pmatrix}, \Sigma_k\right]$$

is the random effect term for antecedent agriculture land-use group, and

$$\begin{pmatrix} \delta_a^{Region_j} \\ \delta_b^{Region_j} \end{pmatrix} \sim MVN\left[\begin{pmatrix} 0 \\ 0 \end{pmatrix}, \Sigma_j\right]$$

is the same for the region group.

Equation 10.18 implies that the relationship between group-level intercepts (slopes) and regional annual average temperature for high and low antecedent agriculture land-use groups is two parallel lines. Fitting this model in R is straightforward:

```
#### R Code ####
ag.full <- as.vector(ag[site])
```

```
ag.cat <- ag.full>0.5

euse.lmer3 <- lmer(richtol ~ nuii+temp.full+nuii:temp.full+
                   (1+nuii|city)+(1+nuii|ag.cat),
                   data=rtol2)
```

The fitted model coefficients are grouped into "fixed effects" (common for all group levels) and "random effects" (group-specific).

```
> round(fixef(euse.lmer3), 4)
   (Intercept)      nuii   temp.full nuii:temp.full
        4.2663    0.0224      0.1143        -0.0006
```

Using the notation of equation 10.18, $\hat{a}_0 = 4.2663$, $\hat{a}_1 = 0.1143$, $\hat{b}_0 = 0.0224$, and $\hat{b}_1 = -0.0006$. The estimated random effects are

```
#### R output ####
> ranef(euse.lmer3)
$site
  (Intercept)        nuii
1   0.070602  0.00174688
2  -0.053474 -0.00132310
3   0.073577  0.00182050
4  -0.010953 -0.00027101
5  -0.065197 -0.00161315
6   0.079344  0.00196320
7  -0.141378 -0.00349809
8   0.157664  0.00390104
9  -0.110185 -0.00272627

$ag.cat
      (Intercept)      nuii
FALSE    -0.82269  0.012530
TRUE      0.82269 -0.012530
```

The difference in intercept between the high and low group is $0.82268 \times 2 = 1.6454$, and the difference in slope between the two groups is 0.02506. Figure 10.10 shows the fitted group level models. Because there are only two levels in the background agriculture group, the estimated between-group variance is less reliable. With only two levels, the model in equation 10.18 can be modified to be:

$$
\begin{array}{rcl}
TOLr_i & \sim & N(\mu_i, \sigma_i^2) \\
\mu_i & = & \beta_{0j[i]} + \beta_{1j[i]} nuii + \epsilon_i \\
\begin{pmatrix} \beta_{0j} \\ \beta_{1j} \end{pmatrix} & \sim & MVN\left[\begin{pmatrix} a_0 + a_1 Temp_j + a_2 Ag_j \\ b_0 + b_1 Temp_j + b_2 Ag_j \end{pmatrix}, \begin{pmatrix} \sigma^2_{\beta_0} & \rho\sigma_{\beta_0}\sigma_{\beta_1} \\ \rho\sigma_{\beta_0}\sigma_{\beta_1} & \sigma^2_{\beta_1} \end{pmatrix}\right]
\end{array}
\tag{10.19}
$$

where $Ag_j = 0$ for low antecedent agriculture group and $Ag_j = 1$ for the high

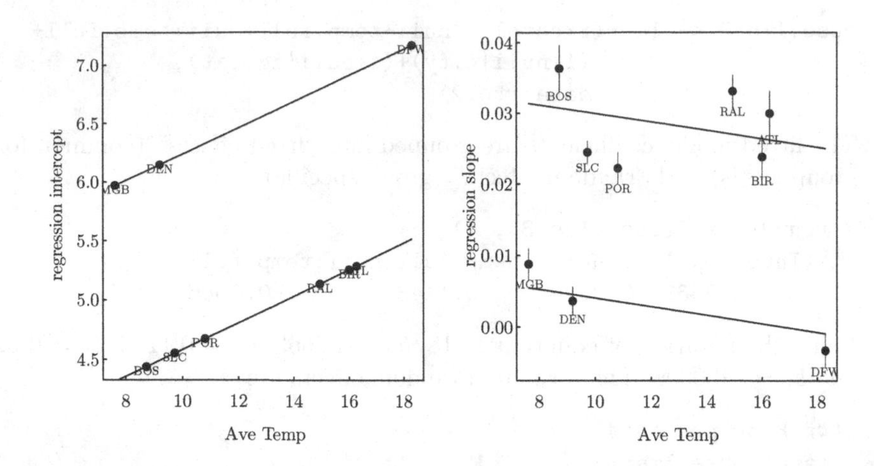

FIGURE 10.10: Antecedent agriculture land-use and temperature as group-level predictors – The antecedent agriculture land-use is used as a categorical predictor. The antecedent agriculture land-use and region form two nonnested groups.

antecedent agriculture land-use group. The coefficients a_2, b_2, representing the antecedent agriculture land-use effect, are now common to all regions. In R, the model of equation 10.19 is fitted by adding *Ag* and *Ag*:*nuii* interaction term:

```
#### R Code ####
euse.lmer4 <- lmer(richtol ~ nuii+temp.full+nuii:temp.full+
                             as.numeric(ag.cat)+
                             as.numeric(ag.cat):nuii +
                             (1+nuii|site), data=rtol2)
```

The estimated $\hat{a}_2 = 1.6555, \hat{b}_2 = -0.025334$ are very close to the effects estimated using the model form in equation 10.18.

This fitting method is more general than equation 10.19, which is limited to a binary group (Ag).

The nonnested model can also include an interaction term to avoid the additive assumption. The interaction can be directly added to equation 10.19:

$$
\begin{array}{rcl}
TOLr_i & \sim & N(\mu_i, \sigma_i^2) \\
\mu_i & = & \beta_{0j[i]} + \beta_{1j[i]} nuii + \epsilon_i \\
\left(\begin{array}{c} \beta_{0j} \\ \beta_{1j} \end{array}\right) & \sim & MVN\left[\left(\begin{array}{c} a_0 + a_1 Temp_j + a_2 Ag_j + a_3 Ag_j Temp_j \\ b_0 + b_1 Temp_j + b_2 Ag_j + b_3 Ag_j Temp_j \end{array}\right),\right. \\
 & & \left.\left(\begin{array}{cc} \sigma^2_{\beta_0} & \rho\sigma_{\beta_0}\sigma_{\beta_1} \\ \rho\sigma_{\beta_0}\sigma_{\beta_1} & \sigma^2_{\beta_1} \end{array}\right)\right]
\end{array}
\tag{10.20}
$$

The coefficient a_3 is the difference in slopes of the two lines in the left panel of Figure 10.11, and b_3 is the difference in slopes of the lines in the right panel of Figure 10.11.

```
#### R Code ####
> euse.lmer5 <-
+     lmer(richtol ~ nuii+temp.full+nuii:temp.full+
+          as.numeric(ag.cat)+ as.numeric(ag.cat):nuii +
+          as.numeric(ag.cat):temp.full +
+          as.numeric(ag.cat):temp.full:nuii +
+          (1+nuii|site), data=rtol2)
```

The estimated differences in slopes are not significantly different from 0 (Figure 10.11):

```
#### R Output ####
Fixed effects:
                                  Estimate Std. Error t value
(Intercept)                       3.102764   0.302048   10.27
nuii                              0.036322   0.011794    3.08
temp.full                         0.140127   0.022961    6.10
as.numeric(ag.cat)                2.210949   0.386754    5.72
nuii:temp.full                   -0.000657   0.000915   -0.72
nuii:as.numeric(ag.cat)          -0.024566   0.014796   -1.66
temp.full:as.numeric(ag.cat)     -0.044174   0.029523   -1.50
nuii:temp.full:as.numeric(ag.cat) -0.000109  0.001157   -0.09
```

Alternatively, we use equation 10.18 and replace the antecedent agriculture land-use group random effect term with

$$
\left(\begin{array}{c} \delta_a^{Ag_k} \\ \delta_b^{Ag_k} \end{array}\right) \sim MVN\left[\left(\begin{array}{c} \delta_{a0} + \delta_{a1} Temp_j \\ \delta_{b0} + \delta_{b1} Temp_j \end{array}\right), \Sigma_k\right]
$$

and fit in R as follows:

```
#### R Code ####
euse.lmer6 <- lmer(richtol ~ nuii+temp.full +
                   nuii:temp.full+
                  (1+nuii|site)+
```

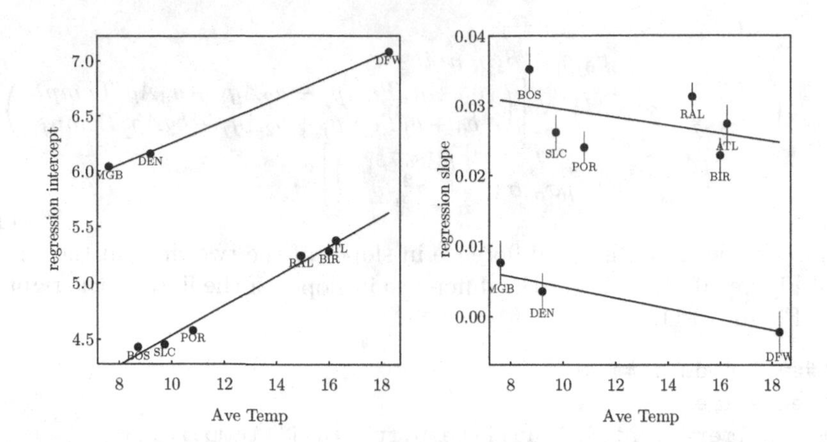

FIGURE 10.11: Antecedent agriculture land-use and temperature interaction – The slopes of the two lines in both panels are only slightly different. The interaction between regional annual average temperature and antecedent agriculture land-use is not obvious.

```
                        (1+nuii+temp.full+nuii:temp.full|ag.cat),
                         data=rtol2)
```

The difference in estimated coefficients is small, except the difference in slopes of the lines in the right panel in Figure 10.11.

```
#### R Output ####
> ranef(euse.lmer6)
$site
...
...
$ag.cat
      (Intercept)       nuii temp.full nuii:temp.full
FALSE     -1.0848  0.011253  0.020961     0.00011844
TRUE       1.0848 -0.011253 -0.020961    -0.00011844
```

These alternative means of fitting the same model resulted in calling the same differences random effects some times and fixed effects some other times, which reinforces the argument for disregarding the differences between the two. The important thing is to know when to use which number in reporting a model's output.

10.4.2 Multiple Regression Problems

The EUSE example has one continuous data level predictor (nuii). In the Finnish lakes example (Section 5.3.7), both total phosphorus and total nitrogen are used as data level predictors. Although the model fitting process is the same, graphical presentations of model results are more complicated. Malve and Qian [2006] compared, in our terminology, the no pooling models and the partial pooling models. One important question from a lake manager is whether phosphorus or nitrogen (or both) is (or are) the limiting nutrient. If a lake is limited by phosphorus, reducing phosphorus input to the lake is a cost-effective means of controlling lake eutrophication, and vice versa. Empirical evidence and limnology theory suggest that inland freshwater lakes are most likely limited by phosphorus. As a result, many lake eutrophication models include only phosphorus as the driving force of eutrophication. Studies have shown that under certain conditions, nitrogen can be the limiting nutrient; including nitrogen in lake eutrophication models often leads to a better model. However, nitrogen and phosphorus concentrations are usually highly correlated. Including both in a multiple regression model will lead to large estimation uncertainty in model coefficients and ambiguous model interpretation. In the Finnish lakes example, the Finnish government conducted studies to classify lakes in Finland into nine types, based on expert assessments on lake morphological and chemical metrics such as depth, surface area, and color. Lakes of the same type are likely to behave similarly. As a result, these lakes are often pooled to increase the sample size for more robust statistical inference. In Section 5.3.7, we discussed the problem of collinearity and the scientific problem of identifying the limiting nutrient. We concluded that when a lake is limited by both nitrogen and phosphorus, the interaction term of the model is often statistically different from 0. When one of the two nutrients is limiting, the interaction effect is usually not different from 0. Identification of the limiting nutrient is largely based on conditional plots or plots of the fitted model (e.g., Figure 5.15 on page 180). In this section, we use data from all nine types to fit type-specific models using a multilevel approach. We use the model in equation 5.11 in a multilevel setting:

$$\log(Chla_{ij}) = \beta_{0j} + \beta_{1j}\log(TP) + \beta_{2j}\log(TN) + \beta_{3j}\log(TN)\log(TP) + \epsilon_{ij} \tag{10.21}$$

where the regression coefficients $(\beta_{0j}, \beta_{1j}, \beta_{2j}, \beta_{3j})$ are for the jth lake type. The model fitting process is straightforward:

```
#### R Code ####
Finn.M3 <- lmer(y ~ lxp+lxn+lxp:lxn+(1+lxp+lxn+lxp:lxn|type),
                data=summer.All)
```

where y is the log chlorophyll a concentration, and lxp and lxn are the standardized log total phosphorus and log total nitrogen, respectively. The multilevel model assumes the four regression coefficients for all lake types are from a common á priori multivariate normal distribution. The `summary` function returns the basic information necessary for generating predictions:

```
#### R Output ####
> summary(Finn.M3)
Linear mixed model fit by REML
Formula: y ~ lxp + lxn + lxp:lxn +
        (1 + lxp + lxn + lxp:lxn | type)
   Data: summer.All
   AIC   BIC logLik deviance REMLdev
 29374 29492 -14672    29325   29344
Random effects:
 Groups Name        Variance Std.Dev. Corr
 type   (Intercept) 0.0139   0.118
        lxp         0.0177   0.133    -0.694
        lxn         0.0631   0.251     0.534 -0.828
        lxp:lxn     0.0326   0.181    -0.831  0.451 -0.511
Residual            0.2635   0.513
Number of obs: 19427, groups: type, 9

Fixed effects:
            Estimate Std. Error t value
(Intercept)   2.2305     0.0400    55.8
lxp           0.7641     0.0459    16.7
lxn           0.7082     0.0863     8.2
lxp:lxn      -0.0129     0.0617    -0.2

Correlation of Fixed Effects:
        (Intr) lxp    lxn
lxp     -0.666
lxn      0.517 -0.818
lxp:lxn -0.811  0.424 -0.487
```

The "fixed effects" section provides the estimated mean regression coefficients, and the "random effects" section shows the variance-covariance matrix. The residual standard deviation is the estimated σ. What we are interested in is whether phosphorus or nitrogen (or both) is (are) limiting the growth of phytoplankton. Our discussion in Section 5.3.7 suggested that inference on the limiting nutrient can be made by comparing regression model coefficients, especially the interaction effect. Because the initial assumption in this example is that lakes of the same type are similar, the resulting lake-type specific models are to be seen as a reference. Lake-specific models should be used in managing a specific lake.

Figure 10.12 shows the estimated type-level model coefficients. These estimates are based on the standardized log TP and log TN. As a result, the intercept (β_0) is the average log chlorophyll a concentration when TP and TN

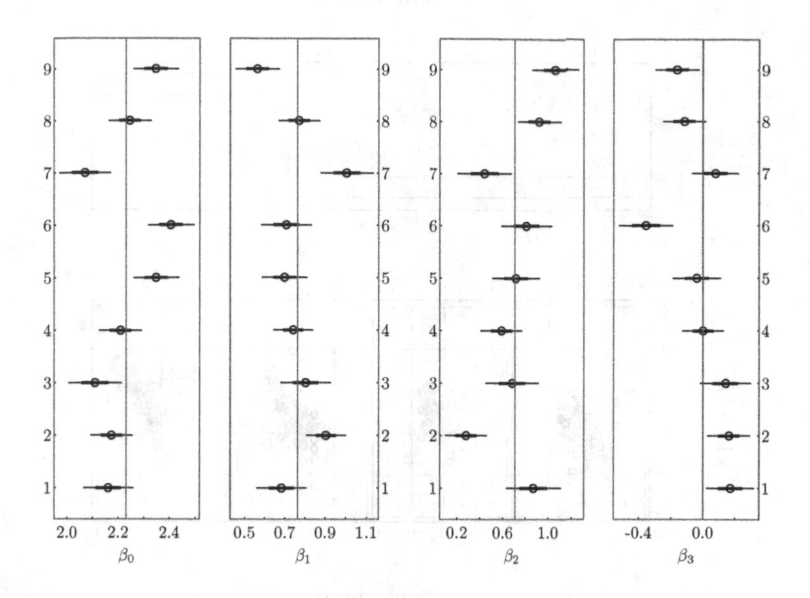

FIGURE 10.12: Lake type-level multilevel model coefficients – Multilevel model coefficients estimated for each lake type are represented by the dots. The thin and thick line segments are the mean ± one and two standard errors.

TABLE 10.1: Finnish lake type definition – Geomorphological typology of Finnish lakes specified by Finnish Environment Institute (SA=Surface Area, D=Depth)

Lake Type	*Name*	*Characteristics*
I	Large, nonhumic lakes	SA > 4,000 Ha, color < 30
II	Large, humic lakes	SA > 4,000 Ha, color > 30
III	Medium and small, nonhumic lakes	SA: 50 - 4,000 Ha, color < 30
IV	Medium area, humic deep lakes	SA: 500 - 4,000 Ha, color: 30-90, D > 3 m
V	Small, humic, deep lakes	SA: 50 - 500 Ha, color: 30-90, D > 3 m
VI	Deep, very humic lakes	Color > 90, D > 3 m
VII	Shallow, nonhumic lakes	Color < 30, D < 3
VIII	Shallow, humic lakes	Color: 30-90, D < 3 m
IX	Shallow, very humic lakes	Color > 90, D < 3 m

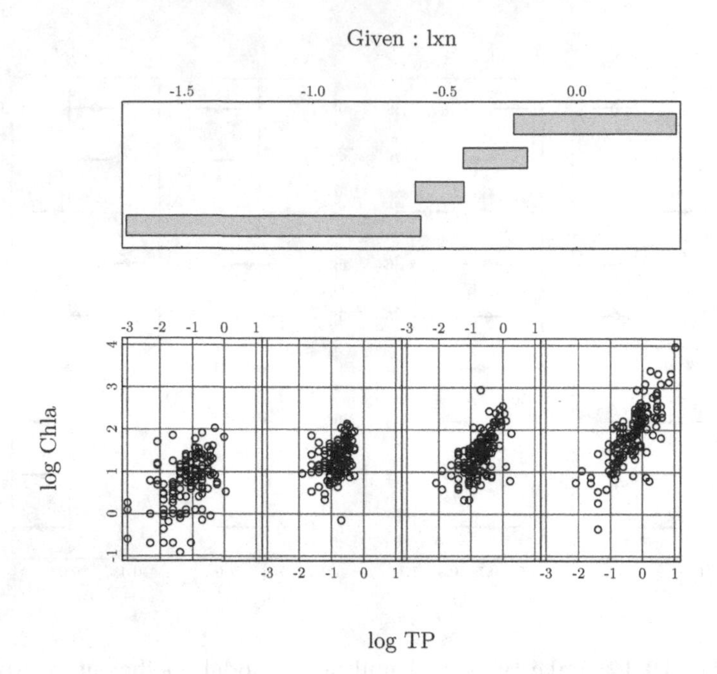

FIGURE 10.13: Conditional plots of oligotrophic lakes (TP) – Scatter plots of log chlorophyll a against centered log TP concentration (conditional on TN) show increased response to phosphorus as nitrogen level increases (from left to right). Data are from large nonhumic (type 1), possibly oligotrophic, lakes.

are at their overall geometric means (calculated using data from all lakes). The intercept can be a measure of the average lake primary productivity. The slopes (β_1 and β_2) are the % changes in chlorophyll a for every one percent increase in TP and TN, respectively (see Section 5.4).

Comparing the lake-type specific intercepts and lake-type definition shown in Table 10.1, it seems that lake-type average chlorophyll a concentrations are related to the average humic level of the lakes. The higher the humic level, the higher the type average chlorophyll a tend to be when the TP and TN concentrations are the same. The signs of the interaction term for lake types 1 and 2 (large lakes) are positive, suggesting that both nitrogen and phosphorus are likely limiting the growth of phytoplankton. For type 1 (large, nonhumic) lakes, slopes of TP and TN (β_1, β_2) are comparable, and conditional plots (Figures 10.13 and 10.14) also show a typical *colimiting* pattern – when one nutrient is low the effect of the other is weak, and vice versa.

The average chlorophyll a level for type 1 lakes is low. When both nitrogen and phosphorus are limiting, the overall nutrient level of the lake is usually very low (oligotrophic). The opposite of oligotrophic is eutrophic, where a lake's overall nutrient level is high. Type 6 lakes are examples of eutrophic

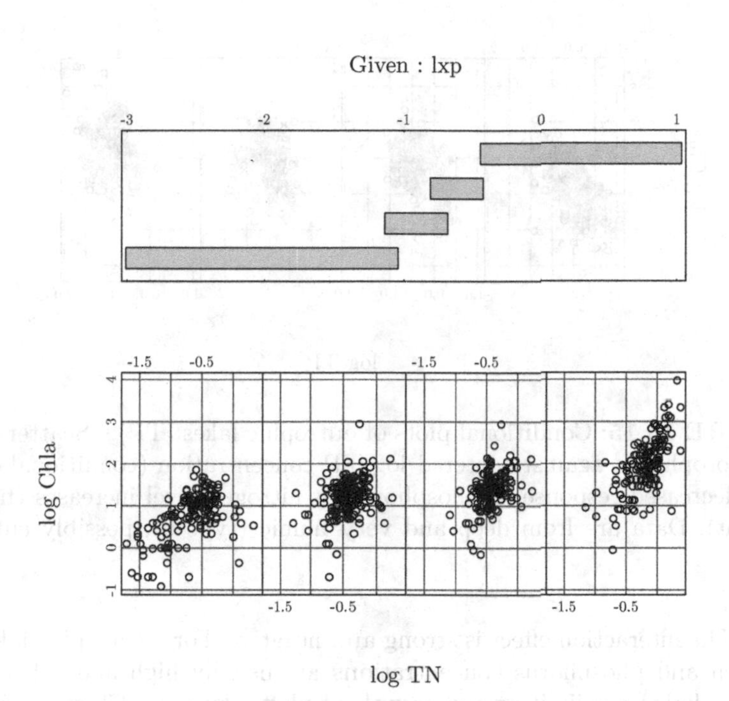

FIGURE 10.14: Conditional plots of oligotrophic lakes (TN) – Scatter plots of log chlorophyll a against centered log TN concentration (conditional on TP) show increased response to nitrogen as phosphorus level increases (from left to right). Data are from large nonhumic (type 1) lakes.

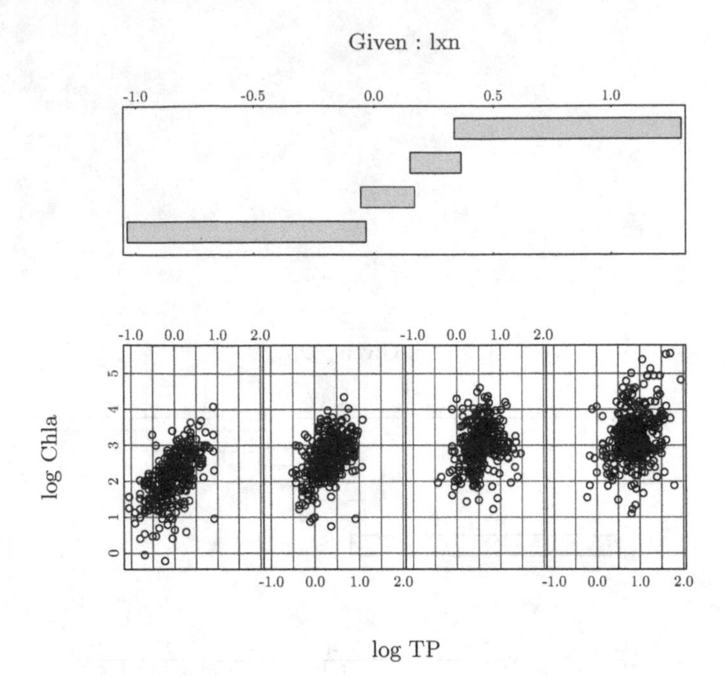

FIGURE 10.15: Conditional plots of eutrophic lakes (TP) – Scatter plots of log chlorophyll a against centered log TP concentration (conditional on TN) show decreased response to phosphorus as nitrogen level increases (from left to right). Data are from deep and very humic (type 6), possibly eutrophic, lakes.

lakes. The interaction effect is strong and negative. For a eutrophic lake, both nitrogen and phosphorus concentrations are usually high and other factors (such as light) are limiting the growth of phytoplankton. Changes in one or both nutrient concentrations will not change the level of chlorophyll *a* by much. Only when nutrient concentrations drop below a certain level will the growth of phytoplankton respond to the changes in nutrient concentration. Figures 10.15 and 10.16 show typical conditional plots of eutrophic lakes.

Shallow nonhumic lakes (type 7) are also oligotrophic. These lakes seem to be limited only by phosphorus, indicated by a weak interaction effect, a large $\hat{\beta}_1$ and a small $\hat{\beta}_2$. Conditional plots for this group of lakes show typical phosphorus limiting patterns (Figure 10.17 and 10.18).

Large humic (type 2) lakes are somewhat complicated. Although the low $\hat{\beta}_0$ and positive $\hat{\beta}_3$ suggest oligotrophic lakes, the large difference between $\hat{\beta}_1$ and $\hat{\beta}_2$ seems to suggest that only phosphorus is limiting. The lake examined in Section 5.3.7 belongs to this type, and it is likely to be limited only by phosphorus. Conditional plots of these lakes (Figure 10.19 and 10.20) are not as clear as the conditional plots for shallow nonhumic lakes (Figure 10.17 and

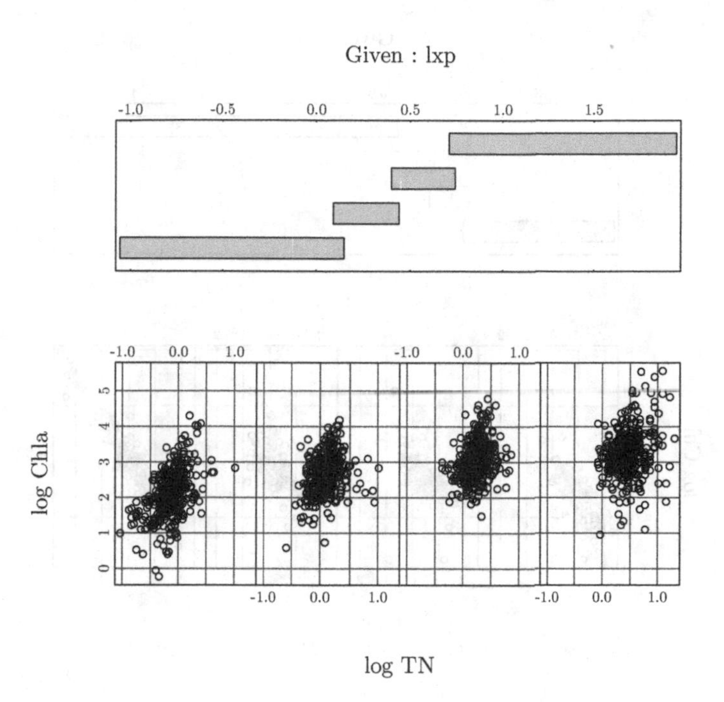

FIGURE 10.16: Conditional plots of eutrophic lakes (TN) – Scatter plots of log chlorophyll a against centered log TN concentration (conditional on TP) show decreased response to nitrogen as phosphorus level increases (from left to right). Data are from deep and very humic (type 6) lakes.

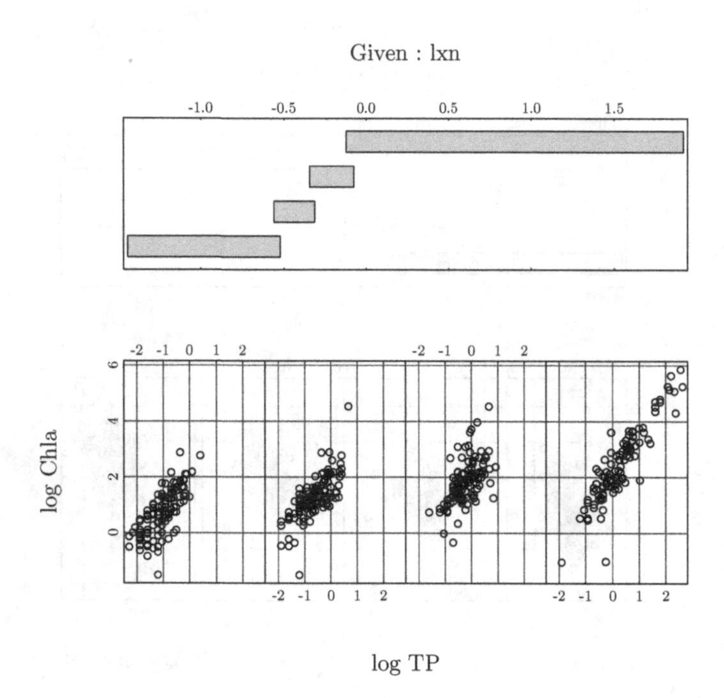

FIGURE 10.17: Conditional plots of oligotrophic (P limited) lakes (TP) – Scatter plots of log chlorophyll a against centered log TP concentration (conditional on TN) show a stable response to phosphorus as nitrogen level increases (from left to right). Data are from shallow nonhumic (type 7), possibly oligotrophic, lakes.

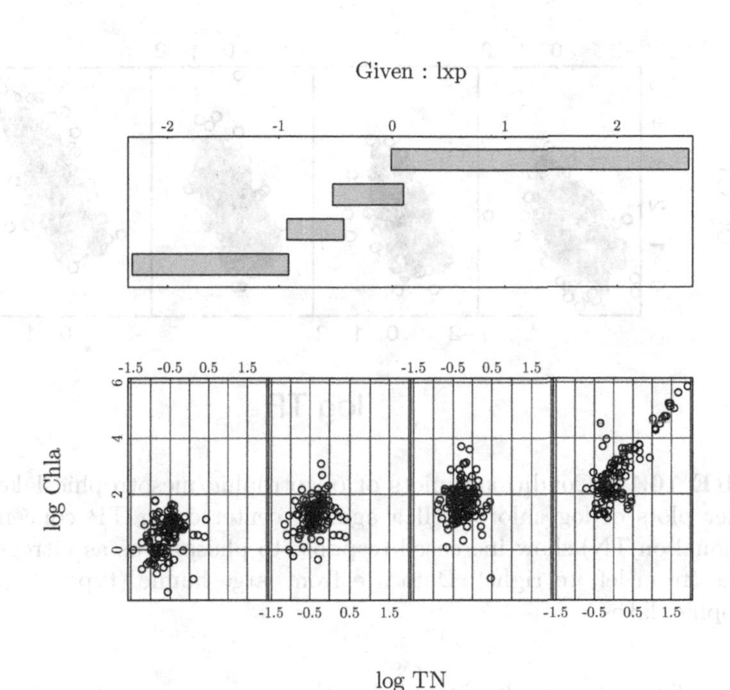

FIGURE 10.18: Conditional plots of oligotrophic (P limited) lakes (TN) – Scatter plots of log chlorophyll a against centered log TN concentration (conditional on TP) show generally no response to nitrogen as phosphorus level increases (from left to right) until phosphorus reaches a very high level. Data are from shallow nonhumic (type 7) lakes.

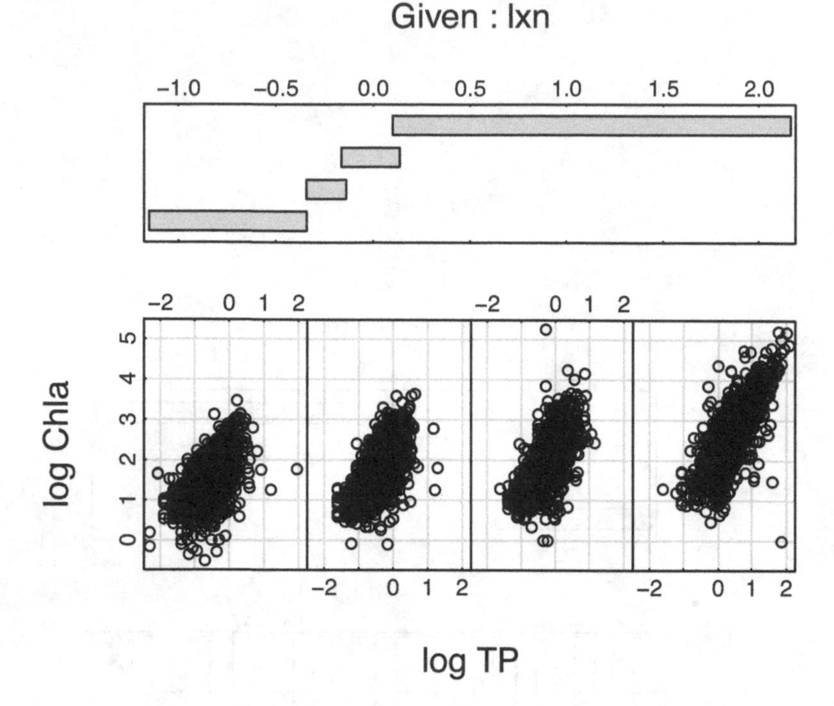

FIGURE 10.19: Conditional plots of oligotrophic/mesotrophic lakes (TP) – Scatter plots of log chlorophyll a against centered log TP concentration (conditional on TN) show increased response to phosphorus as nitrogen level increases (from left to right). Data are from large humic (type 2), possibly mesotrophic, lakes.

10.18). We find that the lakes included in this group are more heavily sampled and are more diverse. Lumping them together may not be appropriate. Further studies are necessary to reclassify lakes within this group so that group-level models can be useful.

The lake typology developed by the Finnish Environment Institute provides a broad division of lakes with different characteristics. Management strategies for improving a lake's water quality should thereby be lake-group specific. However, the classification scheme is not specifically designed for managing eutrophication. Lakes within a group may vary in many aspects and therefore each lake may require a lake-specific management plan.

General conclusions from the lake-type models are:

- Large and nonhumic lakes in Finland tend to be oligotrophic, either limited by phosphorus or by both nitrogen and phosphorus.

- Humic or very humic lakes tend to be eutrophic, limited by neither phosphorus nor nitrogen, with a high primary productivity.

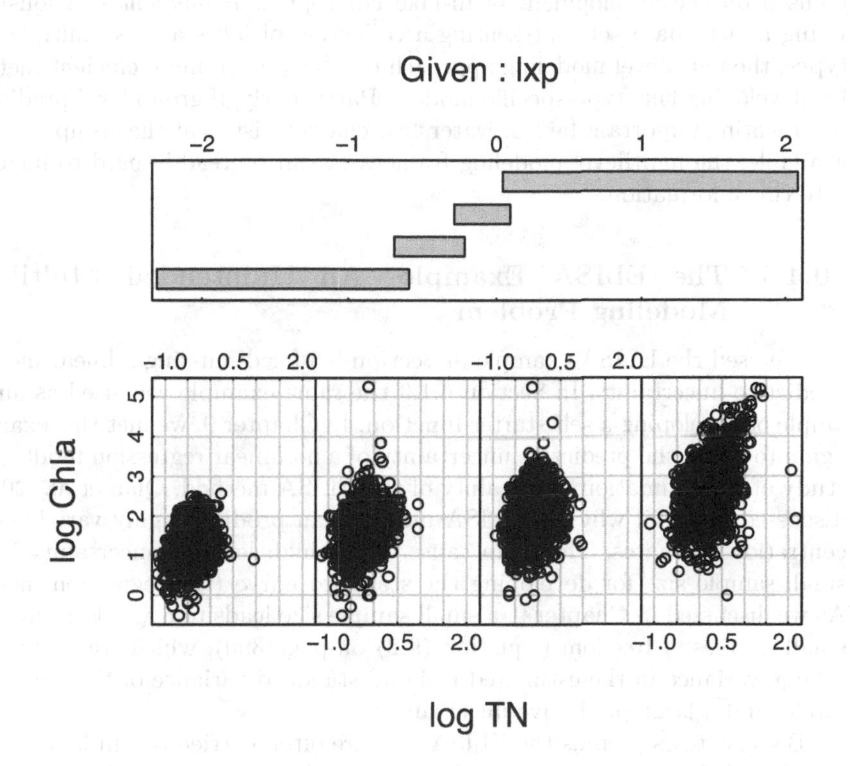

FIGURE 10.20: Conditional plots of oligotrophic/mesotrophic lakes (TN) – Scatter plots of log chlorophyll a against centered log TN concentration (conditional on TP) show increased response to nitrogen as phosphorus level increases (from left to right). Data are from large humic (type 2) lakes.

- Mesotrophic lakes are most likely limited by phosphorus.
- Using equation 10.21, the estimated interaction effect can often be used to identify a lake's trophic status: a negative interaction suggests an eutrophic lake, a positive interaction suggests an oligotrophic lake, and a statistically insignificant interaction may suggest a mesotrophic lake.

Traditionally, modeling for a lake eutrophication management problem is focused on the development of in-lake chlorophyll *a*–nutrient relationships. Using a large data set representing a collection of lakes across multiple lake types, the multilevel modeling approach can be a much more efficient method for developing lake type-specific models. Particularly, if group level predictors representing important lake or watershed characteristics at the group level are available, the multilevel modeling framework can be readily used to incorporate this information.

10.4.3 The ELISA Example—An Unintended Multilevel Modeling Problem

We used the ELISA example in Section 5.5 for discussing a linear model's predictive uncertainty. In Section 6.1.3 the same example was used as an example of developing a self-starter function. In Chapter 9, we met this example again to study the predictive uncertainty of a nonlinear regression model. In a study of the estimation uncertainty of the ELISA method, Qian et al. [2015a] discussed in detail why the ELISA method can produce highly variable concentration estimates. The main cause of the high level of uncertainty is the small sample size for developing the standard curve (the regression model). As we discussed in Chapter 9, a small sample size leads to a χ^2 distribution of small degrees of freedom (equation (9.2) on page 390), which translates into a large variance in the estimated residual standard variance of the regression model and a large predictive uncertainty.

Because tests such as the ELISA test are often carried out in laboratories that handle large number of samples to make the test cost effective, these labs are likely to have access to data from many ELISA tests. Although we often consider a test to be unique because of different water samples and other environmental conditions, the data used for developing the standard curve are generated by using standard solutions with known concentrations under uniform experimental conditions. Consequently, we should expect that the resulting standard curves be more or less consistent, except for the expected fluctuation due to the small degrees of freedom and measurement error. Such random fluctuation can be reduced by using the multilevel modeling approach through partial pooling data from multiple ELISA tests.

The standard curve in an ELISA test has always been developed for one test at a time. However, the small number of data points used for deriving the standard curve (a regression model) resulted in a high level of prediction error, as discussed in Section 5.5. Pooling data from multiple ELISA tests is a natural consequence of the Stein's paradox. That is, tests carried out in the past have relevant information about the standard curve. These sources of information should be properly utilized to help reduce the uncertainty we have about the standard curve. The multilevel model is a shrinkage estimator of the standard curve coefficients. As a result, the multilevel model estimated standard curve should be superior to the standard curve based on one test data alone.

The implementation of the multilevel model for the log-linear model (Section 5.5.1) is straightforward: pooling data from multiple tests using a multilevel model will improve the standard curve thereby improving the measurement accuracy. Qian et al. [2015a] analyzed data from 21 ELISA tests and compared the multilevel model results to the results from using the conventional approach.

10.5 Nonlinear Multilevel Models

In addition to the log-linear model, the standard curve in an ELISA test can also be parameterized using a four-parameter logistic model (Section 6.1.3). As discussed in the previous sections, pooling data from multiple tests is advisable. For a nonlinear regression model, the concept of multilevel can be easily borrowed from equation (10.14), where a multilevel linear model is expressed by the addition of a common prior distribution on model coefficients. Likewise, for a nonlinear regression mean function, the multilevel model includes a common prior distribution of model coefficients. The multilevel version of the four-parameter logistic regression model of equation (6.4) will include a prior model for the four parameters:

$$\begin{array}{rcl} y_{ij} & = & \alpha_{4j} + \frac{\alpha_{1j}-\alpha_{4j}}{1+\left(\frac{x}{\alpha_{3j}}\right)^{\alpha_{2j}}} + \epsilon_{ij} \\ \begin{pmatrix} \alpha_{1j} \\ \alpha_{2j} \\ \alpha_{3j} \\ \alpha_{4j} \end{pmatrix} & \sim & MVN\left[\begin{pmatrix} \mu_1 \\ \mu_2 \\ \mu_3 \\ \mu_4 \end{pmatrix}, \Sigma\right] \end{array} \tag{10.22}$$

where subscript ij represents the ith observation from the jth ELISA test. As in the EUSE example, the regression model coefficients are test-specific, but share the same prior distribution, which pulls test-specific coefficients towards the overall mean. The shrinkage effect will improve the overall predictive accuracy. In other words, when using a multilevel modeling approach for ELISA testing, a lab can improve its overall test accuracy.

The nonlinear multilevel model is implemented in the R function `nlmer`. I will use data from the six ELISA tests carried out from August 1 to 3, 2014 in the City of Toledo's Water Department as an example. The data were published in a report released shortly after the Toledo water crisis. Standard solutions used in all six tests have the same concentrations. Data from these tests are in Section 9.2.4.

For this analysis, the response variable is the observed optical density (`Abs`) and the predictor variable is the known microcystin concentration (`stdConc`). As in all R models, the function `nlmer` requires a formula. In addition to the usual "two-part" formula (`y~x`), `nlmer` takes a "three-part" formula: `y ~ Nonlin(...) ~ fixed + random`. In the current version of R (v3.2.3), the nonlinear model function `Nonlin(...)` must return a numeric vector and a "gradient" attribute. The gradient attribute is a matrix of first order partial derivatives of model coefficients. In other words, the model function needs to be something similar to the self-starter function described in Section 6.1.3. Using the self-starter function `SSfpl2`, the multilevel model formula is:

```
Abs ~ SSfpl2(stdConc,al1,al2,al3,al4) ~ (al1+al2+al3+al4|Test)
```

We also need to supply a set of starting values for model coefficients. In this example, I used the nonlinear regression results in Section 6.1.3:

```
#### R Code ####
tm1 <- nls(Abs ~ SSfpl2(stdConc, al1, al2, al3, al4),
           data=toledo[toledo$Test==1,])

tm1.nlmer <- nlmer(Abs ~ SSfpl2(stdConc, al1, al2, al3, al4)~
                        (al1+al2+al3+al4|Test), data=toledo,
                    start=c(al1=1, al2=1, al3=0.5, al4=0.2))

#### R Output ####

> print(tm1.nlmer)

Nonlinear mixed model fit by maximum likelihood  ['nlmerMod']
Formula: Abs ~ SSfpl2(stdConc, al1, al2, al3, al4) ~
    (al1 + al2 + al3 + al4 | Test)
   Data: toledo
      AIC       BIC    logLik  deviance  df.resid
-214.5027 -180.3527  122.2514 -244.5027        57
Random effects:
 Groups   Name Std.Dev. Corr
 Test     al1  0.04227
          al2  0.06635   0.67
          al3  0.07759  -0.94 -0.62
          al4  0.05976   0.96  0.71 -0.99
 Residual      0.04003
Number of obs: 72, groups:  Test, 6
Fixed Effects:
   al1     al2     al3     al4
1.1237  1.0171  0.4504  0.1695
```

As in a linear multilevel model, the fitted model returns a set of estimated

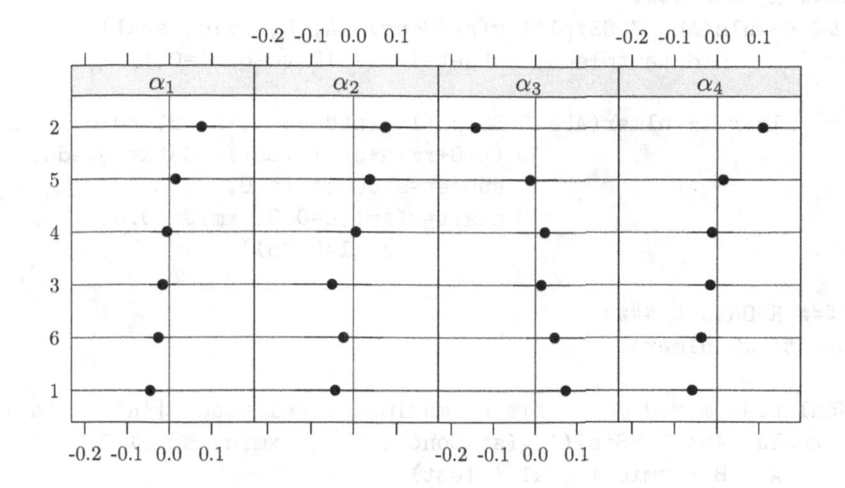

FIGURE 10.21: ELISA standard curve coefficients were estimated using the self-starter function `SSfpl2` (defined on the concentration scale). All four coefficients vary among the six tests.

model coefficients $\hat{\alpha}_{1j}, \cdots, \hat{\alpha}_{4j}$ for all tests (j). The estimated coefficients are expressed as the sum of a "fixed" effect and a random effect: $\hat{\alpha}_{ij} = \hat{\mu}_i + \hat{\delta}_{ij}$, where $i = 1, \cdots, 4$ representing the four parameters of the logistic function, and j is the index of tests. For each coefficient, the fixed effects $\hat{\mu}_i$ are the estimated μ_i ($i = 1, \cdots, 4$) in equation (10.22) (representing the mean of α_{ij} over all tests j, or the among tests mean), while the random effects are the differences between test-specific coefficients and the respective among test mean. The functions `fixef` and `ranef` can be used to extract the estimated fixed and random effects, respectively.

To graphically compare the differences among the six tests, we can directly use the function `dotplot` from the `lattice` package (Figure 10.21).

```
#### R Code ####
temp <- ranef(tm1.nlmer, condVar=T)
dotplot(temp, layout=c(4,1), main=F)
```

As we discussed in Section 6.1.3, the four-parameter logistic function is perhaps better defined in the logarithmic scale of the microcystin concentration based on model residual characteristics. The fitted multilevel model `tm1.nlmer` is based on the function defined in the concentration scale (using `SSfpl2`). After trying several optimization options, I cannot avoid the convergence problem (a warning message). When removing the 0 concentration observations and fitting the model in the log-concentration scale (using `SSfpl`), the process converged without incident.

```
#### R Code ####
tm2 <- nls(Abs ~ SSfpl(log(stdConc), A, B, xmid, scal),
           data=toledo, subset=Test==1&stdConc!=0,])

tm2.nlmer <- nlmer(Abs ~ SSfpl(log(stdConc), A, B, xmid, scal) ~
                          (A+B+xmid+scal|Test), data=toledo,
                          subset=stdConc != 0,
                          start=c(A=1,B=0.2, xmid=-0.6,
                                  scal=0.75))


#### R Output ####
print(tm2.nlmer)

Nonlinear mixed model fit by maximum likelihood  ['nlmerMod']
Formula: Abs ~ SSfpl(log(stdConc), A, B, xmid, scal) ~
   (A + B + xmid + scal | Test)
   Data: toledo1
      AIC       BIC    logLik  deviance  df.resid
-175.4859 -144.0708  102.7430 -205.4859        45
Random effects:
 Groups   Name Std.Dev. Corr
 Test     A    0.01725
          B    0.06041  -0.44
          xmid 0.10768   0.37 -0.82
          scal 0.07584   0.45 -0.88  0.47
 Residual      0.03998
Number of obs: 60, groups:  Test, 6
Fixed Effects:
      A         B      xmid       scal
 1.4041    0.1111   -1.3191    1.3427
```

Compared to the model defined in the concentration scale (using `SSfpl2`), the model fit in the log-concentration scale (using `SSfpl`, Figure 10.22) has a very stable parameter A (the upper bound of the observed optical density at the low end of concentration spectrum) when microcystin concentration approaches 0. When defining the logistic model in the concentration scale (model `tm1.nlmer`), the optical density upper bound is the parameter α_1. The observed optical densities from standard solutions with a concentration of 0 in each test serve as an anchor for α_1. As a result, observation error can be easily reflected in the estimated α_1.

This parameter affects the estimated concentrations near the lower concentration bound of 0. In the logarithmic scale, the distance between two small concentrations (concentration value below 1) increases (e.g., between 0.1 and 0.01 versus log(0.1) [−2.3] and log(0.01) [−4.6]), allowing the model to have increased flexibility to fit the low end of the curve better. Combined with the reduced variability in the estimated parameter A, I believe that the model

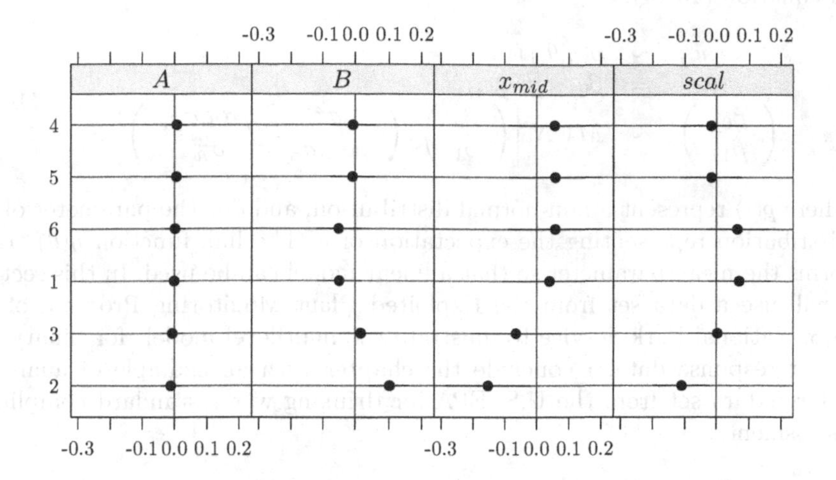

FIGURE 10.22: ELISA standard curve coefficients were estimated using the self-starter function `SSfpl` (defined on the log concentration scale). The coefficient A (the upper bound of the curve near the low end of the concentration spectrum) is stable.

should be fit in the logarithmic scale. When fitting the curve in the logarithmic scale, the 0 concentration standard solution is not used. Replacing the 0 concentration standard solution with a small concentration one would greatly improve the model fitting process.

10.6 Generalized Multilevel Models

When the response variable distribution cannot be approximated by the normal distribution, we move from the multilevel regression model to the generalized multilevel models. In Chapter 8, we discussed the transition from linear regression models to the generalized linear models, with two important concepts – the maximum likelihood estimator and the link function. In multilevel modeling, the maximum likelihood estimator (and its variations) is always used for both normal and non-normal response variables. The generalized multilevel model is the same.

A generalized multilevel regression can be expressed in a similar form as in equation (10.14):

$$
\begin{array}{rcl}
y_{ij} & \sim & p(y|\theta_{ij}) \\
\eta(\theta_{ij}) & = & \beta_{0j} + \beta_{1j}x_{ij} \\
\left(\begin{array}{c} \beta_{0j} \\ \beta_{1j} \end{array}\right) & \sim & MVN\left[\left(\begin{array}{c} \beta_0 \\ \beta_1 \end{array}\right), \left(\begin{array}{cc} \sigma^2_{\beta_0} & \rho\sigma_{\beta_0}\sigma_{\beta_1} \\ \rho\sigma_{\beta_0}\sigma_{\beta_1} & \sigma^2_{\beta_1} \end{array}\right)\right]
\end{array}
\tag{10.23}
$$

where $p(\cdot)$ represent a non-normal distribution, and θ is the parameter of the distribution representing the expectation of y. The link function $\eta(\theta)$ transforms the mean parameter so that a linear model can be used. In this section, I will use a data set from the Exploited Plant Monitoring Program of the U.S. National Park Service to illustrate the multilevel models for binary and count response data. I conclude the chapter with an example of analyzing a large data set from the U.S. EPA for drinking water standard compliance assessment.

10.6.1 Exploited Plant Monitoring—Galax

The Blue Ridge Parkway is a 750-km long scenic highway stretching between the Great Smoky Mountains National Park and the Shenandoah National Park. It is part of the U.S. National Parks system. As in many well-preserved nature areas, illegal harvesting of plants is often a concern. Along the Blue Ridge Parkway there are many species of plants that are targeted by poachers and illegal harvesting activities are increasing. One plant species of concern is galax (*Galax urceolata*). The targeted populations of galax belong to a genetically unique form that produces very large leaves ideal for floral arrangements. The goal of the monitoring effort is to determine whether galax populations are declining. Uncontrolled harvesting of large leaves will inevitably lead to a wild population of mostly young plants. An initial sampling was carried out by the National Park Service and NatureServe to develop a monitoring protocol for long-term monitoring. Specifically, the monitoring program is focused on the detection of changes in the "poachable" plant population, defined both in terms of habitat location (close to roads) and sizes (large patch size and large leaf size).

Galax leaves are usually classified into two size groups (large and small). Currently, plant data collected include leaf densities of two size classes (number of leaves per unit area in a patch). Of interest is the ratio of large leaves (poaching target) and the change of the ratio over time. Data were collected along pre-determined transects using point-intercept method. Each observation consists of counts of small and large leaves at a point. These counts are then transformed into density using the area covered by the "points." A ratio is also calculated.

10.6.1.1 A Multilevel Poisson Model

The first response variable of interest is the density of large leaves. The response variable data are counts in a survey area. As in the generalized linear model, the count response variable is typically modeled by the Poisson distribution:

$$\begin{array}{rcl} y_{ij} & \sim & Pois(\lambda_{ij}) \\ log(\lambda_{ij}) & = & \log(A_{ij}\mu_j) \end{array} \tag{10.24}$$

where ij represents the ith observation observed from site j, A is the area covered by the point associated with the ith observation, y is the counts of large leaves, and μ is the site-specific areal density. Multiple sampling points were used in several sites along the Blue Ridge Parkway.

The initial surveys were carried out in some of the eight sites from 2007 to 2009. Based on these initial data, a complete survey was conducted in 2010 with all eight sites visited. The 2010 results are used for analysis (data file `Galax_2010.csv`):

```
survey.data <- read.csv("Galax_2010.csv", header=T)
```

With the data, we first want to learn whether large leaf density varies by site. A multilevel model imposes the exchangeability assumption on site-specific densities:

$$\mu_j \sim N(\theta, \sigma^2) \tag{10.25}$$

The multilevel model formula for the model specified in equations (10.24) and (10.25) is:

```
CountL ~ 1 + (1|Site)
```

The model is fit using function `glmer`:

```
#### R Code ####
G.lmer1 <- glmer(CountL~1+(1|Site), family="poisson",
                 offset=log(TotalPoints), data=survey.data)
```

The summary of this model can be shown by using the function `display` from the `arm` package:

```
#### R Output ####
display(G.lmer1)

glmer(formula = CountL ~ 1 + (1 | Site), data = survey.data,
    family = "poisson", offset = log(TotalPoints))
coef.est  coef.se
   -4.43     0.49

Error terms:
 Groups   Name         Std.Dev.
```

```
 Site     (Intercept) 1.08
 Residual             1.00
---
number of obs: 57, groups: Site, 8
AIC = 151.3, DIC = 147.3
deviance = 147.3
```

The second line of the model in equation 10.24 can be rewritten as:

$$\log(\lambda_{ij}) = \log(A_{ij}) + \mu + \delta_j \quad (10.26)$$

which breaks a site-specific mean (μ_j) into two parts: the overall mean (μ) and the difference in mean for a specific site. These two ways of expression are mathematically identical. The R function `glmer` uses the overall mean plus difference. For applications, extracting the overall mean and the among site differences are more important. We use function `fixef` and `ranef` to extract these two pieces of information:

```
#### R Output ####
fixef(G.lmer1)
(Intercept)
  -4.433006

ranef(G.lmer1)
$Site
  (Intercept)
A -1.04262497
B -0.52022349
C  0.89850091
D  1.33406236
E  1.03905486
F  0.30419124
G -1.18924836
H -0.05435251
```

That is, the overall mean density (in logorithm) is -4.433, and the site A has a mean 1.04 units *below* the overall mean. In the density scale, the overall mean density is $e^{-4.433}$ or 0.012, and the mean for site A is $e^{-4.433-1.042} = 0.0042$ or 35% ($e^{-1.04}$) of the overall mean density.

Graphically, this model can be shown using a dot plot. If only the differences among sites are of interest, we can plot the estimated "random effects" and the associated uncertainty:

```
dotplot(ranef(G.lmer1, condVar=T))
```

The estimated random effects show a large site to site variation (Figure 10.23). The estimated random effects are in the logarithmic scale. To view the site effects in the density scale, we can add the estimated fix effect to each of the random effects and plot them in the original scale:

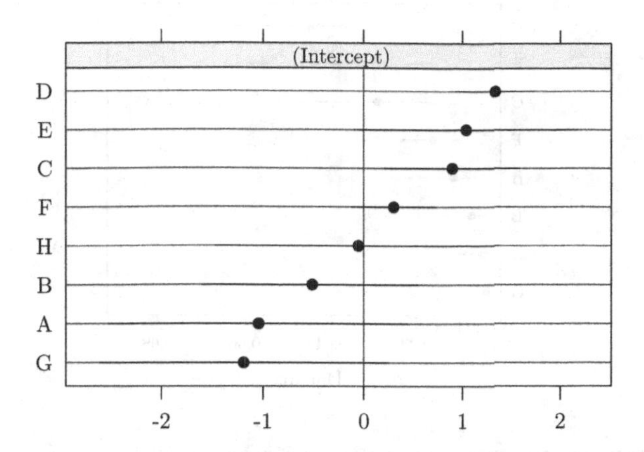

FIGURE 10.23: Estimated log site effects show a wide site to site variation of large leaf density.

```
#### R Code ####
new.data <- data.frame(x = fixef(lmer1)[1] +
                         ranef(lmer1)$Site[,1],
                     y=row.names(ranef(lmer1)$Site),
                     sd = sqrt(se.ranef(lmer1)$Site[,1]^2+
                               se.fixef(lmer1)[1]^2))
new.data$y <- ordered(new.data$y,
                     levels=new.data$y[order(new.data$x)])

dotplot(y ~ exp(x), data = new.data,
        aspect = 0.8,
        xlim=c(0,1.2*range(exp(new.data$x-new.data$sd),
          exp(new.data$x+new.data$sd))[2]),
        panel = function (x, y) {
          panel.xyplot(x, y, pch = 16, col = "black")
          panel.segments(exp(new.data$x-new.data$sd),
                         as.numeric(y),
                         exp(new.data$x+new.data$sd),
                         as.numeric(y), lty = 1},
        xlab="Density")
```

The estimated random effect standard errors are in logarithmic scale. Directly converting them back to the original scale is subject to the retransformation bias. As a result, I often use simulation to estimate the es-

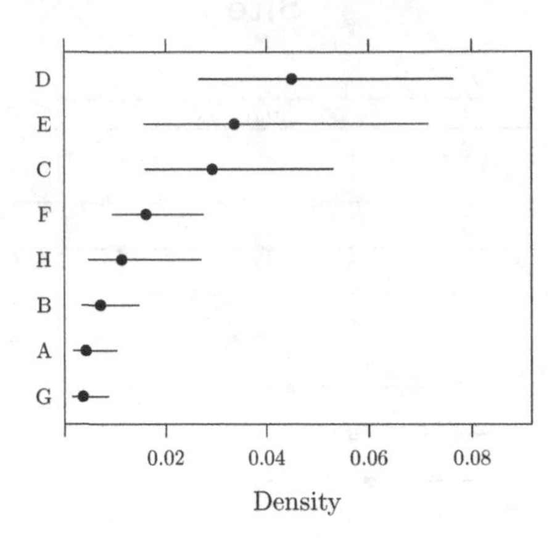

FIGURE 10.24: Estimated large leaf density using 2010 survey data

timation uncertainty in the original scale. The simulation will be left as an exercise.

10.6.1.2 A Multilevel Logistic Regression Model

The Galax survey data recorded the leaf counts for both large and small leaves. As discussed earlier, another variable of interest is the proportion of large leaves as a measure of poaching severeness. Ideally, the ratio should be stable over time without poaching. Using the 2010 data, I will illustrate the model fitting process and the use of simulation to estimate the large leaf proportion and its uncertainty.

The statistical model for describing the variation in the large and small leaf counts is the binomial distribution. When modeling the proportion of large leaves, we are modeling the probability of observing a large leaf. In this case, we are interested in the among site variation of the proportion, which will provide information for deciding the likely site characteristics that made a site vulnerable for poaching. The statistical model is the same as the Poisson multilevel model, except that the response variable distribution is now a binomial distribution and the link function is the logistic function. In R, we use the function `glmer`:

```
#### R Code ####
G.lmer2 <- glmer(cbind(CountL, CountS) ~ 1+(1|Site),
                 family="binomial", data=survey.data)
```

The model output can be displayed by either the function `display` from package `arm` or `summary`.

```
#### R Output ####
```

```
display(lmer3)
glmer(formula = cbind(CountL, CountS) ~ 1 + (1 | Site),
     data = survey.data, family = "binomial")
coef.est  coef.se
   -1.66     0.61

Error terms:
 Groups    Name        Std.Dev.
 Site      (Intercept) 1.43
 Residual              1.00
---
number of obs: 57, groups: Site, 8
AIC = 129.8, DIC = 125.8
deviance = 125.8
```

The estimated coefficient (**coef.est**) of -1.66 is the estimated "fixed effect." This value is the overall average of the logit of the large leaf proportion. In the original scale, the estimated large leaf proportion is 0.16. The "random effects" (differences between the proportion of a site and the overall mean) are obtained by using the function **ranef**:

```
#### R Output ####
ranef(lmer3)
$Site
  (Intercept)
A  -1.1515331
B  -1.4932040
C   2.0069025
D   1.1916822
E   0.6586113
F  -0.9770496
G   0.0000000
H   0.3341795
```

and, graphically, we can use **dotplot** to display the estimated random effects. Again, we see a large among-site variation (Figure 10.25). However, the logit scale is nonlinear with respect to the proportion (Figure 8.2). The same level of uncertainty (estimated standard error) can be very different in the original scale for different levels of proportion. For example, the lowest large leaf proportion is from site B ($-1.66 - 1.49$, or about 4%) and the largest proportion is observed at site C ($-1.66 + 2.01$, or about 57%). The estimated standard error for both sites is about 0.55 in the logit scale (use function **se.ranef**). Using the 95% confidence interval (approximately the estimated proportion plus/minus 2 times standard error), the width of the interval in the logit scale is similar $[(-4.25, -2.05)$ versus $(-0.75, 1.45)]$. When converting these intervals into the original scale, they are very different [(0.014, 0.114) versus

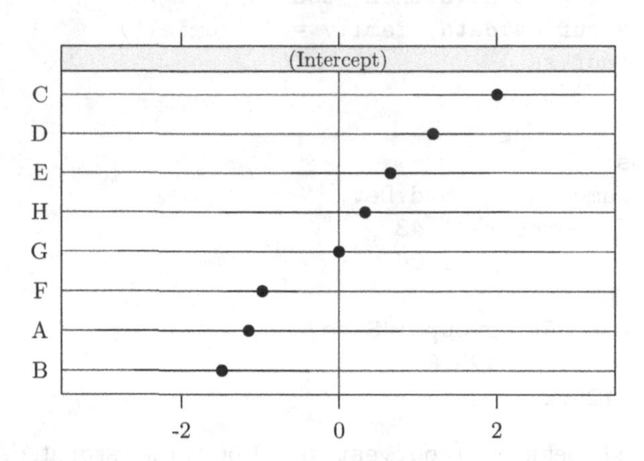

FIGURE 10.25: Estimated large leaf proportions (model random effects) are presented in the logit scale.

(0.321, 0.810), respectively]. Such re-transformation can be misleading. Simulation should be used to properly convert estimation uncertainty from the logit scale to the original scale.

The R function sim (from package arm) uses a Monte Carlo simulation algorithm similar to the ones discussed in Chapter 9 to draw random samples of the estimated model coefficients (the fixed and random effects). Each set of coefficients represents a likely model, in our case, a set of likely large leaf proportions for the eight sites. Using these random samples, we can construct a 95% interval representing the range of the middle 95% likely proportions (Figure 10.26).

```
#### R Code ####

lmer3.sim <- sim(lmer3, 2000) ## 2000 random samples

## predicted proportion and convert to rv object:
pred3 <- rvsims(lmer3.sim@fixef[,1] +
                lmer3.sim@ranef$Site[,,1])
pred3.sum <- summary(invlogit(pred3))

new.data <- data.frame(x = pred3.sum[,2],
                       y = labels(pred3),
                       q2.5=pred3.sum[,5],
                       q25= pred3.sum[,6],
```

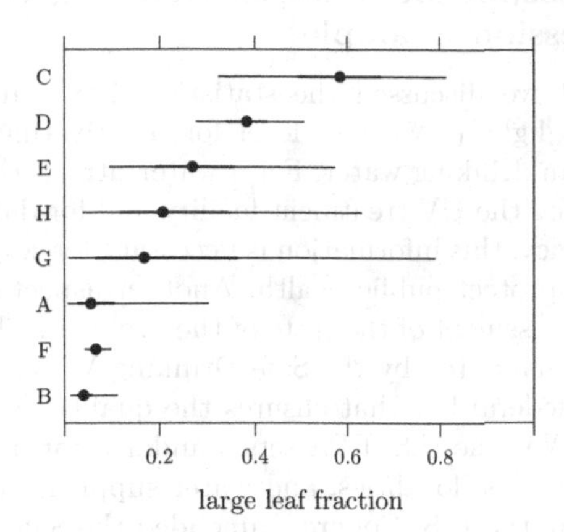

FIGURE 10.26: Estimated large leaf proportions (dots) along with the middle 50% (thick line segments) and 95% (thin line segments) are presented in the scale of proportion.

```
                          q50= pred3.sum[,7],
                          q75 =pred3.sum[,8],
                          q975=pred3.sum[,9])
new.data$y <- ordered(new.data$y,
             levels=new.data$y[order(new.data$x)])

dotplot(y~x, data=new.data,xlim=c(0,1),
        panel=function(x,y){
          panel.points(x=new.data$q50, y=as.numeric(y),
                               pch=16)
          panel.segments(new.data$q2.5, as.numeric(y),
                               new.data$q975, as.numeric(y))
          panel.segments(new.data$q25, as.numeric(y),
                               new.data$q75, as.numeric(y),
                               lwd=2.5)
        }, xlab="large leaf fraction")
```

The 2010 survey will be used for refining the monitoring protocol. In addition to understand the spatial pattern of likely poaching activities, researchers also are interested in estimating changes in large leaf density and proportion over time. Temporal trends will enable the Park Service to evaluate the effectiveness of various management strategies. These initial analyses also provide estimation on the number of sites and number of years necessary for detecting a given magnitude temporal trend.

10.6.2 Cryptosporidium in U.S. Drinking Water—A Poisson Regression Example

In Section 8.1 we discussed the statistical issues in estimating the required ultraviolet light (UV) dose level for inactivating a given amount of cryptosporidium in drinking water. For a water utility, this information is essential for designing the UV treatment facility and for daily management. For a government agency, this information is necessary for setting the standard for UV treatment to protect public health. Another aspect of the cryptosporidium work is the assessment of the state of the problem. The assessment, in the United States, is mandated by the Safe Drinking Water Act (SDWA), which is the main U.S. federal law that ensures the quality of Americans' drinking water. Under SDWA, the U.S. EPA sets standards for drinking water quality and oversees the states, localities, and water suppliers who implement those standards. In 1996, the U.S. Congress amended the Safe Drinking Water Act to emphasize sound science and risk-based standard setting. Two requirements set in the 1996 amendment resulted in a cyclical review on the national distribution of system mean concentrations of cryptosporidium. These two main points are:

- Cost-Benefit Analysis: the U.S. EPA must conduct a thorough cost-benefit analysis for every new standard to determine whether the benefits of a drinking water standard justify the costs.

- Microbial Contaminants and Disinfection By-products: the U.S. EPA is required to strengthen protection for microbial contaminants, including cryptosporidium, while strengthening control over the by-products of chemical disinfection.

The EPA's strategy is to develop numerical distribution functions of system mean concentrations for all contaminants. These distribution functions provide basic information on the level of standard compliance. The method developed in Qian et al. [2004] provides such distributions stratified based on the source water type (groundwater or surface water) and the population served of a public drinking water system. When a new (more protective) standard is proposed, the EPA can use these distributions to estimate the number of drinking water systems that will be affected by the change as well as the number of people who will benefit. The number of systems affected can be translated into cost, and the number of people benefited can be translated into benefit.

The model developed in Qian et al. [2004] is, however, not appropriate for estimating the distribution of cryptosporidium concentration. This is because the datum reported for cryptosporidium is the number of oocysts detected in a given volume of water sample. In this section, we illustrate the use of a multilevel Poisson regression model for estimating the system mean distribution of cryptosporidium in the U.S. public drinking water systems, using the data collected by the EPA's Data Collection and Tracking System (DCTS) for

Cryptosporidium, *E. coli*, and turbidity data in source waters of U.S. public drinking water systems.

The basic idea of a model-based national distribution of system mean contaminant is the hierarchical modeling principle discussed in Section 10.3. Let us describe the problem using a normally distributed variable first. In the United States, drinking water quality is regulated under the same law. If we assume the contaminant concentration distribution in a public drinking water system is a log-normal distribution, we can use the normal distribution to describe the log-concentration variation:

$$y_{ij} \sim N(\theta_i, \sigma^2)$$

Because all drinking water systems are regulated under the same law, there is no reason to believe that the mean of one system θ_i is different from the means of other systems. Therefore, it is reasonable to assume that system means are exchangeable and can be modeled as from the same a priori distribution:

$$\theta_i \sim N(\mu, \tau^2)$$

The distribution $N(\mu, \tau^2)$ is the distribution of (log) system means. The difficulty of this model applied in drinking water data is that most of the reported concentration data are below the detection limits of measurement methods (or method detection limit, MDL). The work of Qian et al. [2004] resolved this problem.

For the cryptosporidium mean distribution, the problem is somewhat different. In theory, there is no detection limit. If a cryptosporidium oocyst is in the water sample, the detection method will be able to detect its presence about 44% of the time based on "spiked" tests conducted by many EPA certified labs. Because the cryptosporidium data reported to the EPA's DCTS database were analyzed by the same group of certified labs, we will use this 44% recovery rate in our model. To set up the model, we first consider the probabilistic distribution of the reported cryptosporidium oocyst. Suppose the true concentration in the water is c and a volume of v was analyzed. On average, the number of oocyst in the sample is $n_o = cv$. Because the sample is taken at random, the actual number of oocyst included in the sample is random. The most commonly used probability distribution for this count random variable is the Poisson distribution. The number of oocyst observed y_{ij} will have a Poisson distribution:

$$y_{ij} \sim Pois(\lambda_{ij}) \tag{10.27}$$

The Poisson intensity λ_{ij} is the expected number of y, which is $0.44c_iv_{ij}$. The parameter of interest is the distribution of c_i. Because the number of systems is large, the commonly used approach is to fit a Poisson regression model using $\log(0.44v_{ij})$ as the offset and system identification as the categorical predictor. In this data set, the measured cryptosporidium oocyst (the response variable) is named `n.cT` and the system identification is in a variable named `PWSID`:

```
#### R Code ####
crypto.glm <- glm(n.cT ~ factor(PWSID), data=dcts.data,
                  offset=log(0.44*volume),
                  family="poisson")
```

This model is the same as equation 10.27 and the log of c_i is estimated. Using the multilevel modeling terminology, the Poisson regression with a factor predictor is the no pooling model, where system means are estimated separately. The model would be fine given that there are a large number of systems in the data. Because the waters tested are source water for drinking water supply, they generally have good water quality and most of the samples have no cryptosporidium in them. In the data used here, 68% of systems reported all 0 counts. When all the observed counts are 0s, the `glm` estimate of the mean concentration is undefined (log of 0), which is reflected in the extremely large standard error of these estimates. For example,

```
> display(crypto.glm)
glm(formula = n.cT ~ factor(PWSID) - 1,
    family = "poisson", data = dcts.data,
    offset = log(volume * 0.44))
                          coef.est coef.se
factor(PWSID)               -22.48 15541.86
factor(PWSID)010106001      -21.86 15541.86
factor(PWSID)104121115      -21.78 15541.86
factor(PWSID)AK2210906       -4.72     0.58
factor(PWSID)AK2260309       -3.97     0.71
factor(PWSID)AL0000133      -21.78  2590.31

......

factor(PWSID)WV3304005       -3.90     1.00
factor(PWSID)WV3304104       -4.05     1.00
factor(PWSID)WY5600011      -21.79  3047.51
factor(PWSID)WY5600029      -21.78  7770.93
factor(PWSID)WY5600050      -21.78 15541.86
---
  n = 13103, k = 884
  residual deviance = 6789.9, null
  deviance = 142338.8 (difference = 135548.9)
```

The estimated coefficients are the estimated mean concentrations for the systems included in the data. When estimating the mean concentration, we don't really believe that the true mean c can be 0. After all, cryptosporidium is a naturally occurring microbe. As a result, even with 884 public drinking water systems in the data, we cannot use the estimated means to estimate the empirical distribution of system means.

Using the multilevel modeling, the system means c_i are further assumed to have a common á priori distribution:

$$\log(c_i) \sim N(\mu, \tau^2) \tag{10.28}$$

In R, the function `glmer` with option `family="poisson"` is used for a multilevel Poisson regression:

```
crypto.lmer1 <- glmer(n.cT ~ 1+(1|PWSID),
                      data=dcts.data, family="poisson",
                      offset=log(volume*0.44))
```

The fitted model provides estimates of μ and τ^2, as well as system means. The estimated system means are model coefficients:

```
#### R Output ####
> coef(crypto.lmer)
$PWSID
    (Intercept)
1          -5.53
2          -5.47
3          -5.47
4          -4.77
5          -4.12
6          -6.46
7          -5.65
8          -6.16
9          -6.17
10         -6.17
11         -6.18
12         -3.46
......
```

Figure 10.27 shows the estimated system means and their standard errors. Systems with all 0s in the observed data are those with the lowest mean concentrations and large standard error. The figure also shows the relative amount of pulling toward the overall mean for these systems – a system with a smaller sample size is pulled toward the overall mean more. As sample size increases, the amount of pulling of the estimated system mean (and the standard error) decreases.

When considering the system mean distributions, we can either use the empirical distribution of the estimated 884 system means or directly use the estimated μ and τ^2. The empirical cumulative distribution function (CDF) can be estimated using equation 3.2:

```
#### R Code ####
mus <- coef(icr.lmer1)[[1]][,1]
```

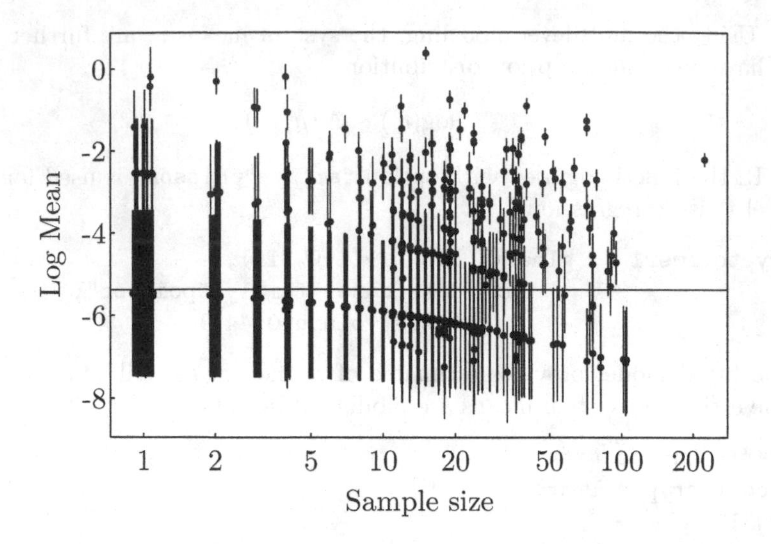

FIGURE 10.27: System means of cryptosporidium in U.S. drinking water systems – Estimated system mean concentrations of cryptosporidium (dots) and their standard errors (line segments) are plotted against the respective sample sizes for the systems. The horizontal line is the overall mean concentration.

```
n.sys <- length(mus)
f = ((1:n.sys)-0.5)/n.sys
```

Comparing the model-based cumulative distribution of system means and empirical CDF of the estimated system means, we see obvious differences and the difference is largely contributed by those systems with all 0 counts (Figure 10.28). If the interest is in correctly quantifying the fraction of systems with cryptosporidium concentration about certain high levels (e.g., 0.5 oocyst/L), both model-based CDF and empirical CDF will yield similar results.

10.6.3 Model Checking Using Simulation

How well does the model fit the data? A quick way to answer this question is to let the model recreate the data set and compare the replicated data to the real data. Replicated data from the multilevel model are random numbers generated from the fitted model. As in the linear regression model, we use the model output to generate random numbers of μ and τ^2, from which plant level crypto concentrations c_i are generated. The generated c_i's are then used to generate a number of oocysts. Using the generated replicates, we can calculate important statistics and make assessments of a model's performance. One quantity in which the EPA is interested is the number of systems exceeding

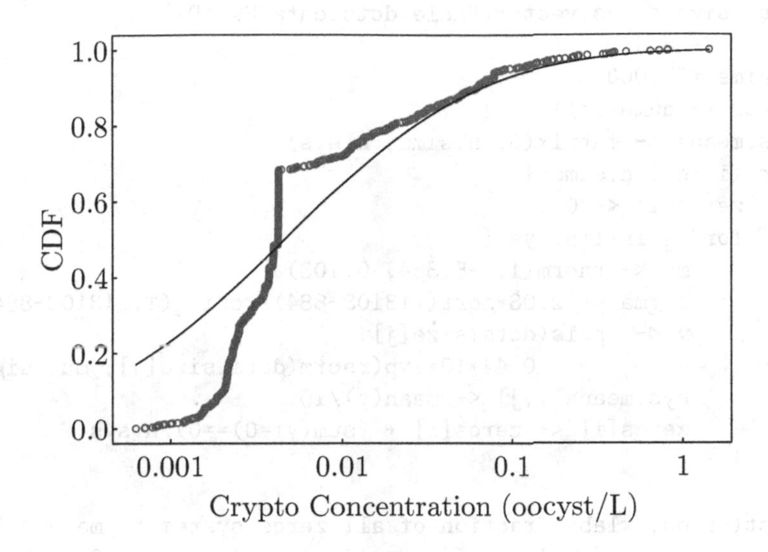

FIGURE 10.28: System mean distributions of cryptosporidium in the United States – The system mean distribution can be estimated empirically by using the empirical CDF of the multilevel model estimated system means (gray open circles) or by the normal distribution parameters of the log system means (μ and τ^2 in equation 10.28, solid line).

the water quality standard. The current standard for cryptosporidium is 0, an impossible value to verify. As a result, the EPA requires public drinking water systems to inactivate 99.9% of any cryptosporidium present. For the model developed in the previous section to be useful, it must be able to reproduce the fraction of systems with all 0 counts and systems with extremely high concentration values.

To assess these two features, we use a simple program to repeatedly generate random samples of μ and τ and generate a c_i for each pair of μ and τ. The generated c_i's are then used to generate possible counts y:

```
#### R Code ####
dcts.size <- as.vector(table(dcts.data$PWSID))
n.sys <- 884
n.sims <- 10000
zeros <- numeric()
sys.means <- matrix(0, n.sims, n.sys)
for (i in 1:n.sims){
    zeros[i] <- 0
    for (j in 1:n.sys){
        mu <- rnorm(1, -5.384, 0.103)
        sigma <- 2.08*sqrt((13103-884)/rchisq(1, 13103-884))
        y <- rpois(dcts.size[j],
                   0.44*10*exp(rnorm(dcts.size[j], mu, sigma)))
        sys.means[i,j] <- mean(y)/10
        zeros[i] <- zeros[i] + (sum(y!=0)==0)/n.sys
    }
}
hist(zeros, xlab="fraction of all zeros systems", main="")
hist(as.vector(apply(sys.means, 1, quantile, prob=0.99))
```

The simulation algorithm used the typical volume of 10 liters. The actual sample volumes in the data range from 2 to 10,000 liters. The 25th percentile and the median of the volumes are both 10 liters. The simulation results (Figure 10.29) suggest that the model is adequate in replicating the extreme highs, but underpredicts the number of systems with all 0 observations.

As in the Cape Sable seaside sparrow example (Figure 9.7 on page 404), the model is unable to replicate the number of all 0 systems. When using a Poisson regression model, we imply that the "true" mean is nonzero. If cryptosporidium is ubiquitous in source water, this assumption is scientifically sound. Because cryptosporidium is transmitted by environmentally hardy cysts (closed sacs having a distinct membrane and division on the nearby tissue) and their presence is associated with the presence of mammals, the ubiquitous assumption is reasonable. Human influence is always present in an environment where a public drinking water system is needed. The discrepancy between the model replicated and observed percentage of all 0 systems can have important consequences. Because more systems would be predicted to have a detectable pres-

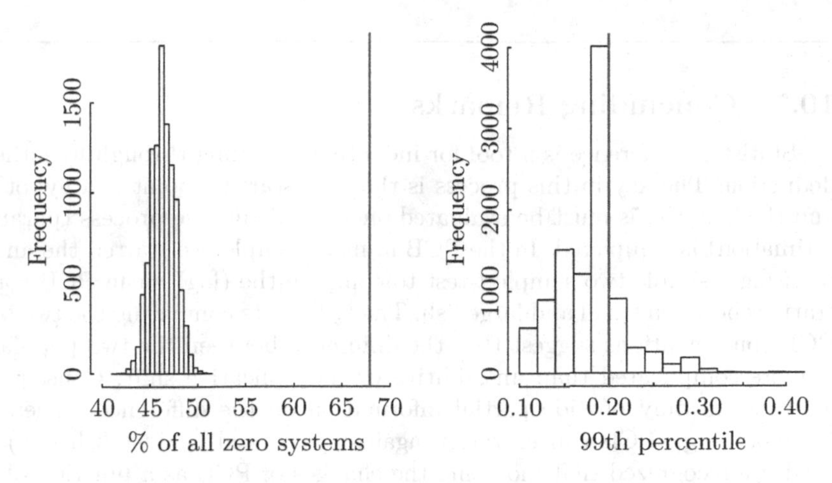

FIGURE 10.29: Simulating cryptosporidium in U.S. drinking water systems – The observed percentage of systems with all 0 observed cryptosporidium count is 68% (the gray vertical line), well above the simulated percentages (left panel); while the observed 99th percentile of system means (0.2 oocyst/L, the gray line) is well represented by the simulation (right panel).

ence of cryptosporidium, the use of this model would somewhat overstate the severity of the problem. One reason for observing more than expected number of zeroes may be the use of an average recovery rate, an oversimplification of the process of data generation. If there is a significant between-lab variation in the recovery rate, this model may have underestimated the concentration for some systems. There are many possible directions for the revision of this model.

First, when a water sample is taken to the lab for measuring cryptosporidium concentration, the sample may contain no oocyst even when the true concentration in the water is nonzero. As a result, a reported 0 is truly 0. When there is one oocyst in the sample, the chance that it will be detected is only 0.44. In other words, the probability of reporting a 0, a function of the recovery rate and the true concentration, is always larger than the probability of 0.56 calculated based on the measured recovery rate of 0.44. The smaller the true concentration, the larger the probability of observing a 0.

Second, there are many labs certified by the U.S. EPA for measuring cryptosporidium. Although they are all certified, these labs may have contributed additional variations in the reported number of cryptosporidium detections. These labs are required to report the results of "spiked samples," where a known number of oocysts are spiked into a sample and the number recovered is reported. This information is lab-specific. The analysis of the drinking water data should include this information to better quantify the recovery rate.

10.7 Concluding Remarks

Statistical inference is a tool for inductive reasoning through hypothetical deduction. The key in this process is that the starting point is a hypothesis, and the hypothesis must be evaluated once the deductive process (parameter estimation) is completed. In the PCB in fish example, we started the analysis by using a simple two sample t-test to compare the (log) mean PCB concentrations between small and large fish. The Q-Q plots comparing the two sets of PCB concentrations suggest that the difference between the two populations is more complicated than an additive or multiplicative shift. Consequently, a t-test can only provide partial information on the difference. When scatter plots (e.g., PCB concentration against year, and against fish size) were used, we recognized that modeling the changes of PCB as a function of both year and fish length is more informative. But when examining the residuals of the resulting model, we further noticed that the linearity assumption on the fish length is problematic. As a result, we used a hockey stick model. While preparing Chapter 9 of this book, the fish length imbalance was first discovered. Through subsequent inquiries, I learned the change in sampling (fish collection) method.

In the crypto example of Section 10.6.2, an implicit assumption of using the Poisson regression model is that the underlying true mean crypto concentration is nonzero. This assumption may well be true for a large body of water. But when a sample of 10 liters of water is taken from a lake or river, there is a chance that there will be no crypto in it (hence the concentration will be underestimated). When the water sample did capture one crypto oocyst (and recovered in the measuring process), the concentration for that particular sample may be overestimated if the true concentration is less than 0.1 oocyst per liter. As a result, a shrinkage estimator such as the multilevel model is desirable. Through the shrinkage effect, we move low concentrations up and high concentrations down.

These two examples illustrate the nature of statistics and the difficulty of learning statistics. When facing a data analysis or modeling problem, we are interested in the underlying process (the science) that resulted in the data. We observed the results (particular) and want to make an inference about the underlying model (general). A scientific problem is almost always an induction problem. As a result, a formula for a successful scientific inquiry does not exist. However, inductive reasoning can also be seen as the process of ascertaining the likelihood of a theory, a model, or a hypothesis being "true." Many important contributions from R.A. Fisher are related to the facilitation of the hypothetico-deductive reasoning in science. The hypothesis testing using p-value is such an example. As an analytical tool for induction, we should not expect a unique formula for a successful application of statistics. However, statistics provides a means for probabilistic inference on the likelihood of a proposed theory (or model) being true. Using statistics, we almost always follow the hypothetico-deductive reasoning method, where a model is proposed and the fitted model is evaluated either using scientific knowledge or new data. This feature is most clearly demonstrated in Fisher's hypothesis testing process and the use of the p-value. The null hypothesis is the theory from which we derive the expected behavior of a test statistic should the null hypothesis be true. The null hypothesis is then evaluated by calculating the p-value. The Neyman–Pearson paradigm of hypothesis testing removed the connection between statistics and science and led to the use of the null hypothesis as a straw man to be shot down. Although the Neyman–Pearson approach is useful in decision-making situations, its place in scientific research is questionable.

Traditionally, statistics is taught through a sequence of topics that appears to have a system. But such a sequence often leads students to confuse statistics with mathematics. Because mathematics is a tool for deductive reasoning, this confusion is not helpful. In a typical environmental/ecological curriculum, we teach t-test, ANOVA, and linear regression in a graduate level statistics course. This approach of teaching statistics emphasizes the problems of parameter estimation and distribution but neglects the problem of specification (see Chapter 1). The problem of specification is, however, the most important step in science. In this regard, Fisher's narrative of the three problems in statistics is a concise summary of the scientific methods: an iterative process of proposing hypothesis (a model), fitting data to the hypothesized model, checking the discrepancy between the hypothesis and the data, and revising the hypothesis.

Fitting a model is important, but assessing the validity of assumptions is more important. But unlike the model fitting process, there is no rule to follow in model evaluation. As a result, I find the process of learning statistics is both difficult and easy. The easy part is the procedures for carrying out a hypothesis test and the routines used for fitting specific models. The difficult part is model assessment, especially the assessment of model assumptions. We use model residuals for checking important assumptions and for searching for discrepancies between data and the fitted model. Because the fitted model is optimized with respect to the data, residuals are less efficient in revealing some misfits in the model. Chamberlin [1890] recommended using "multiple working hypotheses" to develop rational explanations of new phenomena. In Chamberlin's words, when only one hypothesis is proposed to describe a new phenomenon, the investigator may have "the partiality of intellectual parentage." Proposing multiple working hypotheses will likely neutralize the partiality as the investigator is now "the parent of the family of hypotheses." In fitting a statistical model, model parameters are estimated so that the model fits the data the best. Often we cannot readily detect a model's flaws using standard model evaluation methods. The multiple working hypotheses approach would suggest that we propose multiple alternative models. Qian and Cuffney [2012] used four alternative models to learn about the pattern of macroinvertebrate response to watershed urbanization and suggested that a threshold response pattern is unlikely. Qian [2014b] used the same approach and reexamined the estimated phosphorus threshold of the Everglades wetland. Both studies suggested that the simple step function model is unlikely the best model.

Many examples in this book include a simulation component. Using simulation, we can evaluate a model by comparing model predictions to many unique features in data or important criteria based on knowledge of the subject matter. Discrepancies revealed from simulations are often the starting point of a revised model that is likely to result in improved understanding of the subject. The basic principle of simulation is simple and easy to understand: we draw random samples (simulated data) from a model. These simulated data are compared to observed data. These simulated data can also be used to calculate parameters representing important features of the data or known criteria based on knowledge of the subject matter. In practice, the difficulty is the choice of the evaluation criteria. What to simulate or what to compare is not only a statistical problem, but also a scientific problem. Without subject matter knowledge, effective simulation can be difficult. I used the percentage of observed zeros in both the seaside sparrow and the cryptosporidium examples. In both cases, I explained the possible reasons for observing a 0 in the experiment or survey. Once the problem is clearly described, it is obvious that some zeroes are "real" in the sense that there was really no bird or pathogen in the area or sample when a 0 was recorded. The Poisson regression model cannot account for such a problem because the link function is a logarithmic transformation. In both examples, the detection probability (the probability of detecting a bird and the probability of capturing a cryptosporidium oocyst) is a conditional probability, that is, conditional on the presence of at least one bird or one oocyst. When the recovery rate (the estimated detection probability) was used in the Poisson regression model, we made a statistical mistake by ignoring the conditional nature of the probability of detecting an oocyst. The simulation results suggest that the model is adequate in predicting the 99th percentile of system means. The 99th percentile is represented by systems with a very high concentration and zero inflation is less likely than in systems with a lower concentration.

Peters [1991] suggested that the most common weaknesses of the methods section of an ecological paper are likely to be statistical. This is not because of the lack of statistical training of environmental and ecological scientists, but because of the disconnect between statistics and the applied disciplines. Learning statistics and applying statistics are two very different processes. In learning statistics, we learn separate methods for different types of data. In applying statistics, we usually don't know what would be the appropriate model until we try and explore. Peters further suggested that "statistics are better learned from direct applications of statistics in the context of one's own research, supplemented whenever possible with appropriate readings, texts and courses." But such continued learning should be guided by a clear understanding that a statistical analysis/model is built upon a set of assumptions. As a result, a critical mindset is necessary in learning how to ask questions about these assumptions.

10.8 Bibliography Notes

The subject of multilevel models is discussed by Andrew Gelman and his group at Columbia University, summarized with an emphasis on application in Gelman and Hill [2007]. A more theoretical discussion on the use of multilevel models for ANOVA problems can be found in Gelman [2005]. Applications of multilevel models along with some theoretical background are in Pinheiro and Bates [2000].

10.9 Exercises

1. Routine water quality data are used in the U.S. by state agencies for assessing environmental standard compliance. Frey et al. [2011] collected water quality and biological monitoring data from wadeable streams in watersheds surrounding the Great Lakes to understand the impact of nutrient enrichment on stream biological communities. Because sample sizes for different streams vary greatly, assessment uncertainty also fluctuates. Qian et al. [2015b] recommended that similar sites be partially pooled using multilevel models for improving assessment accuracy. Water quality monitoring data from Frey et al. [2011] are in file `greatlakes.csv`. The data file includes information on sites (e.g., location), sampling dates, and various nutrient concentrations. Of interest is the total phosphorus concentration (`Tpwu`). Detailed site descriptions are in file `greatlakessites.csv`, including level III ecoregrion, drainage area, and other calculated nutrient loading information.

   When assessing a water's compliance to a water quality standard, we compare the estimated concentration distribution to the water quality standard. The U.S. EPA recommended TP standard for this area is 0.02413 mg/L. We can use monitoring data from a site to estimate the log-mean and log-variance to approximate the TP concentration distribution (a log-normal distribution) and can calculate the probability of a site exceeding the standard.

   - Use linear regression to estimate site means simultaneously (with site as the only predictor variable) and estimate the probability of each site exceeding the standard assuming a common within-site variance.

- Use the multilevel model to estimate site means and estimate the probability of each site exceeding the standard. Compare the multilevel model results to the linear regression result and discuss the difference.

2. A frequently used indicator of stream ecosystem conditions is the species richness (number of species or taxa found in a sample). In the EUSE example, we often use (1) total number of macroinvertebrate taxa (richness) and (2) the number of taxa belonging to three orders known to be sensitive to pollution. These three orders are Ephemeroptera (mayflies), Plecoptera (stone flies), and Trichoptera (caddisflies). Taxa in these three orders are collectively known as EPT taxa. EPT taxa richness is often more indicative of water quality than total taxa richness. The data file posted on USGS web page (`http://pubs.usgs.gov/sir/2009/5243/data/EUSE_USGSReportData.csv`) includes total richness (`RICH`) and EPT taxa richness (`EPTRICH`).

   (a) Develop a multilevel model to model the response of total taxa richness (`RICH`) to changes in watershed urbanization represented by the national urban intensity index (`NUII`), as well as to regional level climate conditions.

   (b) Develop a multilevel model to study the response of EPT taxa richness (`EPTRICH`) to changes in urban intensity, as well as to regional level variables such as mean temperature and precipitation.

   (c) Another way to examine the changes in biological community is to examine the changes in the relative EPT taxa richness (i.e., EPT taxa richness as a fraction of total number of macroinvertebrate taxa) along the urban gradient. Develop a multilevel logistic regression model to study the changes of relative EPT taxa richness and compare the results to the EPT taxa richness model.

   (d) Considering the connection between multinomial and Poisson models discussed in Chapter 8, discuss how the connection can be used in a multilevel setting.

3. In the Galax example, a simulation is used to present the uncertainty of the estimated large leaf fraction (Figure 10.26). Use the same simulation method to estimate the uncertainty of large leaf density (model `G.lmer1`) and compare the results to Figure 10.24.

4. Many institutions around Lake Erie have long-term water quality monitoring programs. These institutions, however, use several different water sampling methods and several chemical analytical methods for measuring water quality variables. The data file `Eriecombined.csv` includes

monitoring data from six institutions collected from 2010 to 2013. Figure 3.4 shows data from NOAA. One way to examine institutional differences is to use a multilevel model with institution as a factor variable, after other factors affecting water quality (TP and chlorophyll a concentrations in particular) are accounted for. These factors include (1) distance to the source of phosphorus (the Maumee River), (2) year, and (3) season (months).

(a) Use exploratory data analysis tools (e.g., Q-Q plots) to determine the nature of the difference (e.g., multiplicative or additive differences). Based on the exploratory analysis, recommend an appropriate transformation for the two water quality variables of interest (TP and chlorophyll a concentrations).

(b) Fit multilevel models for TP and chlorophyll a concentrations (`TP` and `CHLA`, respectively), using distance to Maumee River mouth (`DISTANCE`) as a continuous predictor and `INSTITUTION`, `YEAR`, and `SEASON` as three factor variables. Describe model outputs in plain language.

(c) Present the differences among institutions graphically.

Chapter 11

Evaluating Models Based on Statistical Significance Testing

11.1	Introduction	493
11.2	Evaluating TITAN	495
	11.2.1 A Brief Description of TITAN	496
	11.2.2 Hypothesis Testing in TITAN	498
	11.2.3 Type I Error Probability	499
	11.2.4 Statistical Power	503
	11.2.5 Bootstrapping	511
	11.2.6 Community Threshold	512
	11.2.7 Conclusions	513
11.3	Exercises	514

11.1 Introduction

Applications of statistics can be grouped into two categories: model development and model evaluation. Model development is to propose a hypothesis or model and model evaluation is to assess the validity of the model. We discussed the differences between Fisher and Neyman-Pearson in Chapter 4. Throughout the book, I followed Fisher's hypothetical deductive approach because I believe the Fisherian approach is consistent with scientific methods. The hypothetical deductive approach, however, requires knowledge in both statistics and subject matter knowledge. Statistics helps us to propose feasible probabilistic distribution assumptions, while the subject matter knowledge ensures that the proposed model is reasonable.

When developing a hypothesis, I consult subject matter experts on the likely model forms. In addition, I follow Tukey's advice in conducting a thorough exploratory data analysis (EDA). Some EDA tools have been described in this book. A thorough EDA often allows me to provide a good summary of the data. But more importantly, a model derived based on EDA results is more likely consistent with data. After necessary revision of the initial model based on discussions with subject matter experts, the model will be fit to the data. Once a model is developed, we start the process of model evaluation. In addition to the usual model assessment steps described in, for example,

Chapter 5, I often use simulation to evaluate whether a mode can capture known features of the data.

In this chapter, I present a detailed critique of a model proposed for detecting ecological thresholds. This example illustrates how statistical simulation can be used to characterize a hypothesis test-based model, and how such characterization can be used for model assessment. With readily available computing power, computation intensive methods such as CART and bootstrapping are increasingly accessible. However, hypothesis testing using these computation intensive methods can often lead to an unintended "multiple comparison trap," where the result is selected among many comparisons or repeated hypothesis testing. For example, I presented a threshold estimation method borrowing the CART strategy of searching for the break point resulting in the largest reduction of response variable deviance [Qian et al., 2003a]. In that paper, I suggested that the statistical significance of the resulting "threshold" can be tested using a χ^2 test. The null hypothesis of the test is that the response variable distribution does not change along the gradient. The χ^2 test can be simplified to a t-test when the response variable distribution is normal. However, the process of finding the change point searches all possible splits to find the one resulting in the largest difference (in deviance) between the two groups. In other words, the estimated threshold is based on repeated comparisons of two groups. A t-test on the threshold is then misleading because the true type I error probability is much larger than the declared significance level. The type I error probability can be estimated by using a simple simulation.

To estimate the probability of making a type I error using simulation, we repeatedly draw data from a distribution specified by the null hypothesis and carry out the test for each set of simulated data. The probability of making a type I error is the fraction of time when the null is rejected. For this case, we can use the function `chngp` in Section 9.3.2 to calculate the change point and conduct a t-test to see if the difference in means is statistically different. Using a significance level of 0.05, we should reject the null hypothesis about 5% of the time.

```
#### R Code ####
set.seed(123)
reject <- 0
n.sims <- 50000
for (i in 1:n.sims){
   temp <- data.frame(X=runif(30), Y=rnorm(30))
   split <- chngp(temp)
   if (split==min(temp$X) | split==max(temp$X))
        p.value=0.5
   else
        p.value <- t.test(Y~I(X<split), data=temp,
                              var.equal=T)$p.value
   reject <- reject + (p.value<0.05)/n.sims
```

```
}
print(reject)
> [1] 0.1568
```

When using a sample size of 30, the type I error probability is 0.1568, not the declared 0.05. The inflated type I error probability is expected; and the larger the sample size is, the higher the type I error probability is. In other words, using a significance test on a CART model result can be misleading. I did not recognize the problem while writing the paper because my emphasis was on another change point model described in the same paper. The significance test of the change point was added to the paper during the review process.

A type I error in the context of identifying ecological threshold is a false positive result. With the increased popularity of computation intensive methods, the likelihood of inflated type I error probability becoming a problem is also increasing. In the rest of this chapter, I will focus on a more complicated model for detecting ecological threshold. The statistical issue behind the model is the same: the significance level of a test is not what is declared because of the multiple comparison trap. Furthermore, the rejection of the null hypothesis does not imply support to a specific alternative hypothesis. When we are to conclude a step function as the alternative model, rejecting the null of no change is not enough.

11.2 Evaluating TITAN

Model evaluation is a difficult task requiring knowledge in both statistics and the subject matter science. We have discussed model evaluation extensively in this book, with a focus on evaluating model assumption compliance. In this example, I will focus on the assessment of a model or method based on statistical significance test. A statistical significance test is often characterized by the probabilities of making type I and type II errors. A properly designed test (based on the Neyman–Pearson lemma) sets the probability of type I error to be a small constant (the significance level α) and minimizes the type II error probability. Briefly, a test starts with a definition of the null hypothesis and a test statistic. The test statistic has a known probability distribution under the null hypothesis, known as the null distribution. A rejection region is defined based on the null hypothesis such that the type I error probability is limited to α. The lemma shows that a test following this procedure has the smallest type II error probability among all possible tests.

Accordingly, when evaluating a significant test-based method, we should present (1) the null hypothesis, (2) the test statistic, (3) the null distribution, and (4) the rejection region. If this information can be clearly presented and verified, the Newman–Pearson lemma guarantees that the test is optimal, in that the probability of making a type I error is limited to α and the probability of making a type II error is minimized. In scientific research, questions facing a scientist are often more complicated than the questions addressed by typical significant tests described in statistics textbooks. As a result, significant test-based methods developed to address scientific questions are often a combination of more than one "textbook test." Consequently, the evaluation of these methods can be difficult. However, the basic concepts of a significant test apply. We should still be able to characterize a test-based method using the probabilities of making type I and type II errors.

As in the simulation for comparing a nonparametric test and the t-test in Section 4.5.4 (page 106), we examine the test by evaluating the type I error probability under the condition of the null hypothesis model. Comparing this probability with the declared significance level (α) we can describe the performance of the test when the null hypothesis is true. Furthermore, we examine the behavior of the test under selected alternative models by estimating the test's statistical power.

The concept of ecological threshold is appealing to environmental managers as it implies that a critical point, beyond which the ecosystem in question may be irrevocably changed, can be identified. Consequently, knowing the threshold would help the manager to set a goal for the protection of the ecosystem. In the Everglades example, we see the use of the concept in developing an environmental standard for total phosphorus. For example, Richardson et al. [2007] used a step function model to study how several ecological indicators change as a function of total phosphorus concentration. The step function is a type of threshold model. It assumes that the specific feature of the Everglades ecosystem measured by the ecological indicator does not change as phosphorus concentration increases. Once the concentration exceeds a threshold the indicator will jump to a different level and stay constant again as phosphorus concentration continues to increase. In Fisher's hypothetical deduction framework, the model is a hypothesis to be tested. I discussed issues related to model evaluation of this type of models in Qian [2014a]. In this section, I present an evaluation of a more complicated hypothesis testing based model for estimating ecological threshold.

11.2.1 A Brief Description of TITAN

Baker and King [2010] presented a program known as the threshold indicator taxa analysis (TITAN) for calculating community "thresholds." TITAN is often applied to find the level of disturbance beyond which a significant change at a community level is expected. Data used by TITAN are species abundances along an environmental or disturbance gradient. For each species (or taxon), the program finds a splitpoint along the gradient to divide the samples into two groups. The splitpoint is selected based on an indicator value that describes a taxon's association with a number of existing clusters

[Dufrêne and Legendre, 1997]. When there are only two clusters, the indicator value (IV) is the product of the taxon's relative abundance and its frequency of occurrence:

$$IV_i = A_i B_i \tag{11.1}$$

where $i = 1, 2$ is the cluster index, A_i is the relative abundance (fraction of individuals of the taxon in cluster i) calculated as the ratio of the mean abundance in cluster i (a_i) over the sum of the cluster means ($A_i = a_i/(a_1 + a_2)$), and B_i is the frequency of occurrence (fraction of non-zero observations) in cluster i.

While IV was developed to describe the association of a given taxon to an existing cluster, TITAN uses IV to define clusters along a disturbance gradient. Observations along the gradient are successively divided into two groups by moving a dividing line along the gradient; an IV is calculated for each group at each potential splitpoint. Baker and King [2010] defines the IV for each potential splitpoint as the larger of the two IVs and selects the splitpoint as the one with the maximum IV. The splitpoint selected under this definition is the same as the splitpoint with the largest difference between the two IVs based on the definition of Dufrêne and Legendre [1997]. This process searches for a maximum of the indicator value differences between the two groups. The gradient value associated with the maximum IV value is identified as an ecological threshold. Because the calculation of IV requires two clusters (or groups in this case), TITAN starts the search at some distance from the low end of the gradient to allow a pre-determined number of data points to be included in the "left group" and ends also at a distance from the upper end of the gradient to allow the same minimum number of data points in the "right group." This calculation is equivalent to a truncated variable transformation (truncating the data at both ends of the disturbance gradient and transforming the total abundance data into IV).

Two statistical inference methods were used on the maximum. One is the permutation test for "statistical significance" of the identified threshold. When the maximum is statistically "significant," the location of the maximum along the gradient is used as the estimate of the threshold. The other inference is about the uncertainty of the estimated threshold, using the bootstrapping method. This process of threshold identification (the permutation test) and estimation (bootstrapping) is repeatedly applied to each taxon separately. Thresholds for individual taxa are combined through the use of normalized IV values to derive the "community" threshold. I will focus on the evaluation of the permutation test because it is the basis of the subsequent analyses.

An obvious problem of TITAN is the use of the permutation test on the maximum of IV along the gradient. This problem is similar to the multiple comparisons in an ANOVA problem. That is, when multiple variables are drawn from the same population and we only compare the pair of variables with the largest sample mean difference, the t-test will reject the null hypothesis of no difference more often than the declared significance level α (the probability of making a type I error). This is because the two samples are no longer simple random samples. Likewise, when applying the permutation test on the two groups with the largest IV value, the data are not simple random samples. Often the violation of the independence assumption is not obvious. As a result, we should evaluate the method by its probabilities of making type I and type II errors. A type I error is erroneously rejecting the null hypothesis when the null is true. A type II error is the failure to reject the null hypothesis when the alternative hypothesis is true. Type I error probability can be estimated by carrying out the test using data from the null hypothesis model, which requires that we know the null hypothesis model and are able to draw random data from the model. As in Section 4.5.4 (page 106), we will use simulation to evaluate the probability of making a type I error. A type II error (and the power) is associated with a specific alternative hypothesis. As a result, I will specify a number of relevant alternative models. Unlike the simulations in Section 4.5.4 where the null hypothesis model is a simple normal distribution, the null hypothesis model in this case is not defined in the TITAN literature. Instead, a vague description of one class of potential alternative hypothesis models is mentioned.

11.2.2 Hypothesis Testing in TITAN

TITAN derives the threshold based on a series of hypothesis tests. First, for a given taxon at a given splitpoint, TITAN uses a permutation test. The data distribution assumption was not clearly defined, hence the preference of a "distribution-free" test. The data used to calculate the IVs are the taxon abundance values observed along an environmental gradient. Using hypothesis testing, the TITAN authors want to learn whether the estimated IV is "statistically significant." When a splitpoint is "significant," the respective gradient value is deemed as a potential threshold. When we used the term "statistically significant" in Chapter 4, we mean that the observed data show strong evidence against the null hypothesis model. That is, statistical significance is relative to the null hypothesis model. What is, then, the null hypothesis in the permutation test incorporated in the TITAN program?

The null hypothesis was never clearly stated in a way that can be used to formulate a proper model. But based on what a permutation test does, I will try to deduce what is the null. The observed IV is calculated from two subsets of taxon abundance data separated by the splitpoint along the disturbance gradient. For a specific splitpoint, the sample sizes of the two subsets (n_1 and n_2) are known. The permutation test, in theory, tabulates all possible permutations of splitting the data into two subsets of sample sizes n_1 and n_2. For each permutation, an IV is calculated. The collection of these IVs is used to form an empirical distribution, and the p-value is calculated as the fraction of these IVs exceeding the observed IV. In other words, the distribution of these permutation-derived IVs is taken as the null distribution. The null hypothesis, with respect to taxon abundance, must be that taxon abundance distribution is not affected by the disturbance gradient. Because abundance is a count variable, we can approximate its distribution using the Poisson distribution or its overdispersion variants; the null hypothesis can be simplified to be a constant mean abundance along the gradient.

Because TITAN uses IV as a test statistic, the distribution of IV under the null hypothesis is unknown. Using a permutation test, we can avoid the derivation of a theoretical null distribution of IV. As in all hypothesis testing situations, the null hypothesis is a specific statistical model and the alternative is undefined. Two questions arise:

1. If the permutation test of the IV of one splitpoint has a significance level of $\alpha = 0.05$, what is the type I error probability of the test of the maximum IV?

2. Why is the rejection of the null hypothesis equivalent to a threshold response model?

We address these questions using simulation.

11.2.3 Type I Error Probability

Characterizing a test's type I error probability can be easily done by using simulation. That is, we repeatedly carry out the test on data drawn from a model specified by the null hypothesis. The frequency of rejection is an estimate of the type I error probability. If the frequency is close to the declared significance level of α, the test behaves as expected. If the frequency is very different from α, the test is problematic. A frequency much lower than α suggests that the test has a probability of making a type I error much lower than the declared significance level. A lower than expected type I error probability implies a higher type II error probability (hence a lower statistical power). A test with a type I error probability higher than α is associated with a higher than expected power. At the same time, it is also more likely to reject the null hypothesis when the null is true.

As in Chapter 4, we can easily evaluate a test's type I error probability by using simulation: repeatedly drawing data from the null hypothesis model and carrying out tests on these fake data to record the frequency of rejection. In this example, TITAN's null hypothesis is that a taxon's mean abundance does not change along the gradient of interest. To simulate data likely to happen under the null hypothesis, I will assign a number of sampling points and draw taxon abundances from a Poisson distribution with mean of 20:

1. The simulation starts with a predetermined number (`ns`) of sampling points along a gradient between 0 and 1: `x <- seq(0,1,,ns)`.

2. For each sampling point, a taxon abundance is drawn from a Poisson distribution: `y <- dpois(ns, 20)`.

3. Assuming the resulting `ns` Poisson random variates are taxon abundance

data along the gradient, equation (11.1) is used to calculate IV values for all potential splitpoints, using TITAN's default setup. This step leads to the test statistic – the maximum of these IV values, and the gradient value at which IV is the maximum.

4. A permutation test is carried out (at the splitpoint with the maximun IV) to derive the null distribution and a p-value is calculated. When the p-value is less than $\alpha = 0.05$, the null is rejected.

5. The process is repeated 5000 times to record the number of times the null is rejected.

I repeated the simulation using `ns = 15, 25, 51, 101,` and `201` to evaluate whether the type I error probability is a function of sample size. The estimated type I error probabilities are 0.14, 0.23, 0.31, 0.31, and 0.30, respectively. These numbers are considerably larger than the significance level of $\alpha = 0.05$. Furthermore, it seems that the type I error probability increases as the number of sampling points increases. When discussing ANOVA, we mentioned the difference between family-wise and test-wise type I errors. In an ANOVA setting, a family-wise type I error concerns the rejection of one of many possible comparisons. This concern arises when multiple comparisons are of interest. In a multiple comparisons problem, we used the Tukey's method, where the null hypothesis distribution of the largest difference is derived. TITAN is similar to a multiple comparisons problem. The permutation test used in testing the significance of the maximum IV has a comparison-wise significance level of 0.05. As such, the "family-wise" type I error probability is always higher than 0.05.

Perhaps in an attempt to correct for the considerably higher than expected type I error probability, TITAN also uses a normalized IV calculated based on the permutation test. The normalized IV is called the z-score.

In the permutation test, a number of IVs are calculated at each potential splitpoint, one for each random permutation. These IV values form the null hypothesis distribution of IV. The mean ($\hat{\mu}_i$) and standard deviation ($\hat{\sigma}_i$) of these IV values are used to normalize the observed IV value:

$$z_i = \frac{IV_i - \hat{\mu}_i}{\hat{\sigma}_i}$$

Although the methods section of Baker and King [2010] stated that the gradient associated with the highest IV is used as the estimated threshold, the gradient value associated with the highest z_i value is clearly used as the threshold in the accompanying computer code. In other words, the test statistic is the z-score. Because it is the normalized IV, using z-score as the test statistic appears to assume that the z-score is a standard normal random variable under the null hypothesis. In the previous simulation, I also calculated the z-score and the type I error probability using the z-score as the test statistics, which are 0.14, 0.23, 0.31, 0.31, and 0.30, for ns = 15, 25, 51, 101, and 201, respectively. In other words, the change of test statistic did not make any difference in terms of the type I error probability.

There was little discussion on whether there would be a difference in the estimated threshold value when the test statistic is switched from IV to the z-score. It seems that the authors of TITAN assumed that the two test statistics will result in the same outcome (the same p-value and the same estimated threshold value). But this assumption was not obvious to me. To compare the two statistics under the null hypothesis, I carried out another simulation with `ns=101`. This time, I calculate IV and the z-score for all potential splitpoints. That is, a permutation test is carried out at each potential splitpoint. As before, the gradient is between 0 and 1 and the 101 sampling points are evenly spaced along the gradient. Using this simulation, I illustrate the pattern of IV and z-score along the gradient.

The null hypothesis assumes a constant taxon abundance along the gradient, which is simulated by drawing random count variables along the gradient from the same Poisson distribution with a mean of 20 (Figure 11.1(a)). For each iteration of the simulation, I calculate IV, as well as $\hat{\mu}, \hat{\sigma}$, and z-score, for all potential splitpoints with a minimum of five data points in each group. There are a total of 92 potential splitpoints. After the process is repeated 5000 times, there are 5000 simulated $IV, \hat{\mu}, \hat{\sigma}$, and z-scores at each potential splitpoint to approximate the distributions of these statistics along the gradient. The simulated IV distributions show a distinct pattern (Figure 11.1(b)); the means and standard deviations near both ends of the gradient are higher than the same near the middle of the gradient. The pattern in IV along the gradient suggests that we are more likely to identify split points near both ends as thresholds using IV even when the underlying taxon abundances are the same across the gradient.

Permutation means, as well as standard deviations, show a similar pattern as IV values, high near both ends of the gradient and low near the middle of the gradient (Figure 11.2). However, the estimated z-scores show no discernable pattern along the gradient (Figure 11.1 (c)). The distribution of these simulated z-scores is very close to the standard normal distribution at all potential splitpoints. The type I error probability would be around 0.05 at any given potential splitpoint if we reject the null when $z > 1.96$.

The z-score distributions for all potential splitpoints are the same ($N(0,1)$). Yet, our first simulation resulted in a type I error probability much larger than the significance level. Figure 11.1(c) explains the cause of the inflated type error probability: a test on the maximum IV along the gradient is more likely to be statistically significant, just like the comparison of the maximum difference among the differences of multiple pairs of means in an ANOVA problem.

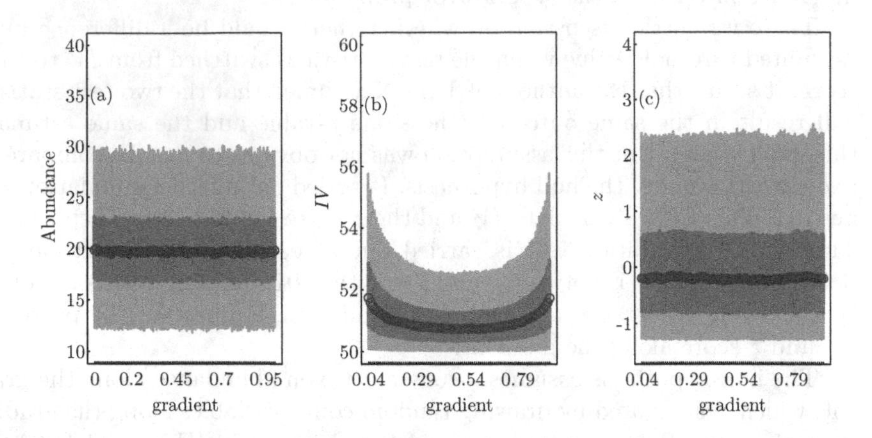

FIGURE 11.1: Simulated data under the null hypothesis model (a), IV (b), and z-score (c) distributions are shown along the gradient. Taxon abundance data at each location along the gradient were drawn from the same Poisson distribution with mean 20. The box plots show a summary of 5000 estimated quantities at each gradient location.

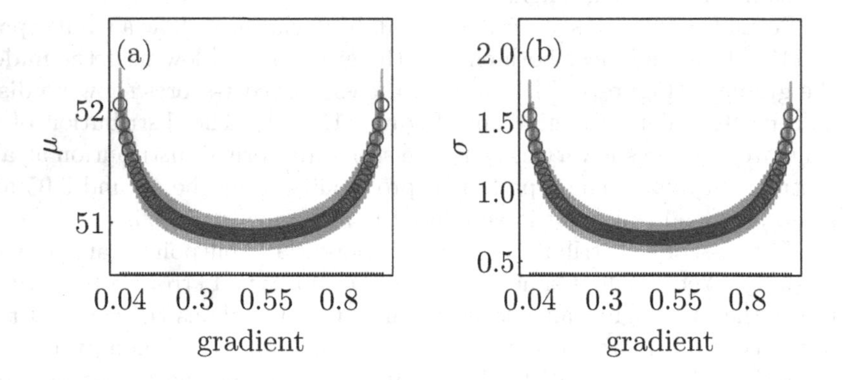

FIGURE 11.2: Permutation estimated mean (μ) and standard deviation (σ) of IV change along the gradient. The box plots show a summary of 5000 estimated quantities μ and σ at each potential splitpoint.

11.2.4 Statistical Power

The objective of TITAN is to find a threshold value. The stated goal of TITAN was "exploring and identifying abrupt changes in both the occurrence frequency and relative abundance of individual taxa along an environmental, spatial or temporal gradient." Using IV as the measure of such changes, Baker and King [2010] assume that the threshold value is the point along the gradient corresponding to the maximum IV value, as they described that when the maximum of IV along a gradient is statistically "significant," a threshold is identified.

A statistically significant result suggests that the null hypothesis is rejected. However, rejecting the null is not the same as evidence supporting a specific alternative hypothesis. There are many "threshold" models, among other possible alternatives. Only one is the "useful" model. The question we need to address after the null hypothesis is rejected should be "which model is supported by the data?" In a t-test, when the null hypothesis of no difference is rejected, we report the observed difference as an estimate, together with the estimated confidence interval of the difference. The confidence interval represents a range of the difference (alternative hypothesis) that is supported by the data. In other words, the confidence interval narrows down the range of all alternative hypotheses. Because the taxon abundance model is a function of the gradient, a likely model should be explored when the null hypothesis is rejected.

The stated goal is vague in that it does not allow a user to write a specific alternative hypothesis model with respect to taxon abundance data. The term abrupt change is not defined. One way to elucidate the meaning is to try several typically used mathematical forms for modeling abrupt change. For each potential alternative model, I will calculate the IV for each potential splitpoint on a set of simulated data without error and see which model results in a peak IV at the known "threshold." These alternative models are:

- Step function (SF) model

  This is the simplest model, stating that the response variable stays constant as we move along the gradient (the x-axis) until reaching a threshold (change point). Once crossing the change point the response variable jumps to a different value and stays constant again (Figure 11.3(a)).

  Mathematically, a step function model for a normal response variable is

  $$y_i = \beta_0 + \delta_1 I(x_i - \phi) + \varepsilon_i \tag{11.2}$$

  We can assume a constant variance ($\varepsilon \sim N(0, \sigma^2)$) or allow the variance to change when crossing the threshold (i.e., $\varepsilon_i \sim N(0, \sigma^2 + \delta_2 I(x_i - \phi))$). The function $I(\theta)$ is a unit step function, taking value 0 when $\theta \leq 0$ and 1 otherwise. The SF model has one discontinuity in the function itself, which is the threshold of interest.

- Hockey stick (HS) model

  This function assumes that the response variable changes as a linear function of the gradient with the slope changes at a threshold. The model resembles two line segments joined at the threshold (Figure 11.3(b)).

$$y_i = \beta_0 + (\beta_1 + \delta_1 I(x_i - \phi))(x_i - \phi) + \varepsilon_i \tag{11.3}$$

  The threshold of interest for the HS model is location of the discontinuity in slope (or the first derivative of the function).

- Disjointed broken stick (dBS) model

  A generalization of both the SF and HS models is the broken stick model. It is a model of two line segments with no constraints on the slope (SF model has two line segments with a common slope of 0) nor on intercept (HS model has two line segments joined at the threshold, or the same intercept with respect to the threshold) (Figure 11.3(c)).

$$y_i = (\beta_0 + \delta_0 I(x_i - \phi)) + (\beta_1 + \delta_1 I(x_i - \phi))(x_i - \phi) + \varepsilon_i \tag{11.4}$$

  The dBS model has discontinuities in the function and the first derivative of the function, and they coincide in the same location (hence the threshold of interest).

- Sigmoidal (SM) model

  A sigmoidal model is a continuous nonlinear model with lower and upper bounds, but without a change point (no parameter change along the gradient) (Figure 11.3(d)). I include the SM model because a threshold is often defined as a rapid change in the response over a short distance of the gradient. A change in one or more model parameters is not required. In other words, an abrupt change does not necessarily imply a discontinuity in the function or its derivatives. It can be simply a "rapid" (but smooth) change. I will use a simple logistic model as an example.

$$y_i = \frac{1}{1 + e^{-(\beta_0 + \beta_1 x_i)}} \tag{11.5}$$

  The SM model is the "smooth" version of the SF model. It is a continuous function with a continuous first derivative. The slope of the curve (first derivative) reaches the maximum (or minimum) at the inflection point. It is, therefore, natural to consider the inflection point as the threshold of interest.

To show how the test statistic changes along the gradient as a function of taxon abundances, I use simulated data without error. That is, I will assume that the pattern of change in taxon abundances along the gradient can be described by one of the four models and data are observed without error. Using data without error provides information on the behavior of IV as a

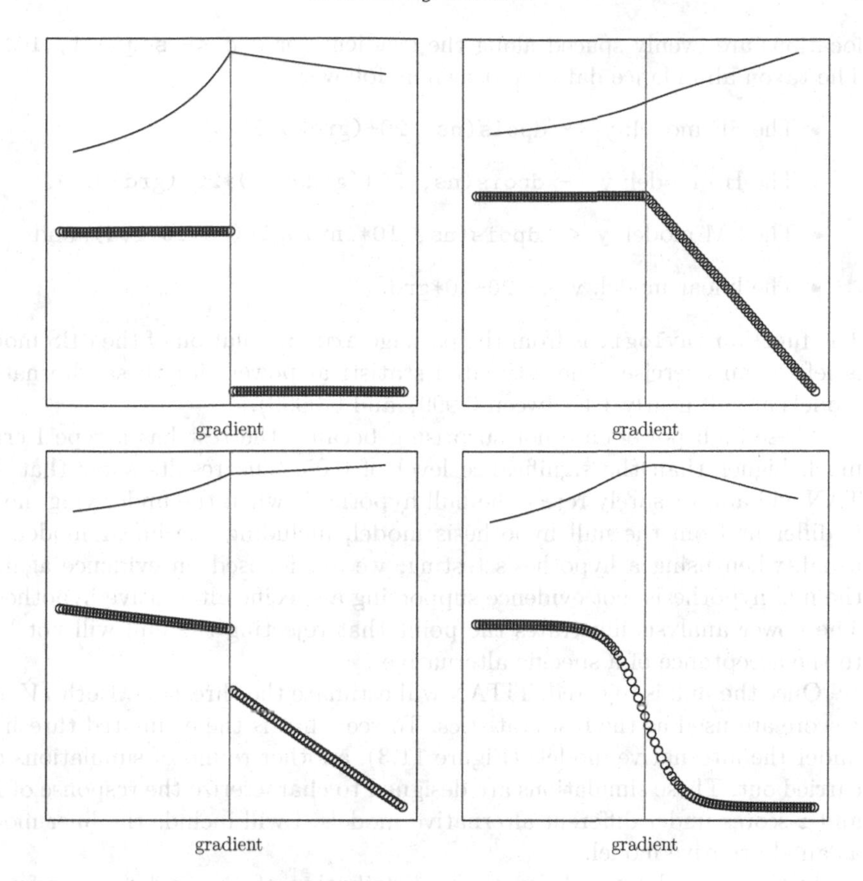

FIGURE 11.3: Data from four potential alternative models describing "abrupt" changes in abundance (open circles) are compared to the calculated *IV*s along the gradient (solid line).

function of the gradient under the four models. I am particularly interested in the location of the peak *IV* in comparison to the known "threshold." The simulated data are compared to the calculated *IV*s (Figure 11.3). It appears that the peak of *IV* coincides with the known threshold when the underlying model is the SF model only. For the HS model, the estimate *IV* is a monotonic function of the gradient. For the dBS model, the peak *IV* is either close to the known threshold or at one end of the gradient, depending on the difference in the slopes and intercepts of the two line segments. When slopes of both line segments are approaching 0, the peak is approaching the known threshold. For the SM model, *IV* peak is close to, but not at, the inflection point.

To estimate the statistical power of the test, I repeat the simulations in Section 11.2.3, except that the taxon abundance data are drawn from Poisson distributions with means calculated by the alternative models. The `ns` sampling locations are evenly spaced along the gradient (or `grd <- seq(0,1,,101)`). The taxon abundance data are drawn as follows:

- The SF model: `y <- dpois(ns, 20+(grd>0.5)*10,`
- The HS model: `y <- dpois(ns, 20+(grd>0.5)*20*(grd-0.5),`
- The SM model: `y <- dpois(ns, 10*invlogit(-5+10*grd)`, and
- The linear model: `y <- 20+10*grd`.

The function `invlogit` is from the package `arm`. Simulation of the dBS model is left as an exercise. The estimated statistical powers for these alternative models are all nearly 1 (between 0.9997 and 0.9999).

These high powers are not surprising because the test has a type I error much higher than the significance level of 0.05. The results show that TITAN will almost surely reject the null hypothesis when the underlying model is different from the null hypothesis model, including the linear model. As usual, when using a hypothesis testing, we are focused on evidence against the null hypothesis, not evidence supporting a specific alternative hypothesis. The power analysis illustrates the point that rejecting the null will not lead to the acceptance of a specific alternative.

Once the null is rejected, TITAN will estimate the threshold. Both IV and z-score are used as the test statistics. To see what is the estimated threshold under the alternative models (Figure 11.3), another round of simulations are carried out. These simulations are designed to characterize the response of IV and z-scores under different alternative models. I will include the liner model as an alternative model.

In these simulations, I derive the distribution of IVs and z-scores for all potential splitpoints. These simulated distributions are presented along the gradient using boxplots.

The linear model assumes that the abundance increases linearly along the gradient (Figure 11.4(a)). This model is different from the null hypothesis model, but without an abrupt change along the gradient. Under the linear model, the IV decreases along the gradient (Figure 11.4(a)), while the permutation estimated means and standard deviations still show the same pattern as before. The resulting z-scores are now shown a clear pattern and we have a statistically "significant" splitpoint near the middle of the gradient (Figure 11.4(c)).

At the low end of the gradient, the range of the middle 95% of the abundance data is between 12 and 30, and the same range at the upper end of the gradient is between 20 and 42. In other words, the null model (e.g., a constant abundance of 25) is within the 95% range of the alternative model. Nevertheless, the statistical power of TITAN for this linear model is still nearly 1, an outcome of the inflated type I error probability due to the multiple comparison trap.

A statistically significant result is desirable because the null hypothesis

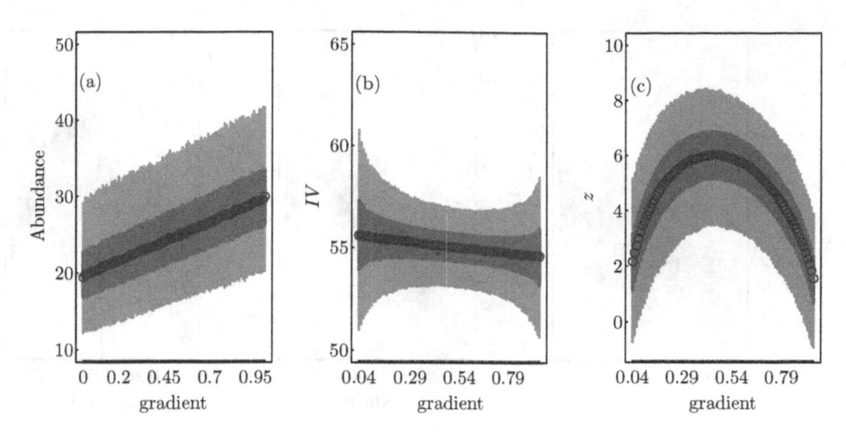

FIGURE 11.4: As in Figure 11.1, but the mean taxon abundance increases along the gradient linearly as shown in panel (a).

model is different from the data (generated from a linear model). But the result does not lead to the conclusion of an abrupt change along the gradient. The data used for the test were drawn from a model with a steady rate of increase. Just as rejecting the null hypothesis in a t-test does not support a specific alternative hypothesis, the rejection of the null hypothesis model of a constant abundance along the gradient cannot be equated to a support for a threshold response model.

Another issue of using TITAN is that the estimated "threshold" value based on IV will be different from the estimated value based on the z-score. The IV peak for the linear model is near the low end of the gradient, while the z-score peak is near the middle of the gradient. This discrepancy is never shown in all applications of TITAN. In a statistical change point problem, if the estimated change point is located near one end of the gradient, we conclude that the change point does not exist. In TITAN, IV is used as an indicator of the presence of a "threshold." If the peak of IV is located near one end of the gradient, we should conclude that no "threshold" is present. In this case, the IV peak is at the low end of the gradient. The standardized IV (the z-score) is calculated by subtracting $\hat{\mu}$ from IVs calculated for a splitpoint and the difference is divided by $\hat{\sigma}$. Because $\hat{\mu}$ is smaller in the middle of the gradient, the difference of $IV - \hat{\mu}$ will be inflated near the middle. Likewise, $\hat{\sigma}$ is lower in the middle, further inflating the difference near the middle of the gradient. As a result, the peak of z-score in this case is likely an artifact of the permutation-based standardization. From a hypothesis testing perspective, this discrepancy is inconsequential because the goal is to evaluate the null hypothesis model of a constant taxon abundance along the gradient. However, because TITAN's authors equate a significant result to the existence of a threshold, the use of the z-score is now misleading.

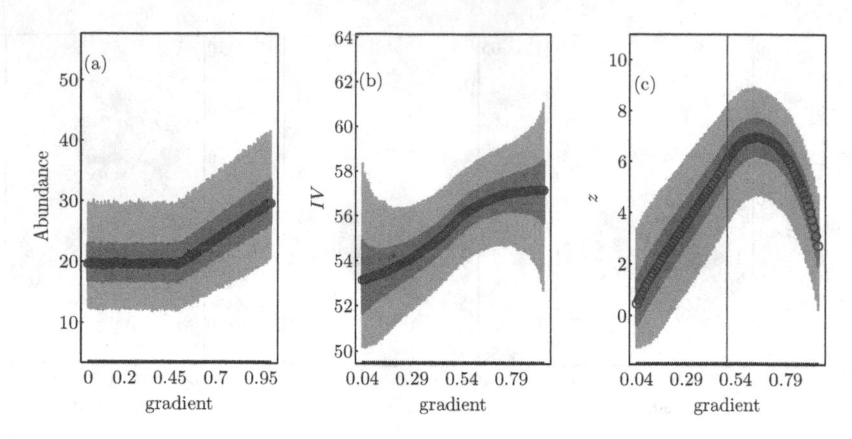

FIGURE 11.5: As in Figure 11.1, but the mean taxon abundance is modeled by a HS model (a).

When the alternative is the HS model (a constant abundance when the gradient is below 0.5 and a linearly increasing abundance after the change point, Figure 11.5(a)), IV increases monotonically along the gradient (Figure 11.5 (b)), while the z-scores show a peak near 0.65 (Figure 11.5 (c)).

Again, the power of the test is practically 1. Although the HS model does have a change point, the estimated "threshold" based on the z-score is not what we expected (0.5); neither is the location of the peak IV.

When the underlying model is the SF model, we have both a power of near 1 and a correctly identified threshold (Figure 11.6).

When the SM model is the underlying model, TITAN will also result in a significant result almost surely. The threshold of interest for the SM model is the inflection point located at 0.5. But the estimated threshold based on the z-score depends on whether the abundance is increasing or decreasing. If the abundance is increasing (as in our simulation), the estimated threshold is below the inflection point. If the abundance is decreasing along the gradient, the estimated threshold is above the inflection point. Furthermore, the test result is also influenced by the rate of change in an SM model. The larger the (maximum) slope is, the closer is the estimated threshold from the inflection point (Figures 11.8 and 11.9).

These simulations suggest that TITAN is effective in terms of rejecting the null hypothesis when the null is not true. However, TITAN has a type I error probability much larger than the declared significance level of 0.05. As a result, it can be overly sensitive to meaningless deviations from the null and a significant result may be practically meaningless. Because a statistical significance test is focused on evidence against the null hypothesis model, a significant result does not imply a threshold response of taxon abundance to disturbance represented by the gradient. Furthermore, TITAN estimated

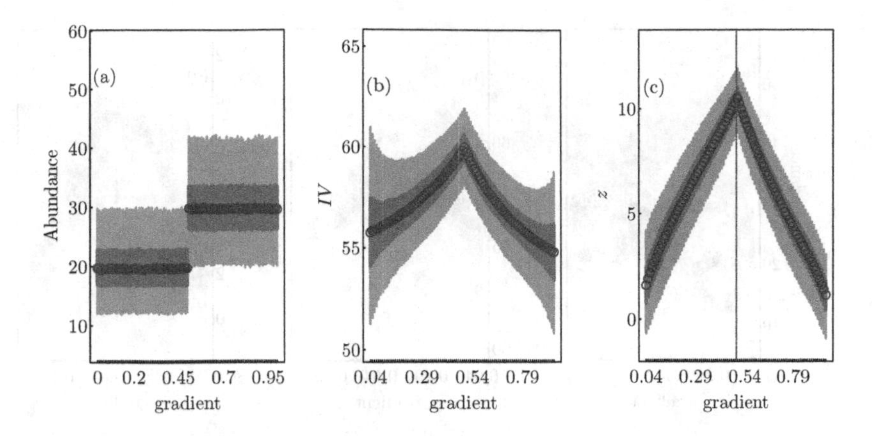

FIGURE 11.6: As in Figure 11.1, but the mean taxon abundance is modeled by a SF model (a).

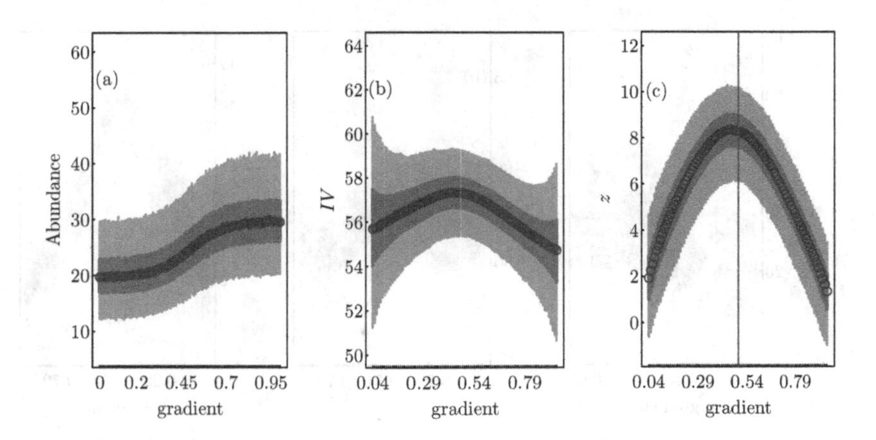

FIGURE 11.7: Same as in Figure 11.1, except that the mean taxon abundance is modeled by a SM model (a).

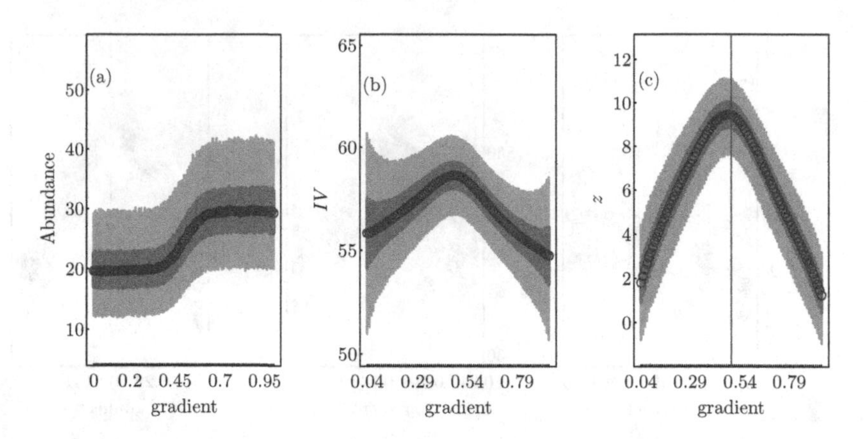

FIGURE 11.8: Same as in Figure 11.7, with a maximum slope twice as large as the same in Figure 11.7.

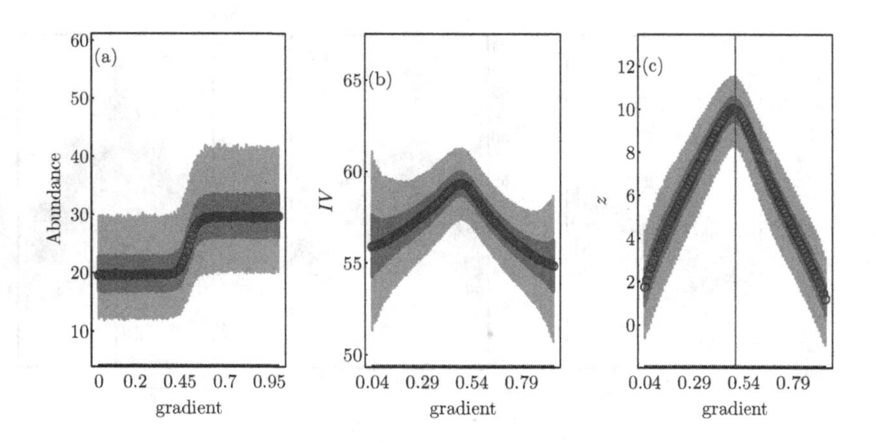

FIGURE 11.9: Same as in Figure 11.7, with a maximum slope 4 times larger than in Figure 11.7.

threshold is correct only when the underlying response pattern is consistent with the SF model.

Because the objective of TITAN is to quantify a specific threshold response model, a significance testing approach is not appropriate. The threshold response model is a specific alternative model. As a result, rejecting the null hypothesis of a constant abundance along the gradient does not translate to evidence supporting any specific alternative model. When a specific model is of interest, we fit the specific model to data and perform model evaluation as outlined in Chapter 9.

In these simulations, the estimated μ and σ have a consistent pattern along the gradient. They tend to be higher near both ends of the gradient than they are in the middle. Because the z-scores are the normalized IVs, z-scores near both ends of the gradient will be lower than z-scores near the middle if we have a constant IV along the gradient. When the null model is true, IVs near both ends are also higher than those near the middle. The net result is a more or less constant z-score along the gradient. When the null model is not the underlying model, for example, when the abundance changes as a linear function of the gradient, the IV will also change monotonically. The patterns in the permutation estimated μ and σ will result in the location of the peak of the z-score being different from the location of the peak of IV, resulting in contradictory estimates of the "threshold." In general, the peak z-score location will be closer to the middle of the gradient than the peak IV location. Apparently, the authors of TITAN did not recognize the potential contradictory results. In the R program accompanying [Baker and King, 2010], the permutation test is used to report the p-value (associated with the maximum of IV) and the peak z-score is used to identify the threshold.

11.2.5 Bootstrapping

TITAN also used bootstrap resampling to calculate the confidence interval of the selected split point. Bootstrapping is a commonly used resampling method for estimating standard deviation and confidence intervals of statistics [Efron and Tibshirani, 1993]. As we discussed in Chapter 9, bootstrapping is a Monte Carlo simulation procedure aimed at obtaining an approximate sampling distribution of the parameter of interest. It substitutes random samples from the target population with random samples of the same size (with replacement) from the existing data. As the sample size of the data increases, bootstrap samples become increasingly closer to random samples from the population. As a result, an empirical distribution of variable calculated from bootstrap samples approximates the true sampling distribution of the variable of interest as sample size increases. The bootstrap method is, however, not appropriate for a splitpoint problem [Bühlmann and Yu, 2002, Banerjee and McKeague, 2007]. Bühlmann and Yu [2002] have shown that the bootstrap estimated standard deviation of the splitpoint is always smaller than the true standard deviation, leading to a narrower confidence interval. In a splitpoint problem, a sample with k unique gradient values has $k-1$ potential splitpoints. Potential splitpoints in a bootstrap sample are a subset of the same $k-1$ potential splitpoints. In other words, the bootstrapping process repeatedly selects the splitpoints from the same pool of $k-1$ potential splitpoints with the result that the bootstrap estimated standard deviation is much smaller than it should be (and the estimated confidence interval is much narrower than it should be). The simulation carried out in Chapter 9 demonstrates this problem.

In addition to the problem of estimated confidence interval, using bootstrapping may also have an "edge effect." When using bootstrapping, only a subset of the individual data points is represented in a bootstrap sample. When data points near one or both ends of the gradient are not included, the range of the gradient is further truncated as we still must maintain the minimum number of data points in the two groups separated by a splitpoint. As the permutation test is not used when running bootstrapping in TITAN's R program, TITAN uses IV in each bootstrap simulation to identify the threshold. If the null model is true, IVs tend to be higher on both ends of the gradient. As a result, bootstrapping is likely to identify one or the other end of the gradient as the threshold. Because the range of the gradient in a bootstrap sample is frequently narrower than the range of the data, a bootstrap estimated distribution of the threshold would be shifted towards the middle of the gradient, resulting in an apparent "threshold" distribution slightly away from the end of a gradient.

11.2.6 Community Threshold

The last step of TITAN is to derive the community threshold based on the sum of z-scores calculated for all taxa at all potential splitpoints. This step implies a hypothesis test of the existence of a community threshold. TITAN calculates the sum of z-scores for every potential splitpoint x_i, which is equivalent to testing whether all taxa share the same splitpoint. TITAN's community threshold is implicitly defined as the splitpoint shared by all or most taxa and is selected as the splitpoint with the largest sum of z-scores, through repeated tests for all potential splitpoints. Statistically, such a test is meaningful only if this community threshold definition is meaningful. Under this definition, the sum of z-scores is a test statistic. Z-scores of individual taxa are random samples from the standard normal distribution. If we assume that these z-scores are independent of each other, the sum of these z-scores is a random variable with mean 0 and standard deviation $\sqrt{N}$, where N is the number of taxa. That is, we use the sum as a composite test statistic for the common null hypothesis for all taxa. We note that the sum of squares of the z-score, which has a χ^2 distribution with degrees of freedom of N, is more frequently used. Because this test is repeated for all potential split points, the resulting "community threshold" is likely a result of the type I error (the multiple comparison trap). However, the meaning of the test is moot as the community threshold defined by this test violates basic ecological principles. Species coexistence forms the conceptual basis for much of community ecology [MacArthur, 1972, Chesson, 2000, Hubbell, 2001] and there is a large body of theoretical work that validates the conclusion of Hutchinson [1959] (species must be different in some way to coexist [Chesson, 1991]). The synchronicity in thresholds reported by King et al. [2011] suggests that coexisting species do not differ in their response (e.g., optima) to changes in environmental resources. Ecological theory (e.g., competition, species packing, resource utilization) suggests that co-occurring species should exhibit differences in species optima, tolerances, and peak abundances across the environmental gradient [Gauch, 1982, Jongman et al., 1995], particularly for species responses that are associated with changes along natural environmental gradients (e.g., elevation, temperature, nutrients, prey abundance). This partitioning of resources should result in species that exhibit different rather than similar response thresholds. Synchronicity might be expected if the environmental change is associated with a toxicant. However, low levels of urbanization or eutrophication are generally not associated with high levels of toxicants; therefore, the synchronicity of thresholds reported by King et al. [2011] cannot be attributed to toxicity. It is far more likely that the apparent synchronicity of thresholds is an artifact of the method used to extract the threshold (e.g., the z-score) rather than an ecological attribute of the community.

11.2.7 Conclusions

TITAN is intended for uncovering discontinuous jumps in taxa abundance data along a disturbance gradient. Instead of formulating specific models about abundance, TITAN's authors used the clustering indicator IV. The resulting program is ambiguous in terms of what kind of threshold was detected. The misuse of the permutation test resulted in a systematic bias in the selected splitpoint based on the z-score towards the center of the data cloud along the gradient. TITAN is written to process large data sets with hundreds of taxa from many sites. As a result, the behavior of the program is opaque. Furthermore, Baker and King [2010] did not give the mathematical and ecological definition of a community threshold, nor the threshold concept at the individual taxon level.

From these simulations, we learned that a statistical test is to assess the evidence against the null hypothesis. In a simple two sample t-test, rejecting the null hypothesis (of a 0 difference of two means) does not provide any evidence in favor of a specific value of the difference. To conclude a specific alternative, evidence supporting the specific alternative model must be provided. Because TITAN's objective is to estimate a threshold, a threshold model is the assumed pattern. We should seek evidence supporting the specific threshold response model. But no specific threshold model was provided. The method packaged in the program implies a null hypothesis model of a constant abundance along the gradient. Rejecting the null model gives us no evidence of supporting any specific alternative model.

As we discussed in Chapter 4, a hypothesis test using a null of no difference should be used as a "devil's advocate." That is, we present our evidence in supporting the hypothesis of interest (in this case, a specific threshold model) and use the null hypothesis of no change as a last step to show that the data cannot be logically attributed to a model of no change. The null hypothesis test alone is not enough.

11.3 Exercises

1. In evaluating TITAN we used several alternative models to show the pattern of the IVs along an environmental gradient. Another natural pattern discussed by Cuffney and Qian [2013] is the Gaussian response model, where the response curve is similar to a bell-shaped curve. This response pattern is often used to represent the "subsidy-stress" response of a taxon. The initial increase in a pollutant (e.g., nutrient) provides subsidy to the growth of the organism, but the organism is stressed after the pollutant exceeds a threshold. The response pattern can be expressed as a parabola function of the gradient in log-abundance scale:

$$\log(y) = \alpha + \beta x + \gamma x^2$$

   where y is the taxon abundance and x is the environmental gradient. A logical threshold is the peak of the quadratic curve. Draw the response curve of IV similar to the curves in Figure 11.3 and discuss whether TITAN is appropriate for this type of threshold response.

2. Often organisms sensitive to pollution are used for setting an environmental standard. Because they are sensitive to environmental disturbance, their abundance can often be 0 near one or the other end of the gradient (e.g., linearly decline and reaching 0 halfway across the gradient). Design a simulation to study the behavior of TITAN when there are multiple zero abundance observations near one or the other end of the gradient.

Bibliography

R.P. Abelson. *Statistics as Principled Argument.* Psychology Press, New York, 1995.

A. Agresti. *Categorical Data Analysis.* Wiley, Hoboken, NJ, 2002.

C.W. Anderson, T.M. Wood, and J.L. Morace. Distribution of dissolved pesticides and other water quality constituents in small streams, and their relation to land use, in the Willamette River Basin, Oregon. Technical report, U.S. Geological Survey, Water-Resources Investigations Report 97-4268, Portland Oregon, 1997.

D.R. Anderson, K.P. Burnham, and W.L. Thompson. Null hypothesis testing: Problems, prevalence, and an alternative. *Journal of Wildlife Management*, 64:912–923, 2000.

E. Anderson. The irises of the Gaspe Peninsula. *Bulletin of the American Iris Society*, 59:2–5, 1935.

S.J. Arnold and M.J. Wade. On the measure of natural and sexual selection: Applications. *Evolution*, 38(4):720–734, 1984.

C.A. Bache, J.W. Serum, W.D. Youngs, and D.J. Lisk. Polychlorinated biphenyl residuals: Accumulation in Cayuga Lake trout with age. *Science*, 117:1192–1193, 1972.

M.E. Baker and R.S. King. A new method for detecting and interpreting biodiversity and ecological community thresholds. *Methods in Ecology and Evolution*, 1(1):25–37, 2010.

M. Banerjee and I. W. McKeague. Confidence sets for split points in decision trees. *The Annals of Statistics*, 35(2):543–574, 2007.

D.M. Bates. *lme4: Mixed-effects modeling with R.* Springer, 2010. URL `http://lme4.r-forge.r-project.org/book/`.

D.M. Bates and D.G. Watts. *Nonlinear Regression Analysis and Its Applications.* Wiley Series in Probability and Statistics. Wiley, New York, 2007.

J.H. Bennett, editor. *Collected Papers of R.A. Fisher.* Adelaide: University of Adelaide, 1971.

P. Bloomfield, A. Royle, and Q. Yang. Accounting for meteorological effects in measuring urban ozone levels and trends. Technical report, Technical Report # 1, National Institute of Statistical Sciences, Research Triangle Park, NC, 1993.

M. E. Borsuk, D. Higdon, C. A. Stow, and K. H. Reckhow. A Bayesian hierarchical model to predict benthic oxygen demand from organic matter loading in estuaries and coastal zones. *Ecological Modelling*, 143(3):165–181, 2001.

G.E.P. Box. Science and statistics. *Journal of the American Statistical Association*, 71(356):791–799, 1976.

G.E.P. Box and D.R. Cox. An analysis of transformations (with discussion). *Journal of the Royal Statistical Society, B*, 26:211–246, 1964.

L. Breiman. Bagging predictors. *Machine Learning*, 24:123–140, 1996.

L. Breiman. Random forests. *Machine Learning*, 45(1):5–32, 2001.

L. Breiman, J.H. Friedman, R. Olshen, and C.J. Stone. *Classification and Regression Trees.* Wadsworth International Group, Belmont, CA, 1984.

P. Bühlmann and B. Yu. Analyzing bagging. *The Annals of Statistics*, 30(4): 927–961, 2002.

R.K. Carey. Modeling N_2O emissions from agricultural soils using a multilevel linear regression. Master's thesis, Nicholas School of the Environment, Duke University, Durham, NC, 2007.

W.W. Carmichael. Cyanobacteria secondary metabolites—the cyanotoxins. *Journal of Applied Bacteriology*, 72(6):445–459, 1992.

T. C. Chamberlin. The method of multiple working hypotheses. *Science*, 15 (old series):92, 1890.

J.M. Chambers and T.J. Hastie, editors. *Statistical Models in S.* CRC Press, Inc., Boca Raton, FL, USA, 1991. ISBN 0412052911.

C.J. Chen, Y.C. Chuang, T.M. Lin, and H.Y. Wu. Malignant neoplasms among residents of a blackfoot disease endemic area in Taiwan: High-arsenic artesian well water and cancers. *Cancer Research*, 45:5895–5899, 1985.

H. Chen, D. Ivanoff, and K. Pietro. Long-term phosphorus removal in the Everglades stormwater treatment areas of South Florida in the United States. *Ecological Engineering*, 79:158–168, 2015.

P. Chesson. A need for niches? *Trends in Ecology and Evolution*, 6:26–28, 1991.

P. Chesson. Mechanisms of maintenance of species diversity. *Annual Review in Ecology and Systemantics*, 31:343–366, 2000.

L.A. Clark and D. Pregibon. Tree-based models. In J.M. Chambers and T.J. Hastie, editors, *Statistical Models in S.* Wadsworth & Brooks, Pacific Grove, CA, 1992.

R.B. Cleveland, W.S. Cleveland, J.E. Mcrae, and I. Terpenning. STL: A eeasonal-trend decomposition procedure based on loess. *Journal of Official Statistics*, 6(1):3–73, 1990.

W.S. Cleveland. *The Elements of Graphing Data.* Hobart Press, Summit, NJ, 1985.

W.S. Cleveland. *Visualizing Data.* Hobart Press, Summit, NJ, 1993.

J. Cohen. *Statistical Power Analysis for the Behavioral Sciences.* Lawrence Erlbaum Associates, Hillsdale, NJ, 1988.

J.J. Cole, B.L. Peierls, N.F. Caraco, and M.L. Pace. Nitrogen loading of rivers as a human-driven process. In M.J. McDonnell and S.T.A. Picket, editors, *Humans as Components of Ecosystems: The Ecology of Subtle Human Effects and Population Areas*, pages 141–157, New York, 1993. Springer-Verlag.

T.F. Cuffney and J.A. Falcone. Derivation of nationally consistent indices representing urban intensity within and across nine metropolitan areas of the conterminous United States. Technical report, U.S. Geological Survey, Scientific Investigations Report 2008-5095, 36 pp., 2008.

T.F. Cuffney and S.S. Qian. A critique of the use of indicator species scores for identifying thresholds in species responses. *Freshwater Science*, 32(2): 471–488, 2013.

T.F. Cuffney, H. Zappia, E.M.P. Giddings, and J.F. Coles. Effects of urbanization on benthic macroinvertebrate assemblages in contrasting environmental settings: Boston, Massachusetts; Birmingham, Alabama; and Salt Lake City, Utah. *American Fisheries Society Symposium*, 47:361–407, 2005.

C.C. Daehler and D.R. Strong. Can you bottle nature? The roles of microcosms in ecological research. *Ecology*, 77:663–664, 1996.

J.H. Davis. The natural features of southern Florida, especially the vegetation, and the Everglades. Technical report, Florida Geological Survey Bulletin, No. 25, 1943.

S.M. Davis and J.C. Ogden, editors. *Everglades: The Ecosystem and Its Restoration.* St. Lucie Press, Delray Beach, FL, 1994.

R.D. De Veaux and P.F. Velleman. Math is music; statistics is literature. *Amstat News*, pages 54–58, September 2008.

G. De'ath and K.E. Fabricius. Classification and regression trees: A powerful yet simple technique for the analysis. *Ecology*, 81(11):3178–3192, 2000.

S. Dodson. Predicting crustacean zooplankton species richness. *Limnology and Oceanography*, 37(4):848–856, 1992.

M. Dufrêne and P. Legendre. Species assemblages and indicator species: The need for a flexible asymmetrical approach. *Ecological Monographs*, 67(3), 1997.

R. Eckhardt. Stan Ulam, John von Neumann, and the Monte Carlo method. *Los Alamos Science*, 15:131–143, 1987.

B Efron. Biased versus unbiased estimation. *Advances in Mathematics*, 16: 259–277, 1975.

B Efron and C. Morris. Stein's estimation rule and its competitors – an empirical Bayes approach. *Journal of the American Statistical Association*, 68(341):117–130, 1973a.

B Efron and C. Morris. Combining possibly related estimation problems. *Journal of the Royal Statistical Society. Series B (Methodological)*, 35(3): 379–421, 1973b.

B Efron and C. Morris. Data analysis using Stein's estimator and its generalizations. *Journal of the American Statistical Association*, 70(350):311–319, 1975.

B. Efron and C. Morris. Stein's paradox in statistics. *Scientific American*, 236:119–127, 1977.

B. Efron and R.J. Tibshirani. *An Introduction to the Bootstrap.* Chapman and Hall, New York, 1993.

A.M. Ellison, E.J. Farnsworth, and R.R. Twilley. Facultative mutualism between red mangroves androot-fouling sponges in Belizean mangal. *Ecology*, 77(8):2431–2444, 1996.

G.R. Finch, C.W. Daniels, E.K. Black, F.W. Schaefer, and M. Belosevic. Dose response of *Cryptosporidium parvum* in outbred neonatal cd-1 mice. *Applied and Environmental Microbiology*, 59(11):3661–3665, 1993.

R.A. Fisher. On the mathematical foundations of theoretical statistics. *Philosophical Transactions of the Royal Society of London, Series A*, 222:309–368, 1922.

R.A. Fisher. *Statistical Methods for Research Workers.* Oliver and Boyd, Edinburgh. (14th edition reprinted in 1970), 1st edition, 1925.

R.A. Fisher. The use of multiple measurements in taxonomic problems. *Annals of Eugenics*, 7, Part II:179–188, 1936.

R.A. Fisher. Statistical methods and scientific induction. *Journal of the Royal Statistical Society, B*, 17:69–78, 1955.

J.F. Fraumeni. Cigarette smoking and cancers of the urinary tract: Geographic variations in the United States. *Journal of the National Cancer Institute*, 41, 1968.

J.W. Frey, A.H. Bell, J.A. Hambrook-Berkman, and D.L. Lorenz. *Assessment of nutrient enrichment by use of algal-, invertebrate-, and fish-community attributions in wadeable streams in ecoregions surrounding the Great Lakes.* Scientific Investigations Report 2011-5009. National Water-Quality Assessment Program, U.S. Geological Survey, Reston, Virginia, 2011.

A.S. Friedlaender, P.N. Halpin, S.S. Qian, G.L. Lawson, P.H. Wiebe, D. Thiele, and A.J. Read. Whale distribution in relation to prey abundance and oceanographic processes in shelf waters of the Western Antarctic Peninsula. *Marine Ecology Progress Series*, 317:297–310, 2006.

J.H.G. Gauch. *Multivariate Analysis in Community Ecology.* Cambridge University Press, New York, 1982.

A. Gelman. Analysis of variance – why it is more important than ever (with discussions). *The Annals of Statistics*, 33(1):1–53, 2005.

A. Gelman. Letter to the editors regarding some papers of Dr. Satoshi Kanazawa. *Journal of Theoretical Biology*, 245(3):597–599, 2007.

A. Gelman and J. Hill. *Data Analysis Using Regression and Multilevel/Hierarchical Models.* Cambridge University Press, New York, 2007.

A. Gelman, J. Hill, and M. Yajima. Why we (usually) don't have to worry about multiple comparisons. *Journal of Research on Educational Effectiveness*, 5:189–211, 2012.

A. Gelman, J.B. Carlin, H.S. Stern, David B. Dunson, Aki Vehtari, and D.B. Rubin. *Bayesian Data Analysis.* CRC Press, Boca Raton, Florida, 3rd edition, 2014.

R.J. Gilliom and D.R. Helsel. Estimation of distribution parameters for censored trace level water quality data 1: Estimation techniques. *Water Resources Research*, 22:135–146, 1986.

A. Gleit. Estimation of small normal data sets with detection limits. *Environmental Science and Technology*, 19:1201–1206, 1985.

J.J. Goeman and S. le Cessie. A goodness-of-fit test for multinomial logistic regression. *Biometrics*, 62:980–985, 2006.

N.J. Gotelli and A.M. Ellison. *A Primer of Ecological Statistics.* Sinauer Associates, Inc. Publishers, Sunderland, MA, 2004.

A. Guisan and N.E. Zimmermann. Predictive habitat distribution models in ecology. *Ecological Modelling*, 135(2-3):147–186, 2000.

W. Härdle. *Smoothing Techniques: With Implementation in S.* Springer-Verlag, New York, 1991.

R.D. Harmel, S. Potter, P. Casebolt, K.H. Reckhow, C. Gree, and R. Haney. Compilation of measured nutrient load data for agricultural land uses in the United States. *Journal of the American Water Resources Association*, 42(5):1163–1178, 2006.

T.J. Hastie and R.J. Tibshirani. *Generalized Additive Models.* Chapman and Hall, London, 1990.

J.P. Hayes and R.J. Steidl. Statistical power analysis and amphibian population trends. *Conservation Biology*, 11(1):273–275, 1997.

D.R. Helsel. Less than obvious. *Environmental Science and Technology*, 24 (12):1767–1774, 1990.

D.R. Helsel and R.J. Gilliom. Estimation of distribution parameters for censored trace level water quality data 2: Verification and applications. *Water Resources Research*, 22:147–155, 1986.

J.M. Hoenig and D.M. Heisey. The abuse of power: The pervasive fallacy of power calculations for data analysis. *The American Statistician*, 55(1):1–6, 2001.

D. W. Hosmer and S. Lemeshow. *Applied Logistic Regression.* Wiley, New York, 2nd edition, 2000.

S.P. Hubbell. *The Unified Neurtral Theory of Biodiversity and Biogeography.* Princeton University Press, Princeton, New Jersey, 2001.

R.B. Huey, G.W. Gilchrist, M.L. Carlson, D. Berringan, and L. Serra. Rapid evolution of geographic cline in size in an introduced fly. *Science*, 287(5451): 308–309, 2000.

D. Hume. *Philosophical Essays Concerning Human Understanding.* A. Millar, London, UK, 1st edition, 1748.

D. Hume. *An Inquiry Concerning Human Understanding.* The Clarendon Press, Oxford, UK, 1777.

G.E. Hutchinson. Homage to Santa Rosalia or why are there so many kinds of animals? *American Maturalist*, 93:145–159, 1959.

J.P.A. Ioannidis. Why most published research findings are false. *PLoS Medicine*, 2(8):e124 doi:10.1371/journal.pmed.0020124, 2005.

W. James and Charles Stein. Estimation with quadratic loss. In *Proceedings of the 4th Berkeley Symposium Mathematics, Statistics and Probability*, volume 1, pages 361–379. University of California Press, Berkeley, California, 1961.

M.P Johnson and P.H. Raven. Species number and endemism: The Galapagos archipelago revisited. *Science*, 179:893–895, 1973.

R.H.G Jongman, C.J.F. ter Braak, and O.F.R. Van Tongeren. *Data Analysis in Community and Landscape Ecology.* Cambridge University Press, New York, 1995.

G.G. Judge and M.E. Bock. *The Statistical Implications of Pre-test and Stein-rule Estimators in Econometrics.* North-Holland, Amsterdam, 1978.

S. Kanazawa and G. Vandermassen. Engineers have more sons, nurses have more daughters: An evolutionary psychological extension of Baron-Cohen's extreme male brain theory of autism. *Journal of Theoretical Biology*, 233 (4):589–599, 2005.

J. Kerman and A. Gelman. Manipulating and summarizing posterior simulations using random variable objects. *Statistics and Computing*, 17(3): 235–244, 2007.

R.S. King, M.E. Baker, P.F. Kazyak, and D.E. Weller. How novel is too novel? Stream community thresolds at exceptionally low levels of catchment urbanization. *Ecological Applications*, 21:1659–1678, 2011.

D.G. Korich, M.M. Marshall, H.V. Smith, J. O'Grady, C.R. Bukhari, Z. Fricker, J.P. Rosen, and J.L. Clancy. Inter-laboratory comparison of the cd-1 neonatal mouse logistic dose-response model for *Cryptosporidium parvum* oocysts. *Journal of Eukaryotic Microbiology*, 47(3):294–298, 2000.

P. Kuhnert and B. Venables. An Introduction to R: Software for Statistical Modelling & Computing. Technical report, CSIRO Mathematical and Information Sciences, Cleveland, Australia, 2005. URL `http://cran.r-project.org/doc/contrib/Kuhnert+Venables-R_Course_Notes.zip`.

D. Lambert. Zero-inflated Poisson regression, with an application to defects in manufacturing. *Technometrics*, 34(1):1–14, 1992.

T.R. Lange, H.E. Royals, and L.L. Connor. Influence of water chemistry on mercury concentration in largemouth bass from Florida lakes. *Transactions of the American Fisheries Society*, 122(1):74–84, 1993.

S. Le Cessie and J. C. Van Houwelingen. A goodness-of-fit test for binary regression models based on smoothing methods. *Biometrics*, 47:1267–1282, 1991.

E.L. Lehmann and G. Casella. *Theory of Point Estimation.* Springer, New York, 2nd edition, 1998.

J. Lenhard. Models and statistical inference: The controversy between Fisher and Neyman-Pearson. *The British Journal for the Philosophy of Science*, 57(1):69–91, 2006.

E. Lesaffre and A. Albert. Multiple-group logistic regression diagnostics. *Applied Statistics*, 38:425–440, 1989.

S.S. Light and J.W. Dineen. Water control in the Everglades: A historical perspective. In S.M. Davis and J.C. Ogden, editors, *Everglades: The Ecosystem and Its Restoration*, pages 47–84. St. Lucie Press, Delray Beach, FL, 1994.

R. MacArthur. *Geographical Ecology.* Princeton University Press, Princeton, New Jersey, 1972.

C.P. Madenjian, R.J. Hesselberg, T.J. Desorcie, L.J. Schmidt, Stedman. R.M., L.J. Begnoche, and D.R. Passino-Reader. Estimate of net trophic transfer efficiency of PCBs to Lake Michigan lake trout from their prey. *Environmental Science and Technology*, 32:886–891, 1998.

O. Malve and S.S. Qian. Estimating nutrients and chlorophyll a relationships in Finnish lakes. *Environmental Science and Technology*, 40(24):7848–7853, 2006.

T.G. Martin, B.A. Wintle, J.R. Rhodes, P.M. Kuhnert, S.A. Field, Samantha J. Low-Choy, A.J. Tyre, and H.P. Possingham. Zero tolerance ecology: Improving ecological inference by modelling the source of zero observations. *Ecology Letters*, 8(11):1235–1246, 2005.

P. McCullagh and J.A. Nelder. *Generalized Linear Models.* Chapman & Hall, London, 1989.

G.C. McDonald and R.C. Schwing. Instabilities of regression estimates relating air pollution to mortality. *Technometrics*, 15(3):463–481, 1973.

G. McMahon and T.F. Cuffney. Quantifying urban intensity in drainage basins for assessing stream ecological conditions. *Journal of the American Water Resources Association*, 36(6):1247–1261, 2000.

K.H. Morales, L. Ryan, T.L. Kuo, M.M. Wu, and C.J. Chen. Risk of internal cancers from arsenic in drinking water. *Environmental Health Perspectives*, 108:655–661, 2000.

V.M.R. Muggeo. Estimating regression models with unknown break-points. *Statistics in Medicine*, 22(19):3055–3071, 2003.

National Research Council. *Arsenic in Drinking Water.* National Academy Press, Washington, DC, 1999.

J. Oksanen and P.R. Minchin. Continuum theory revisited: What shape are species responses along ecological gradients? *Ecological Modelling*, 157(2-3): 119–129, 2002.

W.R. Ott. *Environmental Statistics and Data Analysis.* Lewis Publishers, Boca Raton, 1995.

R.H. Peters. *A Critique for Ecology.* Cambridge University Press, 1991.

J. G. Pigeon and J. F. Heyse. An improved goodness-of-fit statistic for probability prediction models. *Biometrical Journal*, 41:71–82, 1999.

S.L. Pimm, H.L. Jones, and J. Diamond. On the risk of extinction. *The American Naturalist*, 132:757–785, 1988.

J.C. Pinheiro and D.M. Bates. *Mixed-Effects Models in S and S-PLUS.* Springer-Verlag, New York, 2000.

K.P. Popper. *The Logic of Scientific Discovery.* Hutchinson Education (reprinted 1992 by Routledge), London, 1959.

S.S. Qian. *A nonparametric Bayesian model of phosphorus retention.* PhD thesis, Nicholas School of the Environment, Duke University, 1995.

S.S. Qian. Ecological threshold and environmental management: A note on statistical methods for detecting thresholds. *Ecological Indicators*, 38:192–197, 2014a.

S.S. Qian. Statistics in ecology is for making a "principled" argument. *Landscape Ecology*, 29(6):937–939, 2014b.

S.S. Qian and C.W. Anderson. Exploring factors controlling variability of pesticide concentrations in the Willamette River Basin using tree-based models. *Environmental Science and Technology*, 33:3332–3340, 1999.

S.S. Qian and T. F. Cuffney. To threshold or not to threshold? That's the question. *Ecological Indicators*, 15(1):1–9, 2012.

S.S. Qian and T.F. Cuffney. A hierarchical zero-inflated model for species compositional data – from individual taxon responses to community response. *Limnology and Oceanography: Methods*, 12:498–506, 2014.

S.S. Qian and M. Lavine. Setting standards for water quality in the Everglades. *Chance*, 16(3):10–16, 2003.

S.S. Qian and Y. Pan. Historical soil total phosphorus concentration in the Everglades. In A.R. Burk, editor, *Focus on Ecological Research*, pages 131–150. Nova Science, 2006.

S.S. Qian and C.J. Richardson. Estimating the long-term phosphorus accretion rate in the Everglades: A Bayesian approach with risk assessment. *Water Resources Research*, 33(7):1681–1688, 1997.

S.S. Qian and Z. Shen. Ecological applications of multilevel analysis of variance. *Ecology*, 88(10):2489–2495, 2007.

S.S. Qian, M.E. Borsuk, and C. A. Stow. Seasonal and long-term nutrient trend decomposition along a spatial gradient in the Neuse River watershed. *Environmental Science and Technology*, 34:4474–4482, 2000a.

S.S. Qian, M. Lavine, and C.A. Stow. Univariate Bayesian nonparametric binary regression with application in environmental management. *Environmental and Ecological Statistics*, 7:77–91, 2000b.

S.S. Qian, W. Warren-Hicks, J. Keating, D.R.J. Moore, and R.S. Teed. A predictive model of mercury fish tissue concentrations for the southeastern United States. *Environmental Science and Technology*, 35(5):941–947, 2001.

S.S. Qian, R.S. King, and C.J. Richardson. Two statistical methods for the detection of environmental thresholds. *Ecological Modelling*, 166:87–97, 2003a.

S.S. Qian, C.A. Stow, and M.E. Borsuk. On Monte Carlo methods for Bayesian inference. *Ecological Modelling*, 159:269–277, 2003b.

S.S. Qian, A. Schulman, J. Koplos, A. Kotros, and P. Kellar. A hierarchical modeling approach for estimating national distributions of chemicals in public drinking water systems. *Environmental Science and Technology*, 38 (4):1176–1182, 2004.

S.S. Qian, K. Linden, and M. Donnelly. A Bayesian analysis of mouse infectivity data to evaluate the effectiveness of using ultraviolet light as a drinking water disinfectant. *Water Research*, 39:4229–4239, 2005a.

S.S. Qian, K.H. Reckhow, J. Zhai, and G. McMahon. Nonlinear regression modeling of nutrient loads in streams: A Bayesian approach. *Water Resources Research*, 41:W07012, 2005b.

S.S. Qian, T.F. Cuffney, and G. McMahon. Multinomial regression for analyzing macroinvertebrate assemblage composition data. *Freshwater Sciences*, 31(3):681–694, 2012.

S.S. Qian, J.D. Chaffin, M.R. DuFour, J.J. Sherman, P.C. Golnick, C.D. Collier, S.A. Nummer, and M.G. Margida. Quantifying and reducing uncertainty in estimated microcystin concentrations from the ELISA method. *Environmental Science and Technology*, 2015a. doi: 10.1021/acs.est.5b03029.

S.S. Qian, C.A. Stow, and Y.K. Cha. Implications of Stein's Paradox for environmental standard compliance assessment. *Environmental Science and Technology*, 49(10):5913–5920, April 2015b.

F.L. Ramsey and D.W. Schafer. *The Statistical Sleuth, A Course in Methods of Data Analysis.* Duxbury, Pacific Grove, CA, 2002.

K.H. Reckhow and S.S. Qian. Modeling phosphorus trapping in wetland using generalized additive models. *Water Resources Research*, 30(11):3105–3114, 1994.

K.H. Reckhow, J.T. Clements, and R.C. Dodd. Statistical evaluation of mechanistic water quality models. *Journal of Environmental Engineering*, 116 (2):250–268, 1990.

F.J. Richards. A flexible growth function for empirical use. *Journal of Experimental Botany*, 10(2):290–301, 1959.

C.J. Richardson. *The Everglades Experiments: Lessons for Ecosystem Restoration.* Springer, 2008.

C.J. Richardson and S.S. Qian. Long-term phosphorus assimilative capacity in freshwater wetlands: A new paradigm for sustaining ecosystem structure and function. *Environmental Science and Technology*, 33(10):1545–1551, 1999.

C.J. Richardson, R.S. King, S.S. Qian, P. Vaithiyanathan, R.G. Qualls, and C.A. Stow. Estimating ecological thresholds for phosphorus in the Everglades. *Environmental Science and Technology*, 41(23):8084–8091, 2007.

B.D. Ripley. *Pattern Recognition and Neural Networks.* Cambridge University Press, Cambridge, UK, 1996.

C. Ritz and J.C. Streibig. Bioassay analysis using R. *Journal of Statistical Software*, 12(5):1–22, 2005.

K.W. Rizzardi. Alligators and litigators: A recent history of Everglades regulation and litigation. *Florida Bar Journal*, March:18, 2001.

C.P. Robert and G. Casella. *Monte Carlo Statistical Methods.* Springer, 2004.

J.T. Rotenberry and J.A. Wiens. Statistical power analysis and community-wide patterns. *The American Naturalist*, 125(1):164–168, 1985.

C. Ruckdeschel, C.R. Shoop, and R.D. Kenney. On the sex ratio of juvenile *Lepidochelys kempii* in Georgia. *Chelonian Conservation and Biology*, 4(4): 860–863, 2005.

L.M. Ryan. Epidemiologically based environmental risk assessment. *Statistical Science*, 18(4):466–480, 2003.

T.W. Schoener. The anolis lizards of Bimini: Resource partitioning in a complex fauna. *Ecology*, 49(4):704–726, 1968.

M.D. Schwartz and J.M. Caprio. North American First Leaf and First Bloom Lilac Phenology Data. Data contribution series # 2003-078., IGBP PAGES/World Data Center for Paleoclimatology, NOAA/NGDC Paleoclimatology Program, Boulder CO, USA, 2003. URL `ftp://ftp.ncdc.noaa.gov/pub/data/paleo/phenology/north_america_lilac.txt`.

M.D. Schwartz, R. Ahas, and A. Aasa. Onset of spring starting earlier across the Northern Hemisphere. *Global Change Biology*, 12(2):343–351, 2006.

H.M.H. Siersma, C.J. Foley, C.J. Nowicki, S.S. Qian, and D.R. Kashian. Trends in the distribution and abundance of *Hexagenia* spp. in Saginaw Bay, Lake Huron, 1954-2012: Moving towards recovery? *Journal of Great Lakes Research*, 40:156–167, 2014.

A.F.M. Smith. A Bayesian approach to inference about a change-point in a sequence of random variables. *Biometrika*, 62(2):407–416, 1975. doi: 10.1093/biomet/62.2.407.

E.P. Smith, K. Ye, C. Hughes, and L. Shabman. Statistical assessment of violations of water quality standards under section 303(d) of the clean water act. *Environmental Science and Technology*, 35(3):606–612, 2001.

R.A. Smith, G.E. Schwarz, and R.B. Alexander. Regional interpretation of water-quality monitoring data. *Water Resources Research*, 33:2781–2798, 1997.

G.K. Smyth. Nonlinear regression. In *Encyclopedia of Environmentrics*, volume 3, pages 1405–1411. John Wiley and Sons, Ltd., Chichester, 2002.

D. G. Sprugel. Correcting for bias in log-transformed allometric equations. *Ecology*, 64(1):209–210, 1983.

R.J. Steidl, J.P. Hayes, and E. Schauber. Statistical power analysis in wildlife research. *The Journal of Wildlife Management*, 61(2):270–279, 1997.

C. Stein. Inadmissibility of the usual estimator for the mean of a multivariate normal distribution. In *Proceedings of the Third Berkeley Symposium*, volume 1, pages 197–206. University of California Press, 1955.

S.M. Stigler. Napoleonic statistics: The work of Laplace. *Biometrika*, 62(2): 503–517, 1975.

C.A. Stow. Factors associated with PCB concentrations in Lake Michigan salmonids. *Environmental Science and Technology*, 29(2):522–527, 1995.

C.A. Stow and S.S. Qian. A size-based probabilistic assessment of PCB exposure from Lake Michigan fish consumption. *Environmental Science and Technology*, 32(15):2325–2330, 1998.

C.A. Stow and D. Scavia. Modeling hypoxia in the Chesapeake Bay: Ensemble estimation using a Bayesian hierarchical model. *Journal of Marine Systems*, 76(1–2):244–250, 2009.

C.A. Stow, S.R. Carpenter, and J.F. Amrhein. PCB concentration trends in Lake Michigan coho (*Oncorhynchus kisutch*) and chinook salmon (*O. tshawytscha*). *Canadian Journal of Fisheries and Aquatic Sciences*, 51(6): 1384–1390, 1994.

C.A. Stow, S.R. Carpenter, L.A. Eby, J.F. Amrhein, and R.J. Hesselberg. Evidence that PCBs are approaching stable concentrations in Lake Michigan fishes. *Ecological Applications*, 5:248–260, 1995.

C.A. Stow, E.C. Lamon, S.S. Qian, and C.S. Schrank. Will Lake Michigan lake trout meet the Great Lakes strategy 2002 PCB reduction goal? *Environmental Science and Technology*, 38(2):359–363, 2004.

C.A. Stow, K.H. Reckhow, and S.S. Qian. A Bayesian approach to retransformation bias in transformed regression. *Ecology*, 87(6):1472–1477, 2006.

Student. The probable error of a mean. *Biometrika*, 6(1):1–25, 1908.

C.J.F. ter Braak. *Unimodal Models to Relate Species to Environment.* DLO-Agricultureal Mathematics Group, Box 100, NL-6700 AC Wageningen, the Netherland, 1996.

T. Therneau, B. Atkinson, and B. Ripley. *rpart: Recursive Partitioning and Regression Trees*, R package version 4.1-10 edition, 2015. URL `http://CRAN.R-project.org/package=rpart`. R package version 4.1-10.

D.C. Thiele, E.T. Chester, S.E. Moore, A. Sirovic, J.A. Hildebrand, and A.S. Friedlaender. Seasonal variability in whale encounters in the western Antarctic Peninsula. *Deep-Sea Research*, 51:2311–2325, 2004.

J.W. Tukey. The future of data analysis. *The Annals of Mathematical Statistics*, 33(1):1–67, 1962. ISSN 00034851.

J.W. Tukey. *Exploratory Data Analysis.* Addison-Wesley, Reading, MA, 1977.

U.S. EPA. Nutrient criteria technical guidance manual: Lakes and reservoirs. Technical Report EPA 822-B00-001, U.S. Environmental Protection Agency, Office of Water, 2000.

Gerald van Belle. *Statistical Rules of Thumb.* Wiley, 2nd edition, 2002.

R.L. Wasserstein and N.A. Lazar. The ASA's statement on *p*-values: context, process, and purpose. *American Statisticians*, DOI: 10.1080/00031305.2016.1154108, 2016.

S. Weisberg. *Applied Linear Regression.* Wiley, New York, 3rd edition, 2005.

S.N. Wood. *Generalized Additive Models: An Introduction with R.* Chapman and Hall/CRC Press, 2006.

World Health Organization. Guidelines for drinking-water quality. 2nd edition, addendum to volume 2, health criteria and other supporting information. WHO, Geneva, `http://www.who.int/water_sanitation_health/dwq/gdwq2v1/en/index4.html`, 1998.

R. Wu, S.S. Qian, F. Hao, H. Cheng, D. Zhu, and J. Zhang. Modeling contaminant concentration distributions in China's centralized source waters. *Environmental Science and Technology*, 45(14):6041–6048, 2011.

L.L. Yuan and A.I. Pollard. Deriving nutrient targets to prevent excessive cyanobacterial densities in U.S. lakes and reservoirs. *Freshwater Biology*, 60 (9):1901–1916, 2015.

A.F. Zuur, E.N. Ieno, and E.H.W.G. Meesters. *A Beginner's Guide to R.* Springer, Dordrecht, 2009.

Index

additive models, 245–251
additive shift, 59, 60, 69, 153
analysis of variance, 116–127, 137
    dummy variable, 194
    multiple comparisons, 431
    one-way, 116–127, 137
        between group variation, 118
        multiple comparisons, 121
        within group variation, 118
    two-way, 193–200
        interaction, 198

backfitting algorithm, 368
bagging, 410
Bayes risk, 419
Bayesian *p*-value, 387, 397
binomial distribution, 305
boosting, 408
bootstrap, 511
bootstrap aggregation, 410
bootstrap methods, 86–90
    BCa confidence interval, 89
    bootstrap percentile confidence interval, 88
    bootstrap-*t* confidence interval, 88

CART, 271–299
    cross-validation, 281
    distributional assumption, 297
    fitting a tree model, 275
    for variable selection, 371
    Gini impurity, 276
    information index, 276
    plotting options, 289
    variable selection, 293
central limit theorem, 79, 82, 418

χ^2 distribution, 83
χ^2 test, 132
classification and regression tree, *see* CART
climate change, 226
complete pooling, 429, 432
conditional plots, 458
confidence interval, *see* estimation
cumulative probability, 48, 49

data
    air quality in New York, `airquality`, 62
    Atlantic sturgeon
        `sturgeon.csv`, 384
    birds extinction, `birds.csv`, 206
    bullfrog sexual selection
        `bullfrog.csv`, 384
    clines in flies, `flies`, 200
    EUSE
        `EUSE_USGSReportData.csv`, 490
        `usgsmultinomial.csv`, 355
    Galapagos Islands
        `galapagos`, 382
    Lake Erie water quality
        `Eriecombined.csv`, 491
    lizards
        `lizards.txt`, 383
    Maine BCG data, `MaineBCG.csv`, 301
    Maumee River water quality, `MaumeeData`, 45
    mayfly in Saginaw Bay, `SagBayHex.csv`, 301
    North American Wetlands Database, `nadb`, 64

PCB in fish, laketrout, 152
PM2.5 in Baltimore, 66
pollution and mortality,
pollution.csv, 204
rodents in NY appartments
rodents.csv, 382
stream water quality
greatlakes.csv, 489
Toledo water crisis, 192, 466
tree blowdowns
blowdown, 383
UV inactivation
crypto.data, 306
zooplankton in North American lakes
lakes, 381
deviance, 156, 276, 297
dose-response model, 306
dummy variable, 150, 171, 194

estimation
confidence interval, 80
coverage probability, 411
mean, 78
sample mean, 79
sample standard deviation, 79
standard deviation, 78
standard error, 79
Everglades, 7–13
examples
arsenic in drinking water, 332–333
background N_2O emission, 431–434
Cape Sable seaside sparrows, 401–404
crypto in U.S. drinking water, 478–485
drinking water disinfection, 306–307
ELISA, 17, 191–192, 405–408, 464, 465
EUSE, 14–15, 355, 436–452
Everglades, 7–13, 50, 56, 78, 81, 84, 89, 94–96, 104, 116, 124, 127
exploited plants, 470–477
Finnish lakes, 174–185, 424, 453–464
Kemp's ridley Turtles, 128–133
Lake Erie, 54
lilac first bloom dates, 226–229
mangrove and sponges, 137–142, 194
Neuse River water quality, 261–267
North American Wetlands Database, 250–251
PCB in fish, 16, 152–154, 156–173, 209–225, 396, 400
seaweed grazers, 426–431
seed predation, 319–331
threshold confidence interval, 411–414
water quality, 31, 134–137
whales in Antarctic, 369–380
Willamette River pesticides, 272–275
exchangeable, 421, 422, 424
exploratory data analysis, 56–67, 493
boxplot, 59
conditional plots, 67
histogram, 57
power transformation, 64
Q-Q plot, 60
quantile plot, 58
scatter plot, 62
scatter plot matrix, 62
exponential family, 304
exposure, 340

F-statistic, 118
F-test, 119
Fisher, 3–6

GAM, *see* generalized additive models
Gauss, 48, 418
generalized additive models, 367–380

generalized linear model, 304–367
  binary response, 305
  intercept
    through origin, 321
  link function, 304
  logistic regression, 305–331
    binned residual plot, 316
    interaction, 314
    intercept, 310
    intercept through origin, 324
    overdispersion, 316
    slope, 311
  logit transformation, 306
  multinomial regression, 353–361
    fitting in R, 355
    model assessment, 358
  negative binomial, 351–353
  Poisson regression, 332–353
    exposure, 340
    offset, 340
    overdispersion, 341
    quasi-Poisson, 344
  Poisson-multinomial connection, 361–367
generalized multilevel models, 469
GLM, *see* generalized linear model

histogram, 50
hockey stick model, *see* nonlinear models
Hume, 78
humpback whale, 371
hypothesis testing, 90–116
  α, 109
  β, 109
  nonparametric methods, 102
    Wilcoxon rank sum, 104
    Wilcoxon signed rank, 103
  ordered statistics, 102
  p-value, 109
  permutation test, 498
  power, 109
  procedure of, 101
  t-test, 91–99
    test statistic, 92
    two-sided alternatives, 98
    Welch's t-test, 97
  using confidence intervals, 99

indicator value, 497

James–Stein estimator, 417, 420

Laplace, 48, 418
LD_{50}, 309
likelihood, 48
linear models, 149–200
  additive assumption, 160
  ANOVA, 164
  Box–Cox transformation, 187
  collinearity, 174
  Cook's distance, 169
  diagnostics, 162
  dummy variable, 172, 194
  factor predictor, 170
  interaction, 160, 177
  intercept, 157
    through origin, 173
  least squares, 154
  linear transformation, 185
  log transformation, 186
  multiple regression, 158
  prediction, 189
  residual distribution, 166
  residuals, 158
  rfs plot, 169
  R^2, 164
    adjusted, 164
  simple regression, 154, 156
  slope, 158
log-normal distribution, 48
logistic regression, *see* generalized linear model
logit transformation, 306, 310, 354

maximum likelihood estimator, 367
median polish, 260
mesocosm, 9
minke whale, 371
MLE, *see* maximum likelihood estimator

Monte Carlo simulation, 389
multilevel regression, 417
  ANOVA, 425
  exchangeability, 421
  generalized linear model, 469
    logistic regression, 474
    Poisson model, 471
  group level predictor, 433, 444
  linear model, 436
  multiple regression, 453
  nonlinear models, 465
  nonnested groups, 447
multilevel structure, 421
multinomial distribution, 353
multiple comparison trap, 494, 495
multiplicative shift, 60, 69, 153

negative binomial distribution, 351
Neyman-Pearson lemma, 495
no pooling, 429, 432
nonlinear models, 209–267
  hockey stick model, 221
  limiting coefficient range, 216
  piecewise linear model, 220, 221
  threshold model, 221
nonnested groups, 447
nonparametric regression, 240–267
  additive models, 245
  graphical model, 246
  local regression, 243
  loess, 243
  seasonal decomposition of time series, 259
normal distribution, 48
normal Q-Q plot, 51
null distribution, 92
null hypothesis, 498

offset, 340
overdispersion, 316, 341

p-value, 90–91, 494
partial pooling, 429, 432
phenology, 226
Poisson distribution, 332
Poisson regression, *see* generalized linear model
Popper, 78
predictive distribution, 390

Q-Q plot, 153
quantile, 48, 58

R, 19–44
  assignment, 21
  console, 21
  data management, 27–44
    aggregation and reshaping, 38
    creating data, 29
    data cleaning, 34
    dates, 42
    importing data, 27
    subsetting, 36
    transformation, 38
  data types, 23
  function, 25
  functions
    `abline`, 51, 377
    `aov`, 118, 138, 198
    `apply`, 411, 413, 484
    `arm`, 156
    `as.numeric`, 125, 444
    `as.vector`, 360, 448
    `axis`, 85
    `bcanon`, 89
    `binom.test`, 130
    `bootstrap`, 87–89
    `boxcox`, 187
    `boxplot`, 282
    `c`, 107
    `cbind`, 107, 309, 312, 356
    `co.intervals`, 177
    `coef`, 322, 326, 481
    `coplot`, 66, 67, 177
    `curve`, 322, 326
    `data.frame`, 30, 107, 125, 136, 348, 407, 413, 473, 477, 495
    `dim`, 404
    `display`, 156, 195, 309, 333, 472, 475, 480

dnorm, 49
dotplot, 467, 472, 473, 477
dpois, 499
example, 26
exp, 85, 356, 392, 396, 397
fitted, 318
fixef, 442, 449, 472
for, 82
function, 25, 34, 107, 408, 410
gam, 249–251, 254, 256, 375–380
glht, 139, 140
glm, 307–331, 333–351, 401, 479
glm.nb, 352
glmer, 471–482
head, 30
help, 26
hist, 85, 88, 397, 484
invlogit, 322
layout, 282
legend, 322, 326
length, 25, 81, 391
lm, 121, 137, 156–200, 218, 315, 438
lmer, 426–464
lo, 250
loess, 243
log, 81, 85, 391
matrix, 413, 484
mean, 24, 26, 81, 82, 85, 88, 331, 388, 391, 392
medpolish, 261, 263
mode, 23
multinom, 355
mvrnorm, 394
na.omit, 413
nlmer, 466–469
nls, 210–229, 407, 466, 467
order, 25
ordered, 125, 286, 444
packages, 22
par, 251, 255, 282, 322, 326, 375
pchisq, 318, 342, 377
plot, 51, 136, 251, 255, 289, 293, 322, 326, 375, 377
plot.rpart, 277, 282
plotcp, 281
pnorm, 49, 90, 132
points, 322, 326
post.rpart, 277
posterior, 395
power.t.test, 111, 113, 114
predict, 190, 342, 348, 356, 377
print, 395
printcp, 279
prop.test, 132
prune, 281, 287
prune.rpart, 277
pt, 388
qbinom, 135, 136
qnorm, 31, 33, 34, 49, 51, 61, 84, 85
qqline, 52
qqmath, 61, 119
qqnorm, 52
qqplot, 61
qt, 81, 82, 392
quantile, 85, 88, 89, 395, 396, 413
ranef, 428, 442, 449, 452, 467, 472, 475
rank, 103
rbind, 140, 141
rchisq, 85, 391, 393, 484
read.csv, 29
read.table, 29
rep, 30, 262, 360
rnorm, 32, 82, 85, 391, 394
rpart, 277–299, 374
rpart.control, 277
rpois, 404, 484
rt, 388, 395
runif, 82
rvmatrix, 356
rvnorm, 396, 406
rvsims, 356, 408

sample, 86, 410
sapply, 411
sd, 81, 85, 87, 88, 391, 392
se.fixef, 442
se.ranef, 429, 442
seq, 30
set.seed, 32
setnsims, 395
setwd, 23
sim, 223, 323, 394, 397, 476
simpleKey, 349, 360
snip.rpart, 277
sort, 413
sqrt, 81, 82
SSfpl, 468
sum, 25, 33, 82, 318, 342, 377
summary, 118, 122, 254, 308, 318, 342, 377, 393, 395
summary.aov, 121, 137, 198
summary.lmer, 428, 441, 445, 447, 454
summary.rpart, 277
t.test, 94, 96–98, 125, 129, 495
table, 484
tapply, 125, 434
text, 289, 293
text.rpart, 277
title, 326
trellis.par.set, 349
ts, 262
TukeyHSD, 138
unique, 413
unlist, 29, 360
update, 345–347, 410
wilcox.exact, 104, 105
wilcox.test, 104, 105
xyplot, 66, 119, 120, 349, 360

packages, 22
arm, 309, 323, 394
exactRankTests, 104
gam, 249
glht, 139
lattice, 349
lme4, 421, 426
MASS, 352
mgcv, 251, 375
rpart, 277, 374
rv, 356, 395
tree, 277

prompt, 21
RStudio, 20
user interface, 20

random forest, 410
rank transformation, 102
re-sample, 408
re-transformation bias, 392, 396

S-L plot, 55
sampling distribution, 5, 79, 390
sampling error, 12
self-starter function, 233–240, 466
shrinkage estimator, 429
simulation, 31, 81, 82, 84, 85, 107, 125, 218, 223, 323, 387–414, 482–485
linear model, 392
model checking, 390
Monte Carlo, 388
nls, 408
numerical integration, 389
power, 503–511
prediction uncertainty, 405
re-transformation bias, 396
type I error probability, 495, 499–501

smoothing
moving average, 241
scatter plot smoothing, 240

split point, 497
statistical assumptions, 12, 47–56
constant variance, 55
homoscedasticity, 55
independence, 54
normality, 48

Stein's paradox, 417, 423
STL, 259

t-distribution, 80
threshold, 298–299

TITAN, 496
Tukey, 72
type I error, 494

UV disinfection, 306

zero-inflated count data, 404, 484

Printed in the United States
by Baker & Taylor Publisher Services